Springer Finance

Springer Finance

Springer Finance is a programme of books aimed at students, academics and practitioners working on increasingly technical approaches to the analysis of financial markets. It aims to cover a variety of topics, not only mathematical finance but foreign exchanges, term structure, risk management, portfolio theory, equity derivatives, and financial economics.

Ammann M., Credit Risk Valuation: Methods, Models, and Application (2001)
Back K., A Course in Derivative Securities: Introduction to Theory and Computation (2005)
Barucci E., Financial Markets Theory. Equilibrium, Efficiency and Information (2003)
Bielecki T.R. and Rutkowski M., Credit Risk: Modeling, Valuation and Hedging (2002)
Bingham N.H. and Kiesel R., Risk-Neutral Valuation: Pricing and Hedging of Financial Derivatives (1998, 2nd ed. 2004)
Brigo D. and Mercurio F., Interest Rate Models: Theory and Practice (2001)
Buff R., Uncertain Volatility Models-Theory and Application (2002)
Dana R.A. and Jeanblanc M., Financial Markets in Continuous Time (2002)
Deboeck G. and Kohonen T. (Editors), Visual Explorations in Finance with Self-Organizing Maps (1998)
Delbaen F. and Schachermayer W., The Mathematics of Arbitrage (2005)
Elliott R.J. and Kopp P.E., Mathematics of Financial Markets (1999, 2nd ed. 2005)
Fengler M.R., Semiparametric Modeling of Implied Volatility (200)
Geman H., Madan D., Pliska S.R. and Vorst T. (Editors), Mathematical Finance–Bachelier Congress 2000 (2001)
Gundlach M., Lehrbass F. (Editors), CreditRisk^{+} in the Banking Industry (2004)
Kellerhals B.P., Asset Pricing (2004)
Külpmann M., Irrational Exuberance Reconsidered (2004)
Kwok Y.-K., Mathematical Models of Financial Derivatives (1998)
Malliavin P. and Thalmaier A., Stochastic Calculus of Variations in Mathematical Finance (2005)
Meucci A., Risk and Asset Allocation (2005)
Pelsser A., Efficient Methods for Valuing Interest Rate Derivatives (2000)
Prigent J.-L., Weak Convergence of Financial Markets (2003)
Schmid B., Credit Risk Pricing Models (2004)
Shreve S.E., Stochastic Calculus for Finance I (2004)
Shreve S.E., Stochastic Calculus for Finance II (2004)
Yor M., Exponential Functionals of Brownian Motion and Related Processes (2001)
Zagst R., Interest-Rate Management (2002)
Zhu Y.-L., Wu X., Chern I.-L., Derivative Securities and Difference Methods (2004)
Ziegler A., Incomplete Information and Heterogeneous Beliefs in Continuous-time Finance (2003)
Ziegler A., A Game Theory Analysis of Options (2004)

Freddy Delbaen · Walter Schachermayer

The Mathematics
of Arbitrage

Freddy Delbaen

ETH Zürich
Departement Mathematik, Lehrstuhl für Finanzmathematik
Rämistr. 101
8092 Zürich
Switzerland
E-mail: delbaen@math.ethz.ch

Walter Schachermayer

Technische Universität Wien
Institut für Finanz- und Versicherungsmathematik
Wiedner Hauptstr. 8-10
1040 Wien
Austria
E-mail: wschach@fam.tuwien.ac.at

Mathematics Subject Classification (2000): M13062, M27004, M12066

Library of Congress Control Number: 2005937005

ISBN-10 3-540-21992-7 Springer Berlin Heidelberg New York
ISBN-13 978-3-540-21992-7 Springer Berlin Heidelberg New York

Springer is a part of Springer Science+Business Media
springer.com
© Springer-Verlag Berlin Heidelberg 2006
Printed in Germany

Cover design: *design & production*, Heidelberg
Typesetting by the authors using a Springer LaTeX macro package
Produktion: LE-TeX Jelonek, Schmidt & Vöckler GbR

Printed on acid-free paper   41/3142YL - 5 4 3 2 1 0

To Rita and Christine with love

Preface

In 1973 F. Black and M. Scholes published their pathbreaking paper [BS 73] on option pricing. The key idea — attributed to R. Merton in a footnote of the Black-Scholes paper — is the use of trading in continuous time and the notion of arbitrage. The simple and economically very convincing *"principle of no-arbitrage"* allows one to derive, in certain mathematical models of financial markets (such as the Samuelson model, [S 65], nowadays also referred to as the "Black-Scholes" model, based on geometric Brownian motion), unique prices for options and other contingent claims.

This remarkable achievement by F. Black, M. Scholes and R. Merton had a profound effect on financial markets and it shifted the paradigm of dealing with financial risks towards the use of quite sophisticated mathematical models.

It was in the late seventies that the central role of no-arbitrage arguments was crystallised in three seminal papers by M. Harrison, D. Kreps and S. Pliska ([HK 79], [HP 81], [K 81]) They considered a general framework, which allows a systematic study of different models of financial markets. The Black-Scholes model is just one, obviously very important, example embedded into the framework of a general theory. A basic insight of these papers was the intimate relation between no-arbitrage arguments on one hand, and martingale theory on the other hand. This relation is the theme of the *"Fundamental Theorem of Asset Pricing"* (this name was given by Ph. Dybvig and S. Ross [DR 87]), which is not just a single theorem but rather a general principle to relate no-arbitrage with martingale theory. Loosely speaking, it states that a mathematical model of a financial market is free of arbitrage if and only if it is a martingale under an equivalent probability measure; once this basic relation is established, one can quickly deduce precise information on the pricing and hedging of contingent claims such as options. In fact, the relation to martingale theory and stochastic integration opens the gates to the application of a powerful mathematical theory.

The mathematical challenge is to turn this general principle into precise theorems. This was first established by M. Harrison and S. Pliska in [HP 81] for the case of finite probability spaces. The typical example of a model based on a finite probability space is the "binomial" model, also known as the "Cox-Ross-Rubinstein" model in finance.

Clearly, the assumption of finite Ω is very restrictive and does not even apply to the very first examples of the theory, such as the Black-Scholes model or the much older model considered by L. Bachelier [B 00] in 1900, namely just Brownian motion. Hence the question of establishing theorems applying to more general situations than just finite probability spaces Ω remained open.

Starting with the work of D. Kreps [K 81], a long line of research of increasingly general — and mathematically rigorous — versions of the "Fundamental Theorem of Asset Pricing" was achieved in the past two decades. It turned out that this task was mathematically quite challenging and to the benefit of both theories which it links. As far as the financial aspect is concerned, it helped to develop a deeper understanding of the notions of arbitrage, trading strategies, etc., which turned out to be crucial for several applications, such as for the development of a dynamic duality theory of portfolio optimisation (compare, e.g., the survey paper [S 01a]). Furthermore, it also was fruitful for the purely mathematical aspects of stochastic integration theory, leading in the nineties to a renaissance of this theory, which had originally flourished in the sixties and seventies.

It would go beyond the framework of this preface to give an account of the many contributors to this development. We refer, e.g., to the papers [DS 94] and [DS 98], which are reprinted in Chapters 9 and 14.

In these two papers the present authors obtained a version of the "Fundamental Theorem of Asset Pricing", pertaining to general $\mathbb{R}^d$-valued semimartingales. The arguments are quite technical. Many colleagues have asked us to provide a more accessible approach to these results as well as to several other of our related papers on Mathematical Finance, which are scattered through various journals. The idea for such a book already started in 1993 and 1994 when we visited the Department of Mathematics of Tokyo University and gave a series of lectures there.

Following the example of M. Yor [Y 01] and the advice of C. Byrne of Springer-Verlag, we finally decided to reprint updated versions of seven of our papers on Mathematical Finance, accompanied by a guided tour through the theory. This guided tour provides the background and the motivation for these research papers, hopefully making them more accessible to a broader audience.

The present book therefore is organised as follows. Part I contains the "guided tour" which is divided into eight chapters. In the introductory chapter we present, as we did before in a note in the Notices of the American Mathematical Society [DS 04], the theme of the Fundamental Theorem of Asset Pricing in a nutshell. This chapter is very informal and should serve mainly to build up some economic intuition.

In Chapter 2 we then start to present things in a mathematically rigourous way. In order to keep the technicalities as simple as possible we first restrict ourselves to the case of finite probability spaces Ω. This implies that all the function spaces $L^p(\Omega, \mathcal{F}, \mathbf{P})$ are finite-dimensional, thus reducing the functional analytic delicacies to simple linear algebra. In this chapter, which presents the theory of pricing and hedging of contingent claims in the framework of finite probability spaces, we follow closely the Saint Flour lectures given by the second author [S 03].

In Chapter 3 we still consider only finite probability spaces and develop the basic duality theory for the optimisation of dynamic portfolios. We deal with the cases of complete as well as incomplete markets and illustrate these results by applying them to the cases of the binomial as well as the trinomial model.

In Chapter 4 we give an overview of the two basic continuous-time models, the "Bachelier" and the "Black-Scholes" models. These topics are of course standard and may be found in many textbooks on Mathematical Finance. Nevertheless we hope that some of the material, e.g., the comparison of Bachelier versus Black-Scholes, based on the data used by L. Bachelier in 1900, will be of interest to the initiated reader as well.

Thus Chapters 1–4 give expositions of basic topics of Mathematical Finance and are kept at an elementary technical level. From Chapter 5 on, the level of technical sophistication has to increase rather steeply in order to build a bridge to the original research papers. We systematically study the setting of general probability spaces $(\Omega, \mathcal{F}, \mathbf{P})$. We start by presenting, in Chapter 5, D. Kreps' version of the Fundamental Theorem of Asset Pricing involving the notion of "No Free Lunch". In Chapter 6 we apply this theory to prove the Fundamental Theorem of Asset Pricing for the case of finite, discrete time (but using a probability space that is not necessarily finite). This is the theme of the Dalang-Morton-Willinger theorem [DMW 90]. For dimension $d \geq 2$, its proof is surprisingly tricky and is sometimes called the "100 meter sprint" of Mathematical Finance, as many authors have elaborated on different proofs of this result. We deal with this topic quite extensively, considering several different proofs of this theorem. In particular, we present a proof based on the notion of "measurably parameterised subsequences" of a sequence $(f_n)_{n=1}^{\infty}$ of functions. This technique, due to Y. Kabanov and C. Stricker [KS 01], seems at present to provide the easiest approach to a proof of the Dalang-Morton-Willinger theorem.

In Chapter 7 we give a quick overview of stochastic integration. Because of the general nature of the models we draw attention to general stochastic integration theory and therefore include processes with jumps. However, a systematic development of stochastic integration theory is beyond the scope of the present "guided tour". We suppose (at least from Chapter 7 onwards) that the reader is sufficiently familiar with this theory as presented in several beautiful textbooks (e.g., [P 90], [RY 91], [RW 00]). Nevertheless, we do highlight those aspects that are particularly important for the applications to Finance.

Finally, in Chapter 8, we discuss the proof of the Fundamental Theorem of Asset Pricing in its version obtained in [DS 94] and [DS 98]. These papers are reprinted in Chapters 9 and 14.

The main goal of our "guided tour" is to build up some intuitive insight into the Mathematics of Arbitrage. We have refrained from a logically well-ordered deductive approach; rather we have tried to pass from examples and special situations to the general theory. We did so at the cost of occasionally being somewhat incoherent, for instance when applying the theory with a degree of generality that has not yet been formally developed. A typical example is the discussion of the Bachelier and Black-Scholes models in Chapter 4, which is introduced before the formal development of the continuous time theory. This approach corresponds to our experience that the human mind works inductively rather than by logical deduction. We decided therefore on several occasions, e.g., in the introductory chapter, to jump right into the subject in order to build up the motivation for the subsequent theory, which will be formally developed only in later chapters.

In Part II we reproduce updated versions of the following papers. We have corrected a number of typographical errors and two mathematical inaccuracies (indicated by footnotes) pointed out to us over the past years by several colleagues. Here is the list of the papers.

Chapter 9: [DS 94] A General Version of the Fundamental Theorem of Asset Pricing

Chapter 10: [DS 98a] A Simple Counter-Example to Several Problems in the Theory of Asset Pricing

Chapter 11: [DS 95b] The No-Arbitrage Property under a Change of Numéraire

Chapter 12: [DS 95a] The Existence of Absolutely Continuous Local Martingale Measures

Chapter 13: [DS 97] The Banach Space of Workable Contingent Claims in Arbitrage Theory

Chapter 14: [DS 98] The Fundamental Theorem of Asset Pricing for Unbounded Stochastic Processes

Chapter 15: [DS 99] A Compactness Principle for Bounded Sequences of Martingales with Applications

Our sincere thanks go to Catriona Byrne from Springer-Verlag, who encouraged us to undertake the venture of this book and provided the logistic background. We also thank Sandra Trenovatz from TU Vienna for her infinite patience in typing and organising the text.

This book owes much to many: in particular, we are deeply indebted to our many friends in the functional analysis, the probability, as well as the mathematical finance communities, from whom we have learned and benefitted over the years.

Zurich, November 2005,   *Freddy Delbaen*
Vienna, November 2005   *Walter Schachermayer*

Contents

Part III   Bibliography

Part I

A Guided Tour to Arbitrage Theory

1

The Story in a Nutshell

1.1 Arbitrage

The notion of arbitrage is crucial to the modern theory of Finance. It is the corner-stone of the option pricing theory due to F. Black, R. Merton and M. Scholes [BS 73], [M 73] (published in 1973, honoured by the Nobel prize in Economics 1997).

The idea of arbitrage is best explained by telling a little joke: a professor working in Mathematical Finance and a normal person go on a walk and the normal person sees a 100 € bill lying on the street. When the normal person wants to pick it up, the professor says: don't try to do that. It is absolutely impossible that there is a 100 € bill lying on the street. Indeed, if it were lying on the street, somebody else would have picked it up before you. (end of joke)

How about financial markets? There it is already much more reasonable to assume that there are no arbitrage possibilities, i.e., that there are no 100 € bills lying around and waiting to be picked up. Let us illustrate this with an easy example.

Consider the trading of $ versus € that takes place simultaneously at two exchanges, say in New York and Frankfurt. Assume for simplicity that in New York the $/€ rate is 1 : 1. Then it is quite obvious that in Frankfurt the exchange rate (at the same moment of time) also is 1 : 1. Let us have a closer look why this is the case. Suppose to the contrary that you can buy in Frankfurt a $ for 0.999 €. Then, indeed, the so-called "arbitrageurs" (these are people with two telephones in their hands and three screens in front of them) would quickly act to buy $ in Frankfurt and simultaneously sell the same amount of $ in New York, keeping the margin in their (or their bank's) pocket. Note that there is no normalising factor in front of the exchanged amount and the arbitrageur would try to do this on a scale as large as possible.

It is rather obvious that in the situation described above the market cannot be in equilibrium. A moment's reflection reveals that the market forces triggered by the arbitrageurs will make the $ rise in Frankfurt and fall in

New York. The arbitrage possibility will disappear when the two prices become equal. Of course, "equality" here is to be understood as an approximate identity where — even for arbitrageurs with very low transaction costs — the above scheme is not profitable any more.

This brings us to a first — informal and intuitive — definition of arbitrage: an arbitrage opportunity is the possibility to make a profit in a financial market *without risk* and *without net investment of capital*. The *principle of no-arbitrage* states that a mathematical model of a financial market should not allow for arbitrage possibilities.

1.2 An Easy Model of a Financial Market

To apply this principle to less trivial cases than the Euro/Dollar example above, we consider a still extremely simple mathematical model of a financial market: there are two assets, called the bond and the stock. The bond is riskless, hence by definition we know what it is worth tomorrow. For (mainly notational) simplicity we neglect interest rates and assume that the price of a bond equals $1 \, €$ today as well as tomorrow, i.e.,

$$B_0 = B_1 = 1$$

The more interesting feature of the model is the stock which is risky: we know its value today, say (w.l.o.g.)

$$S_0 = 1,$$

but we don't know its value tomorrow. We model this uncertainty stochastically by defining S_1 to be a random variable depending on the random element $\omega \in \Omega$. To keep things as simple as possible, we let Ω consist of two elements only, g for "good" and b for "bad", with probability $\mathbf{P}[g] = \mathbf{P}[b] = \frac{1}{2}$. We define $S_1(\omega)$ by

$$S_1(\omega) = \begin{cases} 2 & \text{for } \omega = g \\ \frac{1}{2} & \text{for } \omega = b. \end{cases}$$

Now we introduce a third financial instrument in our model, an *option on the stock* with strike price K: the buyer of the option has the right — but not the obligation — to buy one stock at time $t = 1$ at a predefined price K. To fix ideas let $K = 1$. A moment's reflexion reveals that the price C_1 of the option at time $t = 1$ (where C stands for "call") equals

$$C_1 = (S_1 - K)_+,$$

i.e., in our simple example

$$C_1(\omega) = \begin{cases} 1 & \text{for } \omega = g \\ 0 & \text{for } \omega = b. \end{cases}$$

Hence we know the value of the option at time $t = 1$, *contingent on the value of the stock*. But what is the price of the option today?

The classical approach, used by actuaries for centuries, is to price contingent claims by taking expectations. In our example this gives the value $C_0 := \mathbf{E}[C_1] = \frac{1}{2}$. Although this simple approach is very successful in many actuarial applications, it is not at all satisfactory in the present context. Indeed, the rationale behind taking the expected value is the following argument based on the law of large numbers: in the long run the buyer of an option will neither gain nor lose in the average. We rephrase this fact in a more financial lingo: the performance of an investment into the option would in average equal the performance of the bond (for which we have assumed an interest rate equal to zero). However, a basic feature of finance is that an investment into a risky asset should in average yield a better performance than an investment into the bond (for the sceptical reader: at least, these two values should not necessarily coincide). In our "toy example" we have chosen the numbers such that $\mathbf{E}[S_1] = 1.25 > 1 = S_0$, so that in average the stock performs better than the bond. This indicates that the option (which clearly is a risky investment) should not necessarily have the same performance (in average) as the bond. It also shows that the old method of calculating prices via expectation is not directly applicable. It already fails for the stock and hence there is no reason why the price of the option should be given by its expectation $\mathbf{E}[C_1]$.

1.3 Pricing by No-Arbitrage

A different approach to the pricing of the option goes like this: we can buy at time $t = 0$ a *portfolio Π* consisting of $\frac{2}{3}$ of stock and $-\frac{1}{3}$ of bond. The reader might be puzzled about the negative sign: investing a negative amount into a bond — "going short" in the financial lingo — means borrowing money.

Note that — although normal people like most of us may not be able to do so — the "big players" can go "long" as well as "short". In fact they can do so not only with respect to the bond (i.e. to invest or borrow money at a fixed rate of interest) but can also go "long" as well as "short" in other assets like shares. In addition, they can do so at (relatively) low transaction costs, which is reflected by completely neglecting transaction costs in our present basic modelling.

Turning back to our portfolio Π one verifies that the value Π_1 of the portfolio at time $t = 1$ equals

$$\Pi_1(\omega) = \begin{cases} 1 & \text{for } \omega = g \\ 0 & \text{for } \omega = b. \end{cases}$$

The portfolio "replicates" the option, i.e.,

$$C_1 \equiv \Pi_1, \tag{1.1}$$

or, written more explicitly,

$$C_1(g) = \Pi_1(g), \tag{1.2}$$
$$C_1(b) = \Pi_1(b). \tag{1.3}$$

We are confident that the reader now sees why we have chosen the above weights $\frac{2}{3}$ and $-\frac{1}{3}$: the mathematical complexity of determining these weights such that (1.2) and (1.3) hold true, amounts to solving two linear equations in two variables.

The portfolio Π has a well-defined price at time $t = 0$, namely $\Pi_0 = \frac{2}{3}S_0 - \frac{1}{3}B_0 = \frac{1}{3}$. Now comes the "pricing by no-arbitrage" argument: equality (1.1) implies that we also must have

$$C_0 = \Pi_0 \tag{1.4}$$

whence $C_0 = \frac{1}{3}$. Indeed, suppose that (1.4) does not hold true; to fix ideas, suppose we have $C_0 = \frac{1}{2}$ as we had proposed above. This would allow an arbitrage by buying ("going long in") the portfolio Π and simultaneously selling ("going short in") the option C. The difference $C_0 - \Pi_0 = \frac{1}{6}$ remains as arbitrage profit at time $t = 0$, while at time $t = 1$ the two positions cancel out *independently of whether the random element ω equals g or b.*

Of course, the above considered size of the arbitrage profit by applying the above scheme to one option was only chosen for expository reasons: it is important to note that you may multiply the size of the above portfolios with your favourite power of ten, thus multiplying also your arbitrage profit.

At this stage we see that the story with the 100 € bill at the beginning of this chapter did not fully describe the idea of an arbitrage: The correct analogue would be to find instead of a single 100 € bill a "money pump", i.e., something like a box from which you can take one 100 € bill after another. While it might have happened to some of us, to occasionally find a 100 € bill lying around, we are confident that nobody ever found such a "money pump".

Another aspect where the little story at the beginning of this chapter did not fully describe the idea of arbitrage is the question of information. We shall assume throughout this book that all agents have the same information (there are no "insiders"). The theory changes completely when different agents have different information (which would correspond to the situation in the above joke). We will not address these extensions.

These arguments should convince the reader that the *"no-arbitrage principle"* is economically very appealing: in a liquid financial market there should be no arbitrage opportunities. Hence a mathematical model of a financial market should be designed in such a way that it does not permit arbitrage.

It is remarkable that this rather obvious principle yielded a unique price for the option considered in the above model.

1.4 Variations of the Example

Although the preceding "toy example" is extremely simple and, of course, far from reality, it contains the heart of the matter: the possibility of replicating a contingent claim, e.g. an option, by trading on the existing assets and to apply the no-arbitrage principle.

It is straightforward to generalise the example by passing from the time index set $\{0, 1\}$ to an arbitrary finite discrete time set $\{0, \ldots, T\}$, and by considering T independent Bernoulli random variables. This binomial model is called the Cox-Ross-Rubinstein model in finance (see Chap. 3 below).

It is also relatively simple — at least with the technology of stochastic calculus, which is available today — to pass to the (properly normalised) limit as T tends to infinity, thus ending up with a stochastic process driven by Brownian motion (see Chap. 4 below). The so-called geometric Brownian motion, i.e., Brownian motion on an exponential scale, is the celebrated *Black-Scholes model* which was proposed in 1965 by P. Samuelson, see [S 65]. In fact, already in 1900 L. Bachelier [B 00] used Brownian motion to price options in his remarkable thesis "Théorie de la spéculation" (a member of the jury and rapporteur was H. Poincaré).

In order to apply the above no-arbitrage arguments to more complex models we still need one additional, crucial concept.

1.5 Martingale Measures

To explain this notion let us turn back to our "toy example", where we have seen that the unique arbitrage free price of our option equals $C_0 = \frac{1}{3}$. We also have seen that, by taking expectations, we obtained $\mathbf{E}[C_1] = \frac{1}{2}$ as the price of the option, which was a "wrong price" as it allowed for arbitrage opportunities. The economic rationale for this discrepancy was that the expected return of the stock was higher than that of the bond.

Now make the following mind experiment: suppose that the world were governed by a different probability than $\mathbf{P}$ which assigns different weights to g and b, such that under this new probability, let's call it $\mathbf{Q}$, the expected return of the stock equals that of the bond. An elementary calculation reveals that the probability measure defined by $\mathbf{Q}[g] = \frac{1}{3}$ and $\mathbf{Q}[b] = \frac{2}{3}$ is the unique solution satisfying $\mathbf{E}_{\mathbf{Q}}[S_1] = S_0 = 1$. Mathematically speaking, the process S is a *martingale* under $\mathbf{Q}$, and $\mathbf{Q}$ is a *martingale measure* for S.

Speaking again economically, it is not unreasonable to expect that in a world governed by $\mathbf{Q}$, the recipe of taking expected values should indeed give a price for the option which is compatible with the no-arbitrage principle. After all, our original objection, that the average performance of the stock and the bond differ, now has disappeared. A direct calculation reveals that in our "toy example" these two prices for the option indeed coincide as

$$\mathbf{E_Q}[C_1] = \tfrac{1}{3}.$$

Clearly we suspect that this numerical match is not just a coincidence. At this stage it is, of course, the reflex of every mathematician to ask: what is precisely going on behind this phenomenon? A preliminary answer is that the expectation under the new measure $\mathbf{Q}$ defines a linear function of the span of B_1 and S_1. The price of an element in this span should therefore be the corresponding linear combination of the prices at time 0. Thus, using simple linear algebra, we get $C_0 = \tfrac{2}{3}S_0 - \tfrac{1}{3}B_0$ and moreover we identify this as $\mathbf{E_Q}[C_1]$.

1.6 The Fundamental Theorem of Asset Pricing

To make a long story very short: for a general stochastic process $(S_t)_{0 \le t \le T}$, modelled on a filtered probability space $(\Omega, (\mathcal{F}_t)_{0 \le t \le T}, \mathbf{P})$, the following statement *essentially* holds true. For any "contingent claim" C_T, i.e. an $\mathcal{F}_T$-measurable random variable, the formula

$$C_0 := \mathbf{E_Q}[C_T] \qquad (1.5)$$

yields precisely the arbitrage-free prices for C_T, when $\mathbf{Q}$ runs through the probability measures on $\mathcal{F}_T$, which are equivalent to $\mathbf{P}$ and under which the process S is a martingale (*"equivalent martingale measures"*). In particular, when there is precisely one equivalent martingale measure (as it is the case in the Cox-Ross-Rubinstein, the Black-Scholes and the Bachelier model), formula (1.5) gives the unique arbitrage free price C_0 for C_T. In this case we may "replicate" the contingent claim C_T as

$$C_T = C_0 + \int_0^T H_t dS_t, \qquad (1.6)$$

where $(H_t)_{0 \le t \le T}$ is a predictable process (a *"trading strategy"*) and where H_t models the holding in the stock S during the infinitesimal interval $[t, t + dt]$.

Of course, the stochastic integral appearing in (1.6) needs some care; fortunately people like K. Itô and P.A. Meyer's school of probability in Strasbourg told us very precisely how to interpret such an integral.

The mathematical challenge of the above story consists of getting rid of the word "essentially" and to turn this program into precise theorems.

The central piece of the theory relating the no-arbitrage arguments with martingale theory is the so-called Fundamental Theorem of Asset Pricing. We quote a general version of this theorem, which is proved in Chap. 14.

Theorem 1.6.1 (Fundamental Theorem of Asset Pricing). *For an $\mathbf{R}^d$-valued semi-martingale $S = (S_t)_{0 \le t \le T}$ t.f.a.e.:*

(i)   *There exists a probability measure* **Q** *equivalent to* **P** *under which S is a* sigma-martingale.

(ii) *S does not permit a* free lunch with vanishing risk.

This theorem was proved for the case of a probability space Ω consisting of finitely many elements by Harrison and Pliska [HP 81]. In this case one may equivalently write *no-arbitrage* instead of *no free lunch with vanishing risk* and *martingale* instead of *sigma-martingale*.

In the general case it is unavoidable to speak about more technical concepts, such as *sigma-martingales* (which is a generalisation of the notion of a local martingale) and *free lunches*. A *free lunch* (a notion introduced by D. Kreps [K 81]) is something like an arbitrage, where — roughly speaking — agents are allowed to form integrals as in (1.6), to subsequently "throw away money" (if they want do so), and finally to pass to the limit in an appropriate topology. It was the — somewhat surprising — insight of [DS 94] (reprinted in Chap. 9) that one may take the topology of *uniform convergence* (which allows for an economic interpretation to which the term "with vanishing risk" alludes) and still get a valid theorem.

The remainder of this book is devoted to the development of this theme, as well as to its remarkable scope of applications in Finance.

2

Models of Financial Markets
on Finite Probability Spaces

2.1 Description of the Model

In this section we shall develop the theory of pricing and hedging of derivative securities in financial markets.

In order to reduce the technical difficulties of the theory of option pricing to a minimum, we assume throughout this chapter that the probability space Ω underlying our model will be finite, say, $\Omega = \{\omega_1, \omega_2, \ldots, \omega_N\}$ equipped with a probability measure $\mathbf{P}$ such that $\mathbf{P}[\omega_n] = p_n > 0$, for $n = 1, \ldots, N$. This assumption implies that all functional-analytic delicacies pertaining to different topologies on $L^\infty(\Omega, \mathcal{F}, \mathbf{P})$, $L^1(\Omega, \mathcal{F}, \mathbf{P})$, $L^0(\Omega, \mathcal{F}, \mathbf{P})$ etc. evaporate, as all these spaces are simply $\mathbb{R}^N$ (we assume w.l.o.g. that the σ-algebra $\mathcal{F}$ is the power set of Ω). Hence all the functional analysis, which we shall need in later chapters for the case of more general processes, reduces in the setting of the present chapter to simple linear algebra. For example, the use of the Hahn-Banach theorem is replaced by the use of the separating hyperplane theorem in finite dimensional spaces.

Nevertheless we shall write $L^\infty(\Omega, \mathcal{F}, \mathbf{P})$, $L^1(\Omega, \mathcal{F}, \mathbf{P})$ etc. (knowing very well that in the present setting these spaces are all isomorphic to $\mathbb{R}^N$) to indicate, which function spaces we shall encounter in the setting of the general theory. It also helps to see if an element of $\mathbb{R}^N$ is a contingent claim or an element of the dual space, i.e. a price vector.

In addition to the probability space $(\Omega, \mathcal{F}, \mathbf{P})$ we fix a natural number $T \geq 1$ and a filtration $(\mathcal{F}_t)_{t=0}^T$ on Ω, i.e., an increasing sequence of σ-algebras. To avoid trivialities, we shall always assume that $\mathcal{F}_T = \mathcal{F}$; on the other hand, we shall *not* assume that $\mathcal{F}_0$ is trivial, i.e. $\mathcal{F}_0 = \{\emptyset, \Omega\}$, although this will be the case in most applications. But for technical reasons it will be more convenient to allow for general σ-algebras $\mathcal{F}_0$.

We now introduce a model of a financial market in not necessarily discounted terms. The rest of Sect. 2.1 will be devoted to reducing this situation to a model in discounted terms which, as we shall see, will make life much easier.

Readers who are not so enthusiastic about this mainly formal and elementary reduction might proceed directly to Definition 2.1.4. On the other hand, we know from sad experience that often there is a lot of myth and confusion arising in this operation of discounting; for this reason we decided to devote this section to the clarification of this issue.

Definition 2.1.1. *A model of a* financial market *is an* $\mathbb{R}^{d+1}$*-valued stochastic process* $\widehat{S} = (\widehat{S}_t)_{t=0}^T = (\widehat{S}_t^0, \widehat{S}_t^1, \dots, \widehat{S}_t^d)_{t=0}^T$*, based on and adapted to the filtered stochastic base* $(\Omega, \mathcal{F}, (\mathcal{F}_t)_{t=0}^T, \mathbf{P})$*. We shall assume that the zero coordinate* $\widehat{S}^0$ *satisfies* $\widehat{S}_t^0 > 0$ *for all* $t = 0, \dots, T$ *and* $\widehat{S}_0^0 = 1$.

The interpretation is the following. The prices of the assets $0, \dots, d$ are measured in a fixed money unit, say Euros. For $1 \le j \le d$ they are not necessarily non-negative (think, e.g., of forward contracts). The asset 0 plays a special role. It is supposed to be strictly positive and will be used as a numéraire. It allows us to compare money (e.g., Euros) at time 0 to money at time $t > 0$. In many elementary models, $\widehat{S}^0$ is simply a bank account which in case of constant interest rate r is then defined as $\widehat{S}_t^0 = e^{rt}$. However, it might also be more complicated, e.g. $\widehat{S}_t^0 = \exp(r_0 h + r_1 h + \dots + r_{t-1} h)$ where $h > 0$ is the length of the time interval between $t - 1$ and t (here kept fixed) and where r_{t-1} is the stochastic interest rate valid between $t - 1$ and t. Other models are also possible and to prepare the reader for more general situations, we only require $\widehat{S}_t^0$ to be strictly positive. Notice that we only require that $\widehat{S}_t^0$ to be $\mathcal{F}_t$-measurable and that it is not necessarily $\mathcal{F}_{t-1}$-measurable. In other words, we assume that the process $\widehat{S}^0 = (\widehat{S}_t^0)_{t=0}^T$ is adapted, but not necessarily predictable.

An economic agent is able to buy and sell financial assets. The decision taken at time t can only use information available at time t which is modelled by the σ-algebra $\mathcal{F}_t$.

Definition 2.1.2. *A* trading strategy *$(\widehat{H}_t)_{t=1}^T = (\widehat{H}_t^0, \widehat{H}_t^1, \dots, \widehat{H}_t^d)_{t=1}^T$ is an* $\mathbb{R}^{d+1}$*-valued process which is predictable, i.e.* $\widehat{H}_t$ *is* $\mathcal{F}_{t-1}$*-measurable.*

The interpretation is that between time $t - 1$ and time t, the agent holds a quantity equal to $\widehat{H}_t^j$ of asset j. The decision is taken at time $t - 1$ and therefore, $\widehat{H}_t$ is required to be $\mathcal{F}_{t-1}$-measurable.

Definition 2.1.3. *A strategy* $(\widehat{H}_t)_{t=1}^T$ *is called* self financing *if for every* $t = 1, \dots, T - 1$, *we have*

$$\left(\widehat{H}_t, \widehat{S}_t\right) = \left(\widehat{H}_{t+1}, \widehat{S}_t\right) \tag{2.1}$$

or, written more explicitly,

$$\sum_{j=0}^d \widehat{H}_t^j \widehat{S}_t^j = \sum_{j=0}^d \widehat{H}_{t+1}^j \widehat{S}_t^j. \tag{2.2}$$

The initial investment required for a strategy is $\widehat{V}_0 = (\widehat{H}_1, \widehat{S}_0) = \sum_{j=0}^d \widehat{H}_1^j \widehat{S}_0^j.$

The interpretation goes as follows. By changing the portfolio from $\widehat{H}_{t-1}$ to $\widehat{H}_t$ there is no input/outflow of money. We remark that we assume that changing a portfolio does not trigger transaction costs. Also note that $\widehat{H}_t^j$ may assume negative values, which corresponds to short selling asset j during the time interval $]t_{j-1}, t_j]$.

The $\mathcal{F}_t$-measurable random variable defined in (2.1) is interpreted as the value $\widehat{V}_t$ of the portfolio at time t defined by the trading strategy $\widehat{H}$:

$$\widehat{V}_t = (\widehat{H}_t, \widehat{S}_t) = (\widehat{H}_{t+1}, \widehat{S}_t).$$

The way in which the value $(\widehat{H}_t, \widehat{S}_t)$ evolves can be described much easier when we use discounted prices using the asset $\widehat{S}^0$ as numéraire. Discounting allows us to compare money at time t to money at time 0. For instance we could say that $\widehat{S}_t^0$ units of money at time t are the "same" as 1 unit of money, e.g., Euros, at time 0. So let us see what happens if we replace prices $\widehat{S}$ by discounted prices $\left(\frac{\widehat{S}}{\widehat{S}^0}\right) = \left(\frac{\widehat{S}^0}{\widehat{S}^0}, \frac{\widehat{S}^1}{\widehat{S}^0}, \ldots, \frac{\widehat{S}^d}{\widehat{S}^0}\right)$. We will use the notation

$$S_t^j := \frac{\widehat{S}_t^j}{\widehat{S}_t^0}, \quad \text{for } j = 1, \ldots, d \text{ and } t = 0, \ldots, T. \tag{2.3}$$

There is no need to include the coordinate 0, since obviously $S_t^0 = 1$. Let us now consider $(\widehat{H}_t)_{t=1}^T = (\widehat{H}_t^0, \widehat{H}_t^1, \ldots, \widehat{H}_t^d)_{t=1}^T$ to be a self financing strategy with initial investment $\widehat{V}_0$; we then have

$$\widehat{V}_0 = \sum_{j=0}^d \widehat{H}_1^j \widehat{S}_0^j = \widehat{H}_1^0 + \sum_{j=1}^d \widehat{H}_1^j \widehat{S}_0^j = \widehat{H}_1^0 + \sum_{j=1}^d \widehat{H}_1^j S_0^j,$$

since by definition $\widehat{S}_0^0 = 1$.

We now write $(H_t)_{t=1}^T = (H_t^1, \ldots, H_t^d)_{t=1}^T$ for the $\mathbb{R}^d$-valued process obtained by discarding the 0'th coordinate of the $\mathbb{R}^{d+1}$-valued process $(\widehat{H}_t)_{t=1}^T = (\widehat{H}_t^0, \widehat{H}_t^1, \ldots, \widehat{H}_t^d)_{t=1}^T$, i.e., $H_t^j = \widehat{H}_t^j$ for $j = 1, \ldots, d$. The reason for dropping the 0'th coordinate is, as we shall discover in a moment, that the holdings $\widehat{H}_t^0$ in the numéraire asset S_t^0 will be no longer of importance when we do the book-keeping in terms of the numéraire asset, i.e., in discounted terms.

One can make the following easy, but crucial observation: for *every* $\mathbb{R}^d$-valued, predictable process $(H_t)_{t=1}^T = (H_t^1, \ldots, H_t^d)_{t=1}^T$ there exists a unique *self financing* $\mathbb{R}^{d+1}$-valued predictable process $(\widehat{H}_t)_{t=1}^T = (\widehat{H}_t^0, \widehat{H}_t^1, \ldots, \widehat{H}_t^d)_{t=1}^T$ such that $(\widehat{H}_t^j)_{t=1}^T = (H_t^j)_{t=1}^T$ for $j = 1, \ldots, d$ and $\widehat{H}_1^0 = 0$. Indeed, one determines the values of $\widehat{H}_{t+1}^0$, for $t = 1, \ldots, T-1$, by inductively applying (2.2). The strict positivity of $(\widehat{S}_t^0)_{t=0}^{T-1}$ implies that there is precisely one function $\widehat{H}_{t+1}^0$ such that equality (2.2) holds true. Clearly such a function $\widehat{H}_{t+1}^0$ is

$\mathcal{F}_t$-measurable. In economic terms the above argument is rather obvious: for any given trading strategy $(H_t)_{t=1}^T = (H_t^1, \ldots, H_t^d)_{t=1}^T$ in the "risky" assets $j = 1, \ldots, d$, we may always add a trading strategy $(\widehat{H}_t^0)_{t=1}^T$ in the numéraire asset 0 such that the total strategy becomes self financing. Moreover, by normalising $\widehat{H}_1^0 = 0$, this trading strategy becomes unique. This can be particularly well visualised when interpreting the asset 0 as a cash account, into which at all times $t = 1, \ldots, T-1$, the gains and losses occurring from the investments in the d risky assets are absorbed and from which the investments in the risky assets are financed. If we normalise this procedure by requiring $\widehat{H}_1^0 = 0$, i.e., by starting with an empty cash account, then clearly the subsequent evolution of the holdings in the cash account is uniquely determined by the holdings in the "risky" assets $1, \ldots, d$. From now on we fix two processes $(\widehat{H}_t)_{t=1}^T = (\widehat{H}_t^0, \widehat{H}_t^1, \ldots, \widehat{H}_t^d)_{t=1}^T$ and $(H_t)_{t=1}^T = (H_t^1, \ldots, H_t^d)_{t=1}^T$ corresponding uniquely one to each other in the above described way.

Now one can make a second straightforward observation: the investment $(\widehat{H}_t^0)_{t=1}^T$ in the numéraire asset does not change the *discounted* value $(V_t)_{t=0}^T$ of the portfolio. Indeed, by definition — and rather trivially — the numéraire asset remains constant in discounted terms (i.e., expressed in units of itself).

Hence the discounted value V_t of the portfolio

$$V_t = \frac{\widehat{V}_t}{\widehat{S}_t^0}, \quad t = 0, \ldots, T,$$

depends only on the $\mathbb{R}^d$-dimensional process $(H_t)_{t=1}^T = (H_t^1, \ldots, H_t^d)_{t=1}^T$.

More precisely, in view of the normalisation $\widehat{S}_0^0 = 1$ and $\widehat{H}_1^0 = 0$ we have

$$\widehat{V}_0 = V_0 = \sum_{j=1}^d H_1^j S_0^j.$$

For the increment $\Delta V_{t+1} = V_{t+1} - V_t$ we find, using (2.2),

$$\Delta V_{t+1} = V_{t+1} - V_t = \frac{\widehat{V}_{t+1}}{\widehat{S}_{t+1}^0} - \frac{\widehat{V}_t}{\widehat{S}_t^0}$$

$$= \sum_{j=0}^d \widehat{H}_{t+1}^j \frac{\widehat{S}_{t+1}^j}{\widehat{S}_{t+1}^0} - \sum_{j=0}^d \widehat{H}_{t+1}^j \frac{\widehat{S}_t^j}{\widehat{S}_t^0}$$

$$= \widehat{H}_{t+1}^0 (1 - 1) + \sum_{j=1}^d \widehat{H}_{t+1}^j \left(S_{t+1}^j - S_t^j \right)$$

$$= \left(H_{t+1}^j, \Delta S_{t+1}^j \right),$$

where $(.,.)$ now denotes the inner product in $\mathbb{R}^d$.

In particular, the final value V_T of the portfolio becomes (in discounted units)

$$V_T = V_0 + \sum_{t=1}^{T} (H_t, \Delta S_t) = V_0 + (H \cdot S)_T,$$

where $(H \cdot S)_T = \sum_{t=1}^{T} (H_t, \Delta S_t)$ is the notation for a stochastic integral familiar from the theory of stochastic integration. In our discrete time framework the "stochastic integral" is simply a finite Riemann sum.

In order to know the value V_T of the portfolio in real money, we still would have to multiply by $\widehat{S}_T^0$, i.e., we have $\widehat{V}_T = V_T \widehat{S}_T^0$. This, however, is rarely needed.

We can therefore replace Definition 2.1.2 by the following definition in discounted terms, which will turn out to be much easier to handle.

Definition 2.1.4. *Let* $S = (S^1, \ldots, S^d)$ *be a model of a financial market in discounted terms. A* trading strategy *is an* $\mathbb{R}^d$-*valued process* $(H_t)_{t=1}^{T} = (H_t^1, H_t^2, \ldots, H_t^d)_{t=1}^{T}$ *which is predictable, i.e., each* H_t *is* $\mathcal{F}_{t-1}$-*measurable. We denote by* $\mathcal{H}$ *the set of all such trading strategies.*

We then define the stochastic integral $H \cdot S$ *as the* $\mathbb{R}$-*valued process* $((H \cdot S)_t)_{t=0}^{T}$ *given by*

$$(H \cdot S)_t = \sum_{u=1}^{t} (H_u, \Delta S_u), \quad t = 0, \ldots, T, \tag{2.4}$$

where $(.\,,.)$ *denotes the inner product in* $\mathbb{R}^d$. *The random variable*

$$(H \cdot S)_t = \sum_{u=1}^{t} (H_u, \Delta S_u)$$

models — when following the trading strategy H *— the gain or loss occurred up to time* t *in discounted terms.*

Summing up: by following the good old actuarial tradition of discounting, i.e. by passing from the process $\widehat{S}$, denoted in units of money, to the process S, denoted in terms of the numéraire asset (e.g., the cash account), things become considerably simpler and more transparent. In particular the value process V of an agent starting with initial wealth $V_0 = 0$ and subsequently applying the trading strategy H, is given by the stochastic integral $V_t = (H \cdot S)_t$ defined in (2.4).

We still emphasize that the choice of the numéraire is not unique; only for notational convenience we have fixed it to be the asset indexed by 0. But it may be chosen as any traded asset, provided only that it always remains strictly positive. We shall deal with this topic in more detail in Sect. 2.5 below.

From now on we shall work in terms of the discounted $\mathbb{R}^d$-valued process, denoted by S.

2.2 No-Arbitrage and the Fundamental Theorem of Asset Pricing

Definition 2.2.1. *We call the subspace K of $L^0(\Omega, \mathcal{F}, \mathbf{P})$ defined by*

$$K = \{(H \cdot S)_T \mid H \in \mathcal{H}\},$$

the set of contingent claims attainable at price 0.

We leave it to the reader to check that K is indeed a vector space.

The economic interpretation is the following: the random variables $f = (H \cdot S)_T$ are precisely those contingent claims, i.e., the pay-off functions at time T, depending on $\omega \in \Omega$, that an economic agent may replicate with zero initial investment by pursuing some predictable trading strategy H.

For $a \in \mathbb{R}$, we call the *set of contingent claims attainable at price a* the affine space $K_a = a + K$, obtained by shifting K by the constant function a, in other words, the space of all the random variables of the form $a + (H \cdot S)_T$, for some trading strategy H. Again the economic interpretation is that these are precisely the contingent claims that an economic agent may replicate with an initial investment of a by pursuing some predictable trading strategy H.

Definition 2.2.2. *We call the convex cone C in $L^\infty(\Omega, \mathcal{F}, \mathbf{P})$ defined by*

$$C = \{g \in L^\infty(\Omega, \mathcal{F}, \mathbf{P}) \mid \text{there exists } f \in K \text{ with } f \geq g\}.$$

the set of contingent claims super-replicable at price 0.

Economically speaking, a contingent claim $g \in L^\infty(\Omega, \mathcal{F}, \mathbf{P})$ is *super-replicable at price* 0, if we can achieve it with zero net investment by pursuing some predictable trading strategy H. Thus we arrive at some contingent claim f and if necessary we "throw away money" to arrive at g. This operation of "throwing away money" or "free disposal" may seem awkward at this stage, but we shall see later that the set C plays an important role in the development of the theory. Observe that C is a convex cone containing the negative orthant $L^\infty_-(\Omega, \mathcal{F}, \mathbf{P})$. Again we may define $C_a = a + C$ as the *contingent claims super-replicable at price a*, if we shift C by the constant function a.

Definition 2.2.3. *A financial market S satisfies the* no-arbitrage condition *(NA) if*

$$K \cap L^0_+(\Omega, \mathcal{F}, \mathbf{P}) = \{0\}$$

or, equivalently,

$$C \cap L^0_+(\Omega, \mathcal{F}, \mathbf{P}) = \{0\}$$

where 0 denotes the function identically equal to zero.

Recall that $L^0(\Omega, \mathcal{F}, \mathbf{P})$ denotes the space of all $\mathcal{F}$-measurable real-valued functions and $L^0_+(\Omega, \mathcal{F}, \mathbf{P})$ its positive orthant.

We now have formalised the concept of an arbitrage possibility: it means the existence of a trading strategy H such that — starting from an initial investment zero — the resulting contingent claim $f = (H \cdot S)_T$ is non-negative and not identically equal to zero. Such an opportunity is of course the dream of every arbitrageur. If a financial market does not allow for arbitrage opportunities, we say it satisfies the *no-arbitrage condition (NA)*.

Proposition 2.2.4. *Assume S satisfies (NA) then*

$$C \cap (-C) = K.$$

Proof. Let $g \in C \cap (-C)$ then $g = f_1 - h_1$ with $f_1 \in K$, $h_1 \in L^\infty_+$ and $g = f_2 + h_2$ with $f_2 \in K$ and $h_2 \in L^\infty_+$. Then $f_1 - f_2 = h_1 + h_2 \in L^\infty_+$ and hence $f_1 - f_2 \in K \cap L^\infty_+ = \{0\}$. It follows that $f_1 = f_2$ and $h_1 + h_2 = 0$, hence $h_1 = h_2 = 0$. This means that $g = f_1 = f_2 \in K$. $\qquad\square$

Definition 2.2.5. *A probability measure $\mathbf{Q}$ on $(\Omega, \mathcal{F})$ is called an* equivalent martingale measure *for S, if $\mathbf{Q} \sim \mathbf{P}$ and S is a martingale under $\mathbf{Q}$, i.e., $\mathbf{E_Q}[S_{t+1}|\mathcal{F}_t] = S_t$ for $t = 0, \dots, T - 1$.*

We denote by $\mathcal{M}^e(S)$ the set of equivalent martingale measures and by $\mathcal{M}^a(S)$ the set of all (not necessarily equivalent) martingale probability measures. The letter a stands for "absolutely continuous with respect to $\mathbf{P}$" which in the present setting (finite Ω and $\mathbf{P}$ having full support) automatically holds true, but which will be of relevance for general probability spaces $(\Omega, \mathcal{F}, \mathbf{P})$ later. Note that in the present setting of a finite probability space Ω with $\mathbf{P}[\omega] > 0$ for each $\omega \in \Omega$, we have that $\mathbf{Q} \sim \mathbf{P}$ iff $\mathbf{Q}[\omega] > 0$, for each $\omega \in \Omega$. We shall often identify a measure $\mathbf{Q}$ on $(\Omega, \mathcal{F})$ with its Radon-Nikodým derivative $\frac{d\mathbf{Q}}{d\mathbf{P}} \in L^1(\Omega, \mathcal{F}, \mathbf{P})$. In the present setting of finite Ω, this simply means

$$\frac{d\mathbf{Q}}{d\mathbf{P}}(\omega) = \frac{\mathbf{Q}[\omega]}{\mathbf{P}[\omega]}.$$

In statistics this quantity is also called the likelihood ratio.

Lemma 2.2.6. *For a probability measure $\mathbf{Q}$ on $(\Omega, \mathcal{F})$ the following are equivalent:*

(i) $\mathbf{Q} \in \mathcal{M}^a(S)$,
(ii) $\mathbf{E_Q}[f] = 0$, *for all $f \in K$,*
(iii) $\mathbf{E_Q}[g] \le 0$, *for all $g \in C$.*

Proof. The equivalences are rather trivial. (ii) is tantamount to the very definition of S being a martingale under $\mathbf{Q}$, i.e., to the validity of

$$\mathbf{E_Q}[S_t \mid \mathcal{F}_{t-1}] = S_{t-1}, \quad \text{for } t = 1, \dots, T. \tag{2.5}$$

Indeed, (2.5) holds true iff for each $\mathcal{F}_{t-1}$-measurable set A we have $\mathbf{E_Q}[\chi_A(S_t - S_{t-1})] = 0 \in \mathbb{R}^d$, in other words $\mathbf{E_Q}[(x\chi_A, \Delta S_t)] = 0$, for each x. By linearity this relation extends to K which shows (ii).

The equivalence of (ii) and (iii) is straightforward.   □

After having fixed these formalities we may formulate and prove the central result of the theory of pricing and hedging by no-arbitrage, sometimes called the "Fundamental Theorem of Asset Pricing", which in its present form (i.e., finite Ω) is due to M. Harrison and S.R. Pliska [HP 81].

Theorem 2.2.7 (Fundamental Theorem of Asset Pricing). *For a financial market S modelled on a finite stochastic base $(\Omega, \mathcal{F}, (\mathcal{F}_t)_{t=0}^T, \mathbf{P})$, the following are equivalent:*

(i) *S satisfies (NA),*
(ii) *$\mathcal{M}^e(S) \neq \emptyset$.*

Proof. (ii) $\Rightarrow$ (i): This is the obvious implication. If there is some $\mathbf{Q} \in \mathcal{M}^e(S)$ then by Lemma 2.2.6 we have that

$$\mathbf{E_Q}[g] \leq 0, \quad \text{for } g \in C.$$

On the other hand, if there were $g \in C \cap L_+^\infty$, $g \neq 0$, then, using the assumption that $\mathbf{Q}$ is equivalent to $\mathbf{P}$, we would have

$$\mathbf{E_Q}[g] > 0,$$

a contradiction.

(i) $\Rightarrow$ (ii) This implication is the important message of the theorem which will allow us to link the no-arbitrage arguments with martingale theory. We give a functional analytic existence proof, which will be extendable — in spirit — to more general situations.

By assumption the space K intersects L_+^∞ only at 0. We want to separate the disjoint convex sets $L_+^\infty \setminus \{0\}$ and K by a hyperplane induced by a linear functional $\mathbf{Q} \in L^1(\Omega, \mathcal{F}, \mathbf{P})$. In order to get a strict separation of K and $L_+^\infty \setminus \{0\}$ we have to be a little careful since the standard separation theorems do not directly apply.

One way to overcome this difficulty (in finite dimension) is to consider the convex hull of the unit vectors $\left(\mathbf{1}_{\{\omega_n\}}\right)_{n=1}^N$ in $L^\infty(\Omega, \mathcal{F}, \mathbf{P})$ i.e.

$$P := \left\{ \sum_{n=1}^N \mu_n \mathbf{1}_{\{\omega_n\}} \,\middle|\, \mu_n \geq 0, \sum_{n=1}^N \mu_n = 1 \right\}.$$

This is a convex, compact subset of $L_+^\infty(\Omega, \mathcal{F}, \mathbf{P})$ and, by the *(NA)* assumption, disjoint from K. Hence we may strictly separate the convex compact set P from the convex closed set K by a linear functional $\mathbf{Q} \in L^\infty(\Omega, \mathcal{F}, \mathbf{P})^* = L^1(\Omega, \mathcal{F}, \mathbf{P})$, i.e., find $\alpha < \beta$ such that

$$(\mathbf{Q}, f) \leq \alpha, \quad \text{for } f \in K,$$
$$(\mathbf{Q}, h) \geq \beta, \quad \text{for } h \in P.$$

Since K is a linear space, we have $\alpha \geq 0$ and may replace α by 0. Hence $\beta > 0$. Defining by I the constant vector $I = (1, \ldots, 1)$, we have $(\mathbf{Q}, I) > 0$, where I denotes the constant function equal to one, and we may normalise $\mathbf{Q}$ such that $(\mathbf{Q}, I) = 1$. As $\mathbf{Q}$ is strictly positive on each $\mathbf{1}_{\{\omega_n\}}$, we therefore have found a probability measure $\mathbf{Q}$ on $(\Omega, \mathcal{F})$ equivalent to $\mathbf{P}$ such that condition (ii) of Lemma 2.2.6 holds true. In other words, we found an equivalent martingale measure $\mathbf{Q}$ for the process S. □

The name "Fundamental Theorem of Asset Pricing" was, as far as we are aware, first used in [DR 87]. We shall see that it plays a truly fundamental role in the theory of pricing and hedging of derivative securities (or, synonymously, contingent claims, i.e., elements of $L^0(\Omega, \mathcal{F}, \mathbf{P})$) by no-arbitrage arguments.

It seems worthwhile to discuss the intuitive interpretation of this basic result: a martingale S (say, under the original measure $\mathbf{P}$) is a mathematical model for a *perfectly fair* game. Applying any strategy $H \in \mathcal{H}$ we always have $\mathbf{E}[(H \cdot S)_T] = 0$, i.e., an investor can neither win nor lose in expectation.

On the other hand, a process S allowing for arbitrage, is a model for an utterly unfair game: choosing a good strategy $H \in \mathcal{H}$, an investor can make "something out of nothing". Applying H, the investor is sure not to lose, but has strictly positive probability to gain something.

In reality, there are many processes S which do not belong to either of these two extreme classes. Nevertheless, the above theorem tells us that there is a sharp dichotomy by allowing to *change the odds*. Either a process S is utterly unfair, in the sense that it allows for arbitrage. In this case there is no remedy to make the process fair by changing the odds: it never becomes a martingale. In fact, the possibility of making an arbitrage is not affected by changing the odds, i.e., by passing to an equivalent probability $\mathbf{Q}$. On the other hand, discarding this extreme case of processes allowing for arbitrage, we can always pass from $\mathbf{P}$ to an equivalent measure $\mathbf{Q}$ under which S is a martingale, i.e., a perfectly fair game. Note that the passage from $\mathbf{P}$ to $\mathbf{Q}$ may change the probabilities (the "odds") but not the impossible events (i.e. the null sets).

We believe that this dichotomy is a remarkable fact, also from a purely intuitive point of view.

Corollary 2.2.8. *Let S satisfy (NA) and let $f \in L^\infty(\Omega, \mathcal{F}, \mathbf{P})$ be an attainable contingent claim. In other words f is of the form*

$$f = a + (H \cdot S)_T, \tag{2.6}$$

for some $a \in \mathbb{R}$ and some trading strategy H. Then the constant a and the process $(H \cdot S)_t$ are uniquely determined by (2.6) and satisfy, for every $\mathbf{Q} \in \mathcal{M}^e(S)$,

$$a = \mathbf{E_Q}[f], \quad and \quad a + (H \cdot S)_t = \mathbf{E_Q}[f \mid \mathcal{F}_t], \quad for \quad 0 \leq t \leq T. \quad (2.7)$$

Proof. As regards the uniqueness of the constant $a \in \mathbb{R}$, suppose that there are two representations $f = a^1 + (H^1 \cdot S)_T$ and $f = a^2 + (H^2 \cdot S)_T$ with $a^1 \neq a^2$. Assuming w.l.o.g. that $a^1 > a^2$ we find an obvious arbitrage possibility by considering the trading strategy $H_2 - H_1$. We have $a^1 - a^2 = ((H^2 - H^1) \cdot S)_T$, i.e. the trading strategy $H^2 - H^1$ produces a strictly positive result at time T, a contradiction to *(NA)*.

As regards the uniqueness of the process $H \cdot S$, we simply apply a conditional version of the previous argument: assume that $f = a + (H^1 \cdot S)_T$ and $f = a + (H^2 \cdot S)_T$ and suppose that the processes $H^1 \cdot S$ and $H^2 \cdot S$ are not identical. Then there is $0 \leq t \leq T$ such that $(H^1 \cdot S)_t \neq (H^2 \cdot S)_t$ and without loss of generality we may suppose that $A := \{(H^1 \cdot S)_t > (H^2 \cdot S)_t\}$ is a non-empty event, which clearly is in $\mathcal{F}_t$. Hence, using the fact hat $(H^1 \cdot S)_T = (H^2 \cdot S)_T$, the trading strategy $H := (H^2 - H^1)\mathbf{1}_A \cdot \mathbf{1}_{]t,T]}$ is a predictable process producing an arbitrage, as $(H \cdot S)_T = 0$ outside A, while $(H \cdot S)_T = (H^1 \cdot S)_t - (H^2 \cdot S)_t > 0$ on A, which again contradicts *(NA)*.

Finally, the equations in (2.7) result from the fact that, for every predictable process H and every $\mathbf{Q} \in \mathcal{M}^a(S)$, the process $H \cdot S$ is a $\mathbf{Q}$-martingale. $\qquad\square$

We denote by $\mathrm{cone}(\mathcal{M}^e(S))$ and $\mathrm{cone}(\mathcal{M}^a(S))$ the cones generated by the convex sets $\mathcal{M}^e(S)$ and $\mathcal{M}^a(S)$ respectively. The subsequent Proposition 2.2.9 clarifies the polar relation between these cones and the cone C.

Let $\langle E, E' \rangle$ be two vector spaces in separating duality. This means that there is a bilinear form $\langle . , . \rangle : E \times E' \to \mathbb{R}$, so that if $\langle x, x' \rangle = 0$ for all $x \in E$, we must have $x' = 0$. Similarly if $\langle x, x' \rangle = 0$ for all $x' \in E'$, we must have $x = 0$. Recall (see, e.g., [Sch 99]) that, for a pair (E, E') of vector spaces in separating duality via the scalar product $\langle . , . \rangle$, the polar C^0 of a set C in E is defined by

$$C^0 = \{g \in E' \mid \langle f, g \rangle \leq 1 \text{ for all } f \in C\}.$$

In the case when C is closed under multiplication by positive scalars (e.g., if C is a convex cone) the polar C^0 may equivalently be defined as

$$C^0 = \{g \in E' \mid \langle f, g \rangle \leq 0 \text{ for all } f \in C\}.$$

The *bipolar theorem* (see, e.g., [Sch 99]) states that the bipolar $C^{00} := (C^0)^0$ of a set C in E is the $\sigma(E, E')$-closed convex hull of C.

In the present, finite dimensional case, $E = L^\infty(\Omega, \mathcal{F}_T, \mathbf{P}) = \mathbb{R}^N$ and $E' = L^1(\Omega, \mathcal{F}_T, \mathbf{P}) = \mathbb{R}^N$ the bipolar theorem is easier. In this case there is only one topology on $\mathbb{R}^N$ compatible with its vector space structure, so that we don't have to speak about different topologies such as $\sigma(E, E')$. However, the proof of the bipolar theorem is in the finite dimensional case and in the infinite dimensional case almost the same and follows from the separating hyperplane resp. the Hahn-Banach theorem.

After these general observations we pass to the concrete setting of the cone $C \subseteq L^\infty(\Omega, \mathcal{F}, \mathbf{P})$ of contingent claims super-replicable at price 0. Note that in our finite dimensional setting this convex cone is closed as it is the algebraic sum of the closed linear space K (a linear space in $\mathbb{R}^N$ is always closed) and the closed polyhedral cone $L^\infty_-(\Omega, \mathcal{F}, \mathbf{P})$ (the verification, that the algebraic sum of a space and a polyhedral cone in $\mathbb{R}^N$ is closed, is an easy, but not completely trivial exercise). We deduce from the bipolar theorem, that C equals its bipolar C^{00}.

Proposition 2.2.9. *Suppose that S satisfies (NA). Then the polar of C is equal to* cone($\mathcal{M}^a(S)$), *the cone generated by $\mathcal{M}^a(S)$, and $\mathcal{M}^e(S)$ is dense in $\mathcal{M}^a(S)$. Hence the following assertions are equivalent for an element $g \in L^\infty(\Omega, \mathcal{F}, \mathbf{P})$:*

(i) $g \in C$,
(ii) $\mathbf{E}_\mathbf{Q}[g] \le 0$, *for all* $\mathbf{Q} \in \mathcal{M}^a(S)$,
(iii) $\mathbf{E}_\mathbf{Q}[g] \le 0$, *for all* $\mathbf{Q} \in \mathcal{M}^e(S)$.

Proof. The fact that the polar C^0 and the set cone($\mathcal{M}^a(S)$) coincide, follows from Lemma 2.2.6 and the observation that $C \supseteq L^\infty_-(\Omega, \mathcal{F}, \mathbf{P})$ and $C^0 \subseteq L^1_+(\Omega, \mathcal{F}, \mathbf{P})$. Hence the equivalence of (i) and (ii) follows from the bipolar theorem.

As regards the density of $\mathcal{M}^e(S)$ in $\mathcal{M}^a(S)$ we first deduce from Theorem 2.2.7 that there is at least one $\mathbf{Q}^* \in \mathcal{M}^e(S)$. For any $\mathbf{Q} \in \mathcal{M}^a(S)$ and $0 < \mu \le 1$ we have that $\mu\mathbf{Q}^* + (1 - \mu)\mathbf{Q} \in \mathcal{M}^e(S)$, which clearly implies the density of $\mathcal{M}^e(S)$ in $\mathcal{M}^a(S)$. The equivalence of (ii) and (iii) is now obvious. $\qquad\square$

Similarly we can show the following:

Proposition 2.2.10. *Suppose S satisfies (NA). Then for $f \in L^\infty$, the following assertions are equivalent*

(i) $f \in K$, *i.e.* $f = (H \cdot S)_T$ *for some strategy $H \in \mathcal{H}$.*
(ii) *For all $\mathbf{Q} \in \mathcal{M}^e(S)$ we have $\mathbf{E}_\mathbf{Q}[f] = 0$.*
(iii) *For all $\mathbf{Q} \in \mathcal{M}^a(S)$ we have $\mathbf{E}_\mathbf{Q}[f] = 0$.*

Proof. By Proposition 2.2.4 we have that $f \in K$ iff $f \in C \cap (-C)$. Hence the result follows from the preceding Proposition 2.2.9. $\qquad\square$

Corollary 2.2.11. *Assume that S satisfies (NA) and that $f \in L^\infty$ satisfies $\mathbf{E}_\mathbf{Q}[f] = a$ for all $\mathbf{Q} \in \mathcal{M}^e(S)$, then $f = a + (H \cdot S)_T$ for some strategy H.* $\square$

Corollary 2.2.12 (complete financial markets). *For a financial market S satisfying the no-arbitrage condition (NA), the following are equivalent:*

(i) $\mathcal{M}^e(S)$ *consists of a single element $\mathbf{Q}$.*
(ii) *Each $f \in L^\infty(\Omega, \mathcal{F}, \mathbf{P})$ may be represented as*

$$f = a + (H \cdot S)_T \quad \text{for some } a \in \mathbb{R} \text{ and } H \in \mathcal{H}.$$

In this case $a = \mathbf{E}_{\mathbf{Q}}[f]$, *the stochastic integral* $H \cdot S$ *is unique and we have that*

$$\mathbf{E}_{\mathbf{Q}}[f \mid \mathcal{F}_t] = \mathbf{E}_{\mathbf{Q}}[f] + (H \cdot S)_t, \quad t = 0, \ldots, T. \qquad \square$$

The Fundamental Theorem of Asset Pricing 2.2.7 allows us to prove the following proposition, which we shall need soon.

Proposition 2.2.13. *Assume that S satisfies (NA) and let $H \cdot S$ be the process obtained from S by means of a fixed strategy $H \in \mathcal{H}$. Fix $a \in \mathbb{R}$ and define the $\mathbb{R}$-valued process $S^{d+1} = (S_t^{d+1})_{t=0}^T$ by $S^{d+1} = a + H \cdot S$. Then the process $\overline{S} = (S^1, S^2, \ldots, S^d, S^{d+1})$ also satisfies the (NA) property and the sets $\mathcal{M}^e(S)$ and $\mathcal{M}^e(\overline{S})$ (as well as $\mathcal{M}^a(S)$ and $\mathcal{M}^a(\overline{S})$) coincide.*

Proof. If $\mathbf{Q} \in \mathcal{M}^e(S)$ then $H \cdot S$ is a $\mathbf{Q}$-martingale. Consequently $\overline{S}$ satisfies (NA). $\qquad \square$

2.3 Equivalence of Single-period with Multiperiod Arbitrage

The aim of this section is to describe the relation between one-period no-arbitrage and multiperiod no-arbitrage. At the same time we will be able to give somewhat more detailed information on the set of risk neutral measures (this term is often used in the finance literature in a synonymous way for martingale measures). We start off with the following observation. Recall that we did not assume that $\mathcal{F}_0$ is trivial.

Proposition 2.3.1. *If S satisfies the no-arbitrage condition, $\mathbf{Q} \in \mathcal{M}^e(S)$ is an equivalent martingale measure, and $Z_t = \mathbf{E}_{\mathbf{P}}\left[\frac{d\mathbf{Q}}{d\mathbf{P}} \;\middle|\; \mathcal{F}_t\right]$ denotes the density process associated with $\mathbf{Q}$, then the process $L_t = \frac{Z_t}{Z_0}$ defines the density process of an equivalent measure $\mathbf{Q}'$ such that $\frac{d\mathbf{Q}'}{d\mathbf{P}} = L_T$, $\mathbf{Q}' \in \mathcal{M}^e(S)$ and $\mathbf{Q}'|_{\mathcal{F}_0} = \mathbf{P}|_{\mathcal{F}_0}$.*

Proof. This is rather straightforward. Since $\mathbf{Q} \in \mathcal{M}^e(S)$ we have that SZ is a $\mathbf{P}$-martingale. Since $Z_0 > 0$ and since it is $\mathcal{F}_0$-measurable the process $S\frac{Z}{Z_0}$ is still a $\mathbf{P}$-martingale. Since SL is now a $\mathbf{P}$-martingale and since the density $L_T > 0$, we necessarily have $\mathbf{Q}' \in \mathcal{M}^e(S)$. As $L_0 = 1$ we obtain $\mathbf{Q}'|_{\mathcal{F}_0} = \mathbf{P}|_{\mathcal{F}_0}$. $\qquad \square$

Theorem 2.3.2. *Let $S = (S_t)_{t=0}^T$ be a price process. Then the following are equivalent:*

(i) *S satisfies the no-arbitrage property.*
(ii) *For each $0 \le t < T$, we have that the one-period market (S_t, S_{t+1}) with respect to the filtration $(\mathcal{F}_t, \mathcal{F}_{t+1})$ satisfies the no-arbitrage property.*

Proof. Obviously (i) implies (ii), since there are less strategies in each single period market than in the multiperiod market. So let us show that (ii) implies (i). By the fundamental theorem applied to (S_t, S_{t+1}), we have that for each t there is a probability measure $\mathbf{Q}_t$ on $\mathcal{F}_{t+1}$ equivalent to $\mathbf{P}$, so that under $\mathbf{Q}_t$ the process (S_t, S_{t+1}) is a $\mathbf{Q}_t$-martingale. This means that $\mathbf{E}_{\mathbf{Q}_t}[S_{t+1} \mid \mathcal{F}_t] = S_t$. By the previous proposition we may take $\mathbf{Q}_t|_{\mathcal{F}_t} = \mathbf{P}|_{\mathcal{F}_t}$. Let $f_{t+1} = \frac{d\mathbf{Q}_t}{d\mathbf{P}}$ and define $L_t = f_1 \ldots f_{t-1} f_t$ and $L_0 = 1$. Clearly $(L_t)_{t=0}^T$ is the density process of an equivalent measure $\mathbf{Q}$ defined by $\frac{d\mathbf{Q}}{d\mathbf{P}} = L_T$. One can easily check that, for all $t = 0, \ldots, T-1$ we have $\mathbf{E}_{\mathbf{Q}}[S_{t+1} \mid \mathcal{F}_t] = S_t$, i.e., $\mathbf{Q} \in \mathcal{M}^e(S)$. $\qquad\square$

Remark 2.3.3. The equivalence between one-period no-arbitrage and multi-period no-arbitrage can also be checked directly by the definition of no-arbitrage. We invite the reader to give a direct proof of the following: if H is a strategy so that $(H \cdot S)_T \geq 0$ and $\mathbf{P}[(H \cdot S)_T > 0] > 0$ then there is a $1 \leq t \leq T$ as well as $A \in \mathcal{F}_{t-1}$, $\mathbf{P}[A] > 0$ so that $\mathbf{1}_A(H_t, \Delta S_t) \geq 0$ and $\mathbf{P}[\mathbf{1}_A(H_t, \Delta S_t) > 0] > 0$ (compare Lemma 5.1.5 below).

Remark 2.3.4. We give one more indication, why there is little difference between the one-period and the T period situation; this discussion also reveals a nice economic interpretation. Given $S = (S_t)_{t=0}^T$ as above, we may associate a one-period process $\widetilde{S} = (\widetilde{S}_t)_{t=0}^1$, adapted to the filtration $(\widetilde{\mathcal{F}}_0, \widetilde{\mathcal{F}}_1) := (\mathcal{F}_0, \mathcal{F}_T)$ in the following way: choose any collection $(f_1, \ldots, f_m)$ in the finite dimensional linear space K defined in 2.2.1, which linearly spans K. Define the $\mathbb{R}^m$-valued process $\widetilde{S}$ by $\widetilde{S}_0 = 0$, $\widetilde{S}_1 = (f_1, \ldots, f_m)$.

Obviously the process $\widetilde{S}$ yields the same space K of stochastic integrals as S. Hence the set of equivalent martingale measures for the processes S and $\widetilde{S}$ coincide and therefore all assertions, depending only on the set of equivalent martingale measures coincide for S and $\widetilde{S}$. In particular S and $\widetilde{S}$ yield the same arbitrage-free prices for derivatives, as we shall see in the next section.

The economic interpretation of the transition from S to $\widetilde{S}$ reads as follows: if we fix the trading strategies H^j yielding $f_j = (H^j \cdot S)_T$, we may think of f_j as a contingent claim at time $t = T$ which may be bought at price 0 at time $t = 0$, by then applying the trading rules given by H^j. By taking sufficiently many of these H^j's, in the sense that the corresponding f_j's linearly span K, we may represent the result $f = (H \cdot S)_T$ of *any* trading strategy H as a linear combination of the f_j's.

The bottom line of this discussion is that in the present framework (i.e. Ω is finite) — from a mathematical as well as from an economic point of view — the T period situation can easily be reduced to the one-period situation.

2.4 Pricing by No-Arbitrage

The subsequent theorem will tell us what the principle of no-arbitrage implies about the possible prices for a contingent claim f. It goes back to the work of D. Kreps [K 81].

For given $f \in L^\infty(\Omega, \mathcal{F}, \mathbf{P})$, we call $a \in \mathbb{R}$ an *arbitrage-free price*, if in addition to the financial market S, the introduction of the contingent claim f at price a does not create an arbitrage possibility. How can we mathematically formalise this economically intuitive idea? We enlarge the financial market S by introducing a new financial instrument which can be bought (or sold) at price a at time $t = 0$ and yields the random cash flow $f(\omega)$ at time $t = T$. We don't postulate anything about the price of this financial instrument at the intermediate times $t = 1, \ldots, T - 1$. The reader might think of an "over the counter" option where the two parties agree on certain payments at times $t = 0$ and $t = T$. So if we look at the linear space generated by K and the vector $(f - a)$ we obtain an enlarged space $K^{f,a}$ of attainable claims. The price a should be such that arbitrage opportunities are inexistent. Mathematically speaking this means that we still should have $K^{f,a} \cap L_+^\infty = \{0\}$. In this case we say that a is an *arbitrage free price* for the contingent claim f.

Theorem 2.4.1 (Pricing by no-arbitrage). *Assume that S satisfies (NA) and let $f \in L^\infty(\Omega, \mathcal{F}, \mathbf{P})$. Define*

$$\underline{\pi}(f) = \inf \left\{ \mathbf{E}_\mathbf{Q}[f] \mid \mathbf{Q} \in \mathcal{M}^e(S) \right\},$$
$$\overline{\pi}(f) = \sup \left\{ \mathbf{E}_\mathbf{Q}[f] \mid \mathbf{Q} \in \mathcal{M}^e(S) \right\}, \tag{2.8}$$

Either $\underline{\pi}(f) = \overline{\pi}(f)$, in which case f is attainable at price $\pi(f) := \underline{\pi}(f) = \overline{\pi}(f)$, i.e. $f = \pi(f) + (H \cdot S)_T$ for some $H \in \mathcal{H}$ and therefore $\pi(f)$ is the unique arbitrage-free price for f.

Or $\underline{\pi}(f) < \overline{\pi}(f)$, in which case

$$]\underline{\pi}(f), \overline{\pi}(f)[= \left\{ \mathbf{E}_\mathbf{Q}[f] \mid \mathbf{Q} \in \mathcal{M}^e(S) \right\}$$

and a is an arbitrage-free price for f iff a lies in the open interval $]\underline{\pi}(f), \overline{\pi}(f)[$.

Proof. The case $\underline{\pi}(f) = \overline{\pi}(f)$ follows from corollary 2.2.11 and so we only have to concentrate on the case $\underline{\pi}(f) < \overline{\pi}(f)$. First observe that the set $\{\mathbf{E}_\mathbf{Q}[f] \mid \mathbf{Q} \in \mathcal{M}^e(S)\}$ forms a bounded non-empty interval in $\mathbb{R}$, which we denote by I.

We claim that a number a is in I iff a is an arbitrage-free price for f. Indeed, supposing that $a \in I$ we may find $\mathbf{Q} \in \mathcal{M}^e(S)$ s.t. $\mathbf{E}_\mathbf{Q}[f - a] = 0$ and therefore $K^{f,a} \cap L_+^\infty(\Omega, \mathcal{F}, \mathbf{P}) = \{0\}$.

Conversely suppose that $K^{f,a} \cap L_+^\infty = \{0\}$. Then exactly as in the proof of the Fundamental Theorem 2.2.7, we find a probability measure $\mathbf{Q}$ so that $\mathbf{E}_\mathbf{Q}[g] = 0$ for all $g \in K^{f,a}$ and so that $\mathbf{Q}$ is equivalent to $\mathbf{P}$. This, of course, implies that $\mathbf{Q} \in \mathcal{M}^e(S)$ and that $a = \mathbf{E}_\mathbf{Q}[f]$.

Now we deal with the boundary case: suppose that a equals the right boundary of I, i.e., $a = \overline{\pi}(f) \in I$, and consider the contingent claim $f - \overline{\pi}(f)$. By definition we have $\mathbf{E}_\mathbf{Q}[f - \overline{\pi}(f)] \leq 0$, for all $\mathbf{Q} \in \mathcal{M}^e(S)$, and therefore by Proposition 2.2.9, that $f - \overline{\pi}(f) \in C$. We may find $g \in K$ such that $g \geq f - \overline{\pi}(f)$. If the sup in (2.8) is attained, i.e., if there is $\mathbf{Q}^* \in \mathcal{M}^e(S)$ such that $\mathbf{E}_{\mathbf{Q}^*}[f] = \overline{\pi}(f)$, then we have $0 = \mathbf{E}_{\mathbf{Q}^*}[g] \geq \mathbf{E}_{\mathbf{Q}^*}[f - \overline{\pi}(f)] = 0$ which in view of $\mathbf{Q}^* \sim \mathbf{P}$ implies that $f - \overline{\pi}(f) \equiv g$; in other words f is attainable at price $\overline{\pi}(f)$. This in turn implies that $\mathbf{E_Q}[f] = \overline{\pi}(f)$ for all $\mathbf{Q} \in \mathcal{M}^e(S)$, and therefore I is reduced to the singleton $\{\overline{\pi}(f)\}$.

Hence, if $\underline{\pi}(f) < \overline{\pi}(f)$, $\overline{\pi}(f)$ cannot belong to the interval I, which is therefore open on the right hand side. Passing from f to $-f$, we obtain the analogous result for the left hand side of I, which is therefore equal to $I =]\underline{\pi}(f), \overline{\pi}(f)[$. $\qquad\square$

The argument in the proof of the preceding theorem can be recast to yield the following duality theorem. The reader familiar with the duality theory of linear programming will recognise the primal-dual relation.

Theorem 2.4.2 (Superreplication). *Assume that S satisfies (NA). Then, for $f \in L^\infty$, we have*

$$\overline{\pi}(f) = \sup\{\mathbf{E_Q}[f] \mid \mathbf{Q} \in \mathcal{M}^e(S)\}$$
$$= \max\{\mathbf{E_Q}[f] \mid \mathbf{Q} \in \mathcal{M}^a(S)\}$$
$$= \min\{a \mid \text{there exists } k \in K, a + k \geq f\}.$$

Proof. As shown in the previous proof we have $f - \overline{\pi}(f) \in C$ and hence

$$\begin{aligned} f &= \overline{\pi}(f) + g, & \text{for some } g \in C \\ &= \overline{\pi}(f) + k - h, & \text{for some } k \in K \text{ and } h \in L^\infty_+ \\ &\leq \overline{\pi}(f) + k, & \text{for some } k \in K. \end{aligned}$$

This shows that $\overline{\pi}(f) \geq \inf\{a \mid \text{there exists } k \in K, a + k \geq f\}$.

Let now $a < \overline{\pi}(f)$. We will show that there is no element $k \in K$ with $a + k \geq f$. This shows that $\overline{\pi}(f) = \inf\{a \mid \text{there exists } k \in K, a + k \geq f\}$ and moreover establishes that the infimum is a minimum. Since $a < \overline{\pi}(f)$ there is $\mathbf{Q} \in \mathcal{M}^e(S)$ with $\mathbf{E_Q}[f] > a$. But this implies that for all $k \in K$ we have that $\mathbf{E_Q}[a + k] = a < \mathbf{E_Q}[f]$, in contradiction to the relation $a + k \geq f$. $\qquad\square$

Remark 2.4.3. Theorem 2.4.2 may be rephrased in economic terms: in order to superreplicate f, i.e., to find $a \in \mathbb{R}$ and $H \in \mathcal{H}$ s.t. $a + (H \cdot S)_T \geq f$, we need at least an initial investment a equal to $\overline{\pi}(f)$.

We now give a conditional version of the duality theorem that allows us to use initial investments that are not constant and to possibly use the information $\mathcal{F}_0$ available at time $t = 0$. This is relevant when the initial σ-algebra $\mathcal{F}_0$ is not trivial.

Theorem 2.4.4. *Let us assume that S satisfies (NA). Denote by $\mathcal{M}^e(S, \mathcal{F}_0)$ the set of equivalent martingale measures $\mathbf{Q} \in \mathcal{M}^e(S)$ so that $\mathbf{Q}|_{\mathcal{F}_0} = \mathbf{P}$. Then, for $f \in L^\infty$, we have*

$$\sup\{\mathbf{E_Q}[f \mid \mathcal{F}_0] \mid \mathbf{Q} \in \mathcal{M}^e(S, \mathcal{F}_0)\}$$
$$= \min\{h \mid h \text{ is } \mathcal{F}_0\text{-measurable and there exists } g \in K \text{ such that } h + g \geq f\}.$$

Remark 2.4.5. Before we prove the theorem let us remark that the "sup" and the "min" are taken in the space $L^0(\Omega, \mathcal{F}_0, \mathbf{P})$ of $\mathcal{F}_0$-measurable functions. Both sets are lattice ordered. Indeed, if $\mathbf{E}_{\mathbf{Q}_1}[f \mid \mathcal{F}_0]$ and $\mathbf{E}_{\mathbf{Q}_2}[f \mid \mathcal{F}_0]$ are given, where $\mathbf{Q}_1, \mathbf{Q}_2 \in \mathcal{M}^e(S, \mathcal{F}_0)$, then there is an element $\mathbf{Q}_3 \in \mathcal{M}^e(S, \mathcal{F}_0)$ so that $\mathbf{E}_{\mathbf{Q}_3}[f \mid \mathcal{F}_0] = \max\{\mathbf{E}_{\mathbf{Q}_1}[f \mid \mathcal{F}_0], \mathbf{E}_{\mathbf{Q}_2}[f \mid \mathcal{F}_0]\}$. The construction is rather straightforward. Let $A = \{\mathbf{E}_{\mathbf{Q}_1}[f \mid \mathcal{F}_0] > \mathbf{E}_{\mathbf{Q}_2}[f \mid \mathcal{F}_0]\} \in \mathcal{F}_0$ and let $\mathbf{Q}_3[B] = \mathbf{Q}_1[A \cap B] + \mathbf{Q}_2[A^c \cap B]$. Because $\mathbf{Q}_1|_{\mathcal{F}_0} = \mathbf{Q}_2|_{\mathcal{F}_0} = \mathbf{P}$ we get that $\mathbf{Q}_3$ is a probability and that $\mathbf{Q}_3 \in \mathcal{M}^e(S, \mathcal{F}_0)$. Also $\mathbf{E}_{\mathbf{Q}_3}[f \mid \mathcal{F}_0] = \mathbf{E}_{\mathbf{Q}_1}[f \mid \mathcal{F}_0] \vee \mathbf{E}_{\mathbf{Q}_2}[f \mid \mathcal{F}_0]$.

Similarly, the set on the right is stable for the "min" operation. Indeed, let $h_1 + g_1 \geq f$ and $h_2 + g_2 \geq f$. For $A = \{h_1 < h_2\}$, an $\mathcal{F}_0$-measurable set, we define $h = h_1 \mathbf{1}_A + h_2 \mathbf{1}_{A^c}$ and $g_1 \mathbf{1}_A + g_2 \mathbf{1}_{A^c} = g$. The function h is $\mathcal{F}_0$-measurable and $g \in K$ (because $A \in \mathcal{F}_0$). Clearly $h + g \geq f$.

Proof of Theorem 2.4.4. If $f \leq h + g$, where h is $\mathcal{F}_0$-measurable and $g \in K$, then for $\mathbf{Q} \in \mathcal{M}^e(S, \mathcal{F}_0)$ we have $\mathbf{E}_{\mathbf{Q}}[f \mid \mathcal{F}_0] \leq h + 0 = h$. This shows that

$$a_1 := \sup \{\mathbf{E}_{\mathbf{Q}}[f \mid \mathcal{F}_0] \mid \mathbf{Q} \in \mathcal{M}^e(S, \mathcal{F}_0)\}$$
$$\leq \inf \{h \mid h \ \mathcal{F}_0\text{-measurable}, h + g \geq f, \text{ for some } g \in K\}$$
$$=: a_2.$$

To prove the converse inequality, we show that there is $g \in K$ with $a_1 + g \geq f$. If this were not be true then $(a_1 + K) \cap (f + L^\infty_+) = \emptyset$ and we could find, using the separating hyperplane theorem, a linear functional φ and $\varepsilon > 0$, so that $\forall g \in K$, $\forall l \geq 0$ we have $\varepsilon + \varphi(a_1 + g) < \varphi(f + l)$. This implies that $\varphi \geq 0$ and $\varphi(g) = 0$ for all $g \in K$. Of course we can normalise φ so that it comes from a probability measure $\mathbf{Q}$. So we get $\mathbf{E}_{\mathbf{Q}}[a_1] + \varepsilon' < \mathbf{E}_{\mathbf{Q}}[f]$ and $\mathbf{Q} \in \mathcal{M}^a(S)$, where $\varepsilon' > 0$.

By the density of $\mathcal{M}^e(S)$ in $\mathcal{M}^a(S)$ we may perturb $\mathbf{Q}$ a little bit to make it an element of $\mathcal{M}^e(S)$. We still get $\mathbf{E}_{\mathbf{Q}}[a_1] + \varepsilon < \mathbf{E}_{\mathbf{Q}}[f]$, but this time for a measure $\mathbf{Q} \in \mathcal{M}^e(S)$. Let now $Z_t = \frac{d\mathbf{Q}}{d\mathbf{P}}\big|_{\mathcal{F}_t}$ and set $L_t = \frac{Z_t}{Z_0}$. The process $(L_t)_{t=0}^\infty$ defines a measure $\mathbf{Q}^0 \in \mathcal{M}^e(S, \mathcal{F}_0)$ via $\frac{d\mathbf{Q}^0}{d\mathbf{P}} = L_T$. Furthermore

$$\mathbf{E}_{\mathbf{Q}^0}[f \mid \mathcal{F}_0] = \mathbf{E}_{\mathbf{P}}[fL_T \mid \mathcal{F}_0]$$
$$= \frac{\mathbf{E}_{\mathbf{P}}[fZ_T \mid \mathcal{F}_0]}{Z_0} = \mathbf{E}_{\mathbf{Q}}[f \mid \mathcal{F}_0]$$

Therefore $\mathbf{E}_{\mathbf{Q}}[f \mid \mathcal{F}_0] \leq a_1$ and hence $\mathbf{E}_{\mathbf{Q}}[f] \leq \mathbf{E}_{\mathbf{Q}}[a_1]$, contradicting the choice of $\mathbf{Q}$. $\square$

Corollary 2.4.6. *Under the assumptions of Theorem 2.4.4 we have*

$$\{\mathbf{E}_{\mathbf{Q}}[f \mid \mathcal{F}_0] \mid \mathbf{Q} \in \mathcal{M}^e(S)\} = \{\mathbf{E}_{\mathbf{Q}}[f \mid \mathcal{F}_0] \mid \mathbf{Q} \in \mathcal{M}^e(S, \mathcal{F}_0)\}.$$

Hence, for $f \in L^\infty_+(\Omega, \mathcal{F}, \mathbf{P})$, we have $\sup_{\mathbf{Q} \in \mathcal{M}^e(S)} \mathbf{E}_{\mathbf{Q}}[f] = \|a_1\|_\infty$ where

$$a_1 = \sup \{\mathbf{E}_{\mathbf{Q}}[f \mid \mathcal{F}_0] \mid \mathbf{Q} \in \mathcal{M}^e(S, \mathcal{F}_0)\}.$$

Proof. As observed in the proof of Theorem 2.3.2 and Proposition 2.3.1, every $\mathbf{Q} \in \mathcal{M}^e(S)$ can be written as $\frac{d\mathbf{Q}}{d\mathbf{P}} = f_0 \frac{d\mathbf{Q}^0}{d\mathbf{P}}$ where $\mathbf{Q}^0|_{\mathcal{F}_0} = \mathbf{P}|_{\mathcal{F}_0}$, $\mathbf{Q}^0 \in \mathcal{M}^e(S, \mathcal{F}_0)$ and where f_0 is $\mathcal{F}_0$-measurable, strictly positive and $\mathbf{E}_\mathbf{P}[f_0] = 1$. But otherwise f_0 is arbitrary. Now for $\mathbf{Q} \in \mathcal{M}^e(S)$ we have $\frac{d\mathbf{Q}}{d\mathbf{P}} = f_0 \frac{d\mathbf{Q}^0}{d\mathbf{P}}$ and hence

$$\mathbf{E}_\mathbf{Q}[f] = \mathbf{E}_\mathbf{Q}\left[\mathbf{E}_{\mathbf{Q}^0}[f \mid \mathcal{F}_0]\right]$$
$$\leq \mathbf{E}_\mathbf{Q}[a_1] = \mathbf{E}_\mathbf{P}[a_1 f_0].$$

Thus $\sup_{\mathbf{Q} \in \mathcal{M}^e(S)} \mathbf{E}_\mathbf{Q}[f] \leq \|a_1\|_\infty$.

To prove the converse inequality we need some more approximations. First for given $\varepsilon > 0$, we choose f_0, $\mathcal{F}_0$-measurable, $f_0 > 0$, $\mathbf{E}_\mathbf{P}[f_0] = 1$ and so that $\mathbf{E}_\mathbf{P}[f_0 a_1] \geq \|a_1\|_\infty - \varepsilon$. Given f_0 we may take $\mathbf{Q}^1 \in \mathcal{M}^e(S, \mathcal{F}_0)$ so that $\mathbf{E}[f_0(a_1 - \mathbf{E}_{\mathbf{Q}^1}[f \mid \mathcal{F}_0])] \leq \varepsilon$. This is possible since the family $\{\mathbf{E}_\mathbf{Q}[f \mid \mathcal{F}_0] \mid \mathbf{Q} \in \mathcal{M}^e(S, \mathcal{F}_0)\}$ is a lattice and since all these functions are in the L^∞-ball with radius $\|f\|_\infty$. Now take $\mathbf{Q}^0$ defined by $\frac{d\mathbf{Q}^0}{d\mathbf{P}} = f_0 \frac{d\mathbf{Q}^1}{d\mathbf{P}}$. Clearly $\mathbf{Q}^0 \in \mathcal{M}^e(S)$ and we have

$$\mathbf{E}_{\mathbf{Q}^0}[f] = \mathbf{E}_\mathbf{P}\left[f_0 \frac{d\mathbf{Q}^1}{d\mathbf{P}} f\right]$$
$$= \mathbf{E}_\mathbf{P}\left[f_0 \mathbf{E}_{\mathbf{Q}^1}[f \mid \mathcal{F}_0]\right] \text{ since } \mathbf{Q}^1|_{\mathcal{F}_0} = \mathbf{P}$$
$$\geq \mathbf{E}_\mathbf{P}[f_0 a_1] - \varepsilon \text{ by the choice of } \mathbf{Q}^1$$
$$\geq \|a_1\|_\infty - 2\varepsilon \text{ by the choice of } f_0. \qquad \square$$

2.5 Change of Numéraire

In the previous sections we have developed the basic tools for the pricing and hedging of derivative securities. Recall that we did our analysis in a *discounted model* where we did choose one of the traded assets as numéraire.

How do these things change, when we pass to a new numéraire, i.e., a new unit in which we denote the values of the stocks? Of course, the arbitrage free prices should remain unchanged (after denominating things in the new numéraire), as the notion of arbitrage should not depend on whether we do the book-keeping in € or in \$. On the other hand, we shall see that the risk-neutral measures $\mathbf{Q}$ do depend on the choice of numéraire. We will also show how, conversely, a change of risk neutral measures corresponds to a change of numéraire.

Let us analyse the situation in the appropriate degree of generality: the model of a financial market $\widehat{S} = (\widehat{S}_t^0, \widehat{S}_t^1, \ldots, \widehat{S}_t^d)_{t=0}^T$ is defined as in 2.1 above. Recall that we assumed that the traded asset $\widehat{S}^0$ serves as numéraire, i.e., we have passed from the value $\widehat{S}_t^j$ of the j'th asset at time t to its value $S_t^j = \frac{\widehat{S}_t^j}{\widehat{S}_t^0}$, expressed in units of $\widehat{S}_t^0$. This led us in (2.3) to the introduction of the process

$$S = (S^1, S^2, \ldots, S^d) = \left(\frac{\widehat{S}^1}{\widehat{S}^0}, \ldots, \frac{\widehat{S}^d}{\widehat{S}^0} \right).$$

Before we prove the theorem, let us first see what assets can be used as numéraire. The crucial requirement on a numéraire is that it is a *traded asset*. We could of course use one of the assets $1, \ldots, d$ but we want to be more general and also want to accept, e.g., baskets as new numéraires. So we might use the value $(V_t)_{t=0}^T$ of a portfolio as a numéraire. Of course, we need to assume $V_t > 0$ for all t. Indeed, if the numéraire becomes zero or even negative, then we obviously have a problem in calculating the value of an asset in terms of V. Further, for normalisation reasons, it is convenient to assume that $V_0 = 1$, exactly as we did for $\widehat{S}^0$. So we start with a value process $V = 1 + (H^0 \cdot S)$ satisfying $V_t > 0$ a.s. for all t, where H^0 is a fixed element of $\mathcal{H}$. Observe that the processes V and S are denoted in terms of our originally chosen numéraire asset $\widehat{S}^0$.

As we have seen above (Proposition 2.2.13), the extended market

$$S^{\text{ext}} = (S^1, S^2, \ldots, S^d, 1, V) \tag{2.9}$$

is still arbitrage free and $\mathcal{M}^e(S) = \mathcal{M}^e(S^{\text{ext}})$. In real money terms this process is described by the process

$$\widehat{S}^{\text{ext}} = \left(S^1 \widehat{S}^0, \ldots, S^d \widehat{S}^0, \widehat{S}^0, V \widehat{S}^0 \right)$$
$$= \left(\widehat{S}^1, \ldots, \widehat{S}^d, \widehat{S}^0, V \widehat{S}^0 \right).$$

If we now use the last coordinate as numéraire, we obtain the process

$$X = \left(\frac{S^1}{V}, \ldots, \frac{S^d}{V}, \frac{1}{V}, 1 \right). \tag{2.10}$$

In order to keep the notation more symmetric we will drop the dummy entry 1 and use $(d+1)$-dimensional predictable processes as strategies. Similarly we shall also drop in (2.9) the dummy entry 1 for S^{ext}. This allows us to pass more easily from S^{ext} to X.

The next lemma shows the economically rather obvious fact that when passing from S to S^{ext}, the space K of claims attainable at price 0 does not change.

Lemma 2.5.1. *Using the above notation we have*

$$K(S^{\text{ext}}) = \{(H \cdot S^{\text{ext}})_T \mid H \ (d+1)\text{-}dimensional \ predictable\}$$
$$= K(S) = \{(H' \cdot S)_T \mid H' \ d\text{-}dimensional \ predictable\}.$$

Proof. The process V is given by the stochastic integral $(H^0 \cdot S)$ with respect to S, so we expect that nothing new can be created by using the additional

V. It suffices to show that, for a one-dimensional predictable process L, the quantities $L_t \Delta V_t$ are in $K(S)$. This is easy, since

$$L_t \Delta V_t = L_t \left(H_t^0, \Delta S_t \right) = \left(L_t H_t^0, \Delta S_t \right) \in K(S)$$

by definition of $K(S)$. This shows that $K(S^{\text{ext}}) = K(S)$.   □

Lemma 2.5.2. *Fix $0 \le t \le T$, and let $f \in K(S) = K(S^{\text{ext}})$ be $\mathcal{F}_t$-measurable. Then the random variable $\frac{f}{V_t}$ is of the form $\frac{f'}{V_T}$ where $f' \in K(S)$.*

Proof. Clearly

$$\frac{f}{V_t} - \frac{f}{V_T} = \frac{1}{V_T} \left(f \frac{V_T - V_t}{V_t} \right) = \frac{1}{V_T} \sum_{s=t+1}^{T} \frac{f}{V_t} (V_s - V_{s-1}).$$

We see that $f'' = \sum_{s=t+1}^{T} \frac{f}{V_t} (V_s - V_{s-1})$ belongs to $K(S^{\text{ext}})$ because $\frac{f}{V_t}$ is $\mathcal{F}_t$-measurable and the summation is on $s > t$. Hence $f' = f'' + f$ does the job.   □

Proposition 2.5.3. *Assume that X is defined as in (2.10). Then*

$$K(X) = \left\{ \frac{f}{V_T} \,\middle|\, f \in K(S) \right\}.$$

Proof. We have that $g \in K(X)$ if and only if there is a $(d+1)$-dimensional predictable process H, with $g = \sum_{t=1}^{T} (H_t, \Delta X_t) = \sum_{t=1}^{T} \sum_{j=1}^{d+1} H_t^j \Delta X_t^j$. Clearly, for $j = 1, \dots, d$ and $t = 1, \dots, T$,

$$\Delta X_t^j = \left(\frac{S_t^j}{V_t} - \frac{S_{t-1}^j}{V_{t-1}} \right)$$

$$= \frac{\Delta S_t^j}{V_t} + S_{t-1}^j \left(\frac{1}{V_t} - \frac{1}{V_{t-1}} \right)$$

$$= \frac{\Delta S_t^j}{V_t} - \frac{S_{t-1}^j}{V_{t-1}} \frac{\Delta V_t}{V_t}$$

$$= \frac{1}{V_t} \left(\Delta S_t^j - X_{t-1}^j \Delta V_t \right).$$

So we get that $H_t^j \Delta X_t^j = \frac{1}{V_t} \left(H_t^j \Delta S_t^j - \left(H_t^j X_{t-1}^j \right) \Delta V_t \right)$, which is of the form $\frac{f}{V_t}$ for some $f \in K(S^{\text{ext}}) = K(S)$. For $j = d+1$ and $t = 1, \dots, T$ the same argument applies by replacing S_t^j and S_{t-1}^j by 1.

By the previous lemma we have $\frac{f}{V_t} = \frac{f'}{V_T}$ for some $f' \in K(S)$. This shows that $K(X) \subset \frac{1}{V_T} K(S)$.

The converse inclusion follows by symmetry. In the financial market modelled by X we can choose $W_t = \frac{1}{V_t}$ as numéraire. The passage from X to S^{ext} is then done by using W as a new numéraire and the inclusion we just proved then yields

$$K(S) \subset \frac{1}{W_T}K(X) = V_T K(X).$$

This shows that $K(S) = V_T K(X)$ as required.   □

Theorem 2.5.4 (change of numéraire). *Let S satisfy the no-arbitrage condition, let $V = 1 + H^0 \cdot S$ be such that $V_t > 0$ for all t, and let $X = \left(\frac{S^1}{V}, \ldots, \frac{S^d}{V}, \frac{1}{V}\right)$. Then X satisfies the no-arbitrage condition too and $\mathbf{Q}$ belongs to $\mathcal{M}^e(S)$ if and only if the measure $\mathbf{Q}'$ defined by $d\mathbf{Q}' = V_T d\mathbf{Q}$ belongs to $\mathcal{M}^e(X)$.*

Proof. Since $K(X) = \frac{1}{V_T}K(S)$ we have that X satisfies the no-arbitrage property by directly verifying Definition 2.2.3. By Proposition 2.2.10 an equivalent probability measure $\mathbf{Q}$ is in $\mathcal{M}^e(S)$ if and only if, for all $f \in K(S)$, we have $\mathbf{E}_{\mathbf{Q}}[f] = 0$. But this is the same as

$$\mathbf{E}_{\mathbf{Q}}\left[V_T \frac{f}{V_T}\right] = 0, \quad \text{for all } f \in K(S),$$

which is equivalent to $\mathbf{E}_{\mathbf{Q}}[V_T g] = 0$ for all $g \in K(X)$. This happens if and only if the probability measure $\mathbf{Q}'$, defined as $d\mathbf{Q}' = V_T d\mathbf{Q}$, is in $\mathcal{M}^e(X)$. (Note that by the martingale property we have $\mathbf{E}_{\mathbf{Q}}[V_T] = V_0 = 1$.)   □

Remark 2.5.5. The process $(V_t)_{t=0}^T$ is a $\mathbf{Q}$-martingale for every $\mathbf{Q} \in \mathcal{M}^e(S)$. Now if $d\mathbf{Q}' = V_T d\mathbf{Q}$, then we have the following so-called Bayes' rule for $f \in L^\infty(\Omega, \mathcal{F}, \mathbf{P})$:

$$\mathbf{E}_{\mathbf{Q}'}[f \mid \mathcal{F}_t] = \frac{\mathbf{E}_{\mathbf{Q}}[fV_T \mid \mathcal{F}_t]}{\mathbf{E}_{\mathbf{Q}}[V_T \mid \mathcal{F}_t]} = \frac{\mathbf{E}_{\mathbf{Q}}[fV_T \mid \mathcal{F}_t]}{V_t}$$

$$= \mathbf{E}_{\mathbf{Q}}\left[f\frac{V_T}{V_t} \;\middle|\; \mathcal{F}_t\right].$$

The previous equality can also be written as

$$V_t \mathbf{E}_{\mathbf{Q}'}[f \mid \mathcal{F}_t] = \mathbf{E}_{\mathbf{Q}}[fV_T \mid \mathcal{F}_t].$$

From this it follows that $(Z_t)_{t=0}^T$ is a $\mathbf{Q}'$-martingale if and only if $(Z_t V_t)_{t=0}^T$ is a $\mathbf{Q}$-martingale. This statement can also be seen as the martingale formulation of Theorem 2.5.4 above.

2.6 Kramkov's Optional Decomposition Theorem

We now present a dynamic version of Theorem 2.4.2 (superreplication), due to D. Kramkov, who actually proved this theorem in a much more general version (see [K 96a], [FK 98], and Chap. 15 below). An earlier version of this theorem is due to N. El Karoui and M.-C. Quenez [EQ 95]. We refer to Chap. 15 for more detailed references.

Theorem 2.6.1 (Optional Decomposition). *Assume that S satisfies (NA) and let $V = (V_t)_{t=0}^T$ be an adapted process.*
The following assertions are equivalent:

(i) V *is a super-martingale for each* $\mathbf{Q} \in \mathcal{M}^e(S)$.
(i') V *is a super-martingale for each* $\mathbf{Q} \in \mathcal{M}^a(S)$
(ii) V *may be decomposed into* $V = V_0 + H \cdot S - C$, *where* $H \in \mathcal{H}$ *and* $C = (C_t)_{t=0}^T$ *is an increasing adapted process starting at* $C_0 = 0$.

Remark 2.6.2. To clarify the terminology *"optional decomposition"* let us compare this theorem with Doob's celebrated decomposition theorem for non-negative super-martingales $(V_t)_{t=0}^T$ (see, e.g., [P 90]): this theorem asserts that, for a non-negative (adapted, càdlàg) process V defined on a general filtered probability space we have the equivalence of the following two statements:

(i) V is a super-martingale (with respect to the fixed measure $\mathbf{P}$),
(ii) V may be decomposed in a unique way into $V = V_0 + M - C$, where M is a local martingale (with respect to $\mathbf{P}$) and C is an increasing predictable process s.t. $M_0 = C_0 = 0$.

We immediately recognise the similarity in spirit. However, there are significant differences. As to condition (i) the difference is that, in the setting of the optional decomposition theorem, the super-martingale property pertains to *all* martingale measures $\mathbf{Q}$ for the process S. As to condition (ii), the role of the local martingale M in Doob's theorem is taken by the stochastic integral $H \cdot S$.

A decisive difference between the two theorems is that in Theorem 2.6.1, the decomposition is no longer unique and one cannot choose, in general, C to be predictable. The process C can only be chosen to be optional, which in the present setting is the same as adapted.

The economic interpretation of the optional decomposition theorem reads as follows: a process of the form $V = V_0 + H \cdot S - C$ describes the wealth process of an economic agent. Starting at an initial wealth V_0, subsequently investing in the financial market according to the trading strategy H, and consuming as described by the process C where the random variable C_t models the accumulated consumption during the time period $\{1, \ldots, t\}$, the agent clearly obtains the wealth V_t at time t. The message of the optional decomposition theorem is that these wealth processes are characterised by condition (i) (or, equivalently, (i')).

Proof of Theorem 2.6.1. First assume that $T = 1$, i.e., we have a one-period model $S = (S_0, S_1)$. In this case the present theorem is just a reformulation of Theorem 2.4.2: if V is a super-martingale under each $\mathbf{Q} \in \mathcal{M}^e(S)$, then

$$\mathbf{E}_\mathbf{Q}[V_1 - V_0] \leq 0, \quad \text{for all } \mathbf{Q} \in \mathcal{M}^e(S).$$

Hence there is a predictable trading strategy H (i.e., an $\mathcal{F}_0$-measurable $\mathbb{R}^d$-valued function - in the present case $T = 1$) such that $(H \cdot S)_1 \geq V_1 - V_0$. Letting $C_0 = 0$ and writing $\Delta C_1 = C_1 = -V_1 + (V_0 + (H \cdot S)_1)$ we get the desired decomposition. This completes the construction for the case $T = 1$.

For general $T > 1$ we may apply, for each fixed $t \in \{1, \ldots, T\}$, the same argument as above to the one-period financial market (S_{t-1}, S_t) based on $(\Omega, \mathcal{F}, \mathbf{P})$ and adapted to the filtration $(\mathcal{F}_{t-1}, \mathcal{F}_t)$. We thus obtain an $\mathcal{F}_{t-1}$-measurable, $\mathbb{R}^d$-valued function H_t and a non-negative $\mathcal{F}_t$-measurable function ΔC_t such that

$$\Delta V_t = (H_t, \Delta S_t) - \Delta C_t,$$

where again $(.,.)$ denotes the inner product in $\mathbb{R}^d$. This will finish the construction of the optional decomposition: define the predictable process H as $(H_t)_{t=1}^T$ and the adapted increasing process C by $C_t = \sum_{u=1}^t \Delta C_u$. This proves the implication (i) $\Rightarrow$ (ii).

The implications (ii) $\Rightarrow$ (i') $\Rightarrow$ (i) are trivial. $\square$

3

Utility Maximisation
on Finite Probability Spaces

In addition to the model S of a financial market, we now consider a function $U(x)$, modelling the utility of an agent's wealth x at the terminal time T.

We make the classical assumptions that $U : \mathbb{R} \to \mathbb{R} \cup \{-\infty\}$ is *increasing on* $\mathbb{R}$, *continuous* on $\{U > -\infty\}$, *differentiable and strictly concave* on the interior of $\{U > -\infty\}$, and that the marginal utility tends to zero when wealth tends to infinity, i.e.,

$$U'(\infty) := \lim_{x \to \infty} U'(x) = 0.$$

These assumptions make perfect sense economically. Regarding the behaviour of the (marginal) utility at the other end of the wealth scale we shall distinguish two cases.

Case 1 (negative wealth not allowed): in this setting we assume that U satisfies the conditions $U(x) = -\infty$, for $x < 0$, while $U(x) > -\infty$, for $x > 0$, and the so-called *Inada condition*

$$U'(0) := \lim_{x \searrow 0} U'(x) = \infty.$$

Case 2 (negative wealth allowed): in this case we assume that $U(x) > -\infty$, for all $x \in \mathbb{R}$, and that

$$U'(-\infty) := \lim_{x \searrow -\infty} U'(x) = \infty.$$

Typical examples for case 1 are

$$U(x) = \ln(x), \quad x > 0,$$

or

$$U(x) = \frac{x^\alpha}{\alpha}, \quad \alpha \in (-\infty, 1) \setminus \{0\}, \quad x > 0,$$

whereas a typical example for case 2 is

$$U(x) = -e^{-\gamma x}, \quad \gamma > 0, \qquad x \in \mathbb{R}.$$

We note that it is natural from an economic point of view to require that the marginal utility tends to zero, when wealth x tends to infinity, and to infinity when the wealth x tends to the infimum of its allowed values. The infimum of the allowed values, i.e., of the domain $\{U > -\infty\}$ of U, may be finite or equal to $-\infty$. In the former case we have assumed w.l.g. the normalisation that this infimum equals zero.

We can now give a precise meaning to the problem of maximising expected utility of terminal wealth. Define the value function

$$u(x) := \sup_{H \in \mathcal{H}} \mathbf{E_P} \left[U(x + (H \cdot S)_T) \right], \quad x \in \mathrm{dom}(U), \tag{3.1}$$

where H runs through the family $\mathcal{H}$ of trading strategies.

The optimisation of expected utility of wealth at a fixed terminal date T is a typical example of a larger family of portfolio optimisation problems, where one can also include utility of intermediate consumption and many other features. We only consider the prototypical optimisation problem (3.1) above. The duality techniques developed for this case can easily be adapted to variants of it.

The value function $u(x)$ is called the *indirect utility function*. Economically speaking it indicates the expected utility of an economic agent at time T for given initial endowment x, provided she invests optimally in the financial market S.

We shall analyze the problem of finding, for given initial wealth x, the optimiser $\widehat{H}(x) \in \mathcal{H}$ in (3.1) at two levels of difficulty: first we consider the case of an arbitrage-free *complete* financial market S. In a second step, we generalise to arbitrage-free markets S, which are not necessarily complete.

3.1 The Complete Case

We assume that the set $\mathcal{M}^e(S)$ of equivalent probability measures under which S is a martingale, is reduced to a singleton $\{\mathbf{Q}\}$. In this setting consider the *Arrow-Debreu assets* $\mathbf{1}_{\{\omega_n\}}$, which pay 1 unit of the numéraire at time T, when ω_n turns out to be the true state of the world and pay out 0 otherwise. In view of our normalisation of the numéraire $S_t^0 \equiv 1$, we get the following relation for the price of the Arrow-Debreu assets at time $t = 0$:

$$\mathbf{E_Q} \left[\mathbf{1}_{\{\omega_n\}} \right] = \mathbf{Q}[\omega_n] =: q_n,$$

and by Corollary 2.2.12 each such asset $\mathbf{1}_{\{\omega_n\}}$ may be represented as $\mathbf{1}_{\{\omega_n\}} = \mathbf{Q}[\omega_n] + (H^n \cdot S)_T$, for some predictable trading strategy $H^n \in \mathcal{H}$.

Hence, for fixed initial endowment $x \in \mathrm{dom}(U)$, the utility maximisation problem (3.1) above may simply be written as

$$\mathbf{E_P} \left[U(X_T) \right] = \sum_{n=1}^{N} p_n U(\xi_n) \to \max! \tag{3.2}$$

under the constraint

$$\mathbf{E_Q}[X_T] = \sum_{n=1}^{N} q_n \xi_n \ \leq \ x. \tag{3.3}$$

To verify that (3.2) and (3.3) are indeed equivalent to the original problem (3.1) above (in the present finite, complete case), note that by Theorem 2.4.2 a random variable $(X_T(\omega_n))_{n=1}^{N} = (\xi_n)_{n=1}^{N}$ can be dominated by a random variable of the form $x + (H \cdot S)_T = x + \sum_{t=1}^{T} H_t \Delta S_t$ iff $\mathbf{E_Q}[X_T] = \sum_{n=1}^{N} q_n \xi_n \leq x$. This basic relation has a particularly evident interpretation in the present setting, as q_n is simply the price of the asset $\mathbf{1}_{\{\omega_n\}}$.

We have written ξ_n for $X_T(\omega_n)$ to stress that (3.2) is simply a concave maximisation problem in $\mathbb{R}^N$ with one linear constraint which is a rather elementary problem. To solve it, we form the Lagrangian

$$L(\xi_1, \ldots, \xi_N, y) = \sum_{n=1}^{N} p_n U(\xi_n) - y \left(\sum_{n=1}^{N} q_n \xi_n - x \right) \tag{3.4}$$

$$= \sum_{n=1}^{N} p_n \left(U(\xi_n) - y \tfrac{q_n}{p_n} \xi_n \right) + yx. \tag{3.5}$$

We have used the letter $y \geq 0$ instead of the usual $\lambda \geq 0$ for the Lagrange multiplier; the reason is the dual relation between x and y which will become apparent in a moment.

Write

$$\Phi(\xi_1, \ldots, \xi_N) = \inf_{y>0} L(\xi_1, \ldots, \xi_N, y), \quad \xi_n \in \mathrm{dom}(U), \tag{3.6}$$

and

$$\Psi(y) = \sup_{\xi_1, \ldots, \xi_N} L(\xi_1, \ldots, \xi_N, y), \quad y \geq 0. \tag{3.7}$$

Note that we have

$$\sup_{\xi_1, \ldots, \xi_N} \Phi(\xi_1, \ldots, \xi_N) = \sup_{\substack{\xi_1, \ldots, \xi_N \\ \sum_{n=1}^{N} q_n \xi_n \leq x}} \sum_{n=1}^{N} p_n U(\xi_n) = u(x). \tag{3.8}$$

Indeed, if $(\xi_1, \ldots, \xi_N)$ is in the admissible region $\left\{ \sum_{n=1}^{N} q_n \xi_n \leq x \right\}$, then $\Phi(\xi_1, \ldots, \xi_N) = L(\xi_1, \ldots, \xi_N, 0) = \sum_{n=1}^{N} p_n U(\xi_n)$. On the other hand, if $(\xi_1, \ldots, \xi_N)$ satisfies $\sum_{n=1}^{N} q_n \xi_n > x$, then by letting $y \to \infty$ in (3.6) we note that $\Phi(\xi_1, \ldots, \xi_N) = -\infty$.

Regarding the function $\Psi(y)$ we make the following pleasant observation, which is the basic reason for the efficiency of the duality approach: using the form (3.5) of the Lagrangian and fixing $y > 0$, the optimisation problem over $\mathbb{R}^N$ appearing in (3.7) splits into N independent optimisation problems over $\mathbb{R}$

$$U(\xi_n) - y\frac{q_n}{p_n}\xi_n \to \max!, \quad \xi_n \in \mathbb{R}.$$

In fact, these one-dimensional optimisation problems are of a very convenient form: recall (see, e.g., [R 70], [ET 76] or [KLSX 91]) that, for a concave function $U : \mathbb{R} \to \mathbb{R} \cup \{-\infty\}$, the *conjugate function* V of U (which is just the *Legendre-transform* of $x \mapsto -U(-x)$) is defined by

$$V(\eta) = \sup_{\xi \in \mathbb{R}} [U(\xi) - \eta\xi], \quad \eta > 0. \tag{3.9}$$

Definition 3.1.1. *We say that the function $V : \mathbb{R} \to \mathbb{R}$, conjugate to the function U, satisfies the usual regularity assumptions, if V is finitely valued, differentiable, strictly convex on $]0, \infty[$, and satisfies*

$$V'(0) := \lim_{y \searrow 0} V'(y) = -\infty. \tag{3.10}$$

Regarding the behaviour of V at infinity, we have to distinguish between case 1 and case 2 above:

$$\text{case 1:} \quad \lim_{y \to \infty} V(y) = \lim_{x \to 0} U(x) \quad \text{and} \quad \lim_{y \to \infty} V'(y) = 0 \tag{3.11}$$

$$\text{case 2:} \quad \lim_{y \to \infty} V(y) = \infty \quad \text{and} \quad \lim_{y \to \infty} V'(y) = \infty \tag{3.12}$$

We have the following well-known fact (see [R 70] or [ET 76]):

Proposition 3.1.2. *If U satisfies the assumptions made at the beginning of this section, then its conjugate function V satisfies the inversion formula*

$$U(\xi) = \inf_{\eta} [V(\eta) + \eta\xi], \quad \xi \in \text{dom}(U) \tag{3.13}$$

and it satisfies the regularity assumptions in Definition 3.1.1. In addition, $-V'(y)$ is the inverse function of $U'(x)$.

Conversely, if V satisfies the regularity assumptions of Definition 3.1.1, then U defined by (3.13) satisfies the regularity assumptions made at the beginning of this section.

Following [KLS 87] we write $-V' = I$ (for "inverse" function). We then have $I = (U')^{-1}$. Naturally, U' has a nice economic interpretation as the *marginal utility* of an economic agent modelled by the utility function U.

Here are some concrete examples of pairs of conjugate functions:

$$U(x) = \ln(x), \quad x > 0, \quad V(y) = -\ln(y) - 1,$$
$$U(x) = -\frac{e^{-\gamma x}}{\gamma}, \quad x \in \mathbb{R}, \quad V(y) = \frac{y}{\gamma}(\ln(y) - 1), \gamma > 0$$
$$U(x) = \frac{x^\alpha}{\alpha}, \quad x > 0, \quad V(y) = \frac{1-\alpha}{\alpha}y^{\frac{\alpha}{\alpha-1}}, \quad \alpha \in (-\infty, 1) \setminus \{0\}.$$

We now apply these general facts about the Legendre transform to calculate $\Psi(y)$. Using definition (3.9) of the conjugate function V and (3.5), formula (3.7) becomes:

$$\Psi(y) = \sum_{n=1}^{N} p_n V \left(y \frac{q_n}{p_n} \right) + yx$$

$$= \mathbf{E_P} \left[V \left(y \frac{d\mathbf{Q}}{d\mathbf{P}} \right) \right] + yx.$$

Denoting by $v(y)$ the dual value function

$$v(y) := \mathbf{E_P} \left[V \left(y \frac{d\mathbf{Q}}{d\mathbf{P}} \right) \right] = \sum_{n=1}^{N} p_n V \left(y \frac{q_n}{p_n} \right), \quad y > 0, \tag{3.14}$$

the function v has the same qualitative properties as the function V listed in Definition 3.1.1, since it is a convex combination of V calculated on linearly scaled arguments.

Hence by (3.10), (3.11), and (3.12) we find, for fixed $x \in \text{dom}(U)$, a unique $\widehat{y} = \widehat{y}(x) > 0$ such that $v'(\widehat{y}(x)) = -x$, which is therefore the unique minimiser to the dual problem

$$\Psi(y) = \mathbf{E_P} \left[V \left(y \frac{d\mathbf{Q}}{d\mathbf{P}} \right) \right] + yx = \min!$$

Fixing the critical value $\widehat{y}(x)$, the concave function

$$(\xi_1, \ldots, \xi_N) \mapsto L(\xi_1, \ldots, \xi_N, \widehat{y}(x))$$

defined in (3.5) assumes its unique maximum at the point $(\widehat{\xi}_1, \ldots, \widehat{\xi}_N)$ satisfying

$$U'(\widehat{\xi}_n) = \widehat{y}(x) \frac{q_n}{p_n} \quad \text{or, equivalently,} \quad \widehat{\xi}_n = I \left(\widehat{y}(x) \frac{q_n}{p_n} \right),$$

so that we have

$$\inf_{y>0} \Psi(y) = \inf_{y>0} (v(y) + xy) \tag{3.15}$$

$$= v(\widehat{y}(x)) + x\widehat{y}(x)$$

$$= L(\widehat{\xi}_1, \ldots, \widehat{\xi}_N, \widehat{y}(x)).$$

Note that the $\widehat{\xi}_n$ are in the interior of $\text{dom}(U)$, for $1 \leq n \leq N$, so that L is continuously differentiable at $(\widehat{\xi}_1, \ldots, \widehat{\xi}_N, \widehat{y}(x))$, which implies that the gradient of L vanishes at $(\widehat{\xi}_1, \ldots, \widehat{\xi}_N, \widehat{y}(x))$ and, in particular, that $\frac{\partial}{\partial y} L(\xi_1, \ldots, \xi_N, y)|_{(\widehat{\xi}_1, \ldots, \widehat{\xi}_N, \widehat{y}(x))} = 0$. Hence we infer from (3.4) and $\widehat{y}(x) > 0$, that the constraint (3.3) is binding, i.e.,

$$\sum_{n=1}^{N} q_n \widehat{\xi}_n = x, \tag{3.16}$$

and

$$\sum_{n=1}^{N} p_n U(\widehat{\xi}_n) = L(\widehat{\xi}_1, \ldots, \widehat{\xi}_N, \widehat{y}(x)). \tag{3.17}$$

In particular, we obtain that

$$u(x) = \sum_{n=1}^{N} p_n U(\widehat{\xi}_n). \tag{3.18}$$

Indeed, the inequality $u(x) \geq \sum_{n=1}^{N} p_n U(\widehat{\xi}_n)$ follows from (3.16) and (3.8), while the reverse inequality follows from (3.17) and the fact that, for all $\xi_1, \ldots, \xi_N$ verifying the constraint (3.3), we have:

$$\sum_{n=1}^{N} p_n U(\xi_n) \leq L(\xi_1, \ldots, \xi_N, \widehat{y}(x)) \leq L(\widehat{\xi}_1, \ldots, \widehat{\xi}_N, \widehat{y}(x)).$$

We shall write $\widehat{X}_T(x) \in C(x)$ for the optimiser $\widehat{X}_T(x)(\omega_n) = \widehat{\xi}_n$, $n = 1, \ldots, N$.

Combining (3.15), (3.17) and (3.18) we note that the value functions u and v are conjugate:

$$\inf_{y > 0} (v(y) + xy) = v(\widehat{y}(x)) + x\widehat{y}(x) = u(x), \quad x \in \mathrm{dom}(U).$$

Thus the relation $v'(\widehat{y}(x)) = -x$, which was used to define $\widehat{y}(x)$, translates into

$$u'(x) = \widehat{y}(x), \quad \text{for } x \in \mathrm{dom}(U).$$

From Proposition 3.1.2 and the remarks after equation (3.14), we deduce that u inherits the properties of U listed at the beginning of this chapter.

Let us summarise what we have proved so far:

Theorem 3.1.3 (finite Ω, complete market). *Let the financial market $S = (S_t)_{t=0}^{T}$ be defined over the finite filtered probability space $(\Omega, \mathcal{F}, (\mathcal{F})_{t=0}^{T}, \mathbf{P})$ and suppose $\mathcal{M}^e(S) = \{\mathbf{Q}\}$. Let the utility function U satisfy the above assumptions.*

Denote by $u(x)$ and $v(y)$ the value functions

$$\begin{aligned} u(x) &= \sup_{X_T \in C(x)} \mathbf{E}[U(X_T)], & x \in \mathrm{dom}(U), \\ v(y) &= \mathbf{E}\left[V\left(y\frac{d\mathbf{Q}}{d\mathbf{P}}\right)\right], & y > 0. \end{aligned} \tag{3.19}$$

We then have:

(i) *The value functions $u(x)$ and $v(y)$ are conjugate and u inherits the qualitative properties of U listed in the beginning of this chapter.*

(ii) *The optimiser $\widehat{X}_T(x)$ in (3.19) exists, is unique and satisfies*

$$\widehat{X}_T(x) = I\left(y\frac{d\mathbf{Q}}{d\mathbf{P}}\right), \quad \text{or, equivalently,} \quad y\frac{d\mathbf{Q}}{d\mathbf{P}} = U'(\widehat{X}_T(x)), \quad (3.20)$$

where $x \in \text{dom}(U)$ and $y > 0$ are related via $u'(x) = y$ or, equivalently, $x = -v'(y)$.

(iii) *The following formulae for u' and v' hold true:*

$$u'(x) = \mathbf{E_P}[U'(\widehat{X}_T(x))], \quad v'(y) = \mathbf{E_Q}\left[V'\left(y\frac{d\mathbf{Q}}{d\mathbf{P}}\right)\right] \quad (3.21)$$

$$xu'(x) = \mathbf{E_P}\left[\widehat{X}_T(x)U'(\widehat{X}_T(x))\right], \quad yv'(y) = \mathbf{E_P}\left[y\frac{d\mathbf{Q}}{d\mathbf{P}}V'\left(y\frac{d\mathbf{Q}}{d\mathbf{P}}\right)\right]. \quad (3.22)$$

Proof. Items (i) and (ii) have been shown in the preceding discussion, hence we only have to show (iii). The formula for $v'(y)$ in (3.21) and immediately follows by differentiating the relation

$$v(y) = \mathbf{E_P}\left[V\left(y\frac{d\mathbf{Q}}{d\mathbf{P}}\right)\right] = \sum_{n=1}^{N} p_n V\left(y\frac{q_n}{p_n}\right).$$

Of course, the formula for v' in (3.22) is an obvious reformulation of the one in (3.21). But we present both of them to stress their symmetry with the formulae for $u'(x)$.

The formula for u' in (3.21) translates via the relations exhibited in (ii) into the identity

$$y = \mathbf{E_P}\left[y\frac{d\mathbf{Q}}{d\mathbf{P}}\right],$$

while the formula for $u'(x)$ in (3.22) translates into

$$v'(y)y = \mathbf{E_P}\left[V'\left(y\frac{d\mathbf{Q}}{d\mathbf{P}}\right)y\frac{d\mathbf{Q}}{d\mathbf{P}}\right],$$

which we just have verified to hold true. □

Remark 3.1.4. Let us recall the economic interpretation of (3.20)

$$U'\left(\widehat{X}_T(x)(\omega_n)\right) = y\frac{q_n}{p_n}, \quad n = 1,\ldots,N.$$

This equality means that in every possible state of the world ω_n, the *marginal utility* $U'(\widehat{X}_T(x)(\omega_n))$ of the wealth of an optimally investing agent at time T is *proportional to the ratio of the price q_n of the corresponding Arrow security $\mathbf{1}_{\{\omega_n\}}$ and the probability of its success $p_n = \mathbf{P}[\omega_n]$.* This basic relation was analyzed in the fundamental work of K. Arrow and allows for a convincing economic interpretation: consider for a moment the situation where this proportionality relation fails to hold. Then one immediately deduces from a marginal variation argument that the investment of the agent cannot be optimal. Indeed, by investing a little more in the more favourable Arrow asset and a little less in the less favourable one, the economic agent can strictly increase her expected utility under the same budget constraint. Hence for the optimal investment the proportionality must hold true. The above result also identifies the proportionality factor as $y = u'(x)$, where x is the initial endowment of the investor. This marginal utility of the indirect utility function $u(x)$ also allows for a straightforward economic interpretation.

Theorem 3.1.3 indicates an easy way to solve the utility maximisation problem at hand: calculate $v(y)$ using (3.19), which reduces to a simple one-dimensional computation. Once we know $v(y)$, the theorem provides easy formulae to calculate all the other quantities of interest, e.g., $\widehat{X}_T(x)$, $u(x)$, $u'(x)$ etc.

Another message of the previous theorem is that the value function $x \mapsto u(x)$ may be viewed as a utility function as well, sharing all the qualitative features of the original utility function U. This makes sense economically, as the "indirect utility" function $u(x)$ denotes the expected utility of an agent with initial endowment x, when optimally investing in the financial market S.

Let us now give an economic interpretation of the formulae for $u'(x)$ in item (iii) along these lines: suppose the initial endowment x is varied to $x + h$, for some small real number h. The economic agent may use the additional endowment h to finance, in addition to the optimal pay-off function $\widehat{X}_T(x)$, h units of the numéraire asset, thus ending up with the pay-off function $\widehat{X}_T(x) + h$ at time T. Comparing this investment strategy to the optimal one corresponding to the initial endowment $x + h$, which is $\widehat{X}_T(x + h)$, we obtain

$$\lim_{h \to 0} \frac{u(x + h) - u(x)}{h} = \lim_{h \to 0} \frac{\mathbf{E}[U(\widehat{X}_T(x + h)) - U(\widehat{X}_T(x))]}{h} \quad (3.23)$$

$$\geq \lim_{h \to 0} \frac{\mathbf{E}[U(\widehat{X}_T(x) + h) - U(\widehat{X}_T(x))]}{h} \quad (3.24)$$

$$= \mathbf{E}[U'(\widehat{X}_T(x))].$$

Using the fact that u is differentiable and that h may be positive as well as negative, we must have equality in (3.24) and have therefore found another proof of formula (3.21) for $u'(x)$; the economic interpretation of this proof is that the economic agent, who is optimally investing, is indifferent of first order towards a (small) additional investment into the numéraire asset.

Playing the same game as above, but using the additional endowment $h \in \mathbb{R}$ to finance an additional investment into the optimal portfolio $\widehat{X}_T(x)$ (assuming, for simplicity, $x \neq 0$), we arrive at the pay-off function $\frac{x+h}{x} \widehat{X}_T(x)$. Comparing this investment with $\widehat{X}_T(x + h)$, an analogous calculation as in (3.23) leads to the formula for $u'(x)$ displayed in (3.22). The interpretation now is, that the optimally investing economic agent is indifferent of first order towards a marginal variation of the investment into the portfolio $\widehat{X}_T(x)$.

It now becomes clear that formulae (3.21) and (3.22) for $u'(x)$ are just special cases of a more general principle: for each $f \in L^\infty(\Omega, \mathcal{F}, \mathbf{P})$ we have

$$\mathbf{E_Q}[f]u'(x) = \lim_{h \to 0} \frac{\mathbf{E_P}[U(\widehat{X}_T(x) + hf) - U(\widehat{X}_T(x))]}{h}. \qquad (3.25)$$

The proof of this formula again is along the lines of (3.23) and the interpretation is the following: by investing an additional endowment $h\mathbf{E_Q}[f]$ to finance the contingent claim hf, the increase in expected utility is of first order equal to $h\mathbf{E_Q}[f]u'(x)$; hence again the economic agent is of first order indifferent towards an additional investment into the contingent claim f.

3.2 The Incomplete Case

We now drop the assumption that the set $\mathcal{M}^e(S)$ of equivalent martingale measures is reduced to a singleton (but we still remain in the framework of a finite probability space Ω) and replace it by $\mathcal{M}^e(S) \neq \emptyset$.

It follows from Theorem 2.4.2 that a random variable $X_T(\omega_n) = \xi_n$ may be dominated by a random variable of the form $x + (H \cdot S)_T$ iff $\mathbf{E_Q}[X_T] = \sum_{n=1}^N q_n \xi_n \leq x$, for each $\mathbf{Q} = (q_1 \ldots, q_N) \in \mathcal{M}^a(S)$ (or equivalently, for each $\mathbf{Q} \in \mathcal{M}^e(S)$).

In order to reduce these infinitely many constraints, where $\mathbf{Q}$ runs through $\mathcal{M}^a(S)$, to a finite number, make the easy observation that $\mathcal{M}^a(S)$ is a bounded, closed, convex polytope in $\mathbb{R}^N$. Indeed, $\mathcal{M}^a(S)$ is a subset of the probability measures on Ω defined by imposing finitely many linear constraints. Therefore $\mathcal{M}^a(S)$ equals the convex hull of its finitely many extreme points $\{\mathbf{Q}^1, \ldots, \mathbf{Q}^M\}$. For $1 \leq m \leq M$, we identify $\mathbf{Q}^m$ with the probabilities $(q_1^m, \ldots, q_N^m)$.

Fixing the initial endowment $x \in \text{dom}(U)$, we therefore may write the utility maximisation problem (3.1) similarly as in (3.2) as a concave optimisation problem over $\mathbb{R}^N$ with finitely many linear constraints:

$$\mathbf{E_P}[U(X_T)] = \sum_{n=1}^N p_n U(\xi_n) \to \max!$$

$$\mathbf{E_{Q^m}}[X_T] = \sum_{n=1}^N q_n^m \xi_n \ \leq \ x, \quad \text{for} \quad m = 1, \ldots, M.$$

Writing again

$$C(x) = \{X_T \in L^0(\Omega, \mathcal{F}, \mathbf{P}) \mid \mathbf{E_Q}[X_T] \leq x, \ \text{for all } \mathbf{Q} \in \mathcal{M}^a(S)\}$$

we define the value function, for $x \in \text{dom}(U)$,

$$u(x) = \sup_{H \in \mathcal{H}} \mathbf{E}\left[U\left(x + (H \cdot S)_T\right)\right] = \sup_{X_T \in C(x)} \mathbf{E}[U(X_T)].$$

The Lagrangian now is given by

$$L(\xi_1, \ldots, \xi_N, \eta_1, \ldots, \eta_M)$$

$$= \sum_{n=1}^{N} p_n U(\xi_n) - \sum_{m=1}^{M} \eta_m \left(\sum_{n=1}^{N} q_n^m \xi_n - x \right)$$

$$= \sum_{n=1}^{N} p_n \left(U(\xi_n) - \sum_{m=1}^{M} \frac{\eta_m q_n^m}{p_n} \xi_n \right) + \sum_{m=1}^{M} \eta_m x,$$

where $(\xi_1, \ldots, \xi_N) \in \mathrm{dom}(U)^N$, $(\eta_1, \ldots, \eta_M) \in \mathbb{R}_+^M$.

Writing $y = \eta_1 + \cdots + \eta_M$, $\mu_m = \frac{\eta_m}{y}$, $\mu = (\mu_1, \ldots, \mu_M)$ and

$$\mathbf{Q}^\mu = \sum_{m=1}^{M} \mu_m \mathbf{Q}^m,$$

note that, when $(\eta_1, \ldots, \eta_M)$ runs trough $\mathbb{R}_+^M$, the pairs $(y, \mathbf{Q}^\mu)$ run through $\mathbb{R}_+ \times \mathcal{M}^a(S)$. Hence we may write the Lagrangian as

$$L(\xi_1, \ldots, \xi_N, y, \mathbf{Q}) = \mathbf{E_P}[U(X_T)] - y\left(\mathbf{E_Q}[X_T - x]\right) \qquad (3.26)$$

$$= \sum_{n=1}^{N} p_n \left(U(\xi_n) - \frac{y q_n}{p_n} \xi_n \right) + yx,$$

where $\xi_n \in \mathrm{dom}(U)$, $y > 0$, $\mathbf{Q} = (q_1, \ldots, q_N) \in \mathcal{M}^a(S)$.

This expression is entirely analogous to (3.5), the only difference now being that $\mathbf{Q}$ runs through the set $\mathcal{M}^a(S)$ instead of being a fixed probability measure. Defining again

$$\Phi(\xi_1, \ldots, \xi_n) = \inf_{y>0, \mathbf{Q} \in \mathcal{M}^a(S)} L(\xi_1, \ldots, \xi_N, y, \mathbf{Q}),$$

and

$$\Psi(y, \mathbf{Q}) = \sup_{\xi_1, \ldots, \xi_N} L(\xi_1, \ldots, \xi_N, y, \mathbf{Q}),$$

we obtain, just as in the complete case,

$$\sup_{\xi_1, \ldots, \xi_N} \Phi(\xi_1, \ldots, \xi_N) = u(x), \quad x \in \mathrm{dom}(U),$$

and

$$\Psi(y, \mathbf{Q}) = \sum_{n=1}^{N} p_n V\left(\frac{y q_n}{p_n} \right) + yx, \quad y > 0, \quad \mathbf{Q} \in \mathcal{M}^a(S),$$

where $(q_1, \ldots, q_N)$ denotes the probability vector of $\mathbf{Q} \in \mathcal{M}^a(S)$. The minimisation of Ψ will be done in two steps: first we fix $y > 0$ and minimise over $\mathcal{M}^a(S)$, i.e.,

$$\Psi(y) := \inf_{\mathbf{Q} \in \mathcal{M}^a(S)} \Psi(y, \mathbf{Q}), \quad y > 0.$$

For fixed $y > 0$, the continuous function $\mathbf{Q} \to \Psi(y, \mathbf{Q})$ attains its minimum on the compact set $\mathcal{M}^a(S)$ and the minimiser $\widehat{\mathbf{Q}}(y)$ is unique by the strict convexity of V. Writing $\widehat{\mathbf{Q}}(y) = (\widehat{q}_1(y), \ldots, \widehat{q}_N(y))$ for the minimiser, it follows from $V'(0) = -\infty$ that $\widehat{q}_n(y) > 0$, for each $n = 1, \ldots, N$. Indeed, suppose that $\widehat{q}_n(y) = 0$ for some $1 \leq n \leq N$ and fix any equivalent martingale measure $\mathbf{Q} \in \mathcal{M}^e(S)$. Letting $\mathbf{Q}^\varepsilon = \varepsilon \mathbf{Q} + (1 - \varepsilon)\widehat{\mathbf{Q}}$ we have that $\mathbf{Q}^\varepsilon \in \mathcal{M}^e(S)$, for $0 < \varepsilon < 1$, and $\Psi(y, \mathbf{Q}^\varepsilon) < \Psi(y, \widehat{\mathbf{Q}})$ for $\varepsilon > 0$ sufficiently small - a contradiction. In other words, $\widehat{\mathbf{Q}}(y)$ is an equivalent martingale measure for S.

Defining the dual value function $v(y)$ by

$$v(y) = \inf_{\mathbf{Q} \in \mathcal{M}^a(S)} \sum_{n=1}^{N} p_n V\left(y\frac{q_n}{p_n}\right)$$

$$= \sum_{n=1}^{N} p_n V\left(y\frac{\widehat{q}_n(y)}{p_n}\right)$$

we find ourselves in an analogous situation as in the complete case above: defining again $\widehat{y}(x)$ by $v'(\widehat{y}(x)) = -x$ and

$$\widehat{\xi}_n = I\left(\widehat{y}(x)\frac{\widehat{q}_n(y)}{p_n}\right),$$

similar arguments as above apply to show that $(\widehat{\xi}_1, \ldots, \widehat{\xi}_N, \widehat{y}(x), \widehat{\mathbf{Q}}(y))$ is the unique saddle-point of the Lagrangian (3.26) and that the value functions u and v are conjugate.

Let us summarise what we have found in the incomplete case:

Theorem 3.2.1 (finite Ω, incomplete market). *Let the financial market $S = (S_t)_{t=0}^T$ be defined over the finite filtered probability space $(\Omega, \mathcal{F}, (\mathcal{F})_{t=0}^T, \mathbf{P})$ and let $\mathcal{M}^e(S) \neq \emptyset$. Let the utility function U satisfy the above assumptions.*

Denote by $u(x)$ and $v(y)$ the value functions

$$u(x) = \sup_{X_T \in C(x)} \mathbf{E}[U(X_T)], \qquad x \in \operatorname{dom}(U), \qquad (3.27)$$

$$v(y) = \inf_{\mathbf{Q} \in \mathcal{M}^a(S)} \mathbf{E}\left[V\left(y\frac{d\mathbf{Q}}{d\mathbf{P}}\right)\right], \qquad y > 0. \qquad (3.28)$$

We then have:

(i) *The value functions $u(x)$ and $v(y)$ are conjugate and u shares the qualitative properties of U listed in the above assumptions.*

(ii) *The optimisers $\widehat{X}_T(x)$ and $\widehat{Q}(y)$ in (3.27) and (3.28) exist, are unique, $\widehat{Q}(y) \in \mathcal{M}^e(S)$ and satisfy*

$$\widehat{X}_T(x) = I\left(y\frac{d\widehat{Q}(y)}{d\mathbf{P}}\right), \qquad y\frac{d\widehat{Q}(y)}{d\mathbf{P}} = U'(\widehat{X}_T(x)), \qquad (3.29)$$

where $x \in \mathrm{dom}(U)$ and $y > 0$ are related via $u'(x) = y$ or, equivalently, $x = -v'(y)$.

(iii) *The following formulae for u' and v' hold true:*

$$u'(x) = \mathbf{E_P}[U'(\widehat{X}_T(x))], \qquad v'(y) = \mathbf{E}_{\widehat{Q}}\left[V'\left(y\tfrac{d\widehat{Q}(y)}{d\mathbf{P}}\right)\right] \qquad (3.30)$$

$$xu'(x) = \mathbf{E_P}[\widehat{X}_T(x)U'(\widehat{X}_T(x))], \quad yv'(y) = \mathbf{E_P}\left[y\tfrac{d\widehat{Q}(y)}{d\mathbf{P}}V'\left(y\tfrac{d\widehat{Q}(y)}{d\mathbf{P}}\right)\right]. \,(3.31)$$

Remark 3.2.2. Let us again interpret the formulae (3.30), (3.31) for $u'(x)$ similarly as in Remark 3.1.4 above. In fact, the interpretations of these formulae as well as their derivation remain exactly the same in the incomplete case .

But a new and interesting phenomenon arises when we pass to the variation of the optimal pay-off function $\widehat{X}_T(x)$ by a small unit of an arbitrary pay-off function $f \in L^\infty(\Omega, \mathcal{F}, \mathbf{P})$. Similarly as in (3.25) we have the formula

$$\mathbf{E}_{\widehat{Q}(y)}[f]u'(x) = \lim_{h \to 0} \frac{\mathbf{E_P}[U(\widehat{X}_T(x) + hf) - U(\widehat{X}_T(x))]}{h}, \qquad (3.32)$$

the only difference being that $\mathbf{Q}$ has been replaced by $\widehat{Q}(y)$ (recall that x and y are related via $u'(x) = y$).

The remarkable feature of this formula is that it does not only pertain to variations of the form $f = x + (H \cdot S)_T$, i.e, contingent claims attainable at price x, but to arbitrary contingent claims f, for which — in general — we cannot derive the price from no-arbitrage considerations.

The economic interpretation of formula (3.32) is the following: the pricing rule $f \mapsto \mathbf{E}_{\widehat{Q}(y)}[f]$ yields precisely those prices at which an economic agent with initial endowment x, utility function U and investing optimally, is indifferent of first order towards adding a (small) unit of the contingent claim f to her portfolio $\widehat{X}_T(x)$.

In fact, one may turn this around: this was done by M. Davis [D 97] (compare also the work of Sir J.R. Hicks [H 86] and L. Foldes [F 90]): one may *define* $\widehat{Q}(y)$ by (3.32), verify that this is indeed an equivalent martingale measure for S and interpret this pricing rule as "pricing by marginal utility", which is, of course, a classical and basic paradigm in economics.

Let us give a proof for (3.32) (under the hypotheses of Theorem 3.2.1). One possible strategy for the proof, which also has the advantage of providing a nice economic interpretation, is the idea of introducing "fictitious securities" as developed in [KLSX 91]: fix $x \in \mathrm{dom}(U)$ and $y = u'(x)$ and let $(f^1, \ldots, f^k)$ be finitely many elements of $L^\infty(\Omega, \mathcal{F}, \mathbf{P})$ such that the space $K = \{(H \cdot S)_T \mid$

$H \in \mathcal{H}\}$, the constant function 1, and $(f^1, \ldots, f^k)$ linearly span $L^\infty(\Omega, \mathcal{F}, \mathbf{P})$. Define the k processes

$$S_t^{d+j} = \mathbf{E}_{\widehat{\mathbf{Q}}(y)}[f^j | \mathcal{F}_t], \quad j = 1, \ldots, k, \quad t = 0, \ldots, T. \tag{3.33}$$

Now extend the $\mathbb{R}^{d+1}$-valued process $S = (S^0, S^1, \ldots, S^d)$ to the $\mathbb{R}^{d+k+1}$-valued process $\overline{S} = (S^0, S^1, \ldots, S^d, S^{d+1}, \ldots, S^{d+k})$ by adding these new co-ordinates. By (3.33) we still have that $\overline{S}$ is a martingale under $\widehat{\mathbf{Q}}(y)$, which, by our choice of $(f^1, \ldots, f^k)$ and Corollary 2.2.8, is now the unique probability under which $\overline{S}$ is a martingale, by our choice of $(f^1, \ldots, f^k)$ and Corollary 2.2.8.

Hence we find ourselves in the situation of Theorem 3.1.3. By comparing (3.20) and (3.29) we observe that the optimal pay-off function $\widehat{X}_T(x)$ has not changed. Economically speaking this means that in the "completed" market $\overline{S}$ the optimal investment may still be achieved by trading only in the first $d+1$ assets and without touching the "fictitious" securities $S^{d+1}, \ldots, S^{d+k}$.

In particular, we now may apply formula (3.25) to $\mathbf{Q} = \widehat{\mathbf{Q}}(y)$ to obtain (3.32).

Finally we remark that the pricing rule induced by $\widehat{\mathbf{Q}}(y)$ is precisely such that the interpretation of the optimal investment $\widehat{X}_T(x)$ defined in (3.29) (given in Remark 3.1.4 in terms of marginal utility and the ratio of Arrow prices $\widehat{q}_n(y)$ and probabilities p_n) carries over to the present incomplete setting. The above completion of the market by introducing "fictitious securities" allows for an economic interpretation of this fact.

3.3 The Binomial and the Trinomial Model

Example 3.3.1 (The binomial model (one-period)). To illustrate the theory we apply the results to the (very) easy case of a one-period binomial model, as encountered in Chap. 1. The probability measure $\mathbf{P}$ assigns $\mathbf{P}[g] = \mathbf{P}[b] = \frac{1}{2}$ to the two states g and b of $\Omega = \{g, b\}$. In order to make the constants obtained below more easily comparable to the usual notation in the literature (e.g. [LL 96]), we refrain for a moment from our usual condition that we are working with a model in discounted terms. Let

$$\widehat{S}_0^0 = 1, \quad \widehat{S}_1^0 = 1 + r,$$
$$\text{and} \quad \widehat{S}_0^1 = 1, \quad \widehat{S}_1^1 = \begin{cases} 1 + u, & \text{for } \omega = g, \\ 1 + d, & \text{for } \omega = b, \end{cases}$$

where $r > -1$ denotes the riskless rate of interest and $u > r$ stands for "up" and $-1 < d < r$ stands for "down". In discounted terms (see Sect. 2.1 above) the model then becomes

$$S_0^0 = 1, \quad S_1^0 = 1$$
$$\text{and } S_0^1 = 1, \quad S_1^1 = \begin{cases} 1 + \widetilde{u}, & \text{for } \omega = g, \\ 1 + \widetilde{d}, & \text{for } \omega = b, \end{cases}$$

where $1 + \widetilde{u} = \frac{1+u}{1+r} > 1$ and $1 + \widetilde{d} = \frac{1+d}{1+r} < 1$.

We assume w.l.o.g. that $\widetilde{u} \geq -\widetilde{d}$; we do so — mainly for notational convenience — in order to ensure that $\mathbf{E_P}[S_1^1] \geq S_0^1$, so that the optimal portfolios calculated below always have a long position in the stock S^1. If $\widetilde{u} < -\widetilde{d}$ we obtain analogous results, but the position in the stock will be short.

Letting $q = \frac{-\widetilde{d}}{\widetilde{u}-\widetilde{d}} = \frac{r-d}{u-d}$ and defining $\mathbf{Q}[g] = q$ and $\mathbf{Q}[b] = 1 - q = \frac{\widetilde{u}}{\widetilde{u}-\widetilde{d}} = \frac{u-r}{u-d}$ we obtain the unique martingale measure $\mathbf{Q}$ for the process S.

Consider the utility function $U(x) = \frac{x^\alpha}{\alpha}$ for $\alpha \in]-\infty, 1[\backslash\{0\}$ with conjugate function $V(y) = -\frac{y^\beta}{\beta}$, where $\alpha - 1 = (\beta - 1)^{-1}$, i.e., $\beta = \frac{\alpha}{\alpha-1}$. We note that the case of logarithmic and exponential utility (which correspond — after proper renormalisation — to $\alpha = 0$ and to $\alpha = -\infty$) are similar (see 3.3.2 and 3.3.3 below).

Fixing the initial endowment $x > 0$ we want to solve the utility maximisation problem (3.1) by applying the duality theory developed above. Well, this is shooting with canons on pigeons, but we find it instructive to do some explicit calculations exemplifying the abstract formulae.

The dual value function

$$v(y) = \mathbf{E}\left[V\left(y\frac{d\mathbf{Q}}{d\mathbf{P}}\right)\right], \quad y > 0,$$

equals

$$v(y) = \frac{1}{2}\, V(y2q) + \frac{1}{2}\, V(y2(1 - q))$$
$$= c_V\, V(y),$$

where

$$c_V = \frac{1}{2}\left((2q)^\beta + (2(1 - q))^\beta\right).$$

For $\beta < 0$ (which corresponds to $\alpha \in]0, 1[$) we have, for $q \neq \frac{1}{2}$, that $c_V > 1$ by Jensen and the strict convexity of $y \mapsto y^\beta$. Similarly, for $\beta \in]0, 1[$ (which corresponds to $\alpha < 0$) we have $c_V < 1$ (or $c_V = 1$ in the case $q = \frac{1}{2}$). In any case $v(y) \geq V(y)$, as this must hold.

To calculate the primal value function $u(x)$ we use the well-known and easily verified fact that, given a constant $c > 0$ and two conjugate functions $U(x)$ and $V(y)$, the function $c\, U\left(\frac{x}{c}\right)$ is conjugate to $c\, V(y)$. Hence

$$u(x) = c_V\, U\left(\frac{x}{c_V}\right) = c_V^{1-\alpha}\, U(x) = c_U U(x), \tag{3.34}$$

where

$$c_U = c_V^{1-\alpha} = \left(\frac{1}{2}\left((2q)^\beta + (2(1 - q))^\beta\right)\right)^{1-\alpha}. \tag{3.35}$$

For fixed $x > 0$, we obtain as critical Lagrange multiplier $\widehat{y}(x) = u'(x) = c_U U'(x)$ so that

$$\widehat{X}_1(x) = -V'\left(\widehat{y}(x)\frac{d\mathbf{Q}}{d\mathbf{P}}\right)$$

$$= -V'\left(U'(x)\right)c_U^{\frac{1}{\alpha-1}}\left(\frac{d\mathbf{Q}}{d\mathbf{P}}\right)^{\frac{1}{\alpha-1}},$$

$$= xc_V^{-1}\left(\frac{d\mathbf{Q}}{d\mathbf{P}}\right)^{\frac{1}{\alpha-1}},$$

where we have used $-V' = (U')^{-1}$ and $V'(y) = -y^{\frac{1}{\alpha-1}}$. Hence

$$\widehat{X}_1(x) = \begin{cases} xc_V^{-1}\left(\frac{-2\widetilde{d}}{\widetilde{u}-\widetilde{d}}\right)^{\frac{1}{\alpha-1}} = xc_V^{-1}(2q)^{\frac{1}{\alpha-1}}, & \text{for } \omega = g, \\ xc_V^{-1}\left(\frac{2\widetilde{u}}{\widetilde{u}-\widetilde{d}}\right)^{\frac{1}{\alpha-1}} = xc_V^{-1}(2(1-q))^{\frac{1}{\alpha-1}}, & \text{for } \omega = b. \end{cases}$$

Let us explicitly verify that $\widehat{X}_1(x)$ is indeed of the form

$$\widehat{X}_1(x) = x + \widehat{h}\Delta S_1^1, \tag{3.36}$$

for some $\widehat{h} \in \mathbb{R}$, or, equivalently, that $\mathbf{E_Q}[X_1(x)] = x$. Indeed:

$$\mathbf{E_Q}\left[\widehat{X}_1(x)\right]$$

$$= x\left(\frac{1}{2}\left((2q)^{\frac{\alpha}{\alpha-1}} + (2(1-q))^{\frac{\alpha}{\alpha-1}}\right)\right)^{-1}$$

$$\cdot\left[q(2q)^{\frac{1}{\alpha-1}} + (1-q)(2(1-q))^{\frac{1}{\alpha-1}}\right]$$

$$= x$$

To calculate $\widehat{h}$ explicitly we may apply (3.36), e.g., for $\omega = g$ to obtain

$$x + \widehat{h}\widetilde{u} = xc_V^{-1}(2q)^{\frac{1}{\alpha-1}}$$

which yields

$$\widehat{h} = x\left[c_V^{-1}(2q)^{\frac{1}{\alpha-1}} - 1\right]\widetilde{u}^{-1}. \tag{3.37}$$

In the special case of $\alpha = \frac{1}{2}$ (so that $\beta = \frac{\alpha}{\alpha-1} = -1$ and $\beta - 1 = \frac{1}{\alpha-1} = -2$) the constants become somewhat nicer: $(2q)^{\frac{1}{\alpha-1}} = \frac{1}{4}\left(\frac{\widetilde{u}-\widetilde{d}}{\widetilde{d}}\right)^2$, $c_V = \frac{1}{2}\left(\frac{\widetilde{u}-\widetilde{d}}{-2\widetilde{d}} + \frac{\widetilde{u}-\widetilde{d}}{2\widetilde{u}}\right) = -\frac{(\widetilde{u}-\widetilde{d})^2}{4\widetilde{u}\widetilde{d}}$ so that

$$\widehat{h} = x\left[\frac{-4\widetilde{u}\widetilde{d}}{(\widetilde{u}-\widetilde{d})^2}\cdot\frac{1}{4}\frac{(\widetilde{u}-\widetilde{d})^2}{\widetilde{d}^2} - 1\right]\widetilde{u}^{-1}$$

$$= x\frac{\widetilde{u}+\widetilde{d}}{|\widetilde{u}\widetilde{d}|}.$$

Coming back to the case of general $\alpha \in \,] - \infty, 1[\, \backslash \{0\}$, we obtain from (3.37) that the optimal investment in the stock equals $\widehat{h} = \widehat{k}x$ where the constant $\widehat{k} = \widehat{k}(u, d, \alpha)$ given by (3.37) satisfies $0 < \widehat{k} < \infty$. It may be seen as a very elementary version of Merton's result (see [M 90] and example 3.3.5 below), that for the Black-Scholes model and power utility, it is optimal to always invest a constant proportion of your wealth, where the constant may be calculated explicitly as a function of the parameters of the model and the utility function. Observe that this constant may very well be bigger than one, in which case one goes short in the bond.

We now specialise to the case that $\widetilde{u} = \sigma \Delta t^{\frac{1}{2}} + \nu \Delta t$ and $\widetilde{d} = -\sigma \Delta t^{\frac{1}{2}} + \nu \Delta t$ for some $\Delta t > 0$, which corresponds to the notation in the Black-Scholes model below (Sect. 4.4). We determine the different quantities up to the relevant terms of powers of Δt:

$$q = \frac{-\widetilde{d}}{\widetilde{u} - \widetilde{d}} = \frac{\sigma \Delta t^{\frac{1}{2}} - \nu \Delta t}{2\sigma \Delta t^{\frac{1}{2}}} = \frac{1}{2}\left(1 - \frac{\nu}{\sigma}\Delta t^{\frac{1}{2}}\right), \tag{3.38}$$

$$1 - q = \frac{\widetilde{u}}{\widetilde{u} - \widetilde{d}} = \frac{\sigma \Delta t^{\frac{1}{2}} + \nu \Delta t}{2\sigma \Delta t^{\frac{1}{2}}} = \frac{1}{2}\left(1 + \frac{\nu}{\sigma}\Delta t^{\frac{1}{2}}\right),$$

$$c_V = \frac{1}{2}\left((2q)^{\beta} + (2(1 - q))^{\beta}\right) \tag{3.39}$$

$$= \frac{1}{2}\left(1 - \beta\frac{\nu}{\sigma}\Delta t^{\frac{1}{2}} + \frac{\beta(\beta - 1)}{2}\frac{\nu^2}{\sigma^2}\Delta t\right.$$

$$\left. + 1 + \beta\frac{\nu}{\sigma}\Delta t^{\frac{1}{2}} + \frac{\beta(\beta - 1)}{2}\frac{\nu^2}{\sigma^2}\Delta t\right)$$

$$= 1 + \frac{\beta(\beta - 1)\nu^2}{2\sigma^2}\Delta t + o(\Delta t),$$

$$c_U = c_V^{1-\alpha} = \left(1 + \frac{\beta(\beta - 1)\nu^2}{2\sigma^2}\Delta t + o(\Delta t)\right)^{\frac{1}{1-\beta}}$$

$$= 1 - \frac{\beta\nu^2}{2\sigma^2}\Delta t + o(\Delta t). \tag{3.40}$$

For the optimal investment $\widehat{h}$ we obtain

$$\widehat{h} = x\left[c_V^{-1}(2q)^{\frac{1}{\alpha - 1}} - 1\right]\widetilde{u}^{-1} \tag{3.41}$$

$$= x\left[\left(1 - \frac{\beta(\beta - 1)\nu^2}{2\sigma^2}\Delta t\right)\left(1 - \frac{\nu(\beta - 1)}{\sigma}\Delta t^{\frac{1}{2}}\right) - 1\right]$$

$$\cdot \sigma^{-1}(\Delta t)^{-\frac{1}{2}} + O\left(\Delta t^{\frac{1}{2}}\right)$$

$$= \frac{x\nu(1 - \beta)}{\sigma^2} + O\left(\Delta t^{\frac{1}{2}}\right).$$

In the special case $\alpha = \frac{1}{2}$, $\beta = -1$, this yields

$$\widehat{h} = \frac{2x\nu}{\sigma^2} + O\left(\Delta t^{\frac{1}{2}}\right).$$

We observe from (3.41) that, for fixed $\nu > 0$, $\sigma > 0$, the factor

$$\widehat{k} = \frac{(1-\beta)\nu}{\sigma^2} \tag{3.42}$$

ranges between 0 and ∞ as β runs through $]-\infty, 1[\setminus \{0\}$ (the case $\beta = 0$ corresponding to the logarithmic utility $U(x) = \log(x)$).

For the optimal portfolio $\widehat{X}_1(x)$ we thus find

$$\widehat{X}_1(x) = \begin{cases} x\left(1 + \frac{\nu(\beta-1)}{\sigma}\Delta t^{\frac{1}{2}}\right) + O(\Delta t), \text{ for } \omega = g, \\ x\left(1 - \frac{\nu(\beta-1)}{\sigma}\Delta t^{\frac{1}{2}}\right) + O(\Delta t), \text{ for } \omega = b. \end{cases} \tag{3.43}$$

Regarding the logarithmic and exponential utility we only report the results and leave the derivation, which is entirely analogous to the above, as exercises.

Example 3.3.2. Under the same assumptions on S as in 3.3.1, but with letting $U(x) = \ln(x)$, we obtain $V(y) = -\ln(y) - 1$,

$$v(y) = -\ln(y) - 1 + c_1,$$

where $c_1 = -\ln 2 - \frac{1}{2}\ln(q(1-q))$, so that

$$u(x) = \ln(x) + c_1.$$

For the optimal investment, we obtain

$$\widehat{X}_1(x) = \begin{cases} \frac{x}{2q}, & \text{for } \omega = g, \\ \frac{x}{2(1-q)}, & \text{for } \omega = b, \end{cases}$$

so that $\widehat{X}_1(x) = x + \widehat{h}\Delta S_1$, where the optimal trading strategy $\widehat{h}$ is given by

$$\widehat{h} = x\frac{1-2q}{2q\widetilde{u}} = x\frac{\widetilde{u}+\widetilde{d}}{-2\widetilde{d}\widetilde{u}}. \tag{3.44}$$

Example 3.3.3. Using again the assumptions on S as in 3.3.1, but letting now $U(x) = -\exp(-x)$, we obtain $V(y) = y(\ln(y) - 1)$ and

$$v(y) = V(y) - c_2 y, \quad u(x) = -\exp(-(x - c_2))$$

where $c_2 = -[q \ln q + (1-q)\ln(1-q) + \ln 2]$.

For the optimal investment, we obtain

$$\widehat{X}_1(x) = \begin{cases} x + \widehat{h}\widetilde{u}, & \text{for } \omega = g, \\ x + \widehat{h}\widetilde{d}, & \text{for } \omega = b, \end{cases}$$

where the optimal trading strategy $\widehat{h}$ is given by

$$\widehat{h} = -\frac{c_2 + \ln(2q)}{\widetilde{u}} = \frac{1}{\widetilde{u} - \widetilde{d}}\ln\left(\frac{\widetilde{u}}{-\widetilde{d}}\right). \tag{3.45}$$

Note that in this case $\widehat{h}$ does not depend on the initial endowment x.

Example 3.3.4 (The trinomial model (one-period)). We now analyze the simplest model of an incomplete market where we add to the possibilities "good" and "bad" a third possibility "neutral". The probability space Ω now consists of three points, $\Omega = \{g, n, b\}$ where $\mathbf{P}[n] = m$, $\mathbf{P}[g] = \mathbf{P}[b] = \frac{1-m}{2}$ and $m \in]0, 1[$ is a parameter still kept free. We define S (already in discounted terms and dropping the notation for the bond) by $S_0 = 1$ and

$$S_1 = \begin{cases} 1 + \widetilde{u}, & \text{if } \omega = g, \\ 1, & \text{if } \omega = n, \\ 1 + \widetilde{d}, & \text{if } \omega = b, \end{cases}$$

where $1 + \widetilde{u} > 1 > 1 + \widetilde{d} > 0$, similarly as above. Again we assume $\widetilde{u} \geq -\widetilde{d}$.

This model may be viewed as an embryonic version of a stochastic volatility model: the determination of the value of the random variable S_1 can be interpreted as the result of two consecutive steps (taking place, however, at the same time $t = 1$). First one performs an experiment describing an event which happens or not with probability m and $1 - m$ respectively. According to the outcome of this event the volatility is "low" or "high". If it is "low" — in the present embryonic example simply zero — the stock price does not change; if it is "high", a fair coin is tossed similarly as in the binomial model to determine whether the stock price moves to $1 + \widetilde{u}$ or $1 + \widetilde{d}$.

We now again consider power, logarithmic and exponential utility and want to apply theorem 3.2.1 to the present situation. One way to solve the portfolio optimisation problem is to proceed similarly as in the complete case above: first solve the dual problem and then derive the primal problem by using (3.29). This is possible but — in contrast to the complete situation — we now would have to solve an optimisation problem to obtain the dual solution.

In our specific example, it will be more convenient to solve the primal problem directly and then to deduce the solution of the dual problem via (3.29). The solution $\widehat{\mathbf{Q}}(y)$ of the dual problem then allows for an interpretation as a pricing functional (see Remark 3.2.2).

Here is the crucial observation to reduce the present example to the case of 3.3.1 above: the optimal strategy $\widehat{h}$ obtained in (3.37) (resp. (3.44) and (3.45) in the case of logarithmic or exponential utility) for the binomial model is also optimal in the present example. Indeed, distinguish the two cases of "low" and "high" volatility: conditioning on the event that volatility is high, we are in the situation of the binomial model, so that the trading strategy $\widehat{h}$, as calculated in 3.3.1, is the unique optimiser. On the other hand, if volatility is low, the present example is designed such that the volatility vanishes, i.e., the stock price does not move. Hence in this case the choice of the investment h does not influence the result as $h(S_1 - S_2)$ is zero anyhow. Summing up, we conclude that the optimal strategy $\widehat{h}$ obtained for the binomial model is optimal in the present trinomial model as well. Thus we conclude that $\widehat{h}$ given by (3.37) defines the optimal investment also in the present situation

$$\widehat{h} = x \left[c_V^{-1}(2q)^{\beta-1} - 1 \right] \widetilde{u}^{-1},$$

which yields

$$\widehat{h} = x \frac{(1-\beta)\nu}{\sigma^2} + O\left(\Delta t^{\frac{1}{2}}\right)$$

if again, we let $\widetilde{u} = \sigma \Delta t^{\frac{1}{2}} + \nu \Delta t$, $\widetilde{d} = -\sigma \Delta t^{\frac{1}{2}} + \nu \Delta t$, and if c_V is defined as in (3.39).

For the optimal portfolio $\widehat{X}_1(x)$ we find, similarly as in (3.43)

$$\widehat{X}_1(x) = \begin{cases} xc_V^{-1}(2q)^{\beta-1} = x\left(1 + \frac{\nu(1-\beta)}{\sigma}\Delta t^{\frac{1}{2}}\right) + O(\Delta t), & \text{for } \omega = g, \\ x, & \text{for } \omega = n \\ xc_V^{-1}(2(1-q))^{\beta-1} = x\left(1 - \frac{\nu(1-\beta)}{\sigma}\Delta t^{\frac{1}{2}}\right) + O(\Delta t), & \text{for } \omega = b. \end{cases}$$

Denoting by $u^{\text{tri}}(x) = \sup_{h \in \mathbb{R}} \mathbf{E}[U(x + h(S_1 - S_2))]$ the value function for the present trinomial model, we still have $u^{\text{tri}}(x) = c_U^{\text{tri}} U(x)$, for some constant $c_U^{\text{tri}} > 0$, by the scaling property of $U(x) = \frac{x^\alpha}{\alpha}$. The explicit form is given by $c_U^{\text{tri}} - 1 = (1-m)\left(c_U^{\text{bi}} - 1\right)$, where c_U^{bi} is given by the constant c_U in the binomial case above (3.34). Indeed, this relation between the constants of the binomial and trinomial model simply follows from the fact that, conditionally on the event $\{\omega \neq n\}$, which happens with probability $(1-m)$, the trinomial model coincides with the binomial one.

Expanding c_U^{tri} in terms of Δt yields

$$c_U^{\text{tri}} = 1 - \frac{(1-m)\beta\nu^2}{2\sigma^2}\Delta t + o(\Delta t).$$

We now may calculate the dual measure $\widehat{\mathbf{Q}}(y)$ via formula (3.29). Fix the initial endowment x. Then

$$u^{\text{tri}}(x) = c_U^{\text{tri}} U(x)$$

so that

$$y := \left(u^{\text{tri}}\right)'(x) = \left(1 - \frac{(1-m)\beta\nu^2}{2\sigma^2}\Delta t + o(\Delta t)\right) x^{\alpha-1}$$

Applying formula (3.29) from Theorem 3.2.1 we find, for $\omega = n$, that

$$
\begin{aligned}
\frac{d\widehat{\mathbf{Q}}(y)}{d\mathbf{P}}[n] &= y^{-1}U'(\widehat{X}_1(x)(n)) \\
&= y^{-1}U'(x) \\
&= \left(1 + \frac{(1-m)\beta\nu^2}{2\sigma^2}\Delta t + o(\Delta t)\right)x^{-(\alpha-1)}x^{\alpha-1} \\
&= 1 + \frac{(1-m)\beta\nu^2}{2\sigma^2}\Delta t + o(\Delta t).
\end{aligned}
$$

As expected, the initial endowment x cancels out so that $\widehat{\mathbf{Q}}(y)[n]$ does not depend on y and we shall therefore denote it by $\widehat{\mathbf{Q}}[n]$. We find

$$
\widehat{\mathbf{Q}}[n] = \frac{d\widehat{\mathbf{Q}}}{d\mathbf{P}}[n]\mathbf{P}[n] = m + \frac{m(1-m)\beta\nu^2}{2\sigma^2}\Delta t + o(\Delta t)
$$

which gives us a rather complete information how this value depends on the parameters of the model. In fact, $\widehat{\mathbf{Q}}[n]$ determines already $\widehat{\mathbf{Q}}$, also for $\omega = g$ and $\omega = b$. Indeed, the two relations $\widehat{\mathbf{Q}}[g] + \widehat{\mathbf{Q}}[n] + \widehat{\mathbf{Q}}[b] = 1$ as well as $\mathbf{E}_{\widehat{\mathbf{Q}}}[S_1 - S_0] = 0$ yield

$$
\widehat{\mathbf{Q}}[g] = q\left(1 - m - \frac{m(1-m)\beta\nu^2}{2\sigma^2}\Delta t\right) + o(\Delta t),
$$

$$
\widehat{\mathbf{Q}}[b] = (1-q)\left(1 - m - \frac{m(1-m)\beta\nu^2}{2\sigma^2}\Delta t\right) + o(\Delta t),
$$

where q is given by (3.38), which gives

$$
\widehat{\mathbf{Q}}[g] = \frac{1}{2}\left((1-m)\left(1 - \frac{\nu}{\sigma}\Delta t^{\frac{1}{2}}\right) - \frac{m(1-m)\beta\nu}{2\sigma^2}\Delta t\right) + o(\Delta t),
$$

$$
\widehat{\mathbf{Q}}[b] = \frac{1}{2}\left((1-m)\left(1 + \frac{\nu}{\sigma}\Delta t^{\frac{1}{2}}\right) - \frac{m(1-m)\beta\nu}{2\sigma^2}\Delta t\right) + o(\Delta t).
$$

Summing up, we have seen that also in the case of the one step trinomial model all relevant quantities may be calculated explicitly. The fact that we have chosen the parameterisation of the example such that, for $\omega = n$, the stock price simply does not move, made the determination of the primal solution particularly easy. However, one could also abandon this assumption, at the cost of somewhat more cumbersome calculations.

Example 3.3.5 (The binomial model (N periods)). This model is called the Cox-Ross-Rubinstein model [CRR 79] and it is extremely popular in finance, for numerical calculations as well as for pedagogical purposes. It is the discrete version of the Black-Scholes model.

Using the notation of the one step model, example 3.3.1, above and fixing $N \geq 1$ we simply concatenate the one step model in a multiplicative way:

the increments are assumed to be independent. To be formal, let $(\varepsilon_t)_{t=1}^N$ be i.i.d. Bernoulli random variables defined on some $(\Omega, \mathcal{F}, \mathbf{P})$ so that $\mathbf{P}[\varepsilon_t = 1] = \mathbf{P}[\varepsilon_t = -1] = \frac{1}{2}$, for $t = 1, \cdots, N$. The reason why we now use the letter N instead of the previously used T will become apparent after (3.46) below. We denote by $\mathcal{F}_t$ the σ-algebra generated by $(\varepsilon_n)_{n=1}^t$. Let $S_0 = 1$ and, for $t = 1, \cdots, N$, define S_t inductively by

$$S_t = \begin{cases} S_{t-1}(1 + \widetilde{u}) & \text{if } \varepsilon_t = 1, \\ S_{t-1}(1 + \widetilde{d}) & \text{if } \varepsilon_t = -1. \end{cases}$$

Letting $U(x) = \frac{x^\alpha}{\alpha}$, for some fixed $\alpha \in]-\infty, 1[\backslash\{0\}$, we again want to determine the optimal investment strategy and other related quantities.

Our aim is to maximise the expected utility of terminal wealth $\widehat{X}_N(x)$, i.e.

$$\mathbf{E}\left[U\left(x + \sum_{n=1}^N h_n \Delta S_n\right)\right] \to \max !$$

where $(h_n)_{n=1}^T$ runs through all predictable processes.

To do so, we define, for $t = 0, \cdots, N$, the conditional value functions

$$u_t(x) = \sup\left\{\mathbf{E}\left[U\left(x + \sum_{n=t+1}^N h_n \Delta S_n\right)\,\Big|\,\mathcal{F}_t\right]\right\}$$

where the supremum is taken over all collections $(h_n)_{n=t+1}^N$ of $(\mathcal{F}_{n-1})_{n=t+1}^N$-measurable functions. In general $u_t(x)$ will depend on $\omega \in \Omega$ in an $\mathcal{F}_t$-measurable way; but in the present easy example, the i.i.d. assumption on the returns $\left(\frac{S_t}{S_{t-1}}\right)_{t=1}^N$ implies, that $u_t(x)$ does not depend on $\omega \in \Omega$.

In fact, it is straightforward to calculate $u_t(x)$ by backward induction on $t = N, \cdots, 0$: for $t = N$, we obviously have

$$u_N(x) = U(x),$$

and for $t = N - 1$ we are just in the situation of the one step model 3.3.1, so that we find

$$u_{N-1}(x) = c_U U(x)$$

where $c_U = \left(\frac{1}{2}\left((2q)^\beta + (2(1-q))^\beta\right)\right)^{1-\alpha}$, as we have computed in (3.35).

Let us take a closer look why this is indeed the case: the reader might object, that in the present example the value S_{N-1} of the stock at time $N - 1$ as well as the possible gain $\widetilde{u}S_{N-1}$ resp. loss $\widetilde{d}S_{N-1}$, depend on $\omega \in \Omega$ in an $\mathcal{F}_{N-1}$-measurable way, while in the one step example 3.3.1 we had $S_0 = 1$, and the possible gains $\widetilde{u}$ and losses $\widetilde{d}$ were also deterministic. But, of course, this difference is only superficial: if $\widehat{h} \in \mathbb{R}$ denotes the optimal investment in the stock S_0 in the one step example, we now have to choose the optimal investment as the $\mathcal{F}_{N-1}$-measurable function $\widehat{h}_N$ defined by $\widehat{h}_N = \frac{h}{S_{N-1}}$. Economically speaking, the value $\widehat{h}_N S_{N-1}$ of the proportion of the investment in the stock is constant, while the number $\widehat{h}_N$ of stocks in the portfolio depends on the current stock price S_{N-1}: it is simply inversely proportional to S_{N-1}.

To compute $u_{N-2}(x)$, note that this step again is reduced to the analysis of the one step problem at times $\{N-2, N-1\}$, where we now have to replace the original utility function $U(x)$ by the conditional utility function $u_{N-1}(x)$. This is just the principle of dynamic programming which reduces to an obvious fact in the present context. Hence

$$u_{N-2}(x) = \sup\left\{ \mathbf{E}\left[u_{N-1}(x + h\Delta S_{N-1}) \right] \mid h \in \mathbb{R} \right\}$$
$$= c_U^2 U(x).$$

By induction we conclude that

$$u_t(x) = c_U^{N-t} U(x), \quad t = 0, \cdots, N,$$

so that we obtain in particular for the value function $u(x) = u_0(x)$

$$u(x) = c_U^N U(x). \tag{3.46}$$

In order to compute the parameters of the optimal investment strategy, we now assume that there is a fixed horizon $T > 0$ such that $N\Delta t = T$ and we let $N \to \infty$, so that $\Delta t = \frac{T}{N} \to 0$. As above we define $\widetilde{u} = \sigma \Delta t^{\frac{1}{2}} + \nu \Delta t$, $\widetilde{d} = -\sigma \Delta t^{\frac{1}{2}} + \mu \Delta t$. We have found in (3.40) that

$$c_U = 1 - \frac{\beta \nu^2}{2\sigma^2} \Delta t + o(\Delta t)$$

so that

$$c_U^N = c_U^{\frac{T}{\Delta t}} = \exp\left(-\frac{\beta \nu^2 T}{2\sigma^2} \right) + o(1).$$

The optimal investment strategy $\left(\widehat{X}_t(x) \right)_{t=0}^T$ for initial wealth $x > 0$ is given by

$$\widehat{X}_t(x) = x + \sum_{n=1}^{t} \widehat{h}_t \Delta S_t$$

where

$$\widehat{h}_t = \frac{\widehat{X}_{t-1}(x)}{S_{t-1}} \widehat{k}$$

and as in (3.42)

$$\widehat{k} = \frac{\widehat{h}}{x} + O(\Delta t^{\frac{1}{2}}) = \frac{(1-\beta)\nu}{\sigma^2} \tag{3.47}$$

is the ratio of the current wealth $\widehat{X}_{t-1}(x)$ invested into the stock. We thus find the discrete version of the well-known "Merton-line" investment strategy [M 90]; the latter applies to the continuous time limit, i.e., the Black-Scholes model.

Let us have a closer look at the constant $\widehat{k}$: the leading term $\frac{(1-\beta)\nu}{\sigma^2}$ is proportional to ν and inversely proportional to σ^2, which is economically intuitive. As regards $(1-\beta) = (1-\alpha)^{-1}$ we observe that this quantity becomes arbitrary large when α tends to 1. In particular, for $\alpha < 1$ sufficiently close to 1, the proportion $\widehat{k}$ of the wealth held in the stock is bigger than one, and therefore the position in the bond is negative ("short").

Regarding the constant $c_U^N \approx \exp\left(-\frac{\beta\nu^2 T}{2\sigma^2}\right)$ we observe that the right hand side is bigger than one for $\beta \in] - \infty, 0[$, which corresponds to $\alpha \in]0, 1[$. This makes sense, as in this case $U(x) = \frac{x^\alpha}{\alpha}$ takes positive values, so that the value function $u(x)$ defined in (3.46) increases with N. If $\beta \in]0, 1[$, which corresponds to $\alpha \in] - \infty, 0[$, then $\exp\left(-\frac{\beta\nu^2 T}{2\sigma^2}\right)$ is less than 1. This too does make sense economically: for $\alpha \in] - \infty, 0[$ the utility function $U(x) = \frac{x^\alpha}{\alpha}$ takes negative values so that again we find that the utility function $u(x)$ in (3.46) increases with N.

Looking at the constants appearing in $\exp\left(-\frac{\beta\nu^2 T}{2\sigma^2}\right)$, the roles of ν^2, σ^2 and T are quite intuitive. Somewhat more puzzling is the role of β, or rather $-\beta$: while it is intuitive that for $\alpha \to 1$ the factor $\exp\left(-\frac{\beta\nu^2 T}{2\sigma^2}\right)$ tends to infinity $\left(\beta = \frac{\alpha}{\alpha-1}\right)$ so that the problem degenerates for $\alpha \to 1$, the behaviour is less intuitive for $\alpha \to 0$: in this case $\beta \to 0$ too so that $\exp\left(-\frac{\beta\nu^2 T}{2\sigma^2}\right)$ tends to 1, i.e., in the limit there seems no difference between $U(x) = \frac{x^\alpha}{\alpha}$ and the corresponding value function $u(x)$. On the other hand $\lim_{\alpha\to 0} \frac{x^\alpha - 1}{\alpha} = \ln(x)$ so that — after proper normalisation — the utility maximisation problem remains meaningful also as α tends to 0. The point is, that one has to be careful about these renormalisations in the limit, which involves also a term of order α^{-1}.

A good way of dealing with these issues is to recall that the multiplicative term c_U^N in (3.46) pertains to the *utility scale* of the investor. We shall transform it to the *wealth scale* of the investor. We follow B. de Finetti's idea of "certainty equivalent": consider the following two possibilities for an investor. Either she holds an initial endowment $x > 0$ and is allowed to invest in stock and bond in the above model up to time $T = N\Delta t$; or she holds an initial endowment $w_U^N x$, where the letter w stands for "wealth", and is only allowed to invest into the bond (where its value simply remains constant) up to time T. If her goal is to maximise expected utility at time T, what is the value of the constant w_U^N for which the agent is indifferent between these two possibilities. Using (3.46) this leads to the equation

$$c_U^N U(x) = U\left(w_U^N x\right)$$

so that x cancels out (as expected) and we obtain

$$\left(c_U^N\right)^{\frac{1}{\alpha}} = w_U^N$$

so

$$w_U^N = \exp\left(-\frac{\beta\nu^2 T}{2\alpha\sigma^2}\right) + o(1)$$

$$= \exp\left(\frac{\nu^2 T}{(1-\alpha)2\sigma^2}\right) + o(1).$$

In this new scaling the factor $\frac{1}{1-\alpha}$ makes perfect sense economically: when $\alpha \to 1$, i.e., the investor is less and less risk averse, the factor w_U^N becomes large: the investor then appreciates highly the possibility to invest in the financial market. If $\alpha \to -\infty$, i.e., the investor is more and more risk averse and the factor w_U^N tends to one: in this case the investor has little appreciation for the possibilities offered by investments into the risky stock.

Note that the function $\alpha \to \exp\left(\frac{\nu^2 T}{2(1-\alpha)\sigma^2}\right)$ is continuous for $\alpha \in]-\infty, 1[$ so that the limiting behaviour as α tends to zero, is not a puzzle any more; in the case $\alpha = 0$ one easily verifies that $w_U^N = \exp\left(\frac{\nu^2 T}{2\sigma^2}\right)$ indeed yields the certainty equivalent in the case of the logarithmic utility function $U(x) = \log(x)$.

Example 3.3.6 (The trinomial model (N periods)). We now extend the one-period trinomial model, example 3.3.4, to the N period setting similarly as we just have done for the binomial model.

Let $0 < m < 1$ and $(\eta_t)_{t=1}^N$ i.i.d. random variables, defined on some $(\Omega, \mathcal{F}, \mathbf{P})$ such that $\mathbf{P}[\eta_t = -1] = \frac{1-m}{2}$, $\mathbf{P}[\eta_t = 0] = m$, $\mathbf{P}[\eta_t = 1] = \frac{1-m}{2}$. Let $S_0 = 1$ and, for $t = 1, \cdots, N$ define inductively

$$S_t = \begin{cases} S_{t-1}(1 + \widetilde{u}), & \text{if } \eta_t = 1, \\ 0, & \text{if } \eta_t = 0, \\ S_{t-1}(1 + \widetilde{d}), & \text{if } \eta_t = -1. \end{cases}$$

The analysis of the maximisation of expected utility of terminal wealth now is entirely analogous to the situation of the binomial model: using the notation from the preceding examples, the optimal investment strategy again consists in investing the constant proportion $\widehat{k} = \frac{(1-\beta)\nu}{\sigma^2} + O(\Delta t^{\frac{1}{2}})$ of current wealth found in (3.47) into the stock. For the value function $u^{\text{tri}}(x)$ we find

$$u^{\text{tri}}(x) = (c_u^{\text{tri}})^N U(x)$$

$$= \exp\left(-\frac{(1-m)\beta\nu^2 T}{2\sigma^2}\right) U(x) + o(1)$$

and for the "certainty equivalent" $(w_U^{\text{tri}})^N$, defined analogously as in the preceding example, we obtain

$$(w_U^{\text{tri}})^N = \exp\left(\frac{(1-m)\nu^2 T}{(1-\alpha)2\sigma^2}\right) + o(1).$$

The verifications are simply a combination of the arguments of Examples 3.3.4.

4

Bachelier and Black-Scholes

4.1 Introduction to Continuous Time Models

In this chapter we illustrate the theory developed in the previous chapters by analyzing the most basic examples in continuous time. They still play an important role in practice.

The binomial model (Example 3.3.1 and 3.3.5) was already analyzed in the previous Chap. 3. It fits perfectly into the framework developed in Chap. 2, i.e., it is based on a finite probability space Ω. Therefore we could rigorously analyze it in Chap. 3.

If we consider the binomial model on a grid in arithmetic progression and pass to the continuous time limit we arrive, similarly as L. Bachelier in 1900 [B 00] at Brownian motion. If we consider the binomial model on a grid in geometric progression (the "Cox-Ross-Rubinstein" model as in example 3.3.5) we arrive at geometric Brownian motion, similarly as P. Samuelson in 1965. The latter model now is often called the "Black-Scholes" model.

We now pass to the continuous time setting. Strictly speaking, we jump already one step ahead, as we have not yet developed the theory for the case of processes in continuous time. But we believe that it is more important to see the theory in action using these important examples in order to build up some motivation for the formal treatment of the general theory which will be developed later. We shall therefore deliberately use some heuristic arguments which will be rigorously justified by the general theory developed later in this book.

4.2 Models in Continuous Time

To do so, we suppose from now on that the reader is familiar with the notion of Brownian motion and related concepts. We recall the *martingale representation theorem* for Brownian motion, which is the continuous analogue to the elementary considerations on the binomial model above (compare Corollary 2.2.12).

Theorem 4.2.1. *(see, e.g., [RY 91] or Sect. 7.3). Let $(W_t)_{t\in[0,T]}$ be a standard Brownian motion modelled on $(\Omega, (\mathcal{F}_t)_{t\in[0,T]}, \mathbf{P})$, where $(\mathcal{F}_t)_{t\in[0,T]}$ is the natural (right continuous saturated) filtration generated by W.*

Then $\mathbf{P}$ is the unique measure on $\mathcal{F}_T$ which is absolutely continuous with respect to itself, and under which W is a martingale.

Correspondingly, for every function $f \in L^1(\Omega, \mathcal{F}_T, \mathbf{P})$ there is a unique predictable process $H = (H_t)_{t\in[0,T]}$ such that

$$f = \mathbf{E}[f] + (H \cdot W)_T,$$

and

$$\mathbf{E}[f \mid \mathcal{F}_t] = \mathbf{E}[f] + (H \cdot W)_t, \quad 0 \leq t \leq T, \tag{4.1}$$

which implies in particular that $(H \cdot W)$ is a uniformly integrable martingale.

4.3 Bachelier's Model

We formulate Bachelier's model in the framework of the formalism developed above. Let $B_t \equiv 1$ and $S_t = S_0 + \sigma W_t$, $0 \leq t \leq T$, where S_0 is the current stock price, $\sigma > 0$ is a fixed constant, and W is standard Brownian motion on its natural base $(\Omega, (\mathcal{F}_t)_{t\in[0,T]}, \mathbf{P})$.

Fixing the strike price K, we want to price and hedge the contingent claim

$$f(\omega) = (S_T(\omega) - K)_+ \in L^1(\Omega, \mathcal{F}_T, \mathbf{P}). \tag{4.2}$$

Using the martingale representation theorem we may find a trading strategy $\overline{H}$ s.t.

$$\begin{aligned} f &= \mathbf{E}[f] + (\overline{H} \cdot W)_T \tag{4.3} \\ &= \mathbf{E}[f] + (H \cdot S)_T, \end{aligned}$$

where $H = \frac{\overline{H}}{\sigma}$.

Interpreting Theorem 2.4.1 in a liberal way, i.e., transferring its message to the case where Ω is no longer finite and using Theorem 4.2.1 above, we conclude that $C(S_0, T) = \mathbf{E}[f]$ is the unique arbitrage free price for the call option defined in (4.2).

Note that S_T is normally distributed with mean S_0 and variance $\sigma^2 T$. Hence

$$C(S_0, T) = \int_{K-S_0}^{\infty} (x - (K - S_0))g(x)dx, \tag{4.4}$$

where

$$g(x) = \frac{1}{\sigma\sqrt{2\pi T}} e^{-\frac{x^2}{2\sigma^2 T}}. \tag{4.5}$$

It is straightforward to derive from (4.4) an "option pricing formula" by calculating the integral in (4.4) (compare, e.g., [Sh 99]): denoting the standard normal density function by $\varphi(x)$, i.e., $\varphi(x)$ equals (4.5) for $\sigma^2 T = 1$, denoting the corresponding distribution function by $\Phi(x)$ and using the relation $\varphi'(x) = -x\varphi(x)$, an elementary calculation reveals that

$$C(S_0, T) = \mathbf{E}[f] = (S_0 - K)\,\Phi\left(\frac{S_0 - K}{\sigma\sqrt{T}}\right) + \sigma\sqrt{T}\,\varphi\left(\frac{S_0 - K}{\sigma\sqrt{T}}\right). \quad (4.6)$$

By the same token we obtain, for every $0 \le t \le T$, and conditionally on the stock price having the value S_t at time t,

$$C(S_t, T - t) \quad (4.7)$$
$$= \mathbf{E}[f \mid S_t] = (S_t - K)\,\Phi\left(\frac{S_t - K}{\sigma\sqrt{T - t}}\right) + \sigma\sqrt{T - t}\,\varphi\left(\frac{S_t - K}{\sigma\sqrt{T - t}}\right).$$

This solves the pricing problem in Bachelier's model, based on no-arbitrage arguments, as we have the "replication formula" (4.3).

But what is the trading strategy H, in other words, the recipe to replicate the option by trading dynamically? Economic intuition suggests that we have

$$H(S, t) = \frac{\partial}{\partial S} C(S, T - t).$$

Indeed, consider the following heuristic reasoning using infinitesimals: suppose that at time t the stock price equals S_t, so that the value of the option equals $C(S_t, T - t)$. During the infinitesimal interval $(t, t + dt)$ the Brownian motion W_t will move by $dW_t = W_{t+dt} - W_t = \varepsilon_t\sqrt{dt}$, where $\mathbf{P}[\varepsilon_t = 1] = \mathbf{P}[\varepsilon_t = -1] = \frac{1}{2}$, so that S_t will move by $dS_t = S_{t+dt} - S_t = \varepsilon_t\sigma\sqrt{dt}$. Hence the value of the option $C(S_t, T-t)$ will move by $dC_t = C(S_{t+dt}, T-(t+dt)) - C(S_t, T-t) \approx \varepsilon_t\frac{\partial C}{\partial S}(S_t, T - t)\sigma\sqrt{dt}$, where we neglect terms of smaller order than $\sqrt{dt}$. In other words, the ratio between the up or down movement of the underlying stock S and the option is

$$\frac{dC_t}{dS_t} = \frac{\partial C}{\partial S}(S_t, T - t)\frac{\varepsilon_t\sigma\sqrt{dt}}{\varepsilon_t\sigma\sqrt{dt}} \quad (4.8)$$
$$= \frac{\partial C}{\partial S}(S_t, T - t).$$

If we want to replicate the option by investing the proper quantity H of the underlying stock, formula (4.8) suggests that this quantity should equal $\frac{\partial C}{\partial S}(S_t, T - t)$.

After these motivating remarks, let us deduce the equation

$$H(S_t, t) = \frac{\partial C}{\partial S}(S_t, T - t) \quad (4.9)$$

more formally. Consider the stochastic process

$$C(S_t, T - t) = C(S_0 + \sigma W_t, T - t), \quad 0 \le t \le T,$$

of the value of the option. By Itô's formula (see, e.g., [RY 91])

$$dC(S_t, T - t) = \frac{\partial C}{\partial S} dS_t + \left(\frac{\partial C}{\partial t} + \frac{1}{2} \frac{\partial^2 C}{\partial S^2} \sigma^2 \right) dt, \qquad (4.10)$$

where we have used $dS_t = \sigma dW_t$. One readily deduces from formula (4.7) that C verifies the heat equation with parameter $\frac{\sigma^2}{2}$ displayed in (4.11) below (time is running into the negative direction in the present setting). In particular, for the process C defined in (4.7), the drift term in (4.10) vanishes as it must be the case according to the general theory (the option price process is a martingale by (4.1)). Hence (4.10) reduces to the formula

$$C(S_t, T - t) = C(S_0, T) + (H \cdot S)_t,$$

where H is given by (4.9). Rephrasing this result once more we have shown that the trading strategy H, whose existence was guaranteed by the martingale representation (Theorem 4.2.1), is of the form (4.9).

One more word on the fact that $C(S, T - t)$ satisfies the heat equation (4.11) below, which was known to L. Bachelier in 1900 and may be verified by simply calculating the partial derivatives in (4.7). Admitting this calculation, we concluded above that the drift term in (4.10) vanishes. One may also turn the argument around to conclude from (4.1) that the drift term in (4.10) must vanish, which then *implies* that $C(S, T - t)$ must satisfy the heat equation (time running inversely)

$$\frac{\partial C}{\partial t}(S, T - t) = -\frac{\sigma^2}{2} \frac{\partial^2 C}{\partial S^2}(S, T - t). \qquad (4.11)$$

Imposing the boundary condition $C(S, T - T) = C(S, 0) = (S - K)_+$ one may *derive* by standard methods the solution (4.7) of this p.d.e.. This is, in fact, how F. Black and M. Scholes originally proceeded (in the framework of their model) to derive their option pricing formula, which we shall now analyze.

4.4 The Black-Scholes Model

This model of a stock market was proposed by the famous economist P. Samuelson in 1965 ([S 65]), who was aware of Bachelier's work. In fact, triggered by a question of J. Savage, it was P. Samuelson who had rediscovered Bachelier's work for the economic literature some years before 1965.

The model is usually called the Black-Scholes model today and became the standard reference model in the context of option pricing:

$$\widehat{B}_t = e^{rt},$$

$$\widehat{S}_t = S_0 e^{\sigma W_t + \left(\mu - \frac{\sigma^2}{2} \right) t}, \quad 0 \le t \le T. \qquad (4.12)$$

Again W is a standard Brownian motion with natural base $(\Omega, (\mathcal{F}_t)_{t \in [0,T]}, \mathbf{P})$.

The parameter r models the "riskless rate of interest", while the parameter μ models the average increase of the stock price. Indeed using Itô's formula one may describe the model equivalently by the differential equations:

$$\frac{d\widehat{B}_t}{\widehat{B}_t} = rdt,$$

$$\frac{d\widehat{S}_t}{\widehat{S}_t} = \mu dt + \sigma dW_t.$$

The numéraire in this model is just the relevant currency (say €). In order to remain consistent with the above theory, we shall rather follow our usual procedure of taking a traded asset as numéraire, namely the bond, to use discounted terms. We then have

$$B_t = \frac{\widehat{B}_t}{\widehat{B}_t} = 1 \tag{4.13}$$

$$S_t = \frac{\widehat{S}_t}{\widehat{B}_t} = S_0 e^{\sigma W_t + \left(\mu - r - \frac{\sigma^2}{2}\right)t}.$$

We shall write ν for $\mu - r$ which is called the "excess return". The only thing we have to keep in mind when passing to the bond as numéraire is that now quantities have to be expressed in terms of the bond: in particular, if K denotes the strike price of an option at time T (expressed in € at time T), we have to express it as Ke^{-rT} units of the bond.

Contrary to Bachelier's setting, the process

$$S_t = S_0 e^{\sigma W_t + \left(\nu - \frac{\sigma^2}{2}\right)t}, \quad 0 \leq t \leq T,$$

is *not* a martingale under **P** (unless $\nu = 0$, which typically is not the case).

The unique martingale measure **Q** for S (which is absolutely **P**-continuous) is given by Girsanov's theorem (see [RY 91] or any introductory text to stochastic calculus)

$$\frac{d\mathbf{Q}}{d\mathbf{P}} = \exp\left(-\frac{\nu}{\sigma}W_T - \frac{\nu^2}{2\sigma^2}T\right). \tag{4.14}$$

Let us price and hedge the contingent claim $f(\omega) = \left(S_T(\omega) - Ke^{-rT}\right)_+$, which is the pay-off function of the European call option with exercise time T and a strike price of K Euros (expressed in terms of the bond numéraire).

Noting that $(W_t + \nu t)_{t=0}^{\infty}$ is a standard Brownian motion under **Q** and applying Theorem 4.2.1 to the **Q**-martingale S, we may calculate, similarly as in (4.6) above.

$$C(S_0, T) = \mathbf{E_Q}[f] = \mathbf{E_Q}\left[\left(S_0 e^{\sigma(W_T + \nu T) - \frac{\sigma^2}{2}T} - Ke^{-rT}\right)_+\right] \tag{4.15}$$

$$= S_0 \mathbf{E_Q}\left[e^{\sigma\sqrt{T}Z - \frac{\sigma^2 T}{2}}\chi_{\{S_T \geq K\}}\right] - Ke^{-rT}\mathbf{Q}[S_T \geq K],$$

where $Z = \frac{W_T + \nu T}{\sqrt{T}}$ is a $N(0,1)$-distributed random variable under **Q**.

After an elementary calculation (see, e.g., [LL 96]) this yields the famous *Black-Scholes formula*

$$C(S_0, T) = S_0 \Phi \left(\frac{\ln \left(\frac{S_0}{K} \right) + \left(r + \frac{\sigma^2}{2} \right) T}{\sigma \sqrt{T}} \right) \tag{4.16}$$

$$- K e^{-rT} \Phi \left(\frac{\ln \left(\frac{S_0}{K} \right) + \left(r - \frac{\sigma^2}{2} \right) T}{\sigma \sqrt{T}} \right)$$

and, by the same token, for $0 \le t \le T$ and $S_t > 0$,

$$C(S_t, T - t) = S_0 \Phi \left(\frac{\ln \left(\frac{S_t}{K} \right) + \left(r + \frac{\sigma^2}{2} \right) (T - t)}{\sigma \sqrt{T - t}} \right) \tag{4.17}$$

$$- K e^{-rT} \Phi \left(\frac{\ln \left(\frac{S_t}{K} \right) + \left(r - \frac{\sigma^2}{2} \right) (T - t)}{\sigma \sqrt{T - t}} \right).$$

Let us take some time to contemplate on this truly remarkable formula (for which R. Merton and M. Scholes received the Nobel prize in economics in 1997; F. Black unfortunately had passed away in 1995).

1.) As a warm-up consider the limits as $\sigma \to \infty$ (which yields $C(S_0, T) = S_0$) and $\sigma \to 0$ (which yields $C(S_0, T) = (S_0 - K e^{-rT})_+$). The reader should convince herself that this does make sense economically. For an extremely risky underlying S, an option on one unit of S is almost as valuable as one unit of S itself (think, for example, of a call option on a lottery ticket with strike price $K = 100$ and exercise time T, such that T is later than the drawing at which it is decided whether the ticket wins a million or nothing). Intuitively, it is quite obvious that the option on the lottery ticket is almost as valuable as the lottery ticket itself). On the other hand, if the underlying S is (almost) riskless a similar consideration reveals that the value of an option is almost equal to its "inner value" $(S_0 - K e^{-rT})_+$.

This behaviour of the Black-Scholes formula should be contrasted to Bachelier's formula (specialising to the case $S_0 = K$ and $r = 0$)

$$C^{\text{Bachelier}}(S_0, T) = \frac{\sigma}{\sqrt{2\pi}} \sqrt{T} \tag{4.18}$$

obtained in (4.6) above, which tends to infinity as $\sigma \to \infty$; this limiting behaviour is economically absurd and contradicts an obvious no-arbitrage argument which — using the fact that S_T is non-negative — shows that the value of a call option always must be less than the value of the underlying stock.

The reason for this difference in the behaviour of the Black-Scholes formula and Bachelier's one, for large values of σ, is that geometric Brownian motion always remains positive, while Brownian motion may also attain negative values. This fact has strong effects for very large σ or — what amounts roughly to the same — for very large T. Nevertheless we shall presently see that — for reasonable values of σ and T — the Black-Scholes formula and Bachelier's formula (4.18) are very close. This seems to be the essential fact, keeping in mind Keynes' famous dictum telling us, *not* to look at the limit $T \to \infty$: in the long run we are all dead.

2.) Let us compare the Black-Scholes formula (4.16) and Bachelier's formula (4.18) more systematically. To do so we specialise in the Black-Scholes formula to $r = 0$ and $S_0 = K$, and we have to let the volatility in the Black-Scholes formula, which we now denote by σ^{BS}, correspond to the "parameter of nervousness" σ (this wording was used by Bachelier) appearing in Bachelier's formula, which we denote by σ^B. As the former pertains to the relative standard deviation of stock prices and the latter to the absolute standard deviation, we roughly find the correspondence — at least for small values of T —

$$\sigma^B \approx \sigma^{BS} S_0$$

In the subsequent calculations we therefore suppose that $\sigma^B \approx \sigma^{BS} S_0$. In this case, the Black-Scholes and Bachelier option prices to be compared are

$$C^{BS} = S_0 \left[\Phi \left(\frac{\sigma^{BS} \sqrt{T}}{2} \right) - \Phi \left(-\frac{\sigma^{BS} \sqrt{T}}{2} \right) \right],$$

while

$$C^B = \frac{\sigma^B}{\sqrt{2\pi}} \sqrt{T} \approx S_0 \frac{\sigma^{BS}}{\sqrt{2\pi}} \sqrt{T}.$$

The difference of the two quantities is best understood by looking at the shaded area in the subsequent graph involving the density $\varphi(x) = \frac{1}{\sqrt{2\pi}} e^{-\frac{x^2}{2}}$ of the standard normal distribution, and noting that $\varphi(0) = \frac{1}{\sqrt{2\pi}}$. This shaded area, let's call it A, equals $\frac{\sigma^{BS}\sqrt{T}}{\sqrt{2\pi}} - \left[\Phi \left(\frac{\sigma^{BS}\sqrt{T}}{2} \right) - \Phi \left(-\frac{\sigma^{BS}\sqrt{T}}{2} \right) \right]$ which — up to the factor S_0 — is just the difference between C^B and C^{BS}

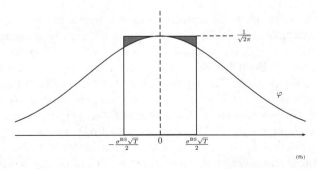

Fig. 4.1. Comparison of the Bachelier with the Black-Scholes formula.

Expanding $\varphi(x)$ into a Taylor series around zero and using $\varphi''(0) = -\frac{1}{\sqrt{2\pi}}$ we get the asymptotic expression

$$C^{\mathrm{B}} - C^{\mathrm{BS}} = S_0 \left[\frac{1}{24\sqrt{2\pi}} \left(\sigma^{\mathrm{BS}} \sqrt{T} \right)^3 \right] + o\left(\left(\sigma^{\mathrm{BS}} \sqrt{T} \right)^3 \right),$$

which indicates a very small difference between the option values C^{B} and C^{BS} in the Bachelier and Black-Scholes model respectively, provided $\sigma^{\mathrm{BS}}\sqrt{T}$ is small. Evaluating this expression for the empirical data reported by Bachelier (see [B 00] or [S 03, Chap. 1]), i.e., $\sigma^{\mathrm{BS}} \approx 2.4\,\%$ on a yearly basis, and $T \approx 2\,\text{months} = \frac{1}{6}\,\text{year}$ (this is a generous upper bound for the periods considered by Bachelier which were ranging between 10 and 45 days) we find

$$C^B - C^{\mathrm{BS}} \approx S_0 \frac{1}{24\sqrt{2\pi}} \left(0.024\sqrt{\frac{1}{6}} \right)^3 \approx 1.56 * 10^{-8} S_0.$$

Hence for this data the difference between the option value obtained using Bachelier's and the Black-Scholes model is of order 10^{-8} times the value S_0 of the underlying; observing that for Bachelier's data, the price of an option was of the order of $\frac{S_0}{100}$, we find that the difference is of the order 10^{-6} of the price of the option.

In view of all the uncertainties involved in option pricing, in particular regarding the estimation of σ, one might be tempted to call this quantity "completely negligible, a priori" (this expression was used by Bachelier when discussing the drawbacks of the normal distribution giving positive probability to negative stock prices).

For more information we refer to the introductory chapter of [S 03, Chap. 1] as well as to [ST 05].

3.) Let us now comment on the role of the riskless rate of interest r, appearing in the Black-Scholes formula and on the reason why this variable does not show up in Bachelier's formula: noting the obvious fact that

$$\ln\left(\frac{S_0}{K} \right) + rT = \ln\left(\frac{S_0}{Ke^{-rT}} \right),$$

one readily observes that this quantity only appears in the Black-Scholes formula (4.16) via the discounting of the strike price, i.e., transforming K units of $\text{€}_{t=T}$ into Ke^{-rT} units of $\text{€}_{t=0}$. When comparing the setting of Black-Scholes to that of Bachelier one should recall that the option premium in Bachelier's days pertained to a payment at time T rather than at time 0. Under the assumption of a constant riskless interest rate — as is the case in the Black-Scholes model — this amounts to considering the present day quantities upcounted by e^{rT}. This was perfectly taken into account by Bachelier, who stressed that the quantities appearing in his formulae have to be understood in terms of forward prices in modern terminology, which amounts to upcounting by e^{rT} in the present setting.

The bottom line of these considerations on the role of r is: when we assumed that $r = 0$ in the above comparison of the Bachelier and Black-Scholes option pricing methodology, this assumption did not restrict the generality of the argument. It also applies to $r \neq 0$ as Bachelier denoted the relevant quantities in terms of "true prices", i.e., forward prices.

4.) What is the partial differential equation satisfied by the solution (4.17) of the Black-Scholes formula? Again we specialise to the case $r = 0$ in order to focus on the heart of the matter, but we note that now we *do restrict the generality* and refer to any introductory text to Mathematical Finance (e.g., [LL 96]) for the Black-Scholes partial differential equation in the case of a riskless rate of interest $r \neq 0$.

From the Martingale Representation Theorem 4.2.1 we know that the Black-Scholes option price *process*

$$C(S_t, T - t)_{t \in [0,T]}$$

is a martingale under the measure $\mathbf{Q}$ defined in (4.14). Hence, denoting by $(\widetilde{W}_t)_{t \in [0,T]}$ a standard Brownian motion under $\mathbf{Q}$, using $dS_t = \sigma S_t d\widetilde{W}_t$, and working under the measure $\mathbf{Q}$, we deduce from Itô's formula

$$dC_t = dC(S_t, T - t) = \frac{\partial C}{\partial S} \sigma S_t d\widetilde{W}_t + \left(\frac{\sigma^2}{2} S_t^2 \frac{\partial^2 C}{\partial S^2} + \frac{\partial C}{\partial t} \right) dt.$$

We first observe, using again $\sigma S_t d\widetilde{W}_t = dS_t$, that — similarly as in the context of Bachelier — the replicating trading strategy H_t is given by $\frac{\partial C}{\partial S}(S_t, T - t)$. In the lingo of finance this quantity is called the "Delta" of the option (which depends on S_t and t) and the trading strategy H is called "delta-hedging".

Next we pass to the drift term: as $C(S_t, T - t)$ is a $\mathbf{Q}$-martingale we infer that it must vanish, which yields the "Black-Scholes partial differential equation"

$$\frac{\partial C}{\partial t}(S, T - t) = -\frac{\sigma^2}{2} S^2 \frac{\partial^2 C}{\partial S^2}(S, T - t), \quad \text{for } S \geq 0, t \geq 0. \tag{4.19}$$

This is the multiplicative analogue of the heat equation (4.11) and may, in fact, easily be reduced to a heat equation (with drift) by passing to logarithmic coordinates $x = \ln(S)$.

Exactly as in Bachelier's case we may proceed by solving the partial differential equation (4.19) for the boundary condition $C(S, T - T) = C(S, 0) = (S - K)_+$ and $C(0, t) = 0$ to obtain the Black-Scholes formula.

In the lingo of finance, the quantity $-\frac{\partial C}{\partial t}$ is called the "Theta" and the quantity $\frac{\partial^2 C}{\partial S^2}$ the "Gamma" of the option. Hence the p.d.e. (4.19) allows for the following economic interpretation: when time to maturity $T - t$ decreases (and S remains fixed), the loss of value of the option is equal to the "convexity" or the "Gamma" of the option price (as a function of S) at time t, normalised by $\frac{\sigma^2}{2}S^2$ (in the case of the Bachelier model the normalisation was simply $\frac{\sigma^2}{2}$). This has a nice economic interpretation and today's option traders think in these terms. They speak about "selling or buying convexity" or rather "going gamma-short or gamma-long" which amounts to the same thing. The interpretation of (4.19) is that, for the buyer of an option, the convexity of the function $C(S, T-t)$ in the variable S corresponds to a kind of insurance with respect to price movements of S. As there is no such thing as a free lunch, this insurance costs (proportional to the second derivative) and a positive $\frac{\sigma^2}{2}S^2\frac{\partial^2 C}{\partial S^2}$ is reflected by a negative partial derivative $\frac{\partial C}{\partial t}$ of $C(S, T-t)$ with respect to time t.

Let us illustrate this fact by reasoning once more heuristically with infinitesimal movements of Brownian motion: we want to explain the infinitesimal change of the option price when "time increases by an infinitesimal while the stock price S remains constant". To do so we apply the heuristic analogue of the Brownian bridge: consider the infinitesimal interval $[t, t+2dt]$ and assume that the driving $\mathbf{Q}$-Brownian motion $\widetilde{W}$ moves in the first half $[t, t+dt]$ from $\widetilde{W}_t$ to $\widetilde{W}_t + \varepsilon_t\sqrt{dt}$, where ε_t is a random variable with $\mathbf{Q}[\varepsilon_t = 1] = \mathbf{Q}[\varepsilon_t = -1] = \frac{1}{2}$, while in the second half $[t+dt, t+2dt]$ it moves back to $\widetilde{W}_t$. What should happen during this time interval to a "hedger" who proceeds according to the Black-Scholes trading strategy H described above, which replicates the option? At time t she holds $\frac{\partial C}{\partial S}(S_t, T-t)$ units of the stock. Following first the scenario $\varepsilon_t = +1$, the stock has a price of $S_t + \sigma S_t\sqrt{dt}$ at time $t+dt$. Apart from being happy about this up movement, the hedger now (i.e., at time $t+dt$) adjusts the portfolio to hold $\frac{\partial C}{\partial S}\left(S_t + \sigma S_t\sqrt{dt}, T-(t+dt)\right)$ units of stock, which results in a net buy of $\frac{\partial^2 C}{\partial S^2}(S_t, T-t)\sigma S_t\sqrt{dt}$ units of stock, where we neglect terms of smaller order than $\sqrt{dt}$. In the next half $[t+dt, t+2dt]$ of the interval the stock price S drops again to the value $S_{t+2dt} = S_t$ and the hedger readjusts at time $t+2dt$ the portfolio by selling again the $\frac{\partial^2 C}{\partial S^2}(S_t, T-t)\sigma S_t\sqrt{dt}$ units of stock (neglecting again terms of smaller order than $\sqrt{dt}$). It seems at first glance that the gains made in the first half are precisely compensated by the losses in the second half, but a closer inspection shows that the hedger did "buy high" and "sell low": the quantity $\frac{\partial^2 C}{\partial S^2}(S_t, T-t)\sigma S_t\sqrt{dt}$ was bought at price $S_t + \sigma S_t\sqrt{dt}$ at time $t+dt$, and sold at price S_t at time $t+2dt$, resulting in a total loss of

$$\left(\frac{\partial^2 C}{\partial S^2}(S_t, T-t)\sigma S_t\sqrt{dt}\right)\left(\sigma S_t\sqrt{dt}\right) = \sigma^2 S_t^2\frac{\partial^2 C}{\partial S^2}(S_t, T-t)dt. \quad (4.20)$$

Going through the scenario $\varepsilon_t = -1$, one finds that the hedger did first "sell low" and then "buy high" resulting in the same loss (where again we neglect infinitesimals resulting in effects (with respect to the final result) of smaller order than dt).

Keeping in mind that this was achieved during an interval of total length $2dt$ (which corresponds to the passage from σ^2 in (4.20) to $\frac{\sigma^2}{2}$ in (4.19)) we have found a heuristic explanation for the Black-Scholes equation (4.19). We also note that the same argument applied to Bachelier's model, yields a heuristic explanation of the heat equation (4.11). The general phenomenon behind this fact is that, in the case of convexity, the "wobbling" of Brownian motion, which is of order $\sqrt{dt}$ in an interval of length dt, causes the hedger to have systematic losses, which are proportional to $\frac{\partial^2 C}{\partial S^2}$ as well as to the increment $d\langle S \rangle_t$ of the quadratic variation process $\langle S \rangle_t = \int_0^t \sigma^2 S_u^2 du$ of the stock price process S.

5.) When deriving the Black-Scholes formula (4.16) we did not go through the (elementary but tedious) trouble of explicitly calculating (4.15). We shall now provide an explicit derivation of the formula which has the merit of yielding an interpretation of the two probabilities appearing in (4.16). It also allows for a better understanding of the formula (for example, for the remarkable fact, that the parameter μ has disappeared) and dispenses us of cumbersome calculation.

As observed in (4.15), the contingent claim $f(\omega) = (S_T(\omega) - Ke^{-rT})_+$ (expressed in terms of the numéraire B_t) splits into

$$
\begin{aligned}
\left(S_T - Ke^{-rT}\right)_+ &= S_T \chi_{\{S_T \geq Ke^{-rT}\}} - Ke^{-rT} \chi_{\{S_T \geq Ke^{-rT}\}} \\
&= f_1 - f_2.
\end{aligned}
$$

We have to calculate $\mathbf{E_Q}[f_1]$ and $\mathbf{E_Q}[f_2]$ under the risk-neutral measure $\mathbf{Q}$ defined in (4.14). This is easy for f_2 and we do not have to use the explicit form of the density (4.14) provided by Girsanov's theorem. It suffices to observe that $S_T = S_0 \exp(\sigma \widetilde{W}_T - \frac{\sigma^2}{2}T)$ where $\widetilde{W}$ is a Brownian motion under $\mathbf{Q}$. So

$$
X := \frac{\ln\left(\frac{S_T}{S_0}\right) + \frac{\sigma^2}{2}T}{\sigma\sqrt{T}} \sim N(0,1) \quad \text{under } \mathbf{Q},
$$

whence

$$
\mathbf{E_Q}[f_2] = e^{-rT} K \, \mathbf{Q}\left[S_T \geq e^{-rT}K\right] \tag{4.21}
$$

$$
= e^{-rT} K \, \mathbf{Q}\left\{ \frac{\ln\left(\frac{S_T}{S_0}\right) + \frac{\sigma^2}{2}T}{\sigma\sqrt{T}} \geq \frac{\ln\left(\frac{e^{-rT}K}{S_0}\right) + \frac{\sigma^2}{2}T}{\sigma\sqrt{T}} \right\}
$$

$$
= e^{-rT} K \, \mathbf{Q}\left\{ X \geq \frac{\ln\left(\frac{e^{-rT}K}{S_0}\right) + \frac{\sigma^2}{2}T}{\sigma\sqrt{T}} \right\}
$$

$$
= e^{-rT} K \, \Phi\left(\frac{\ln\left(\frac{S_0}{K}\right) + \left(r - \frac{\sigma^2}{2}\right)T}{\sigma\sqrt{T}} \right),
$$

which yields the second term of the Black-Scholes formula.

Why was the calculation of $\mathbf{E}_{\mathbf{Q}}[f_2]$ so easy? Simply because the factor Ke^{-rT} appearing in f_2 is a constant (expressed in terms of the present numéraire, namely the bond); hence the calculation of the expectation was reduced to the calculation of the probability of an event, namely the probability that the option will be exercised, with respect to the probability measure $\mathbf{Q}$.

To proceed similarly with the calculation of $\mathbf{E}_{\mathbf{Q}}[f_1]$ we make a change of numéraire, now choosing the risky asset $\widehat{S}$ in the Black-Scholes model (4.12) as numéraire. Under this numéraire the model reads

$$\frac{\widehat{B}_t}{\widehat{S}_t} = \frac{1}{S_t} = S_0^{-1} e^{-\sigma W_t + \left(r - \mu + \frac{\sigma^2}{2}\right)t}$$

$$\frac{\widehat{S}_t}{\widehat{S}_t} \equiv 1$$

where W is a standard Brownian motion under $\mathbf{P}$. The reader has certainly noticed the symmetry with (4.13). But what is the probability measure $\check{\mathbf{Q}}$ under which the process $\frac{\widehat{B}_t}{\widehat{S}_t} = \frac{1}{S_t}$ becomes a martingale? Using Girsanov we can explicitly calculate the density $\frac{d\check{\mathbf{Q}}}{d\mathbf{P}}$; but, in fact, we don't really need this information. All we need is to observe that we may write

$$\frac{1}{S_t} = S_0^{-1} e^{-\sigma \check{W}_t - \frac{\sigma^2}{2}t},$$

where $\check{W}$ is a standard Brownian motion under $\check{\mathbf{Q}}$ (the reader worried by the minus sign in front of $\sigma \check{W}_t$ may note that $-\check{W}$ is also a standard Brownian motion under $\check{\mathbf{Q}}$). We now apply the change of numéraire theorem (in the form of Theorem 2.5.4) to calculate $\mathbf{E}_{\mathbf{Q}}[f_1]$. In fact, we have only proved this theorem for the case of finite Ω, but we rely on the reader's faith that it also applies to the present case (for a thorough investigation for the validity of this theorem for general locally bounded semi-martingale models we refer to Chap. 11 below). Applying this theorem we obtain

$$\mathbf{E}_{\mathbf{Q}}[f_1] = \mathbf{E}_{\mathbf{Q}}\left[S_T \chi_{\{S_T \geq e^{-rT}K\}} \right]$$

$$= S_0 \mathbf{E}_{\check{\mathbf{Q}}}\left[\frac{S_T}{S_T} \chi_{\left\{ \frac{1}{S_T} \leq e^{rT}K^{-1} \right\}} \right]$$

$$= S_0 \mathbf{E}_{\check{\mathbf{Q}}}\left[\chi_{\left\{ S_0^{-1} e^{-\sigma \check{W}_T - \frac{\sigma^2}{2}T} \leq e^{rT}K^{-1} \right\}} \right]$$

$$= S_0 \check{\mathbf{Q}}\left[S_0 e^{\sigma \check{W}_T + \frac{\sigma^2}{2}T} \geq e^{-rT}K \right].$$

Noting that $\frac{\check{W}_T}{\sqrt{T}}$ is $N(0,1)$-distributed under $\check{\mathbf{Q}}$, this expression is completely analogous to the one appearing in (4.21), with the exception of the plus in front of the term $\frac{\sigma^2}{2}T$. Hence we get

$$\mathbf{E_Q}[f_1] = S_0 \Phi \left(\frac{\ln\left(\frac{S_0}{K}\right) + \left(r + \frac{\sigma^2}{2}\right)T}{\sigma\sqrt{T}} \right),$$

which is the first term appearing in the Black-Scholes formula. We now may interpret $\Phi\left(\frac{\ln\left(\frac{S_0}{K}\right) + \left(r + \frac{\sigma^2}{2}\right)T}{\sigma\sqrt{T}}\right)$ as the probability, that the option will be exercised, with respect to the probability measure $\check{\mathbf{Q}}$, under which $\frac{1}{S_t}$ is a martingale.

5

The Kreps-Yan Theorem

Let us turn back to the no-arbitrage theory developed in Chap. 2 to raise again the question: what can we deduce from applying the no-arbitrage principle with respect to pricing and hedging of derivative securities?

While we obtained satisfactory and mathematically rigorous answers to these questions in the case of a finite underlying probability space Ω in Chap. 2, we saw in Chap. 4, that the basic examples for this theory, the Bachelier and the Black-Scholes model, do not fit into this easy setting, as they involve Brownian motion.

In Chap. 4 we overcame this difficulty either by using well-known results from stochastic analysis (e.g., the martingale representation Theorem 4.2.1 for the Brownian filtration), or by appealing to the faith of the reader, that the results obtained in the finite case also carry over — mutatis mutandis — to more general situations, as we did when applying the change of numéraire theorem to the calculation of the Black-Scholes model.

5.1 A General Framework

We now want to develop a "théorie générale of no-arbitrage" applying to a general framework of stochastic processes. Forced by the relatively poor fit of the Black-Scholes model (as well as Bachelier's model) to empirical data (especially with respect to extreme behaviour, i.e., large changes in prices), Mathematical Finance developed towards more general models. In some cases these models still have continuous paths, but processes (in continuous time) with jumps are increasingly gaining importance.

We adopt the following general framework (for more details on the technicalities of stochastic integration we refer to Chap. 7): let $S = (S_t)_{t \geq 0}$ be an $\mathbb{R}^{d+1}$-valued stochastic process based on and adapted to the filtered probability space $(\Omega, \mathcal{F}, (\mathcal{F})_{t \geq 0}, \mathbf{P})$. Again we assume that the zero coordinate S^0, called the bond, is normalised to $S_t^0 \equiv 1$.

We first will make a technical assumption, namely that the process S is *locally bounded*, i.e., that there exists a sequence $(\tau_n)_{n=1}^{\infty}$ of stopping times, increasing a.s. to $+\infty$, such that the stopped processes $S_t^{\tau_n} = S_{t \wedge \tau_n}$ are uniformly bounded, for each $n \in \mathbb{N}$ (We refer to Sect. 7.2 below for unexplained notation). Note that continuous processes — or, more generally, càdlàg processes with uniformly bounded jumps — are locally bounded. This assumption will be very convenient for technical reasons. At the end of Chap. 8 we shall indicate, how to extend this to the general case of processes, which are not necessarily locally bounded.

We have chosen $\mathbb{T} = [0, \infty[$ for the time index set in order to assume full generality; of course this also covers the case of a compact interval $\mathbb{T} = [0, T]$, which is relevant in most applications, when assuming that S_t is constant, for $t \geq T$. The use of $\mathbb{T} = [0, \infty[$ as time index set also covers the case of discrete time (either in its finite version $\mathbb{T} = \{0, 1, \ldots, T\}$, or in its infinite version $\mathbb{T} = \mathbb{N}$). Indeed, it suffices to restrict to processes S which are constant on $[n-1, n[$, for each natural number n and only jump at times $n \in \mathbb{T}$.

We shall always assume that the filtration $(\mathcal{F}_t)_{t=0}^{\infty}$ satisfies the usual conditions i.e. it is right continuous and $\mathcal{F}_0$ contains all null sets of $\mathcal{F}_{\infty}$. Furthermore the process S has a.s. càdlàg trajectories.

How to define the trading strategies H, which played a crucial role in the preceding sections? A very elementary approach, corresponding to the role of step functions in integration theory, is formalised by the subsequent concept.

The reader will notice that the definitions in this chapter are variants of a more general situation to be handled in Chap. 7 and later.

Definition 5.1.1. *(compare, e.g., [P 90]) For a locally bounded stochastic process S we call an $\mathbb{R}^d$-valued process $H = (H_t)_{t=0}^{\infty}$ a simple trading strategy (or, speaking more mathematically, a simple integrand), if H is of the form*

$$H = \sum_{i=1}^{n} h_i \chi_{]\tau_{i-1}, \tau_i]},$$

where $0 = \tau_0 \leq \tau_1 \leq \ldots \leq \tau_n < \infty$ are finite stopping times and h_i are $\mathcal{F}_{\tau_{i-1}}$-measurable, $\mathbb{R}^d$-valued functions. We then may define, similarly as in Definition 2.1.4, the stochastic integral $H \cdot S$ as the stochastic process

$$(H \cdot S)_t = \sum_{i=1}^{n} \left(h_i, S_{\tau_i \wedge t} - S_{\tau_{i-1} \wedge t} \right)$$

$$= \sum_{i=1}^{n} \sum_{j=1}^{d} h_i^j \left(S_{\tau_i \wedge t}^j - S_{\tau_{i-1} \wedge t}^j \right), \quad 0 \leq t < \infty,$$

and its terminal value as the random variable

$$(H \cdot S)_{\infty} = \sum_{i=1}^{n} \left(h_i, S_{\tau_i} - S_{\tau_{i-1}} \right).$$

Throughout this chapter we call H admissible if, in addition, the stopped process S^{τ_n} and the functions $h_1, \ldots, h_n$ are uniformly bounded.

This definition is a well-known building block for developing a stochastic integration theory (see, e.g., [P 90]). It has a clear economic interpretation in the present context: at time τ_{i-1} an investor decides to adjust her portfolio in the assets $S^1, \ldots, S^j, \ldots, S^d$ by fixing her investment in asset S^j to be $h_i^j(\omega)$ units; we allow h_i^j to have arbitrary sign (holding a negative quantity means borrowing or "going short"), and to depend on the random element ω in an $\mathcal{F}_{\tau_{i-1}}$-measurable way, i.e., using the information available at time τ_{i-1}. The funds for adjusting the portfolio in this way are simply financed by taking the appropriate amount from (or putting into) the "cash box", modelled by the numéraire $S^0 \equiv 1$ (compare Sect. 2.1). The investor holds this portfolio fixed up to time τ_i. During this period the value of the risky stocks S^j, $j = 1, \ldots, d$, changed from $S_{\tau_{i-1}}^j(\omega)$ to $S_{\tau_i}^j(\omega)$ resulting in a total gain (or loss) given by the random variable $(h_i, S_{\tau_i} - S_{\tau_{i-1}})$. At time τ_i, for $i < n$, the investor readjusts the portfolio and at time τ_n she liquidates the portfolio, i.e., converts all her positions into the numéraire. Hence the random variable $(H \cdot S)_{\tau_n} = (H \cdot S)_\infty$ models the total gain (in units of the numéraire S_0) which she finally, i.e., at time τ_n, obtained by adhering to the strategy H; the process $(H \cdot S)_t$ models the gains accumulated up to time t.

The concept of a simple trading strategy is designed in a purely algebraic way, avoiding limiting procedures in order to be on safe grounds.

The next crucial ingredient in developing the theory is the proper generalisation of the notion of an equivalent martingale measure.

Definition 5.1.2. *A probability measure* **Q** *on* $\mathcal{F}$ *which is equivalent (resp. absolutely continuous with respect) to* **P** *is called an* equivalent *(resp. absolutely continuous) local martingale measure, if S is a local martingale under* **Q**.

We denote by $\mathcal{M}^e(S)$ *(resp.* $\mathcal{M}^a(S)$*) the family of all such measures, and say that S satisfies the condition of the existence of an equivalent local martingale measure (EMM) if* $\mathcal{M}^e(S) \neq \emptyset$.

Note that, by our assumption of local boundedness of S, we have that S is a local **Q**-martingale iff S^τ is a **Q**-martingale for each stopping time τ such that S^τ is uniformly bounded (compare Chap. 7).

Why did we use the notion of a local martingale instead of the more familiar notion of a martingale? This is simply the natural degree of generality. The subsequent straightforward lemma (whose proof is an obvious consequence of the chosen concepts and left to the reader) shows that this notion does the job just as well as the notion of a martingale for the present purpose of no-arbitrage theory. Last but not least, the restriction to the notion of martingale measures would lead to a different version of the general version of the fundamental theorem of asset pricing (Theorem 2.2.7 above), as may be seen from easy examples (see [DS 94a], compare also [Y 05]).

Lemma 5.1.3. *Let* **Q** *be a probability measure on* $\mathcal{F}$ *which is absolutely continuous w.r. to* **P**. *A locally bounded stochastic process* S *is a local martingale under a probability measure* **Q** *iff*

$$\mathbf{E_Q}\left[(H \cdot S)_\infty\right] = 0, \tag{5.1}$$

for each admissible simple trading strategy H.

Proof. Let $(\tau_n)_{n=1}^\infty$ be a sequence of finitely valued stopping times increasing **P**-a.s. to infinity such that each S^{τ_n} is bounded.

Supposing that (5.1) holds true for each simple admissible integrand we have to show that each S^{τ_n} is a **Q**-martingale. In other words, for each $n \geq 1$ and each pair of stopping times $0 \leq \sigma_1 \leq \sigma_2 \leq \tau_n$ we have to show that

$$\mathbf{E_Q}[S_{\sigma_2} \mid \mathcal{F}_{\sigma_1}] = S_{\sigma_1}.$$

This is tantamount to the requirement that for each $\mathbb{R}^d$-valued $\mathcal{F}_{\sigma_1}$-measurable, bounded function h we have

$$\mathbf{E_Q}[(h, S_{\sigma_2} - S_{\sigma_1})] = 0,$$

which holds true by assumption (5.1). Hence S is a local **Q**-martingale.

The proof of the converse implication, i.e., that the local **Q**-martingale property of S implies (5.1) for each admissible integrand is straightforward (compare Lemma 2.2.6). □

For later use we note that the "=" in (5.1) may equivalently be replaced by "$\leq$" (or "$\geq$"), as H is an admissible simple trading strategy iff $-H$ is so.

We define the subspace K^{simple} of $L^\infty(\Omega, \mathcal{F}, \mathbf{P})$ of contingent claims, available at price zero via an admissible simple trading strategy, by

$$K^{\text{simple}} = \{(H \cdot S)_\infty \mid H \text{ simple, admissible}\}$$

and by C^{simple} the convex cone in $L^\infty(\Omega, \mathcal{F}, \mathbf{P})$ of contingent claims dominated by some $f \in K$

$$C^{\text{simple}} = K^{\text{simple}} - L_+^\infty = \left\{f - k \mid f \in K^{\text{simple}}, f \in L^\infty, k \geq 0\right\}.$$

Definition 5.1.4. S *satisfies the no-arbitrage condition* (*NA*^{simple}) *with respect to simple integrands, if* $K^{\text{simple}} \cap L_+^\infty(\Omega, \mathcal{F}, \mathbf{P}) = \{0\}$ *(or, equivalently,* $C^{\text{simple}} \cap L_+^\infty(\Omega, \mathcal{F}, \mathbf{P}) = \{0\}$*).*

As the following lemma shows there is no difference between an arbitrage opportunity for simple admissible strategies and arbitrage opportunities for admissible "buy and hold" strategies.

Lemma 5.1.5. *Let the process* H *be a simple admissible strategy defined as* $H = \sum_{i=1}^n h_i \chi_{]\!]\tau_{i-1}, \tau_i]\!]}$ *and yielding an arbitrage opportunity. In other words we have* $(H \cdot S)_\infty \geq 0$ *a.s. and* $\mathbf{P}[(H \cdot S)_\infty > 0] > 0$. *Then there is a buy and hold strategy* $K = h\mathbf{1}_{]\!]\sigma_1, \sigma_2]\!]}$ *such that* K *is admissible and* K *yields an arbitrage opportunity.*

Proof. The proof proceeds by induction on n and yields additionally that the stopping times σ_1 and σ_2 can be chosen as $\sigma_1 = \tau_{i-1}$ and $\sigma_2 = \tau_i$, for some $i \le n$. For $n = 1$ the statement is obvious since $H = K$ will do the job. So we only check the inductive step. If $\mathbf{P}\left[(H \cdot S)_{\tau_{n-1}} < 0\right] > 0$ then we put $\sigma_1 = \tau_{n-1}$, $\sigma_2 = \tau_n$ and $h = h_{n-1}\chi_{\{(H\cdot S)_{\tau_{n-1}}<0\}}$. The strategy $K = h\chi_{]\!]\sigma_1,\sigma_2]\!]}$ is an arbitrage opportunity since $(H \cdot S)_{\tau_n} \ge 0$ a.s. and hence $(K \cdot S)_{\tau_n} > 0$ on $\left\{(H \cdot S)_{\tau_{n-1}} < 0\right\}$. If $(H \cdot S)_{\tau_{n-1}} = 0$ a.s. then $K = h_{n-1}\chi_{]\!]\tau_{n-1},\tau_n]\!]}$ must give an arbitrage opportunity. The only remaining case is $(H \cdot S)_{\tau_{n-1}} \ge 0$ a.s. and $\mathbf{P}\left[(H \cdot S)_{\tau_{n-1}} > 0\right] > 0$. It is here that we apply the inductive hypothesis. $\square$

Remark 5.1.6. If there is an arbitrage strategy, can we reduce the strategy even further, e.g. can we take k of the form $\alpha\chi_A\chi_{]\!]\sigma_1,\sigma_2]\!]}$ where $A \in \mathcal{F}_{\sigma_1}$ and α is a constant? The following example shows that, for $d \ge 2$, this is not the case.

Let ϑ and η be independent random variables which are uniformly distributed on $[0, 1[$ and $[0, \frac{1}{2}[$ respectively. Define the $\mathbb{R}^2$-valued process (S_0, S_1) by $S_0 = 0$ and $S_1 = e^{2\pi i(\vartheta+\eta)}$, where we identify $\mathbb{R}^2$ with $\mathbb{C}$. The σ-algebra $\mathcal{F}_0$ is generated by ϑ and $\mathcal{F}_1$ is generated by ϑ and η. This market allows for arbitrage: indeed let h be the $\mathcal{F}_0$-measurable $\mathbb{R}^2$-valued random variable $h = e^{2\pi i(\vartheta+\frac{1}{4})}$ (identifying $\mathbb{R}^2$ with $\mathbb{C}$ once more) so that

$$(h, S_1 - S_0)_{\mathbb{R}^2} > 0, \quad a.s.,$$

where $(\cdot, \cdot)_{\mathbb{R}^2}$ now is the usual inner product on $\mathbb{R}^2$. On the other hand, it is easy to verify that for each h of the form $\alpha\chi_A$, where $\alpha \in \mathbb{R}^2$ is a constant and $A \in \mathcal{F}_0$, $\mathbf{P}[A] > 0$ we have

$$\mathbf{P}[(h, S_1 - S_0)_{\mathbb{R}^2} < 0] > 0.$$

We want to prove a version of the fundamental theorem of asset pricing analogous to Theorem 2.2.7 above. However, things are now more delicate and the notion of *(NA^{simple})* defined above is not sufficiently strong to imply the existence of an equivalent local martingale measure:

Proposition 5.1.7. *The condition (EMM) of existence of an equivalent local martingale measure implies the condition (NA^{simple}) of no-arbitrage with respect to simple integrands, but not vice versa.*

Proof. *(EMM)* $\Rightarrow$ *(NA^{simple})*: this is an immediate consequence of Lemma 5.1.3, noting that for $\mathbf{Q} \sim \mathbf{P}$ and a non-negative function $f \ge 0$, which does not vanish almost surely, we have $\mathbf{E}_{\mathbf{Q}}[f] > 0$.

(NA^{simple}) $\nRightarrow$ *(EMM)*: we give an easy counter-example which is just an infinite random walk.

Let $t_n = 1 - \frac{1}{n+1}$ and define the $\mathbb{R}$-valued process S to start at $S_0 = 1$, and to be constant except for jumps at the points t_n which are defined as

$$\Delta S_{t_n} = 3^{-n}\varepsilon_n, \quad n \ge 1,$$

such that $(\varepsilon_n)_{n=1}^{\infty}$ are independent random variables taking the values $+1$ or -1 with probabilities

$$\mathbf{P}[\varepsilon_n = 1] = \frac{1 + \alpha_n}{2}, \quad \mathbf{P}[\varepsilon_n = -1] = \frac{1 - \alpha_n}{2},$$

where $(\alpha_n)_{n=1}^{\infty}$ is a sequence in $]-1, +1[$ to be specified below.

Clearly this defines a bounded process S, for which there is a unique measure $\mathbf{Q}$ on $(\Omega, \mathcal{F}) = (\{-1, 1\}^{\mathbb{N}}, \mathcal{B}(\{-1, 1\}^{\mathbb{N}}))$, under which S is a martingale in its own filtration $(\mathcal{F}_t)_{t=0}^{\infty}$; this measure is given by

$$\mathbf{Q}[\varepsilon_n = 1] = \mathbf{Q}[\varepsilon_n = -1] = \frac{1}{2},$$

and $(\varepsilon_n)_{n=1}^{\infty}$ are independent under $\mathbf{Q}$.

By a result of Kakutani (see, e.g. [W 91]) we know that $\mathbf{Q}$ is either equivalent to $\mathbf{P}$, or $\mathbf{P}$ and $\mathbf{Q}$ are mutually singular, depending on whether $\sum_{n=1}^{\infty} \alpha_n^2 < \infty$ or not.

Taking, for example, $\alpha_n = \frac{1}{2}$, for all $n \in \mathbb{N}$, we have constructed a process S on $(\Omega, \mathcal{F}, \mathbf{P})$, for which there is no equivalent (local) martingale measure $\mathbf{Q}$. On the other hand, it is an easy and instructive exercise to show that, *for simple trading strategies*, there are no arbitrage opportunities for the process S.

By Lemma 5.1.5 we only need to check for strategies of the form $H = h 1_{]\sigma_1, \sigma_2]}$. Such a strategy gives the outcome $h(S_{\sigma_2} - S_{\sigma_1})$ and h is $\mathcal{F}_{\sigma_1}$-measurable. Suppose that $(H \cdot S)_{\infty} \geq 0$. Of course we may replace h by $\mathrm{sign}(h)$. The choice of 3^{-n} also yields that on the set $\{\sigma_1 = 1 - \frac{1}{n}\} \cap \{\sigma_2 \geq 1 - \frac{1}{n+1}\}$ the sign of $S_{\sigma_2} - S_{\sigma_1}$ is the same as the sign of ε_n. We get that $\mathrm{sign}\left(h(S_{\sigma_2} - S_{\sigma_1})\right)$ is on that same set equal to the sign of $h\varepsilon_n$. The independence of ε_n from $\mathcal{F}_{1-\frac{1}{n}}$ then gives that $h = 0$ on $\{\sigma_1 = 1 - \frac{1}{n}\} \cap \{\sigma_2 \geq 1 - \frac{1}{n+1}\}$. Combining all these facts gives that $h(S_{\sigma_2} - S_{\sigma_1}) = 0$ a.s.. $\qquad\square$

The example in the above proof shows, why the no-arbitrage condition *(NA^{simple})* defined in 5.1.4 is too weak: it is intuitively rather obvious that by a sequence of properly scaled bets on a (sufficiently, i.e., $\sum_{n=1}^{\infty} \alpha_n^2 = \infty$) biased coin one can "produce something like an arbitrage", while a finite number of bets (as formalised by Definition 5.1.1) does not suffice to do so.

But here we are starting to move on thin ice, and it will be the crucial issue to find a mathematically precise framework, in which the above intuitive insight can be properly formalised.

5.2 No Free Lunch

A decisive step in this direction was done in the work of D. Kreps [K 81], who realised that the purely algebraic notion of no-arbitrage with respect to simple integrands has to be complemented with a topological notion:

Definition 5.2.1. *(compare [K 81]) S satisfies the condition of* no free lunch *(NFL) if the closure $\overline{C}$ of C^{simple}, taken with respect to the weak-star topology of $L^\infty(\Omega, \mathcal{F}, \mathbf{P})$, satisfies*

$$\overline{C} \cap L^\infty_+(\Omega, \mathcal{F}, \mathbf{P}) = \{0\}.$$

This strengthening of the condition of no-arbitrage is tailor-made so that the subsequent version of the fundamental theorem of asset pricing holds.

Theorem 5.2.2 (Kreps-Yan). *A locally bounded stochastic process S satisfies the condition of no free lunch (NFL), iff condition (EMM) of the existence of an equivalent local martingale measure is satisfied:*

$$(NFL) \iff (EMM).$$

Proof. (EMM) $\Rightarrow$ (NFL): This is still the easy part. By Lemma 5.1.3 we have $\mathbf{E}_\mathbf{Q}[f] \leq 0$, for each $\mathbf{Q} \in \mathcal{M}^e(S)$ and $f \in C^{\text{simple}}$, and this inequality also extends to the weak-star-closure $\overline{C}$ as $f \mapsto \mathbf{E}_\mathbf{Q}[f]$ is a weak-star continuous functional. On the other hand, if *(EMM)* were true and *(NFL)* were violated, there would exist a $\mathbf{Q} \in \mathcal{M}^e(S)$ and $f \in \overline{C}$, $f \geq 0$ not vanishing almost surely, whence $\mathbf{E}_\mathbf{Q}[f] > 0$, which yields a contradiction.

(NFL) $\Rightarrow$ (EMM): We follow the strategy of the proof for the case of finite Ω, but have to refine the argument:

Step 1 (Hahn-Banach argument): We claim that, for fixed $f \in L^\infty_+$, $f \not\equiv 0$, there is $g \in L^1_+$ which, viewed as a linear functional on L^∞, is less than or equal to zero on $\overline{C}$, and such that $(f, g) > 0$. To see this, apply the separation theorem (e.g., [Sch 99, Theorem II, 9.2]) to the σ^*-closed convex set $\overline{C}$ and the compact set $\{f\}$ to find $g \in L^1$ and $\alpha < \beta$ such that $g|_{\overline{C}} \leq \alpha$ and $(f, g) > \beta$. Since $0 \in C$ we have $\alpha \geq 0$. As $\overline{C}$ is a cone, we have that g is zero or negative on $\overline{C}$ and, in particular, non-negative on L^∞_+, i.e. $g \in L^1_+$. Noting that $\beta > 0$ we have proved step 1.

Step 2 (Exhaustion argument): Denote by $\mathcal{G}$ the set of all $g \in L^1_+$, $g \leq 0$ on C. Since $0 \in \mathcal{G}$ (or by step 1), $\mathcal{G}$ is non-empty.

Let $\mathcal{S}$ be the family of (equivalence classes of) subsets of Ω formed by the supports $\{g > 0\}$ of the elements $g \in \mathcal{G}$. Note that $\mathcal{S}$ is closed under countable unions, as for a sequence $(g_n)_{n=1}^\infty \in \mathcal{G}$, we may find strictly positive scalars $(\alpha_n)_{n=1}^\infty$, such that $\sum_{n=1}^\infty \alpha_n g_n \in \mathcal{G}$. Hence there is $g_0 \in \mathcal{G}$ such that, for $\{g_0 > 0\}$, we have

$$\mathbf{P}[\{g_0 > 0\}] = \sup\{\mathbf{P}[\{g > 0\}] \mid g \in \mathcal{G}\}.$$

We now claim that $\mathbf{P}[\{g_0 > 0\}] = 1$, which readily shows that g_0 is strictly positive almost surely. Indeed, if $\mathbf{P}[\{g_0 > 0\}] < 1$, then we could apply step 1 to $f = \chi_{\{g_0=0\}}$ to find $g_1 \in \mathcal{G}$ with

$$\mathbf{E}[fg_1] = \langle f, g_1 \rangle = \int_{\{g_0=0\}} g_1(\omega)d\mathbf{P}(\omega) > 0$$

Hence, $g_0 + g_1$ would be an element of $\mathcal{G}$ whose support has **P**-measure strictly bigger than $\mathbf{P}[\{g_0 > 0\}]$, a contradiction.

Normalise g_0 so that $\|g_0\|_1 = 1$ and let **Q** be the measure on $\mathcal{F}$ with Radon-Nikodým derivative $\frac{d\mathbf{Q}}{d\mathbf{P}} = g_0$. We conclude from Lemma 5.1.3 that **Q** is a local martingale measure for S, so that $\mathcal{M}^e(S) \neq \emptyset$.   □

Some comments on the Kreps-Yan theorem seem in order: this theorem was obtained by D. Kreps [K 81] in a more abstract setting and under a — rather mild — additional separability assumption; the reason for the need of this assumption was that D. Kreps did not use the above exhaustion argument, but rather some sequential procedure relying on the separability of $L^1(\Omega, \mathcal{F}, \mathbf{P})$. Independently and at about the same time, Ji-An Yan [Y 80] proved in a different context, namely the characterisation of semi-martingales as good integrators (which is the theme of the Bichteler-Dellacherie theorem), and without a direct relation to finance, a general theorem which is similar in spirit to Theorem 5.2.2. Ch. Stricker [Str 90] observed that Yan's theorem may be applied to quickly prove the theorem of Kreps without any separability assumption. We therefore took the liberty to give Theorem 5.2.2 the name of these two authors.

The message of the theorem is, that the assertion of the "fundamental theorem of asset pricing" 2.2.7 is valid for general processes, if one is willing to interpret the notion of "no-arbitrage" in a somewhat liberal way, crystallised in the notion of "no free lunch" above.

What is the economic interpretation of a "free lunch"? By definition S violates the assumption *(NFL)* if there is a function $g_0 \in L_+^\infty(\Omega, \mathcal{F}, \mathbf{P})$, $g_0 \neq 0$, and nets $(g_\alpha)_{\alpha \in I}, (f_\alpha)_{\alpha \in I}$ in $L^\infty(\Omega, \mathcal{F}, \mathbf{P})$, such that $f_\alpha = (H^\alpha \cdot S)_\infty$ for some admissible, simple integrand H^α, $g_\alpha \leq f_\alpha$, and $\lim_{\alpha \in I} g_\alpha = g_0$, the limit converging with respect to the weak-star topology of $L^\infty(\Omega, \mathcal{F}, \mathbf{P})$. Speaking economically: an arbitrage opportunity would be the existence of a trading strategy H such that $(H \cdot S)_\infty \geq 0$, almost surely, and $\mathbf{P}[(H \cdot S)_\infty > 0] > 0$. Of course, this is the dream of each arbitrageur, but we have seen, that — for the purpose of the fundamental theorem to hold true — this is asking for too much (at least, if we only allow for simple admissible trading strategies). Instead, a free lunch is the existence of a contingent claim $g_0 \geq 0$, $g_0 \neq 0$, which may, in general, not be written as (or dominated by) a stochastic integral $(H \cdot S)_\infty$ with respect to a simple admissible integrand H; but there are contingent claims g_α "close to g_0", which can be obtained via the trading strategy H^α, and subsequently "throwing away" the amount of money $f_\alpha - g_\alpha$.

This triggers the question whether we can do somewhat better than the above — admittedly complicated — procedure. Can we find a requirement sharpening the notion of "no free lunch", i.e., being closer to the original notion of "no-arbitrage" and such that a — properly formulated — version of the "fundamental theorem" still holds true?

Here are some questions related to our attempt to make the process of taking the weak-star-closure more understandable:

(i)   Is it possible, in general, to replace the net $(g_\alpha)_{\alpha \in I}$ above by a sequence $(g_n)_{n=0}^\infty$?

(ii) Can we choose the net $(g_\alpha)_{\alpha \in I}$ (or, hopefully, the sequence $(g_n)_{n=0}^\infty$) such that $(g_\alpha)_{\alpha \in I}$ remains bounded in $L^\infty(\mathbf{P})$ (or at least such that the negative parts $((g_\alpha)_-)_{\alpha \in I}$ remain bounded)?

Note that this latter issue is crucial from an economic point of view, as it pertains to the question whether the approximation of f by $(g_\alpha)_{\alpha \in I}$ can be done *respecting a finite credit line*.

(iii) Is it really necessary to allow for the "throwing away of money", i.e., to pass from K^{simple} to C^{simple}?

It turns out that questions (i) and (ii) are intimately related and, in general, the answer to these questions is no. In fact, the study of the pathologies of the operation of taking the weak-star-closure is an old theme of functional analysis. On the very last pages of S. Banach's original book ([B 32]) the following example is given: there is a separable Banach space X such that, for every given fixed number $n \geq 1$ (say $n = 147$), there is a convex cone C in the dual space X^*, such that $C \subsetneq C^{(1)} \subsetneq C^{(2)} \subsetneq \ldots \subsetneq C^{(n)} = C^{(n+1)} = \overline{C}$, where $C^{(k)}$ denotes the sequential weak-star-closure of $C^{(k-1)}$, i.e., the limits of weak-star convergent *sequences* $(x_i)_{i=0}^\infty$, with $x_i \in C^{(k-1)}$, and $\overline{C}$ denotes the weak-star-closure of C. In other words, by taking the limits of weak-star convergent *sequences* in C we do not obtain the weak-star-closure of C immediately, but we have to repeat this operation precisely n times, when finally this process stabilises to arrive at the weak-star-closure $\overline{C}$.

In Banach's book this construction is done for $X = c_0$ and $X^* = \ell^1$ while our present context is $X = L^1(\mathbf{P})$ and $X^* = L^\infty(\mathbf{P})$. Adapting the ideas from Banach's book, it is possible to construct a semi-martingale S such that the corresponding convex cone C^{simple} has the following property: taking the weak-star sequential closure $(C^{\text{simple}})^{(1)}$, the resulting set intersects $L_+^\infty(\mathbf{P})$ only in $\{0\}$; but doing the operation twice, we obtain the weak-star-closure $C^{(2)} = \overline{C}$ and $\overline{C}$ intersects $L_+^\infty(\mathbf{P})$ in a non-trivial way (see Example 9.7.8 below). Hence we cannot — in general — reduce to sequences $(g_n)_{n=0}^\infty$ in the definition of *(NFL)*. The construction of this example uses a process with jumps; for continuous processes the situation is, in fact, nicer, and in this case it is possible to give positive answers to questions (i) and (ii) above (see [Str 90], [D 92], and Chap. 9 below).

Regarding question (iii), the dividing line again is the continuity of the process S (see [Str 90] and [D 92] for positive results for continuous processes, and [S 94] and [S 04a] for counter-examples).

Summing up the above discussion: the theorem of Kreps and Yan is a beautiful and mathematically precise extension of the fundamental theorem of asset pricing 2.2.7 to a general framework of stochastic processes in continuous time. However, in general, the concept of passing to the weak-star-closure does not allow for a clear-cut economic interpretation. It is therefore desirable to prove versions of the theorem, where the closure with respect to the weak-star topology is replaced by the closure with respect to some finer topology (ideally the topology of uniform convergence, which allows for an obvious and convincing economic interpretation).

To do so, let us contemplate once more, where the above encountered difficulties related to the weak-star topology originated from: they are essentially caused by our restriction to consider only *simple, admissible trading strategies*. These nice and simple objects can be defined without any limiting procedure, but we should not forget, that — except for the case of finite discrete time — they are only auxiliary gimmicks, playing the same role as step functions in integration theory. The concrete examples of trading strategies (e.g., replicating a European call option) encountered in Chap. 4 for the case of the Bachelier and the Black-Scholes model led us already out of this class: of course, they are not simple trading strategies. This is similar to the situation in classical integration theory, where the most basic examples, such as polynomials, trigonometric functions etc, of course fail to be step functions.

Hence we have to pass to a suitable class of more general trading strategies then just the simple, admissible ones. Among other pleasant and important features, this will have the following effects on the corresponding sets C and K: these sets will turn out to be "closer to their closures" (ideally they will already be closed in the relevant topology, see Theorem 6.9.2 and 8.2.2 below), than the above considered sets C^{simple} and K^{simple}. The reason is that the passage from simple to more general integrands *involves already a limiting procedure*.

We shall take up the theme of developing a no-arbitrage theory based on general (i.e., not necessarily simple) integrands in Chap. 8 below. But before doing so we shall investigate in Chap. 6 in more detail, the situation of a finite time index set $\mathbb{T} = \{0, 1, \ldots, T\}$, where — as opposed to Chap. 2 — we drop the assumption that $(\Omega, \mathcal{F}, \mathbf{P})$ is finite. This is the theme of the Dalang-Morton-Willinger theorem. In this setting the simple integrands are already the general concept. It turns out that the assumption of *(NA)* (without any strengthening of "no free lunch" type) implies already that the cone C^{simple} *is closed w.r.t. the relevant topologies* which will allow us to directly apply the Kreps-Yan theorem.

In order to prove the Dalang-Morton-Willinger theorem in full generality it will be convenient to state a slightly more general version of the Kreps-Yan theorem. While in Theorem 5.2.2 above we considered the duality between L^∞ and L^1 only, we now state the theorem for the duality between L^p and L^q, for arbitrary $1 \le p \le \infty$ and $\frac{1}{p} + \frac{1}{q} = 1$. We observe that this is still a more restricted degree of generality then the one considered by D. Kreps in his original paper [K 81], who worked with an abstract dual pair $\langle E, E' \rangle$ of vector lattices.

We use the occasion to give a slightly different proof than in Theorem 5.2.2 above, replacing the exhaustion argument by a somewhat more direct reasoning. However, both arguments amount to the same, and the difference is only superficial.

Fix $1 \leq p \leq \infty$, and $1 \leq q \leq \infty$ such that $\frac{1}{p} + \frac{1}{q} = 1$, and let $E = L^p(\Omega, \mathcal{F}, \mathbf{P})$, $E' = L^q(\Omega, \mathcal{F}, \mathbf{P})$. We denote by $E_+ = \{f \in L^p \mid f \geq 0 \text{ a.s.}\}$ (resp. $E_- = \{f \in L^p \mid f \leq 0 \text{ a.s.}\}$) the cone of non-negative (resp. non-positive) random variables in L^p.

Theorem 5.2.3. *With the above notation let $C \subset E$ be a $\sigma(E, E')$-closed convex cone containing E_- and suppose that $C \cap E_+ = \{0\}$. Then there is a probability measure $\mathbf{Q}$ on $\mathcal{F}$, which is equivalent to $\mathbf{P}$, satisfying $\frac{d\mathbf{Q}}{d\mathbf{P}} \in E'$, and so that, for all $f \in C$, we have $\mathbf{E}_{\mathbf{Q}}[f] \leq 0$.*

Conversely, given a probability measure $\mathbf{Q}$ on $\mathcal{F}$, equivalent to $\mathbf{P}$ and satisfying $\frac{d\mathbf{Q}}{d\mathbf{P}} \in E'$, the cone $C = \{f \in E \mid \mathbf{E}_{\mathbf{Q}}[f] \leq 0\}$ is $\sigma(E, E')$-closed and satisfies $C \cap E_+ = \{0\}$.

Proof. For $1 \geq \delta > 0$, let B_δ be the convex set in E

$$B_\delta = \{f \in E \mid 0 \leq f \leq 1, \; \mathbf{E}[f] \geq \delta\}.$$

Clearly B_δ is $\sigma(E, E')$-compact and $C \cap B_\delta = \emptyset$. The Hahn-Banach Theorem in its version as separating hyperplane (see, e.g., [Sch 99]) gives the existence of an element $g_\delta \in E'$ so that

$$\sup_{f \in C} \mathbf{E}[g_\delta f] < \min_{h \in B_\delta} \mathbf{E}[g_\delta h]. \tag{5.2}$$

Because C is a cone this means that for all $f \in C$ we must have $E[g_\delta f] \leq 0$. This implies that the sup on the left hand side of (5.2) equals 0, so that the min on the right hand side must be strictly positive. Since C contains E_- we must have $g_\delta \geq 0$ almost surely. Now g_δ cannot be identically zero since $\mathbb{1} \in B_\delta$ and hence $\mathbf{E}[g_\delta \mathbb{1}] > 0$. We may normalise g_δ to have $\mathbf{E}[g_\delta] = 1$.

For each $n \geq 1$ we consider $\delta_n = 2^{-n}$ and let $\mathbf{Q}_n$ be the probability measure defined as $\frac{d\mathbf{Q}_n}{d\mathbf{P}} = g_{2^{-n}}$. For later use let us denote by α_n the number $\alpha_n = \|g_{2^{-n}}\|_{E'} \in [1, \infty[$. Also remark that for $A \in \mathcal{F}$ with $\mathbf{P}[A] > 2^{-n}$, we must have $\mathbf{Q}_n[A] = \mathbf{E}[1_A g_{2^{-n}}] > 0$.

The element $\mathbf{Q}$ is now defined as

$$\mathbf{Q} = \sum_{n=1}^{\infty} \beta_n \mathbf{Q}_n,$$

where we choose the weights $\beta_n > 0$ such that $\sum_{n=1}^{\infty} \beta_n = 1$ and $\sum_{n=1}^{\infty} \beta_n \alpha_n < \infty$. For instance we could take $\beta_n = \frac{2^{-n}}{c\alpha_n}$ where $c = \sum_{n=1}^{\infty} \frac{2^{-n}}{\alpha_n} < \infty$.

The probability measure $\mathbf{Q}$ satisfies

(i)   for all A, $\mathbf{P}[A] > 0$ we have $\mathbf{Q}[A] = \sum_{n=1}^{\infty} \beta_n \mathbf{Q}_n[A] > 0$.

(ii) $\frac{d\mathbf{Q}}{d\mathbf{P}} \in E'$ since $\left\| \frac{d\mathbf{Q}}{d\mathbf{P}} \right\|_{E'} \leq \sum_{n=1}^{\infty} \beta_n \alpha_n < \infty$.

(iii) For all $f \in C$ we have $\mathbf{E}_{\mathbf{Q}}[f] = \sum_{n=1}^{\infty} \beta_n \mathbf{E}_{\mathbf{Q}_n}[f] \leq 0$.

The final assertion is obvious.   □

We now turn to the question how to characterise the $\sigma(L^p, L^q)$-closedness of a cone $C \subseteq L^p(\Omega, \mathcal{F}, \mathbf{P})$ as considered in the preceding Theorem 5.2.3 and whether this is possible by only considering sequences rather than nets. In the case $1 \leq p < \infty$ this is quite obvious: by the Hahn-Banach theorem a convex set $C \subseteq L^p$ is $\sigma(L^p, L^q)$-closed, where $\frac{1}{p} + \frac{1}{q} = 1$, iff C is closed w.r. to the norm $\|\cdot\|_p$ of L^p. Hence in this case the closedness of C can be characterised by using sequences. The price to pay for this comfortable situation is that — reading Theorem 5.2.3 as an "if and only if" result — we only obtain a probability measure $\mathbf{Q}$ with the additional requirement $\frac{d\mathbf{Q}}{d\mathbf{P}} \in L^q(\Omega, \mathcal{F}, \mathbf{P})$, i.e., $\mathbf{Q}$ must have a finite q-th moment, for some $1 < q \leq \infty$. This additional requirement on $\mathbf{Q}$ is not natural in most applications to finance: for example passing from $\mathbf{P}$ to an equivalent probability measure $\mathbf{P}_1$ (an operation, which we very often apply) the requirement $\frac{d\mathbf{Q}}{d\mathbf{P}} \in L^q(\mathbf{P})$ does not imply $\frac{d\mathbf{Q}}{d\mathbf{P}_1} \in L^q(\mathbf{P}_1)$, if $1 < q \leq \infty$. Only for the case $q = 1$ this difficulty disappears, as for equivalent probability measures $\mathbf{Q}$ and $\mathbf{P}$ we always have $\|\frac{d\mathbf{Q}}{d\mathbf{P}}\|_{L^1(\mathbf{P})} = 1$. This is why, in general, the case $p = \infty$, $q = 1$ is of prime importance and there is no "cheap" way to make the subtleties of the weak-star topology on L^∞ disappear.

There is, however, a notable exception to these considerations: we shall see in the next chapter that, for finite discrete time $\mathbb{T} = \{0, \ldots, T\}$, we obtain a version of the Fundamental Theorem of Asset Pricing, due to R. Dalang, A. Morton, W. Willinger, where we get the additional requirement $\frac{d\mathbf{Q}}{d\mathbf{P}} \in L^\infty(\mathbf{P})$ for the desired equivalent martingale measure $\mathbf{Q}$ "for free". In this case we therefore shall use Theorem 5.2.3 for the case $p = 1$ and $q = \infty$.

The case $p = \infty$ in the above Theorem 5.2.3 is more subtle (and more interesting). The subsequent result, which is essentially based on the Krein-Smulian theorem, goes back to A. Grothendieck [G 54]. We only formulate it for the case of convex cones, but it can be extended to general convex sets in an obvious way.

Proposition 5.2.4. *Let C be a convex cone in $L^\infty(\Omega, \mathcal{F}, \mathbf{P})$. Denote by $\sigma(L^\infty, L^1)$ (resp. $\tau(L^\infty, L^1)$) the weak-star (resp. the Mackey) topology on L^∞ and, for $0 < p \leq \infty$, by $\|\cdot\|_p$ the norm topology induced by $(L^p, \|\cdot\|_p)$ on L^∞ (for $0 < p < 1$ $\|\cdot\|_p$ is only a quasi-norm). We denote by ball_∞ the unit ball of $L^\infty(\Omega, \mathcal{F}, \mathbf{P})$.*

The following assertions are equivalent

(i) C is $\sigma(L^\infty, L^1)$-closed.
(i') C is $\tau(L^\infty, L^1)$-closed.
(ii) $C \cap \mathrm{ball}_\infty$ is $\sigma(L^\infty, L^1)$-closed.
(ii') $C \cap \mathrm{ball}_\infty$ is $\tau(L^\infty, L^1)$-closed.
(iii) $C \cap \mathrm{ball}_\infty$ is $\|\cdot\|_p$-closed, for every $0 < p < \infty$.
(iii') $C \cap \mathrm{ball}_\infty$ is $\|\cdot\|_p$-closed, for some $0 < p < \infty$.

(iv) $C \cap \text{ball}_\infty$ *is closed with respect to the topology of convergence in measure.*

Proof. (ii) $\Rightarrow$ (i): This is the crucial implication. It is a direct consequence of Krein-Šmulian theorem ([S 94, Theorem IV 6.4]).

(i) $\Rightarrow$ (ii) trivial.

(i) $\Leftrightarrow$ (i') and (ii) $\Leftrightarrow$ (ii'): this is the assertion of the Mackey-Arens theorem ([S 94, Theorem IV 3.2]).

(ii) $\Rightarrow$ (iv): Let $(f_n)_{n=1}^\infty$ be a Cauchy sequence in $C \cap \text{ball}_\infty$ with respect to convergence in measure. By passing to a subsequence we may assume that $(f_n)_{n=1}^\infty$ converges a.s. to some $f_0 \in \text{ball}_\infty$. We have to show that $f_0 \in C$.

This follows from Lebesgue's theorem on dominated convergence, which implies that, for each $g \in L^1(\Omega, \mathcal{F}, \mathbf{P})$, we have $\lim_{n \to \infty} \mathbf{E}[f_n g] = \mathbf{E}[f_0 g]$. Hence $(f_n)_{n=1}^\infty$ converges to f_0 in the $\sigma(L^\infty, L^1)$-topology so that by hypothesis (ii) we have that $f_0 \in C$.

(iv) $\Rightarrow$ (iii') $\Rightarrow$ (iii) $\Rightarrow$ (ii'): trivial, noting that the Mackey topology $\tau(L^\infty, L^1)$ is finer than the topology induced by $\|.\|_p$, for any $0 < p < \infty$. $\square$

The Dalang-Morton-Willinger Theorem

6.1 Statement of the Theorem

In Chap. 2 we only dealt with finite probability spaces. This was mainly done because of technical difficulties. As soon as the probability space $(\Omega, \mathcal{F}_T, \mathbf{P})$ is no longer finite, the corresponding function spaces such as $L^1(\Omega, \mathcal{F}_T, \mathbf{P})$ or $L^\infty(\Omega, \mathcal{F}_T, \mathbf{P})$ are infinite dimensional and we have to fall back on functional analysis. In this chapter we will present a proof of the Fundamental Theorem of Asset Pricing, Theorem 2.2.7, in the case of general $(\Omega, \mathcal{F}_T, \mathbf{P})$, but still in finite discrete time. Since discounting does not present any difficulty, we will suppose that the d-dimensional price process S has already been discounted as in Sect. 2.1. Also the notion of the class $\mathcal{H}$ of trading strategies does not present any difficulties and we may adopt Definition 2.1.4 verbatim also for general $(\Omega, \mathcal{F}_T, \mathbf{P})$ as long as we are working in finite discrete time. In this setting we can state the following beautiful version of the Fundamental Theorem of Asset Pricing, due to Dalang, Morton and Willinger [DMW 90].

Theorem 6.1.1. *Let $(\Omega, \mathcal{F}_T, \mathbf{P})$ be a probability space and let $(S_t)_{t=0}^T$ be an $\mathbb{R}^d$-valued stochastic process adapted to the discrete time filtration $(\mathcal{F}_t)_{t=0}^T$. Suppose further that the no-arbitrage (NA) condition holds:*

$$K \cap L_+^0(\Omega, \mathcal{F}_T, \mathbf{P}) = \{0\}, \tag{6.1}$$

where

$$K = \left\{ \sum_{t=1}^T (H_t, \Delta S_t) \,\middle|\, H \in \mathcal{H} \right\}$$

Then there exists an equivalent probability measure $\mathbf{Q}$, $\mathbf{Q} \sim \mathbf{P}$ so that

(i) $S_t \in L^1(\Omega, \mathcal{F}_T, \mathbf{P})$, $t = 0, \ldots, T$,
(ii) $(S_t)_{t=0}^T$ *is a $\mathbf{Q}$-martingale, i.e., $\mathbf{E}[S_t | \mathcal{F}_{t-1}] = S_{t-1}$, for $t = 1, \ldots, T$,*
(iii) $\frac{d\mathbf{Q}}{d\mathbf{P}}$ *is bounded, i.e. $\frac{d\mathbf{Q}}{d\mathbf{P}} \in L^\infty(\Omega, \mathcal{F}_T, \mathbf{P})$.*

The proof of the theorem is not a trivial extension of the case of finite Ω (we remark, however, that, in the case $d = 1$, the verification of the theorem is indeed almost a triviality). Besides the original proof (see [DMW 90]) based on a measurable selection theorem, many authors have tried to give other and perhaps easier proofs. Below we repeat some of these proofs.

We first prove the theorem for $T = 1$. This seemingly easy case contains already the essential difficulties of the proof. We shall develop some abstract notions, which will be convenient for the proof, but which shall also turn out to be of independent interest: the concept of the predictable range of a process and the concept of a random subsequence $(f_{\tau_k})_{k=1}^\infty$ of a sequence $(f_n)_{n=1}^\infty$ of random variables. We also give a different proof, due to C. Rogers [R 94], based on the idea of utility maximisation.

To pass from the case $T = 1$ to the general case $T \geq 1$ we shall again offer two alternative proofs. The first will follow the original argument of [DMW 90] by considering, for each $t = 1, \ldots, T$, the one-step process (S_{t-1}, S_t) to which the previous result applies. The equivalent martingale measure $\mathbf{Q}$ for $(S_0, \ldots, S_T)$ is then obtained by concatenating the densities of the corresponding one step measures, i.e., by multiplicatively composing them (compare Sect. 2.3 above). The second proof will proceed by directly generalising the arguments for the case $T = 1$ to the general case.

We elaborate on these proofs because each of them gives different — and important — insights into the problem: the argument relying on concatenation in particular makes clear that the process $(S_0, \ldots, S_T)$ is free of arbitrage iff each of the one step processes (S_{t-1}, S_t) is so (see also Theorem 2.3.2 and Lemma 5.1.5 above); the direct argument shows the closedness of the cone C of super-replicable claims in $L^0(\Omega, \mathcal{F}_T, \mathbf{P})$, a result which will play a central role in the further development of the theory.

The proofs we present below are a mixture of the proofs in [S 92], [D 92], [R 94], [St 97] and [KS 01].

The proof of the Dalang-Morton-Willinger Theorem for the one-period case $S = (S_0, S_1)$ uses several ingredients that will be explained in the following sections. The first problem we have to solve is the problem of redundancy. The d given assets may have redundancy in the sense that some of them are linear combinations of others. This linear dependence, given the information $\mathcal{F}_0$, may, however, depend on $\omega \in \Omega$. We have to find a way to describe this mathematically. We will do this in Sect. 6.2 where we will introduce the notion of "the predictable range". It is a coordinate-free approach to describe in an $\mathcal{F}_0$-measurable way, the conditional linear (in-)dependence between $S_1^1, \ldots, S_1^d$. The second ingredient we need is the selection principle.

6.2 The Predictable Range

Let us fix two σ-algebras $\mathcal{F}_0 \subset \mathcal{F}_1$ on Ω and a probability measure $\mathbf{P}$ on $\mathcal{F}_1$. For an $\mathcal{F}_1$-measurable mapping $X : \Omega \to \mathbb{R}^d$ we will try to find the smallest

$\mathcal{F}_0$-measurably parameterised subspace of $\mathbb{R}^d$ where X takes its values. This idea was used in [S 92]. The problem is twofold. First we want to describe, in an easy way, how subspaces depend on $\omega \in \Omega$ in a measurable way. Second we have to find in some sense the smallest subspace where X takes its values.

The first problem can be solved in the following way. With each subspace of $\mathbb{R}^d$ we associate the orthogonal projection onto it and then we simply have to ask that the matrices describing these projections (with respect to a fixed orthonormal basis of $\mathbb{R}^d$) depend on ω in an $\mathcal{F}_0$-measurable way. There is a more elegant way of describing subspaces using Grassmannian manifolds but this would bring us too far from our goal.

We start with a measure theoretic result.

Lemma 6.2.1. *Let $(\Omega, \mathcal{F}, \mathbf{P})$ be a probability space and $E \subset L^0(\Omega, \mathcal{F}, \mathbf{P}; \mathbb{R}^d)$ a subspace closed with respect to convergence in probability. We suppose that E satisfies the following stability property. If $f, g \in E$ and $A \in \mathcal{F}$, then $f\mathbf{1}_A + g\mathbf{1}_{A^c} \in E$.*

Under these assumptions there exists an $\mathcal{F}$-measurable mapping P_0 taking values in the orthogonal projections in $\mathbb{R}^d$, so that $f \in E$ if and only if $P_0 f = f$.

Given any $\mathcal{F}$-measurable projection-valued mapping P so that $Pf = f$ for all $f \in E$, we have that $P_0 = P_0 P = P P_0$, meaning that the range of P_0 is a subspace of the range of P.

The proof is rather a formality but requires a lot of technical verifications. We start with a sublemma.

Sublemma 6.2.2. *Under the assumptions of Lemma 6.2.1 we have, for $f \in E$ and a real-valued $\mathcal{F}$-measurable function h, that $hf \in E$.*

Proof. We first prove the statement for elementary functions h. So let $h = \sum_{k=1}^n a_k \mathbf{1}_{A_k}$ where $a_1, \ldots, a_n$ are real numbers and $A_1, \ldots, A_n$ form an $\mathcal{F}$-measurable partition of Ω. Since the stability assumption on E implies that $f\mathbf{1}_{A_k} \in E$ for each k, we obviously have that $hf \in E$ as well.

The general case follows by approximation. If h is real-valued and $\mathcal{F}$-measurable we take a sequence $(h_n)_{n=1}^\infty$ of elementary $\mathcal{F}$-measurable random variables so that $h_n \to h$ in probability. Clearly the closedness of E together with $h_n f \in E$ for each n, implies $hf \in E$. $\qquad \square$

Proof of Lemma 6.2.1. The construction of the projection-valued mapping P_0 will be done through the construction of appropriate orthogonal random vectors having maximal support. The construction is done recursively and we first take $E_1 = E$. We then look at the family of $\mathcal{F}$-measurable sets

$$\{\{f \neq 0\} \mid f \in E_1\}.$$

Since E_1 is closed this system of sets is stable for countable unions and hence there is an element A_1 with maximal probability $\mathbf{P}[A_1]$ and with corresponding function $f \in E_1$, meaning $\{f \neq 0\} = A_1$. By the sublemma the function

$\varphi_1 = \frac{f}{|f|}\mathbf{1}_{A_1}$ is still in E_1 and it has maximal support among all elements of E_1.

We now look at the subspace E_2 of E_1

$$E_2 = \{f \in E_1 \mid (f, \varphi_1) = 0 \text{ a.s.}\}.$$

Clearly E_2 is closed and satisfies the same stability property as E_1. We can therefore find $\varphi_2 \in E_2$, such that $|\varphi_2| = 1$ or 0 and such that φ_2 has maximal support among all elements of E_2. Continuing this way we find random variables $\varphi_1, \varphi_2, \ldots, \varphi_d$ so that $(\varphi_i, \varphi_j) = 0$, $i \neq j$, $|\varphi_i| = 1$ or 0 and φ_i has maximal support among all elements of E_i. We claim that the procedure necessarily stops after at most d steps. So we have to show that, for $g \in E$ and $(g, \varphi_i) = 0$ for all $i = 1, \ldots, d$, we have $g = 0$. Obviously we have by the maximality of φ_1 that $\{g \neq 0\} \subset \{\varphi_1 \neq 0\}$. Also one verifies inductively $g \in E_i$ and by the maximality of φ_i we have $\{g \neq 0\} \subset \{\varphi_i \neq 0\}$. This implies that almost surely we have that $g(\omega) \neq 0$ implies that $\varphi_1(\omega), \ldots, \varphi_d(\omega)$ are all different from zero. But then $\varphi_1(\omega), \ldots, \varphi_d(\omega)$ form an orthogonal basis of $\mathbb{R}^d$ and $g(\omega)$ cannot be orthogonal to $\varphi_1(\omega), \ldots, \varphi_d(\omega)$.

We now claim that

$$E = \left\{ \sum_{k=1}^{d} a_k \varphi_k \; \middle| \; a_1, \ldots, a_d \text{ are } \mathcal{F}\text{-measurable} \right\}. \tag{6.2}$$

Indeed, denoting by F the right hand side of (6.2) we obviously have $F \subset E$ by the sublemma. Conversely if $f \in E$ we compare f with $\sum_{k=1}^{d}(f, \varphi_k)\varphi_k$. Since $g = f - \sum_{k=1}^{d}(f, \varphi_k)\varphi_k$ satisfies $g \in E$ and $(g, \varphi_i) = 0$ for all $i \leq d$, we must have that $g = 0$. We now define $P_0 x = \sum_{i=1}^{d}(x, \varphi_i)\varphi_i$. This is clearly a projection map. Obviously $f \in E$ implies $P_0 f = f$ and conversely we have that $P_0 f = f$ implies $f = \sum(f, \varphi_i)\varphi_i$ which means that $f \in E$.

It remains to check the last statement. So let P be a projection-valued mapping so that $Pf = f$ for all $f \in E$. Then $P\varphi_i = \varphi_i$ for all i and hence $PP_0 = P_0 = P_0 P$.   □

Remark 6.2.3. The reader might wonder why we gave so many details on these rather obvious facts from linear algebra. The reason is that we had to check the measurability. One way of doing this is to give explicit constructions.

Let us now return to the problem described in the beginning of this section. If $X : \Omega \to \mathbb{R}^d$ is $\mathcal{F}_1$-measurable let us look at the space

$$E^X = \{H \mid H : \Omega \to \mathbb{R}^d, \; \mathcal{F}_0\text{-measurable and } (H, X)_{\mathbb{R}^d} = 0 \text{ a.s.}\}. \tag{6.3}$$

Clearly E^X satisfies the assumption of lemma 6.2.1 and hence E^X can be described by a projection-valued mapping P'. The projection-valued mapping $P = \text{Id} - P'$ defines the "orthogonal complement" of E^X. Let us define:

$$\mathcal{H}^X = \{ f : \Omega \to \mathbb{R}^d \mid f \text{ is } \mathcal{F}_0\text{-measurable and } Pf = f \} \tag{6.4}$$

By construction we have for $g \in E^X$ and $f \in \mathcal{H}^X$ that $(f, g)_{\mathbb{R}^d} = 0$ almost surely. Therefore we could say that f is "orthogonal" to E^X a.s.. The space $P(\mathbb{R}^d)$ is a subspace of $\mathbb{R}^d$ which is dependent on $\omega \in \Omega$ in an $\mathcal{F}_0$-measurable way. In some sense (see below for a precise statement) $P(\mathbb{R}^d)$ is the best $\mathcal{F}_0$-measurable prediction of the space spanned by X.

More precisely:

Lemma 6.2.4. *Let $X \in L^0(\Omega, \mathcal{F}_1, \mathbf{P}; \mathbb{R}^d)$ and let $\mathcal{H}^X$ be defined as above. If $h \in L^0(\Omega, \mathcal{F}_1, \mathbf{P}; \mathbb{R})$ is $\mathcal{F}_0$-measurable and $hX \in L^1(\Omega, \mathcal{F}_1, \mathbf{P}; \mathbb{R}^d)$ then $\mathbf{E}[hX \mid \mathcal{F}_0] \in \mathcal{H}^X$.*

Proof. For $\alpha \in E^X$ we have

$$(\alpha, hX)_{\mathbb{R}^d} = h(\alpha, X)_{\mathbb{R}^d} = 0, \quad \text{a.s..}$$

Therefore we have for bounded $\alpha \in E^X$ that $\mathbf{E}[(\alpha, hX)_{\mathbb{R}^d} \mid \mathcal{F}_0] = (\alpha, \mathbf{E}[hX \mid \mathcal{F}_0])_{\mathbb{R}^d} = 0$ almost surely. In other words $\mathbf{E}[hX \mid \mathcal{F}_0] \in \mathcal{H}^X$. $\square$

We shall apply the above results to the situation where $X = \Delta S_1 = S_1 - S_0$ for a one-step financial market (S_t) adapted to the filtration $(\mathcal{F}_t)_{t=0}^1$. For $H \in L^0(\Omega, \mathcal{F}_0, \mathbf{P}; \mathbb{R}^d)$ the random variable $(H, X)_{\mathbb{R}^d}$ then equals the stochastic integral $(H \cdot S)_1$. In general the integration map $I : L^0(\Omega, \mathcal{F}_0, \mathbf{P}; \mathbb{R}^d) \to L^0(\Omega, \mathcal{F}_1, \mathbf{P})$ mapping H to $(H \cdot S)_1$ is not injective; this was precisely the theme of the above considerations. We have to restrict the integration map I to the space $\mathcal{H}^X$ in order to make it injective. This is done in the subsequent Definition 6.2.5 and in Lemma 6.2.6.

Definition 6.2.5. *We say that $H \in L^0(\Omega, \mathcal{F}_0, \mathbf{P}; \mathbb{R}^d)$ is in canonical form for (S_0, S_1) if $H \in \mathcal{H}^X$ where $\mathcal{H}^X$ is defined in (6.3) and (6.4) with $X = S_1 - S_0$.*

Lemma 6.2.6. *The kernel of the mapping $I : L^0(\Omega, \mathcal{F}_0, \mathbf{P}, \mathbb{R}^d) \to L^0(\Omega, \mathcal{F}_1, \mathbf{P})$ equals E^X. The restriction of I to $\mathcal{H}^X$ is injective, linear and has full range.*

Proof. Let H and H' be in $L^0(\Omega, \mathcal{F}_0, \mathbf{P}; \mathbb{R}^d)$ such that $I(H) = I(H')$. Then $(H - H', X) = 0$ a.s. and hence $(H - H') \in E^X$. Since $H - H' \in \mathcal{H}^X$ we necessarily have $H - H' \in \mathcal{H}^X \cap E^X = \{0\}$. $\square$

6.3 The Selection Principle

In this section we consider a probability space $(\Omega, \mathcal{F}, \mathbf{P})$ and a compact metric space $(\mathcal{K}, d)$. In the applications below $(\mathcal{K}, d)$ will typically be the one-point compactification $\mathbb{R}^d \cup \{\infty\}$ of $\mathbb{R}^d$. The Bolzano-Weierstrass Theorem states that, for a sequence $(x_n)_{n=1}^\infty$ in $\mathcal{K}$, we may find a convergent subsequence $(x_{n_k})_{k=1}^\infty$. We want to generalise this theorem to a sequence $(f_n)_{n=1}^\infty$

in $L^0(\Omega, \mathcal{F}, \mathbf{P}; \mathcal{K})$, i.e., a sequence of $\mathcal{K}$-valued functions depending on $\omega \in \Omega$ in a measurable way. Here $(\mathcal{K}, d)$ is equipped with its Borel σ-algebra.

We would like to find a subsequence $(n_k)_{k=1}^\infty$ so that $(f_{n_k}(\omega))_{k=1}^\infty$ converges for each ω (or at least for almost each ω). If Ω is finite, this is clearly possible. But considering a sequence $(f_n)_{n=1}^\infty$ of independent Bernoulli variables i.e., $\mathbf{P}[f_n = 1] = \mathbf{P}[f_n = -1] = \frac{1}{2}$, we see that in general this wish is asking for too much. For each subsequence $(n_k)_{k=1}^\infty$ the sequence $(f_{n_k})_{k=1}^\infty$ still forms an independent sequence of Bernoulli variables and therefore diverges almost surely. On the other hand, for each fixed $\omega \in \Omega$, we may of course extract a subsequence $(n_k(\omega))_{k=1}^\infty$ such that $(f_{n_k}(\omega))_{k=1}^\infty$ converges. The crux is, that the choice of the subsequence depends on ω, as we just have seen. The idea of the subsequent notion of a measurably parameterised subsequence is that this subsequence $(n_k(\omega))_{k=1}^\infty$ may be chosen to depend *measurably* on $\omega \in \Omega$. This nice and simple idea was observed by H.-J. Engelbert and H. v. Weizsäcker and was successfully applied by Y.M. Kabanov and Ch. Stricker ([KS 01]) in the present context.

Definition 6.3.1. *An $\mathbb{N}$-valued, $\mathcal{F}$-measurable function is called a* random time. *A strictly increasing sequence $(\tau_k)_{k=1}^\infty$ of random times is called a* measurably parameterised subsequence *or simply a* measurable subsequence.

Before stating the actual result we first mention the following lemma on random times.

Lemma 6.3.2. *Let $(f_n)_{n=1}^\infty$ be a sequence of $\mathcal{F}$-measurable functions $f_n : \Omega \to \mathcal{K}$. Let $\tau : \Omega \to \{1, 2, 3, \ldots\}$ be an $\mathcal{F}$-measurable random time, then $g(\omega) = f_{\tau(\omega)}(\omega)$ is $\mathcal{F}$-measurable.*

Proof. Let B be a Borel set in $\mathcal{K}$. Then

$$g^{-1}(B) = \bigcup_{n=1}^\infty \left(\{\tau = n\} \cap \{f_n \in B\}\right) \in \mathcal{F}. \qquad \square$$

Proposition 6.3.3. *For a sequence $(f_n)_{n=1}^\infty \in L^0(\Omega, \mathcal{F}, \mathbf{P}; \mathcal{K})$ we may find a measurably parameterised subsequence $(\tau_k)_{k=1}^\infty$ such that $(f_{\tau_k})_{k=1}^\infty$ converges for all $\omega \in \Omega$.*

Proof. It suffices to check that the procedure of finding a convergent subsequence in the proof of the Bolzano-Weierstrass Theorem may be done in a measurable way.

For $n \geq 1$, let $A_1^n, \ldots, A_{N_n}^n$ be finite open coverings of $\mathcal{K}$ by sets of diameter less than n^{-1}. For each fixed $\omega \in \Omega$ we define inductively a sequence $(I^k(\omega))_{k=1}^\infty$ of infinite subsets of $\mathbb{N}$. For $k = 1$ find the smallest number $1 \leq j_1 \leq N_1$ such that $(f_n(\omega))_{n=1}^\infty$ lies infinitely often in $A_{j_1}^1$, and define $I^1(\omega) = \{n \in \mathbb{N} \mid f_n(\omega) \in A_{j_1}^1\}$.

If the set $I^{k-1}(\omega)$ is defined, find the smallest $1 \le j_k \le N_k$ such that $(f_n(\omega))_{n \in I^{k-1}(\omega)}$ lies infinitely often in $A^k_{j_k}$, and define $I^k(\omega) = \{n \in I^{k-1}(\omega) \mid f_n(\omega) \in A^k_{j_k}\}$. Defining $\tau_k(\omega)$ to be the k'th element of $I^k(\omega)$ it is straightforward to check that τ_k is a well-defined $\mathbb{N}$-valued measurable function. Clearly $(f_{\tau_k(\omega)}(\omega))_{k=1}^\infty$ converges for each $\omega \in \Omega$. □

Taking measurable subsequences of $(f_n)_{n=1}^\infty$ in $L^0(\Omega, \mathcal{F}, \mathbf{P}; \mathcal{K})$ works just in the same way as taking subsequences of sequences $(x_n)_{n=1}^\infty$ in the compact space $\mathcal{K}$. For example, consider the usual procedure of taking a subsequence of a subsequence: in the present framework this means that we are given two measurably parameterised subsequences $(\tau_k)_{k=1}^\infty$ and $(\sigma_j)_{j=1}^\infty$. Given $(f_n)_{n=1}^\infty$ we may extract the subsequence $g_k = f_{\tau_k}$ and the subsequence $h_j = g_{\sigma_j} = f_{\tau_{\sigma_j}}$. Similarly, we may take diagonal subsequences etc., just as we are used to do in analysis.

We often shall use the reasoning *"by passing to a measurably parameterised subsequence we may assume that $(f_n)_{n=1}^\infty$ satisfies ..."* which has, as usual, the interpretation: *there exists a measurably parameterised subsequence $(g_k)_{k=1}^\infty = (f_{\tau_k})_{k=1}^\infty$ which has the properties...*

For later use we give some more properties which one may impose on a suitably chosen measurably parameterised subsequence. We place ourselves into the setting of Proposition 6.3.3.

Proposition 6.3.4. *Under the assumptions of Proposition 6.3.3 we have in addition:*

(i) *Let $x_0 \in \mathcal{K}$ and define*

$$B = \{\omega \mid x_0 \text{ is an accumulation point of } (f_n(\omega))_{n=1}^\infty\}.$$

Then the sequence $(\tau_k)_{k=1}^\infty$ in Proposition 6.3.3 may be chosen such that

$$\lim_{k \to \infty} f_{\tau_k(\omega)}(\omega) = x_0, \quad \text{for each } \omega \in B.$$

(ii) *Let $f_0 \in L^0(\Omega, \mathcal{F}, \mathbf{P}; \mathcal{K})$ and define*

$$C = \{\omega \mid f_0(\omega) \text{ is not the limit of } (f_n(\omega))_{n=1}^\infty\},$$

where the above means that either the limit does not exist or, if it exists, it is different from $f_0(\omega)$. Then the sequence $(\tau_k)_{k=1}^\infty$ in Proposition 6.3.3 may be chosen such that

$$\lim_{k \to \infty} f_{\tau_k(\omega)}(\omega) \neq f_0(\omega), \quad \text{for each } \omega \in C.$$

Proof. For (i) it suffices to choose the coverings $(A^n_j)_{j=1}^{N_n}$ in the proof of Proposition 6.3.3 such that $x_0 \in A^n_1$, for each $n \ge 1$.

As regards (ii) define

$$\Delta(\omega) = \limsup_{n \to \infty} d\left(f_n(\omega), f_0(\omega)\right)$$

so that $C = \{\Delta > 0\}$. Modify the construction of the proof of Proposition 6.3.3 in a straightforward way by intersecting each of the subsets $I^k(\omega)$ of $\mathbb{N}$ with $\left\{n \;\middle|\; d(f_n(\omega), f_0(\omega)) \geq \frac{\Delta(\omega)}{2}\right\}$. $\qquad\qquad\square$

6.4 The Closedness of the Cone C

The map $I : H \to (H, \Delta S)$ introduced in Sect. 6.2 above is a linear map from $L^0(\Omega, \mathcal{F}_0, \mathbf{P}; \mathbb{R}^d)$ to $L^0(\Omega, \mathcal{F}_1, \mathbf{P}; \mathbb{R}^d)$ which is continuous for the convergence in measure. We know from Lemma 6.2.6 that I becomes injective when restricted to $\mathcal{H}^X$ where $X = \Delta S = S_1 - S_0$. The relevant feature shown in the subsequent proposition is that its inverse, defined as a function from $I(L^0(\Omega, \mathcal{F}_0, \mathbf{P}; \mathbb{R}^d))$ to the subspace $\mathcal{H}^X$ of $L^0(\Omega, \mathcal{F}_0, \mathbf{P}; \mathbb{R}^d)$ defined in (6.4) above, is continuous too. If, in addition, the process S satisfies (NA), we may even formulate a stronger statement.

Proposition 6.4.1. *Let $S = (S_0, S_1)$ be adapted to $(\Omega, (\mathcal{F}_t)_{t=0}^1, \mathbf{P})$ and let $(H^n)_{n=1}^\infty$ be a sequence in $L^0(\Omega, \mathcal{F}_0, \mathbf{P}; \mathbb{R}^d)$ in canonical form, i.e., $H^n \in \mathcal{H}^{\Delta S}$. Then we have that*

(i) *$(H^n)_{n=1}^\infty$ is a.s. bounded iff $((H^n, \Delta S))_{n=1}^\infty$ is.*
(i') *$(H^n)_{n=1}^\infty$ converges a.s. iff $((H^n, \Delta S))_{n=1}^\infty$ does.*

If we suppose in addition that S satisfies (NA) as defined in (6.1) we also have

(ii) *$(H^n)_{n=1}^\infty$ is a.s. bounded iff $((H^n, \Delta S)_-)_{n=1}^\infty$ is.*
(ii') *$(H^n)_{n=1}^\infty$ converges a.s. to zero iff $((H^n, \Delta S)_-)_{n=1}^\infty$ does so.*

Proof. The "only if" direction is trivial in all the above assertions and holds true without the assumption (NA) and/or that H^n is in canonical form. Let us now show the "if" direction.

(i) and (ii): Suppose that $(H^n)_{n=1}^\infty$ fails to be a.s. bounded and let us show that $((H^n, \Delta S))_{n=1}^\infty$ (resp. $((H^n, \Delta S)_-)_{n=1}^\infty$) fails so too. We apply Proposition 6.3.4 (i) to $\mathcal{K} = \mathbb{R}^d \cup \{\infty\}$ and $x_0 = \infty$: there is a measurably parameterised subsequence $(L^k)_{k=1}^\infty = (H^{\tau_k})_{k=1}^\infty$ such that $(L^k(\omega))$ diverges to ∞ on a set B of positive measure. Note that each L^k is an integrand in canonical form.

Let $\widehat{L}^k = L^k \mathbf{1}_{B \cap \{|L^k| \geq 1\}} / |L^k|$ in order to find a sequence of integrands in canonical form such that $|\widehat{L}^k(\omega)| = 1$, for $\omega \in B$ and k sufficiently large. By passing once more to a measurably parameterised subsequence we may suppose that $(\widehat{L}^k)_{k=1}^\infty$ converges to an integrand $\widehat{L}$, which is in canonical form and satisfies $|\widehat{L}| = 1$ on B.

As regards (i) note that the assumption of the a.s. boundedness of $((H^n, \Delta S))_{n=1}^{\infty}$ implies that $(\widehat{L}^k, \Delta S)$ tends to zero a.s.; indeed, on the set B the ratio $\frac{\widehat{L}^k}{\overline{L}^k}$ tends to zero, while outside of B the integrand $\widehat{L}^k$ vanishes. Hence

$$\left(\widehat{L}, \Delta S\right) = \lim_{k \to \infty} \left(\widehat{L}^k, \Delta S\right) = 0, \quad \text{a.s.,}$$

which by Lemma 6.2.6 implies that $\widehat{L} = 0$, the required contradiction.

As regards (ii) note that

$$\left(\widehat{L}, \Delta S\right)_{-} = \lim_{k \to \infty} \left(\widehat{L}^k, \Delta S\right)_{-} = 0, \quad \text{a.s..}$$

From the no-arbitrage assumption *(NA)* we now may conclude that $\left(\widehat{L}, \Delta S\right) = 0$ a.s., which again implies that $\widehat{L} = 0$ and yields the desired contradiction.

(i'): Suppose now that $(H^n)_{n=1}^{\infty}$ does not converge a.s., assume that $((H^n, \Delta S))_{n=1}^{\infty}$ does so and let us work towards a contradiction. By (i) above we may assume, that $(H^n)_{n=1}^{\infty}$ is a.s. bounded. Applying Proposition 6.3.3 to $\mathcal{K} = \mathbb{R}^d \cup \{\infty\}$, we may find a measurably parameterised subsequence $(H^{\tau_k})_{k=1}^{\infty}$ converging a.s. to some H^0. Applying Proposition 6.3.4 (ii) to $f_0 = H^0$ we may find another measurably parameterised subsequence $(H^{\sigma_k})_{k=1}^{\infty}$ converging a.s. to some $\widehat{H}^0 \in \mathcal{H}$ for which we have $\mathbf{P}\left[\widehat{H}^0 \neq H^0\right] > 0$. Note that H^0 as well as $\widehat{H}^0$ are in canonical form.

As $\left(H^0 - \widehat{H}^0, \Delta S\right) = \lim_{n \to \infty}(H^n, \Delta S) - \lim_{n \to \infty}(H^n, \Delta S) = 0$, we obtain again a contradiction to Lemma 6.2.6.

(ii'): Suppose now that S satisfies *(NA)*, that $(H^n)_{n=1}^{\infty}$ does not converge to zero a.s., while $((H^n, \Delta S)_{-})_{n=1}^{\infty}$ does. Again we may use Proposition 6.3.4 (ii), this time applied to $f_0 = 0$, to find a measurably parameterised subsequence of $(H^n)_{n=1}^{\infty}$ converging a.s. to some H^0 with $\mathbf{P}[H_0 \neq 0] > 0$. We have

$$(H^0, \Delta S)_{-} = \lim_{n \to \infty} (H^n, \Delta S)_{-} = 0, \quad \text{a.s..}$$

From *(NA)* we conclude that $(H^0, \Delta S) = 0$ a.s. so that we get again a contradiction to Lemma 6.2.6. □

We can now formulate a theorem on the closedness of the space (resp. cone) of the stochastic integrals (resp. of functions dominated by stochastic integrals). Assertion (i) is due to Ch. Stricker [Str 90] and (ii) is due to the second named author [S 92].

Theorem 6.4.2. *Let the $\mathbb{R}^d$-valued one-step process $S = (S_0, S_1)$ be adapted to $(\Omega, (\mathcal{F}_t)_{t=0}^{1}, \mathbf{P})$*

(i) *The vector space*

$$K = \{(H, \Delta S) \mid H \in L^0(\Omega, \mathcal{F}_0, \mathbf{P}; \mathbb{R}^d)\}$$

is closed in $L^0(\Omega, \mathcal{F}_1, \mathbf{P})$.

(ii) *If S satisfies (NA) then the cone*

$$C = K - L^0_+(\Omega, \mathcal{F}_1, \mathbf{P})$$

is closed in $L^0(\Omega, \mathcal{F}_1, \mathbf{P})$ too.

Proof. (i): Let $(f_n) = (H^n, \Delta S)_{n=1}^\infty$ be a sequence in K converging to $f_0 \in L^0(\Omega, \mathcal{F}_1, \mathbf{P})$ with respect to convergence in measure. We may suppose that each H^n is in canonical form. Moreover, by passing to a subsequence we may suppose that $(f_n)_{n=1}^\infty$ converges a.s. to f_0. Proposition 6.4.1 implies that $(H^n)_{n=1}^\infty$ converges a.s. to some $H^0 \in L^0(\Omega, \mathcal{F}_1, \mathbf{P}; \mathbb{R}^d)$ so that $f_0 = (H^0, \Delta S)$, whence $f_0 \in K$.

(ii): Let $f_n = g_n - h_n$ be a sequence in C converging in probability to $f_0 \in L^0(\Omega, \mathcal{F}_1, \mathbf{P})$, where $g_n = (H^n, \Delta S)$, where H^n is an integrand in canonical form and $h_n \in L^0_+(\Omega, \mathcal{F}_1, \mathbf{P})$. We have to show that f_0 belongs to C. Again we may suppose that $(f_n)_{n=1}^\infty$ converges a.s. to f_0. As $g_n \geq f_n$ we deduce that $((H^n, \Delta S)_-)_{n=1}^\infty$ is a.s. bounded, so that we may conclude from *(NA)* and Proposition 6.4.1 (ii) that $(H^n)_{n=1}^\infty$ is a.s. bounded too. By passing to a measurably parameterised subsequence $(\tau_k)_{k=1}^\infty$ we may suppose that $g_{\tau_k} = (H^{\tau_k}, \Delta S)$ converges a.s. to $(H^0, \Delta S)$, for some $H^0 \in E$. Note that $(f_{\tau_k})_{k=1}^\infty$ still converges a.s. to f_0 so that $h_{\tau_k} = f_{\tau_k} - g_{\tau_k}$ converges a.s. to some $h_0 \geq 0$. Hence $f_0 = (H^0, \Delta S) - h_0 \in K - L^0_+(\Omega, \mathcal{F}_1, \mathbf{P}) = C$. □

In assertion (ii) of the preceding theorem, the no-arbitrage assumption cannot be dropped. Indeed, consider the following simple example ([S 92]). Let $\Omega = [0,1]$, $\mathcal{F}_0$ trivial, $\mathcal{F}_1$ the Borel σ-algebra and $\mathbf{P}$ Lebesgue measure. Let $S_0 \equiv 0$ and $S_1(\omega) = \omega$, for $\omega \in [0,1]$, and

$$f_n(\omega) = \begin{cases} n\omega & \text{for } 0 \leq \omega \leq n^{-1} \\ 1 & \text{for } n^{-1} \leq \omega \leq 1. \end{cases}$$

As $f_n \leq g_n := n\Delta S$, we have $f_n \in C$, for each n. Clearly $(f_n)_{n=1}^\infty$ converges a.s. to the constant function $f_0 = 1$. But f_0 is not in C, as for each $f \in C$ we may find a constant $M > 0$ s.t. almost surely $f(\omega) \leq M\omega$ so that $f(\omega) \leq \frac{1}{2}$ a.s. for $0 \leq \omega \leq \frac{1}{2M}$.

Summing up, we have an example of a process $S = (S_t)_{t=0}^1$ such that C is not closed in $L^0(\Omega, \mathcal{F}_1, \mathbf{P})$. The crux is, of course, that S does not satisfy *(NA)*.

6.5 Proof of the Dalang-Morton-Willinger Theorem for $T = 1$

As we have proved in Theorem 6.4.2, the cone $C \in L^0(\Omega, \mathcal{F}_1, \mathbf{P})$ is closed if (S_0, S_1) satisfies the *(NA)* condition. The Kreps-Yan Theorem can now be applied in such a way that it yields the existence of an equivalent martingale measure.

Theorem 6.5.1. *Let $S = (S_0, S_1)$ be an $(\mathcal{F}_0, \mathcal{F}_1)$-adapted $\mathbb{R}^d$-valued process satisfying the (NA) condition. Then we have the existence of an equivalent probability measure $\mathbf{Q}$ so that*

(i) $S_0, S_1 \in L^1(\mathbf{Q})$,
(ii) $S_0 = \mathbf{E}_{\mathbf{Q}}[S_1 \mid \mathcal{F}_0]$,
(iii) $\frac{d\mathbf{Q}}{d\mathbf{P}}$ *is bounded.*

Proof. First we take an equivalent probability measure $\mathbf{P}_1$ so that $\frac{d\mathbf{P}_1}{d\mathbf{P}}$ is bounded and $S_0, S_1 \in L^1(\mathbf{P}_1)$. To do so we could take, for example, $\frac{d\mathbf{P}_1}{d\mathbf{P}} = c\exp(-\|S_0\|_{\mathbb{R}^d} - \|S_1\|_{\mathbb{R}^d})$, where c is a suitable normalisation constant.

The next step consists of considering the set

$$C_1 = C \cap L^1(\Omega, \mathcal{F}_1, \mathbf{P}_1)$$

where C is defined as in Theorem 6.4.2. Because C is closed in $L^0(\mathbf{P})$, the set C_1 is closed in $L^1(\mathbf{P}_1)$. Obviously C_1 is a convex cone (since C is a convex cone). The *(NA)* condition implies that $C_1 \cap L^1_+(\Omega, \mathcal{F}_1, \mathbf{P}_1) = \{0\}$. Theorem 5.2.3 now gives the existence of an equivalent probability measure $\mathbf{Q}$ so that $\frac{d\mathbf{Q}}{d\mathbf{P}_1}$ is bounded and so that $\mathbf{E}_{\mathbf{Q}}[f] \le 0$ for all $f \in C_1$. Obviously $S_0, S_1 \in L^1(\mathbf{Q})$ since $\frac{d\mathbf{Q}}{d\mathbf{P}_1}$ is bounded. Since for each coordinate $j = 1, \ldots, d$ and each $A \in \mathcal{F}_0$ we have $\mathbf{1}_A(S_1^j - S_0^j) \in C_1$ and $-\mathbf{1}_A(S_1^j - S_0^j) \in C_1$, we must have $\mathbf{E}_{\mathbf{Q}}[\mathbf{1}_A S_1^j] = \mathbf{E}_{\mathbf{Q}}[\mathbf{1}_A S_0^j]$. This shows $S_0 = \mathbf{E}_{\mathbf{Q}}[S_1 \mid \mathcal{F}_0]$. Since $\frac{d\mathbf{Q}}{d\mathbf{P}} = \frac{d\mathbf{Q}}{d\mathbf{P}_1} \frac{d\mathbf{P}_1}{d\mathbf{P}}$ we also have that $\frac{d\mathbf{Q}}{d\mathbf{P}}$ is bounded. $\qquad\square$

Remark 6.5.2. One may ask whether it is possible to replace in Theorem 6.5.1 assertion (iii) by the assertion that $\frac{d\mathbf{P}}{d\mathbf{Q}}$ is bounded, i.e., by

(iii') there is a constant $c > 0$ such that $\frac{d\mathbf{Q}}{d\mathbf{P}} > c$ almost surely.

A moment's reflexion reveals that, in general, this is not possible. Indeed, if it happens that $\|S_t\|_1 = \mathbf{E}[\|S_t\|] = \infty$, then for each probability measure $\mathbf{Q}$ with $\frac{d\mathbf{Q}}{d\mathbf{P}} \ge c > 0$ we also have $\mathbf{E}_{\mathbf{Q}}[\|S_t\|] = \infty$, so that S cannot be a $\mathbf{Q}$-martingale. But even if S is uniformly bounded, we cannot replace (iii) by (iii') as the subsequent example shows.

Let $\Omega = \mathbb{N}$, the σ-algebra $\mathcal{F}_0$ be generated by the partition of Ω into the sets $(\{2n - 1, 2n\})_{n=1}^\infty$ and $\mathcal{F}_1$ be the power set of Ω. Define the probability measure $\mathbf{P}$ on $\mathcal{F}_1$ by $\mathbf{P}[2n - 1] = \mathbf{P}[2n] = 2^{-(n+1)}$. Let $S_0 \equiv 0$, $S_1 = 1$ on $\{2n - 1\}$ and $S_1 = -2^{-n}$ on $\{2n\}$ to obtain an $\mathbb{R}$-valued adapted process $S = (S_t)_{t=0}^1$ satisfying the *(NA)* condition. If $\mathbf{Q}$ is a probability measure on $\mathcal{F}_1$ with $\frac{d\mathbf{Q}}{d\mathbf{P}} \ge c > 0$, we have $\mathbf{Q}[2n-1] \ge c2^{-(n+1)}$. If in addition $\mathbf{E}_{\mathbf{Q}}[S_1 \mid \mathcal{F}_0] = 0$, we must have $\mathbf{Q}[2n] = 2^n \mathbf{Q}[2n - 1] \ge \frac{c}{2}$, which is a contradiction to the finiteness of $\mathbf{Q}$. Hence, there cannot exist a measure $\mathbf{Q}$ satisfying (i) and (ii) of Theorem 6.5.1 and (iii') above. We refer to [R 04] and [RS 05] for a more thorough analysis of condition (iii').

6.6 A Utility-based Proof of the DMW Theorem for $T = 1$

We give another proof of the theorem of Dalang-Morton-Willinger (for the case $T = 1$) which is based on the ideas of utility maximisation. This proof is due to C. Rogers [R 94]. The basic idea is — transferring the results on utility maximisation, obtained in Chap. 3 for the case of finite Ω, to the present situation — that for the optimal investment $\widehat{X}$, the function $U'(\widehat{X})$ should define the density of an equivalent martingale measure, up to a normalising constant. This was proved in Theorem 3.1.3 above for the case of finite Ω and the hope is, of course, that this result should also hold in a more general context. Rogers' idea was to exploit this basic relation to find an equivalent martingale measure in the context of the theorem of Dalang-Morton-Willinger. Among other features such an approach has the advantage of being more constructive than the mere existence result provided by the theorem of Kreps-Yan. On the other hand we note that the present proof will not yield a bounded density $\frac{d\mathbf{Q}}{d\mathbf{P}}$, i.e., we do not obtain assertion (iii) of Theorem 6.1.1.

Let us fix the $\mathbb{R}^d$-valued process $S = (S_0, S_1)$ based on $(\Omega, (\mathcal{F}_t)_{t=0}^1, \mathbf{P})$ and assume that it is free of arbitrage. As utility function we use $U(x) = -e^{-x}$ and as initial endowment $x_0 = 0$.

Our utility maximisation problem consists of finding the optimal element $\widehat{h} \in L^0(\Omega, \mathcal{F}_0, \mathbf{P}; \mathbb{R}^d)$ solving the maximisation problem

$$\mathbf{E}\left[U\left((h, \Delta S)\right)\right] \to \max!, \quad h \in L^0(\Omega, \mathcal{F}_0, \mathbf{P}, \mathbb{R}^d). \tag{6.5}$$

In order to assure a well-defined solution to this problem, we need some preliminary work. As a first step we may suppose that, for each $h \in L^\infty(\Omega, \mathcal{F}_0, \mathbf{P}; \mathbb{R}^d)$,

$$\mathbf{E}\left[|U\left((h, \Delta S)\right)|\right] < \infty. \tag{6.6}$$

To guarantee that this integrability condition holds, it suffices to pass from $\mathbf{P}$ to the equivalent probability measure $\mathbf{P}'$ defined by

$$\frac{d\mathbf{P}'}{d\mathbf{P}} = c \exp\left(-\|\Delta S\|_{\mathbb{R}^d}^2\right),$$

where $c > 0$ is a normalising constant. Hence we may suppose that the original $\mathbf{P}$ satisfies (6.6).

The idea of the proof is that, for any maximising sequence $(h_n)_{n=1}^\infty$ in $L^\infty(\Omega, \mathcal{F}_0, \mathbf{P}; \mathbb{R}^n)$ for the optimisation problem (6.5), the stochastic integrals $f_n := (h_n, \Delta S)$ automatically converge in measure to the optimal function $\widehat{f}$; if, in addition, $(h_n)_{n=1}^\infty$ is in the predictable range of $S_1 - S_0$, then $(h_n)_{n=1}^\infty$ will also converge in measure to an optimal $\widehat{h} \in L^0(\Omega, \mathcal{F}_0, \mathbf{P}; \mathbb{R}^d)$. Having found $\widehat{h}$ we may conclude that $\widehat{f} = \left(\widehat{h}, \Delta S\right)$ is a.s. finite, and that the formula

$$\frac{d\mathbf{Q}}{d\mathbf{P}} = cU'\left(\widehat{f}\right) \tag{6.7}$$

defines the desired equivalent martingale measure **Q**, where $c > 0$ is a suitable normalising constant. This is roughly the strategy of proof and we now have to work through the details.

We need an auxiliary result (compare [S 92] and Sect. 9.8 below).

Lemma 6.6.1. *Let $U(x) = -e^{-x}$ and let $(f_n)_{n=1}^{\infty}$ be a sequence of real-valued measurable functions defined on $(\Omega, \mathcal{F}, \mathbf{P})$ such that*

$$a := \lim_{n \to \infty} \sup_{g_n \in \text{conv}\{f_n, f_{n+1}, \ldots\}} \mathbf{E}[U(g_n)] > -\infty. \tag{6.8}$$

Then there is a unique element $g_0 \in L^0(\Omega, \mathcal{F}, \mathbf{P};] - \infty, \infty])$ such that, for each sequence $g_n \in \text{conv}\{f_n, f_{n+1}, \ldots\}$ with $\lim_{n \to \infty} \mathbf{E}[U(g_n)] = a$, we have that $(g_n)_{n=1}^{\infty}$ converges to g_0 in probability.

Proof. Let $g_n \in \text{conv}\{f_n, f_{n+1}, \ldots\}$ be such that $\lim_{n \to \infty} \mathbf{E}[U(g_n)] = a$. Fix n, m and $\alpha > 0$ and let

$$A_{n,m,\alpha} = \{|g_n - g_m| > \alpha \text{ and } \min(g_n, g_m) < \alpha^{-1}\}.$$

From the uniform concavity of U on $] - \infty, \alpha^{-1} + \alpha]$ we conclude that there exists a $\beta = \beta(\alpha) > 0$ such that, for $n, m \in \mathbb{N}$ and $\omega \in A_{n,m,\alpha}$, we have

$$U\left(\frac{g_n(\omega) + g_m(\omega)}{2}\right) \geq \frac{U(g_n(\omega)) + U(g_m(\omega))}{2} + \beta.$$

For $\omega \notin A_{n,m,\alpha}$ we only apply the concavity of U to obtain

$$U\left(\frac{g_n(\omega) + g_m(\omega)}{2}\right) \geq \frac{U(g_n(\omega)) + U(g_m(\omega))}{2}.$$

Hence, for $g = \frac{g_n + g_m}{2}$ we have

$$\mathbf{E}[U(g)] > \frac{\mathbf{E}[U(g_n)] + \mathbf{E}[U(g_m)]}{2} + \beta \mathbf{P}[A_{n,m,\alpha}].$$

If $(g_n)_{n=1}^{\infty}$ is a maximising sequence for (6.8) we get therefore for each $\alpha > 0$

$$\lim_{n,m \to \infty} \mathbf{P}[A_{n,m,\alpha}] = 0.$$

This is tantamount to saying that $(g_n)_{n=1}^{\infty}$ is a Cauchy sequence in $L^0(\Omega, \mathcal{F}, \mathbf{P};] - \infty, \infty])$, i.e., with respect to convergence in probability on the half-closed line $] - \infty, \infty]$. Letting $g_0 = \lim_{n \to \infty} g_n$ we have found our desired limiting function. We observe that we cannot exclude the possibility, that g_0 assumes the value $+\infty$ on a set of strictly positive probability.

As to the uniqueness of g_0, let g_n' be any other sequence in $\text{conv}\{f_n, f_{n-1}, \ldots\}$ such that $\lim_{n \to \infty} \mathbf{E}[U(g_n)] = a$. Then by the same argument as above, $(g_n')_{n=1}^{\infty}$ is also Cauchy in $L^0(\Omega, \mathcal{F}, \mathbf{P};]-\infty, +\infty])$. By considering $g_n'' := g_n$ for odd n and $g_n'' := g_n'$ for even n, we may also conclude that $(g_n')_{n=1}^\infty$ converges in probability to g_0.   □

To sketch the idea of Rogers' proof let us first assume that $\mathcal{F}_0$ is trivial so that $L^0(\Omega, \mathcal{F}_0, \mathbf{P}, \mathbb{R}^d)$ may be identified with $\mathbb{R}^d$. Then the problem (6.5) reduces to find the optimal vector $\widehat{H} \in \mathbb{R}^d$.

So, let $(H_n)_{n=1}^\infty$ be a sequence in $\mathbb{R}^d$ such that

$$\lim_{n \to \infty} \mathbf{E}\left[U\left((H_n, \Delta S)\right)\right] = \sup_{H \in \mathbb{R}^d} \mathbf{E}\left[U\left((H, \Delta S)\right)\right].$$

From the previous lemma we know that the sequence $f_n := (H_n, \Delta S)$ converges in measure to a function $g_0 \in L^0(\Omega, \mathcal{F}_1, \mathbf{P},]-\infty, \infty])$; we also know that every sequence $g_n \in \mathrm{conv}\{f_n, f_{n+1}, \ldots\}$ converges to g_0 in measure.

We want to conclude that $(H_n)_{n=1}^\infty$ converges; for this we still need two ingredients: first we have to suppose that each H_n is in the predictable range of S (see 6.4.1 above); in the present case of a trivial σ-algebra $\mathcal{F}_0$ this just means that H_n lies in the smallest subspace R of $\mathbb{R}^d$ such that ΔS takes a.s. its values in R. We may always pass from an arbitrary $H_n \in \mathbb{R}^d$ to $P(H_n)$ where P denotes the orthogonal projection onto R, as $(H_n, \Delta S) = (PH_n, \Delta S)$ almost surely.

The second ingredient is the assumption of no-arbitrage. Using this assumption we claim that $(H_n)_{n=1}^\infty$ is bounded in $\mathbb{R}^d$ and therefore g_0 cannot assume the value $+\infty$.

First we have to isolate the trivial case when $\Delta S = 0$ a.s. so that $R = \{0\}$ and $P = 0$. In this case we may define the martingale measure $\mathbf{Q}$ via $\frac{d\mathbf{Q}}{d\mathbf{P}} = 1$, which — for trivial reasons — coincides with (6.7).

Hence we may suppose that R is a subspace of $\mathbb{R}^d$ of dimension $\dim(R) \geq 1$ and define

$$\gamma = \inf\left\{\mathbf{E}\left[(H, \Delta S)_- \wedge 1\right] \mid H \in R, \|H\|_{\mathbb{R}^d} = 1\right\}. \tag{6.9}$$

As $H \mapsto \mathbf{E}[(H, \Delta S)_- \wedge 1]$ is continuous on the unit sphere of R (by Lebesgue's theorem) and strictly positive (by *(NA)* and the construction of R) we deduce from the compactness of the unit sphere of $\mathbb{R}^d$ that γ is a strictly positive number.

Next we show that $(H_n)_{n=1}^\infty$ is bounded. Indeed, otherwise there is an increasing sequence $(n_k)_{k=1}^\infty$ such that $\|H_{n_k}\|_{\mathbb{R}^d} \geq k$ so that

$$\mathbf{E}\left[U((H_{n_k}, \Delta S))\right] \leq -\|H_{n_k}\|_{\mathbb{R}^d} \mathbf{E}\left[\left(\frac{H_{n_k}}{\|H_{n_k}\|_{\mathbb{R}^d}}, \Delta S\right)_-\right] \leq -k\gamma,$$

in contradiction to the assumption that $(H_n)_{n=1}^\infty$ is a maximising sequence. Hence we obtain that g_0 is a.s. finite.

Finally we show that $(H_n)_{n=1}^\infty$ converges in $\mathbb{R}^d$. Indeed, otherwise there is $\alpha > 0$ and sequences $(n_k)_{k=1}^\infty$, $(m_k)_{k=1}^\infty$ tending to infinity such that

$$\|H_{n_k} - H_{m_k}\|_{\mathbb{R}^d} \geq \alpha,$$

so that

$$\mathbf{E}\left[\left(\frac{H_{n_k} - H_{m_k}}{\alpha}, \Delta S\right)_- \wedge 1\right] \geq \alpha\gamma,$$

contradicting Lemma 6.6.1, which asserts that $(H_n, \Delta S)$ converges a.s. to g_0.

Summing up, we deduce from the *(NA)* assumption and the fact that we choose the maximising sequence $(H_n)_{n=1}^{\infty}$ in the predictable range R that $(H_n)_{n=1}^{\infty}$ converges in $\mathbb{R}^d$. Denoting by $\widehat{H}$ the limit, we deduce from Fatou's lemma that $\widehat{H}$ is the optimiser for (6.5).

Now define the measure $\mathbf{Q}$ on $\mathcal{F}_1$ by

$$\frac{d\mathbf{Q}}{d\mathbf{P}} = cU'\left(\left(\widehat{H}, \Delta S\right)\right),$$

where the normalising constant $c > 0$ is chosen such that $\mathbf{E}\left[\frac{d\mathbf{Q}}{d\mathbf{P}}\right] = 1$ (note that by (6.6) we have $\mathbf{E}\left[\left|U'\left(\left(\widehat{H}, \Delta S\right)\right)\right|\right] < \infty$).

To show that $\mathbf{Q}$ is indeed a martingale measure we have to show that

$$\mathbf{E_Q}\left[(H, \Delta S)\right] = 0, \quad \text{for } H \in \mathbb{R}^d$$

or, equivalently,

$$\mathbf{E_Q}\left[(H, \Delta S)\right] \leq 0, \quad \text{for } H \in \mathbb{R}^d.$$

To do so, we use a variational argument:

$$\mathbf{E_Q}\left[(H, \Delta S)\right]$$
$$= c\mathbf{E_P}\left[(H, \Delta S)U'\left(\left(\widehat{H}, \Delta S\right)\right)\right]$$
$$= c \lim_{\alpha \searrow 0} \left(\frac{\mathbf{E}\left[U\left(\left(\widehat{H} + \alpha H, \Delta S\right)\right)\right] - \mathbf{E}\left[U\left(\left(\widehat{H}, \Delta S\right)\right)\right]}{\alpha}\right) \leq 0$$

where we have used the fact hat $\frac{U((\widehat{H} + \alpha H, \Delta S)) - U((\widehat{H}, \Delta S))}{\alpha}$ is a pointwise increasing function of α (by the concavity of U) so that the monotone convergence theorem applies.

We thus have found the desired martingale measure $\mathbf{Q}$ under the simplifying assumption that $\mathcal{F}_0$ is trivial.

We now extend this argument to the case of an arbitrary σ-algebra $\mathcal{F}_0$.

Proposition 6.6.2. *Suppose that the $\mathbb{R}^d$-valued process $S = (S_0, S_1)$ based on and adapted to $(\Omega, (\mathcal{F}_t)_{t=0}^1, \mathbf{P})$ satisfies (NA). Let $U(x) = -e^{-x}$ and suppose that*

$$\mathbf{E}\left[|U\left((H, \Delta S)\right)|\right] < \infty \tag{6.10}$$

for each $H \in L^{\infty}(\Omega, \mathcal{F}_0, \mathbf{P}; \mathbb{R}^d)$. Denote by P the $\mathcal{F}_0$-measurable predictable projection associated to $\Delta S = S_1 - S_0$. Then the following statements hold true.

(i)   *There is a unique optimiser $\widehat{H} \in L^0(\Omega, \mathcal{F}_0, \mathbf{P}; \mathbb{R}^d)$ for the maximisation problem*

$$\mathbf{E}\left[U\left((H, \Delta S)\right)\right] \to \max, \quad H \in L^0(\Omega, \mathcal{F}_0, \mathbf{P}; \mathbb{R}^d), \qquad (6.11)$$

which is in canonical form (i.e., $\widehat{H} = P(\widehat{H})$).

(ii) *Every maximising sequence $H_n \in L^0(\Omega, \mathcal{F}_0, \mathbf{P}; \mathbb{R}^d)$ for (6.11), verifying $H_n = P(H_n)$, converges to $\widehat{H}$ in measure.*

(iii) *The equation*

$$\frac{d\widehat{\mathbf{Q}}}{d\mathbf{P}} := \frac{\exp\left(-\left(\widehat{H}, \Delta S\right)\right)}{\mathbf{E}\left[\exp\left(-\left(\widehat{H}, \Delta S\right)\right)\right]}$$

defines a measure $\widehat{\mathbf{Q}}$ on $\mathcal{F}_1$ such that S is a martingale under $\widehat{\mathbf{Q}}$.

Before starting the proof of the proposition we remark that condition (6.10) is a "without loss of generality" assumption: the argument (6.6) above also applies to the present setting to yield a measure $\mathbf{P}' \sim \mathbf{P}$ such that $\mathbf{E}_{\mathbf{P}'}\left[|U((H, \Delta S))|\right] < \infty$ for each $H \in L^\infty(\Omega, \mathcal{F}_0, \mathbf{P}; \mathbb{R}^d)$.

Proof. We shall mimic the above argument in an "$\mathcal{F}_0$-parameterised" way. The crucial step is the extension of (6.9) to the present setting. Let P denote the $\mathcal{F}_0$-measurable predictable range projection associated to ΔS. Let $B = \{\mathbf{E}\left[\|\Delta S\|_{\mathbb{R}^d} \wedge 1 \mid \mathcal{F}_0\right] = 0\}$ so that B is the biggest set (modulo null-sets) in $\mathcal{F}_0$ on which $\Delta S = 0$. Note that P vanishes on B.

The case $B = \Omega$ again is trivial as we then have $\Delta S = 0$ a.s. so that $\widehat{H} = 0$ and $\frac{d\widehat{\mathbf{Q}}}{d\mathbf{P}} = 1$.

Excluding this case we define

$$\mathcal{H}_1 = \left\{ H \in \mathcal{H}^{\Delta S_1} \mid \|H\|_{\mathbb{R}^d} \geq \mathbf{1}_{B^c} \text{ a.s.} \right\}.$$

Define the $\mathcal{F}_0$-measurable non-negative function

$$\gamma = \operatorname{ess\,inf}\left\{\mathbf{E}\left[(H, \Delta S)_- \wedge 1 \mid \mathcal{F}_0\right] \mid H \in \mathcal{H}_1\right\}.$$

We claim that the function γ is a.s. strictly positive on $B^c = \Omega \setminus B$. Indeed for $H_1, H_2 \in \mathcal{H}_1$ and $A \in \mathcal{F}_0$, the function $H = H_1 \mathbf{1}_A + H_2 \mathbf{1}_{\Omega \setminus A}$ is in $\mathcal{H}_1$ too; hence we may — similarly as in the proof of Lemma 6.2.1 — find a sequence $(H_n)_{n=1}^\infty \in \mathcal{H}_1$ such that

$$\gamma = \lim_{n \to \infty} \mathbf{E}\left[(H_n, \Delta S)_- \wedge 1 \mid \mathcal{F}_0\right], \quad \text{a.s..}$$

We may suppose that $H_n = PH_n$ and, by multiplying with $\|H_n\|_{\mathbb{R}^d}^{-1}$, that $\|H_n\|_{\mathbb{R}^d} = 1$ almost surely on B^c. We may apply Proposition 6.3.4 above to find an $\mathcal{F}_0$-measurably parameterised subsequence $(H_{\tau_k})_{k=1}^\infty$ converging a.s. to $H \in L^0(\Omega, \mathcal{F}, \mathbf{P}; \mathbb{R}^d)$ such that $(\widetilde{H}, \Delta S)_- \wedge 1 = \liminf_{n \to \infty}(H_n, \Delta S)_- \wedge 1$ a.s., for which we then have $\widetilde{H} = P(\widetilde{H})$, $\|\widetilde{H}\|_{\mathbb{R}^d} = 1$ a.s. on B^c and by Lebesgue's theorem,

$$\gamma = \mathbf{E}\left[(H, \Delta S)_- \wedge 1 \,|\, \mathcal{F}_0\right].$$

Letting $A = \{\gamma = 0\}$ we have $(\mathbf{1}_A \widetilde{H}, \Delta S)_- = 0$ a.s., whence the *(NA)* assumption implies that $(\mathbf{1}_A \widetilde{H}, \Delta S) = 0$ almost surely. As $\mathbf{1}_A \widetilde{H}$ satisfies $\mathbf{1}_A \widetilde{H} = P(\mathbf{1}_A \widetilde{H})$ we must have $\mathbf{1}_A \widetilde{H} = 0$ a.s., so that $A = B$. Hence we have shown that γ is a.s. strictly positive on B^c.

Now let $(H_n)_{n=1}^\infty \in L^0(\Omega, \mathcal{F}_0, \mathbf{P}; \mathbb{R}^d)$ be a maximising sequence for (6.11). By passing to $(P(H_n))_{n=1}^\infty$ we may assume that H_n is in canonical form, i.e., $H_n = P(H_n)$. We infer from Lemma 6.6.1 that $f_n = (H_n, \Delta S)$ converges in measure to some $g_0 \in L^0(\Omega, \mathcal{F}_1, \mathbf{P};]-\infty, \infty])$.

We now show that $(H_n)_{n=1}^\infty$ remains bounded in $L^0(\Omega, \mathcal{F}_0, \mathbf{P}; \mathbb{R}^d)$, i.e., for $\varepsilon > 0$ there is $M > 0$ such that

$$\mathbf{P}\left[\|H_n\|_{\mathbb{R}^d} \geq M\right] < \varepsilon, \quad \text{for } n \geq 1. \tag{6.12}$$

Indeed, otherwise there is $\alpha > 0$ and an increasing sequence $(n_k)_{k=1}^\infty$ such that

$$\mathbf{P}\left[\|H_{n_k}\|_{\mathbb{R}^d} \geq k\right] \geq \alpha. \tag{6.13}$$

On the other hand, the boundedness from below of $(\mathbf{E}[U((H_n, \Delta S))])_{n=1}^\infty$ implies in particular the boundedness of $(\mathbf{E}[(H_n, \Delta S)_-])_{n=1}^\infty$, say by a constant $M > 0$. Hence

$$\widetilde{H}_{n_k} = \frac{H_{n_k}}{k} \mathbf{1}_{\{\|H_{n_k}\| \geq k\}}$$

is in canonical form, satisfies

$$\mathbf{E}\left[\left(\widetilde{H}_{n_k}, \Delta S\right)_-\right] \leq \frac{M}{k}$$

and $\|\widetilde{H}_{n_k}\|_{\mathbb{R}^d} \geq 1$ on $A_k = \{\|H_{n_k}\|_{\mathbb{R}^d} \geq k\}$. It follows that

$$\mathbf{E}\left[\gamma \mathbf{1}_{A_k}\right] \leq \mathbf{E}\left[\left(\widetilde{H}_{n_k}, \Delta S\right)_- \wedge 1\right] \leq \frac{M}{k},$$

which in view of (6.13) leads to a contradiction to the strict positivity of γ.

The L^0-boundedness of $(H_n)_{n=1}^\infty$ given by (6.12) implies in particular that g_0 is a.s. finite.

We now show that $(H_n)_{n=1}^\infty$ is a Cauchy-sequence in $L^0(\Omega, \mathcal{F}, \mathbf{P}; \mathbb{R}^d)$, i.e., with respect to convergence in measure. Indeed, suppose to the contrary that there is $\alpha > 0$ and sequences $(n_k)_{k=1}^\infty$, $(m_k)_{k=1}^\infty$ tending to infinity such that

$$\mathbf{P}\left[\|H_{n_k} - H_{m_k}\|_{\mathbb{R}^d} > \alpha\right] > \alpha.$$

Then on one hand side the strict positivity of γ implies that

$$\mathbf{E}\left[\left(\frac{H_{n_k} - H_{m_k}}{\alpha}, \Delta S\right)_- \wedge 1\right], \quad k \in \mathbb{N},$$

remains bounded away from zero, whence $((H_{n_k} - H_{m_k}, \Delta S))_{k=1}^{\infty}$ does not converge to zero in measure. Combining this fact with the L^0-boundedness (6.12) of $((H_n, \Delta S))_{n=1}^{\infty}$ we deduce that $((H_{n_k}, \Delta S))_{k=1}^{\infty}$ cannot converge in $L^0(\Omega, \mathcal{F}_1, \mathbf{P};] - \infty, +\infty])$, a contradiction to Lemma 6.6.1. Hence $(H_n)_{n=1}^{\infty}$ converges a.s. to some $\widehat{H} \in L^0(\Omega, \mathcal{F}_0, \mathbf{P}; \mathbb{R}^d)$ for which we have $\widehat{H} = P(\widehat{H})$ and $(\widehat{H}, \Delta S) = g_0$. Assertions (i) and (ii) now follow by the same arguments as discussed before.

As regards (iii) let $H \in L^{\infty}(\Omega, \mathcal{F}_0, \mathbf{P}; \mathbb{R}^d)$ and estimate again

$$\mathbf{E}_{\widehat{Q}}\left[(H, \Delta S) \mid \mathcal{F}_0\right]$$
$$= c\mathbf{E}_{\mathbf{P}}\left[(H, \Delta S)U'\left((\widehat{H}, \Delta S)\right) \mid \mathcal{F}_0\right]$$
$$= c\mathbf{E}_{\mathbf{P}}\left[\lim_{\alpha \to 0} \frac{U((\widehat{H} + \alpha H, \Delta S)) - U((\widehat{H}, \Delta S))}{\alpha} \;\middle|\; \mathcal{F}_0\right] \leq 0,$$

where again we have used in the last inequality the monotone convergence theorem, the concavity of U and (6.10). $\qquad \square$

6.7 Proof of the Dalang-Morton-Willinger Theorem for $T \geq 1$ by Induction on T

Proof of Theorem 6.1.1. We proceed by induction on T. For $T = 1$ Theorem 6.5.1 applies. So suppose Theorem 6.1.1 holds true for $T - 1$.

We now consider the process $(S_t)_{t=1}^{T}$ adapted to the filtration $(\mathcal{F}_t)_{t=1}^{T}$. Because of the inductive hypothesis we suppose that there is a probability measure $\mathbf{Q}^1$, defined on $\mathcal{F}_T$, equivalent to $\mathbf{P}$, and so that

(i) $\frac{d\mathbf{Q}^1}{d\mathbf{P}}$ is bounded,
(ii) $S_1, \ldots, S_T$ are in $L^1(\Omega, \mathcal{F}_T, \mathbf{Q}^1)$,
(iii) $(S_t)_{t=1}^{T}$ is a $\mathbf{Q}^1$-martingale, i.e., for all $t \geq 1$, $A \in \mathcal{F}_t$ we have

$$\int_A S_t d\mathbf{Q}^1 = \int_A S_{t+1} d\mathbf{Q}^1.$$

The one-step result in the DMW Theorem (Theorem 6.5.1 or Proposition 6.6.2) applied to the process $(S_t)_{t=0}^{1}$, the probability space $(\Omega, \mathcal{F}_1, \mathbf{Q}^1)$ and the filtration $(\mathcal{F}_t)_{t=0}^{1}$, gives us a bounded function f_1 so that: f_1 is $\mathcal{F}_1$-measurable, $f_1 > 0$, $\mathbf{E}_{\mathbf{Q}^1}[f_1] = 1$, $\mathbf{E}_{\mathbf{Q}^1}[|S_1| f_1] < \infty$, $\mathbf{E}_{\mathbf{Q}^1}[|S_0| f_1] < \infty$ and for all $A \in \mathcal{F}_0$ we have

$$\int_A S_0 f_1 d\mathbf{Q}_1 = \int_A S_1 f_1 d\mathbf{Q}_1.$$

Let us finally define $\mathbf{Q}$ on $\mathcal{F}_T$ by the rule

$$\mathbf{Q}[A] = \int_A f_1 d\mathbf{Q}^1 \quad \text{for all } A \in \mathcal{F}_T.$$

Of course this means that $\frac{d\mathbf{Q}}{d\mathbf{P}} = f_1 \frac{d\mathbf{Q}^1}{d\mathbf{P}}$ and hence this is a bounded random variable. Furthermore $\frac{d\mathbf{Q}}{d\mathbf{P}} > 0$ almost surely and hence $\mathbf{Q}$ and $\mathbf{P}$ are equivalent. Now let us check the integrability properties as well as the martingale properties. For $t = 1, \ldots, T$ we have

$$\int_\Omega |S_t| d\mathbf{Q} = \int_\Omega |S_t| f_1 d\mathbf{Q}^1 < \infty,$$

by construction of $\mathbf{Q}^1$ and the boundedness of f_1.

The martingale property of $(S_t)_{t=0}^T$ with respect to $\mathbf{Q}$ is also an easy calculation. Indeed, for all $A \in \mathcal{F}_0$ we have

$$\int_A S_0 d\mathbf{Q} = \int_A S_0 f_1 d\mathbf{Q}^1 = \int_A S_1 f_1 d\mathbf{Q}^1 = \int_A S_1 d\mathbf{Q}$$

by construction of f_1. For $t \geq 1$ we remark that f_1 was $\mathcal{F}_1$-measurable and bounded, which means that the sequence of the following equalities is easily justified. If $A \in \mathcal{F}_t$, $t \geq 1$ we have

$$\int_A S_t d\mathbf{Q} = \int_A S_t f_1 d\mathbf{Q}^1 = \int_A S_{t+1} f_1 d\mathbf{Q}^1 = \int_A S_{t+1} d\mathbf{Q}.$$

This ends the proof of the induction step. $\qquad\square$

6.8 Proof of the Closedness of K in the Case $T \geq 1$

In this section we extend Stricker's lemma (Theorem 6.4.2 (i)) to the case $T \geq 1$.

Proposition 6.8.1. *Let the process* $S = (S_t)_{t=0}^T$ *be* $\mathbb{R}^d$-*valued and* $(\mathcal{F}_t)_{t=0}^T$-*adapted. The space*

$$K = \left\{ \sum_{t=1}^T (H_t, \Delta S_t) \, \middle| \, (H_t)_{t=1}^T \, \mathbb{R}^d\text{-valued and predictable} \right\}$$

is a closed subspace of $L^0(\Omega, \mathcal{F}_T, \mathbf{P})$.

Proof. The proof is by induction on T. For $T = 1$ the statement is Stricker's lemma, Theorem 6.4.2 (i). The inductive hypothesis reads:

$$K_2 = \left\{ \sum_{t=2}^{T} (H_t, \Delta S_t) \,\Big|\, (H_t)_{t=2}^{T} \text{ predictable and } \mathbb{R}^d\text{-valued} \right\} \tag{6.14}$$

is closed in $L^0(\Omega, \mathcal{F}_T, \mathbf{P})$. The basic ingredient of the proof is the reduction of the integrands H to a canonical form. However, in this multiperiod setting we have to be more careful, since the elements $f \in K$ can, a priori, be represented in many different ways.

Let $P : L^0(\Omega, \mathcal{F}_0, \mathbf{P}; \mathbb{R}^d) \to L^0(\Omega, \mathcal{F}_0, \mathbf{P}; \mathbb{R}^d)$ be the projection on the $\mathcal{F}_0$-predictable range of ΔS_1 as in Lemma 6.2.1 and let

$$\mathcal{H}_1 = \{ H \mid H \text{ is } \mathbb{R}^d\text{-valued } \mathcal{F}_0\text{-measurable and } PH = H \}. \tag{6.15}$$

In other words, the elements of $\mathcal{H}_1$ are in canonical form for ΔS_1. Let I^1 be the linear mapping $I^1 : \mathcal{H}_1 \to L^0(\Omega, \mathcal{F}_T, \mathbf{P})$, $I^1(H_1) = (H_1, \Delta S_1)$. As in Sect. 6.4 above, I^1 is continuous and injective. Let $F_1 \subset \mathcal{H}_1$ be defined by $F_1 = (I^1)^{-1}(K_2 \cap I^1(\mathcal{H}_1))$. Clearly F_1 is a closed subspace of $\mathcal{H}_1$ since K_2 is closed by hypothesis. Also F_1 is stable in the sense of Lemma 6.2.1. This means that there is a projection-valued $\mathcal{F}_0$-measurable map, called P_0 : $L^0(\Omega, \mathcal{F}_0, \mathbf{P}; \mathbb{R}^d) \to L^0(\Omega, \mathcal{F}_0, \mathbf{P}; \mathbb{R}^d)$, so that $f \in F_1$ if and only if $P_0 f = f$ a.s.. Now we take

$$\begin{aligned} E_1 &= \{ H_1 \in \mathcal{H}_1 \mid P_0 H_1 = 0 \} \\ &= \{ H_1 \in L^0(\Omega, \mathcal{F}_0, \mathbf{P}; \mathbb{R}^d) \mid P(\mathrm{Id} - P_0) H_1 = H_1 \}. \end{aligned}$$

The elements H_1 in E_1 are in canonical form and the integrals $(H_1, \Delta S_1)$ cannot be obtained by stochastic integrals on $(S_t)_{t=2}^{T}$ (see (6.14)). We have that

$$K = \left\{ \sum_{t=1}^{T} (H_t, \Delta S_t) \,\Big|\, (H_t)_{t=1}^{T} \text{ is } \mathbb{R}^d\text{-valued, predictable and } H_1 \in E_1 \right\}.$$

Moreover the decomposition of elements $f \in K$ into $f = (H_1, \Delta S_1) + f_2$ where $H_1 \in E_1$ and $f_2 \in K_2$ is unique.

Let now $f^n = (H_1^n, \Delta S_1) + f_2^n$ be a sequence in K with $H_1^n \in E_1$, $f_2^n \in K_2$ so that $f^n \to f$ almost surely. We have to show that $f \in K$. We will show that $(H_1^n)_{n=1}^{\infty}$ is bounded a.s. and the selection principle will do the rest. Let $A = \{\limsup |H_1^n| = \infty\}$. By Proposition 6.3.4 we have that there is a $\mathcal{F}_0$-measurably parameterised subsequence $(\tau_n)_{n=1}^{\infty}$ so that: $|H_1^{\tau_n}| \to \infty$ on A and $H_1^{\tau_n} \to H_1$ on A^c for some $H_1 \in E_1$. We will show that $\mathbf{P}[A] = 0$. If this were not the case we could apply Proposition 6.3.3 and suppose that we have $\frac{H_1^{\tau_n}}{|H_1^{\tau_n}|} \to \psi_1$ a.s. on the set A, where $\psi_1 = \psi_1 \mathbf{1}_A$ is some $\mathcal{F}_0$-measurable function supported by A where it takes values in the unit sphere of $\mathbb{R}^d$. Clearly the functions $\frac{H_1^{\tau_n}}{|H_1^{\tau_n}|}\mathbf{1}_A$ are still in E_1 since E_1 is closed and stable in the sense of lemma 6.2.1. Therefore $\psi_1 \in E_1$. Indeed $H_1^{\tau_n} = \sum_{m=1}^{\infty} \mathbf{1}_{\{\tau_n = m\}} H_1^m$ where $H_1^m \in E_1$ and the τ_n are $\mathcal{F}_0$-measurable. Since $|H_1^{\tau_n}| \to \infty$ on A, we have that a.s.

$$\left(\left(\frac{H_1^{\tau_n}}{|H_1^{\tau_n}|}, \Delta S_1 \right) + \frac{f_2^{\tau_n}}{|H_1^{\tau_n}|} \right) \mathbf{1}_A \longrightarrow 0.$$

It follows that $\frac{f_2^{\tau_n}}{|H_1^{\tau_n}|}\mathbf{1}_A \to -(\psi_1, \Delta S_1)\mathbf{1}_A$ and hence by the closedness of K_2 we have $-(\psi_1, \Delta S_1)\mathbf{1}_A \in K_2$. This implies that $(\psi_1, \Delta S_1) = 0$ since $\psi_1 \in E_1$. But since ψ_1 is in canonical form we must have $\psi_1 = 0$ a contradiction to $|\psi_1| = 1$ on A so that $\mathbf{P}[A] = 0$.

So we get an $\mathcal{F}_0$-measurably parameterised sequence $(H_1^{\tau_n})_{n=1}^{\infty}$ converging a.s. to H_1 on Ω. This implies that $f_2^{\tau_n} \to f - (H_1, \Delta S_1)$ and hence $f_2 = f - (H_1, \Delta S_1) \in K_2$ by the closedness of K_2. Finally $f = (H_1, \Delta S_1) + f_2$ where $H_1 \in E_1$ and $f_2 \in K_2$ i.e. $f \in K$. $\qquad \square$

6.9 Proof of the Closedness of C in the Case $T \geq 1$ under the *(NA)* Condition

We will use the same notation as in the previous section. This means that for the $(\mathcal{F}_t)_{t=0}^{T}$-adapted $\mathbb{R}^d$-valued process $(S_t)_{t=0}^{T}$ we introduce $\mathcal{H}_1$, K_2, and I^1 as in (6.14) and (6.15).

We say that a predictable $\mathbb{R}^d$-valued process $(H_t)_{t=1}^{T}$ is in canonical form, if for each t, H_t is in canonical form for $\Delta S_t = (S_t - S_{t-1})$. The spaces $\mathcal{H}_t$ are defined in the same way as $\mathcal{H}_1$ i.e.

$$\mathcal{H}_t = \left\{ H_t \mid H_t \text{ is } \mathbb{R}^d\text{-valued, } \mathcal{F}_{t-1}\text{-measurable and } P_t H_t = H_t \right\}.$$

Here P_t is the projection in $L^0(\Omega, \mathcal{F}_{t-1}, \mathbf{P}; \mathbb{R}^d)$ associated with the predictable range of $\Delta S_t = S_t - S_{t-1}$.

Proposition 6.9.1. *With the above notation and under the assumption that S satisfies the (NA) condition we have*

(i) $I : \mathcal{H}_1 \times \mathcal{H}_2 \times \ldots \times \mathcal{H}_T \to L^0(\Omega, \mathcal{F}_T, \mathbf{P})$, $I((H_t)_{t=1}^{T}) = \sum_{t=1}^{T}(H_t, \Delta S_t) = (H \cdot S)_T$ *is injective.*

(ii) *If $(H^n)_{n=1}^{\infty}$ is a sequence in $\mathcal{H}_1 \times \mathcal{H}_2 \times \ldots \times \mathcal{H}_T$ so that $I(H^n)^- = (H^n \cdot S)_T^-$ is bounded a.s., then $(H^n)_{n=1}^{\infty} = (H_1^n, \ldots, H_T^n)_{n=1}^{\infty}$ is bounded a.s.*

(iii) *If $(f^n)_{n=1}^{\infty}$ is a sequence in K which is bounded a.s., then there is a $\mathcal{F}_T$-measurably parameterised subsequence σ_n so that $f^{\sigma_n} \to f$ a.s. and $f \in K$.*

Proof. (i): The first statement will follow by induction on T from the fact that $I^1(H_1) \in K_2$ gives $H_1 = 0$ if H_1 is in canonical form. This statement is proved as follows. Let $(H_1, \Delta S_1) + f_2 = 0$ where H_1 is in canonical form and $f_2 \in K_2$. On the set $A = \{(H_1, \Delta S_1) < 0\}$ we have that $f_2 > 0$ and since $1_A f_2 \in K_2$ we have that $\mathbf{P}[A] > 0$ gives an arbitrage opportunity. On the set $\{(H_1, \Delta S_1) > 0\}$ we replace f_2 by $-f_2$ and get the same result. This means that $(H_1, \Delta S_1) = 0$ and since H_1 is in canonical form we get $H_1 = 0$.

(iii): This assertion is immediate from the closedness of K (Proposition 6.8.1) and the measurable sub-sequence principle (Proposition 6.3.3). The reason why we stated it explicitly is to avoid confusion: we only claim that the random subsequence $(\sigma_n)_{n=1}^{\infty}$ is measurably parameterised w.r. to $\mathcal{F}_T$; we do not claim that $H^{\sigma_n} = (H_1^{\sigma_n}, \ldots, H_T^{\sigma_n})$ is predictable.

(ii): To prove (ii) we again use induction on T. For $T = 1$ we refer to Proposition 6.4.1. So let us suppose that assertions (ii) and (iii) hold true for $T - 1$ and fix the horizon T. We now show that the assumption of (ii) implies that $(H_1^n)_{n=1}^{\infty}$ is a.s. bounded. We proceed in the same way as in the proof of 6.8.1. Let $A = \{\limsup |H_1^n| = +\infty\}$ and let τ_n be an $\mathcal{F}_0$-measurably parameterised subsequence selected in such a way that $|H_1^{\tau_n}| \to \infty$ on A and $\frac{H_1^{\tau_n}}{|H_1^{\tau_n}|} \to \psi_1$ on A. On the set A we must have $|\psi_1| = 1$ and we may put $\psi_1 = 0$ on the complement of A. We remark that this is possible by the selection principle and that $\psi_1 \in \mathcal{H}_1$ i.e. ψ_1 is in canonical form. Now $(H^{\tau_n} \cdot S)_T = (H_1^{\tau_n}, \Delta S_1) + f_2^{\tau_n}$, where $f_2^{\sigma_n} \in K_2$. On the set A we get

$$\left(\frac{H^{\tau_n}}{|H_1^{\tau_n}|} \cdot S \right)_T = \left(\frac{H_1^{\tau_n}}{|H_1^{\tau_n}|}, \Delta S_1 \right) + \frac{1}{|H_1^{\tau_n}|} f_2^{\tau_n}.$$

Since the first term on the right hand side is bounded by $|\Delta S_1|$ we get

$$\limsup_{n \to \infty} \left(\frac{1}{|H_1^{\tau_n}|} f_2^{\tau_n} \right)^{-} \leq |\Delta S_1| + \limsup_{n \to \infty} \frac{1}{|H_1^{\tau_n}|} (H^{\tau_n} \cdot S)_T^{-} \leq |\Delta S_1|.$$

By the induction hypothesis this means that the sequence

$$\widetilde{H}_n := \left(0, 1_A \frac{H_2^{\tau_n}}{|H_1^{\tau_n}|}, \ldots, 1_A \frac{H_T^{\tau_n}}{|H_1^{\tau_n}|} \right) = \widetilde{H}^n$$

is a.s. bounded. It follows that the functions $(\widetilde{H}^n \cdot S)_T = f_2^{\tau_n} 1_A \frac{1}{|H_1^{\tau_n}|}$ are also a.s. bounded. By (iii) there is a $\mathcal{F}_T$-measurably parameterised subsequence σ_n of τ_n so that $\left(\widetilde{H}^{\sigma_n} \cdot S \right)_T \to f$ and $f \in K_2$. Since $(\sigma_n)_n$ is a subsequence of τ_n we still have $\frac{H_1^{\sigma_n}}{|H_1^{\tau_n}|} \to \psi_1$ and hence

$$(\psi_1, \Delta S_1) + f = \lim_{n \to \infty} \left(\frac{H_1^{\tau_n}}{|H_1^{\tau_n}|}, \Delta S_1 \right) + (\widetilde{H}^{\sigma_n} \cdot S)_T$$

$$= \lim_{n \to \infty} 1_A \left((H_1^{\sigma_n}, \Delta S_1) + f_2^{\sigma_n} \right) \frac{1}{|H_1^{\tau_n}|}$$

$$= \lim_{n \to \infty} 1_A (H^{\sigma_n} \cdot S)_T \frac{1}{|H_1^{\tau_n}|}.$$

Since $(H^n \cdot S)_T^-$ is a.s. bounded we have that $((\psi_1, \Delta S_1) + f) \geq 0$ and by the *(NA)* condition this implies $(\psi_1, \Delta S_1) + f = 0$. This means $\psi_1 = 0$ by (i). Since $|\psi_1| = 1$ on A we must have $\mathbf{P}[A] = 0$ proving that $(H_1^n)_{n=1}^\infty$ is a bounded sequence. The sums from 2 to T

$$\sum_{t=2}^T H_t^n \Delta S_t = (H^n \cdot S)_T - (H_1^n, \Delta S_1)$$

satisfy

$$\left(\sum_{t=2}^T H_t^n \Delta S_t \right)_- \leq (H^n \cdot S)_T^- + |(H_1^n, \Delta S_1)|$$

and hence are a.s. bounded. The inductive hypothesis shows that $((H_t^n)_{t=2}^T)_{n=1}^\infty$ is then a.s. bounded too.   □

We are now ready to prove the main result of this section.

Theorem 6.9.2. *If $(S_t)_{t=0}^T$ is $\mathbb{R}^d$-valued and adapted with respect to the filtration $(\mathcal{F}_t)_{t=0}^T$, if S satisfies the (NA) condition, then the cone*

$$C = K - L_+^0(\Omega, \mathcal{F}_T, \mathbf{P}) = \{ (H \cdot S)_T - h \mid H \text{ predictable, } h \geq 0\}$$

is closed in $L^0(\Omega, \mathcal{F}_T, \mathbf{P})$.

Proof. Let $f_n = (H^n \cdot S)_T$ and $h_n \geq 0$ be such that $g_n = f_n - h_n \to g$. Clearly $(H^n \cdot S)_T^- = f_n^- \leq g_n^-$ forms a bounded sequence. By Proposition 6.9.1 (ii) and (iii) we have that $(H^n)_{n=1}^\infty$ itself is already bounded and we also have the existence of a $\mathcal{F}_T$-measurably parameterised subsequence σ_n so that $f_{\sigma_n} \to f \in K$. Then necessarily $h_{\sigma_n} = f_{\sigma_n} - g_{\sigma_n}$ tends a.s. to $f - g = h$ and we have therefore $h \geq 0$. Moreover $g = f - h \in C$ a.s. which proves the theorem.   □

6.10 Proof of the Dalang-Morton-Willinger Theorem for $T \geq 1$ using the Closedness of C

Proof of Theorem 6.1.1. The proof is the same as in Sect. 6.5 for Theorem 6.5.1, except for some obvious variations. Let us indicate the changes. First we take an equivalent probability measure $\mathbf{P}_1$ so that $\frac{d\mathbf{P}_1}{d\mathbf{P}}$ is bounded and $S_t \in L^1(\mathbf{P}_1)$ for all $0 \leq t \leq T$. For example we may take $\frac{d\mathbf{P}_1}{d\mathbf{P}} = c \exp(-|S_0| - \ldots - |S_T|)$, where c is the normalisation constant given by $c^{-1} = E[\exp(-|S_0| - \ldots - |S_T|)]$.

The next step consists in considering the set

$$C_1 = C \cap L^1(\Omega, \mathcal{F}_T, \mathbf{P}_1).$$

As C is closed in $L^0(\Omega, \mathcal{F}_T, \mathbf{P})$, the set C_1 is closed in $L^1(\Omega, \mathcal{F}_T, \mathbf{P}_1)$. Obviously C_1 is a convex cone (since C is a convex cone). The *(NA)* condition implies that $C_1 \cap L^1_+(\Omega, \mathcal{F}_T, \mathbf{P}_1) = \{0\}$. The Kreps-Yan Theorem 5.2.2 now gives the existence of an equivalent probability measure $\mathbf{Q}$ so that $\frac{d\mathbf{Q}}{d\mathbf{P}_1}$ is bounded and so that $\mathbf{E}_\mathbf{Q}[f] \leq 0$ for all $f \in C_1$. Obviously all $S_t \in L^1(\mathbf{Q})$ since $\frac{d\mathbf{Q}}{d\mathbf{P}_1}$ is bounded. Since for each coordinate $j = 1, \ldots, d$ and each $A \in \mathcal{F}_t$ we have $\mathbf{1}_A(S^j_{t+1} - S^j_t) \in C_1$ and $-\mathbf{1}_A(S^j_{t+1} - S^j_t) \in C_1$, we must have $\mathbf{E}_\mathbf{Q}[\mathbf{1}_A S^j_{t+1}] = \mathbf{E}_\mathbf{Q}[\mathbf{1}_A S^j_t]$. This shows that $S_t = \mathbf{E}_\mathbf{Q}[S_{t+1} \mid \mathcal{F}_t]$.

Let us finally verify assertion (iii) of Theorem 6.1.1. Since $\frac{d\mathbf{Q}}{d\mathbf{P}} = \frac{d\mathbf{Q}}{d\mathbf{P}_1} \frac{d\mathbf{P}_1}{d\mathbf{P}}$ we have that $\frac{d\mathbf{Q}}{d\mathbf{P}}$ is bounded.   □

6.11 Interpretation of the L^∞-Bound in the DMW Theorem

This section is based on [De 00, Chap. VII]. We will suppose that the process $(S_t)_{t=0}^T$ adapted to $(\mathcal{F}_t)_{t=0}^T$ satisfies *(NA)* and that it is *integrable* with respect to $\mathbf{P}$. The set

$$M^a = \left\{ \frac{d\mathbf{Q}}{d\mathbf{P}} \in L^\infty \;\middle|\; \mathbf{Q} \text{ a probability such that } S \text{ is a } \mathbf{Q}\text{-martingale} \right\}$$

is non-empty by the DMW Theorem. The space

$$W_1 = \left\{ (H \cdot S)_T \mid H \text{ predictable } (H \cdot S)_T \in L^1(\Omega, \mathcal{F}_T, \mathbf{P}) \right\}$$

is closed in $L^1(\Omega, \mathcal{F}_T, \mathbf{P})$ since $W_1 = K \cap L^1(\Omega, \mathcal{F}_T, \mathbf{P})$ and K is L^0-closed. The set M^a is the intersection of $W_1^\perp$ with the set of probability measures. For each $k \geq 1$ we define a utility function $u_k : L^1 \to \mathbb{R}$ as follows. The set $\mathcal{P}_k$ is defined by $\mathcal{P}_k = \left\{ \frac{d\mathbf{Q}}{d\mathbf{P}} \;\middle|\; \mathbf{Q} \text{ a probability, } \frac{d\mathbf{Q}}{d\mathbf{P}} \leq k \right\}$. With this set we define the coherent monetary utility function

$$u_k(Y) = \inf\left\{ \mathbf{E}_\mathbf{Q}[Y] \mid \mathbf{Q} \in \mathcal{P}_k \right\} = \min\left\{ \mathbf{E}_\mathbf{Q}[Y] \mid \mathbf{Q} \in \mathcal{P}_k \right\}.$$

The utility function u_k is concave and L^1-continuous since it is Lipschitz. The set $O_k = \{Y \mid u_k(Y) > 0\}$ is therefore open and convex in $L^1(\Omega, \mathcal{F}_T, \mathbf{P})$. Furthermore it contains the cone of strictly positive integrable random variables. The set $\mathcal{P}_k$ can be described as $\mathcal{P}_k = \{\mathbf{Q} \mid \mathbf{Q} \text{ a probability, } \mathbf{E}_\mathbf{Q}[Y] \geq 0 \text{ for all } Y \in O_k\}$. The interpretation of the L^∞-bound in the DMW Theorem is described in the subsequent result:

Theorem 6.11.1. *Under the above assumptions we have, for each $k \geq 1$,*

$$M^a \cap \mathcal{P}_k \neq \emptyset \iff W_1 \cap O_k = \emptyset.$$

Proof. If $\mathbf{Q}_0 \in M^a \cap \mathcal{P}_k$ then for all $Y \in W_1$ we have $u_k(Y) = \inf\{\mathbf{E}_\mathbf{Q}[Y] \mid \mathbf{Q} \in \mathcal{P}_k\} \le \mathbf{E}_{\mathbf{Q}_0}[Y] = 0$. Therefore $W_1 \cap O_k = \emptyset$.

Conversely if $W_1 \cap O_k = \emptyset$ we may separate the vector space W_1 from the convex open set O_k. This yields an element $f \in L^\infty$ such that for all $Z \in O_k$, all $Y \in W_1$: $\mathbf{E}[fY] < \mathbf{E}_\mathbf{Q}[fZ]$. These inequalities show that f is non-negative and not identically zero. We may normalise f so that $\mathbf{E}[f] = 1$, hereby defining a probability $\mathbf{Q}$ so that $\frac{d\mathbf{Q}}{d\mathbf{P}} = f \in L^\infty$. Since $\mathbf{E}_\mathbf{Q}[Z] \ge 0$ for all $Z \in O_k$ we must have $\mathbf{Q} \in \mathcal{P}_k$. Also $\mathbf{E}_\mathbf{Q}[Y] = 0$ for all $Y \in W_1$ and hence $\mathbf{Q} \in M^a$. $\qquad\square$

Remark 6.11.2. In [HK 79] M. Harrison and D. Kreps related the concept of no-arbitrage to the concept of viability which is based on utility considerations somewhat similar to the above considerations.

Remark 6.11.3. The coherent monetary utility function u_k is essentially the same as the so called *tail expectation* with parameter $\alpha = \frac{1}{k}$.

To avoid trivialities let us suppose that $(\Omega, \mathcal{F}, \mathbf{P})$ is a non-atomic probability space. For $Y \in L^1(\Omega, \mathcal{F}, \mathbf{P})$ define the quantile $q_\alpha(Y)$ as

$$q_\alpha = \inf\{x \mid \mathbf{P}[Y \le x] \ge \alpha\}.$$

If Y has a continuous distribution then we have

$$u_k(Y) = \mathbf{E}[Y \mid Y \le q_\alpha],$$

where the right hand side is called, for obvious reasons, the α-tail expectation of Y. Indeed, it suffices to consider $\frac{d\mathbf{Q}}{d\mathbf{P}} = k\mathbf{1}_{\{Y \le q_\alpha\}}$.

In the general case we have to be slightly more careful as it might happen that $\mathbf{P}[Y = q_\alpha(Y)] > 0$. In this case let $\beta = \mathbf{P}[Y < q_\alpha]$ and verify that

$$u_k(Y) = \mathbf{E}[Y \mid Y < q_\alpha] + (\alpha - \beta)q_\alpha.$$

The above Theorem 6.11.1 tells us that the constant α verifying $1 \ge \frac{1}{k} = \alpha > 0$ is sufficiently small such that $M^a \cap \mathcal{P}_k \ne \emptyset$ iff it is not possible to find an element $(H \cdot S)_T \in W_1$ which yields a strictly higher utility $u_k((H \cdot S)_T)$ than $u_k(0) = 0$. As a trivial illustration consider $\alpha = k = 1$: then this statement boils down to the fact that S is a martingale under $\mathbf{P}$ iff for each $(H \cdot S)_T$ we have that $u_1((H \cdot S)_T) = \mathbf{E}[(H \cdot S)_T] = 0$.

We refer to [De 00] for a more detailed situation in which an interpretation in terms of superhedging and risk measures is given.

7

A Primer in Stochastic Integration

7.1 The Set-up

In the previous chapters we mainly developed the arbitrage theory for models in finite discrete time. In the setting of the previous chapter, where the probability space was not finite, several features of infinite dimensional functional analysis played a role. When trading takes place in continuous time the difficulties increase even more. It is here that we need the full power of stochastic integration theory. Before giving precise definitions, let us give a short overview of the different models and of their mutual relation. In financial problems the following concepts play a dominant role:

(i)   assets to be traded
(ii) trading dates
(iii) trade procedures
(iv) uncertainty

We assume for the moment that the set of *trading dates* $\mathbb{T}$ is a subset of $\mathbb{R}_+$. The following cases are of particular importance

(i) $\mathbb{T} = \{0, 1, \ldots, n\}$ finite discrete time
(ii) $\mathbb{T} = \mathbb{N} = \{0, 1, \ldots\}$ infinite discrete time
(iii) $\mathbb{T} = [0, 1]$ finite horizon continuous time
(iv) $\mathbb{T} = \mathbb{R}_+ = [0, \infty[$ infinite horizon continuous time

The *uncertainty* is modelled using a filtered probability space $(\Omega, (\mathcal{F}_t)_{t \in \mathbb{T}}, \mathbf{P})$. The filtration $(\mathcal{F}_t)_{t \in \mathbb{T}}$ is formed by an increasing family of sub-σ-algebras of $\mathcal{F}_\infty$ where $(\Omega, \mathcal{F}_\infty, \mathbf{P})$ is a probability space. The role of the filtration is very important since it describes the information available at each time t. We suppose that there are finitely many *assets*, indexed by $i = 1, \ldots, d$. An asset to be traded is described by a stochastic process $S^i : \mathbb{T} \times \Omega \to \mathbb{R}$. The collection of assets is therefore described by a finite dimensional stochastic process $S : \mathbb{T} \times \Omega \to \mathbb{R}^d$. There is no need to suppose that prices are positive. It is also understood that there is an asset number 0, that describes "cash".

Cash is convenient to transform money from one date to another. We assume that there are no costs in carrying cash and that interest rate is zero. This results in a constant price process whose value is 1. If interest rate is not zero then we introduce a process that describes the cumulative value of an account earning the instantaneous riskless interest rate and we renormalise all the prices by dividing them by this process. This is exactly what we did in Chap. 2. The choice of a convenient numéraire depends on the application, and the change of numéraire is an important technique in finance. In Chap. 2 we introduced the reader to this technique. We will come back to this later and we will show in Chap. 11 in what cases the change of numéraire can be performed without distorting the model. For the moment we suppose that a traded asset has been chosen as numéraire and that all prices are expressed in units of this numéraire. We will therefore not need the process indexed by the number 0 which is simply identically equal to one. The price S_t^i at time $t \in \mathbb{T}$ is part of the information available at time t. In mathematical terms we translate this by the statement that the processes S^i are adapted, i.e., S_t^i is $\mathcal{F}_t$-measurable, for each $t \in \mathbb{T}$. The filtration $(\mathcal{F}_t)_{t \in \mathbb{T}}$ is not necessarily generated by the process S. This means that other sources of information than prices can be observed (e.g. balance sheets, weather conditions, ...). All agents have access to the same filtration, i.e. information. Agents can buy and sell assets, short selling is allowed. There are *no* transaction costs. In buying and selling assets, only information available from the past is to be used. We cannot buy 100.000 shares of some stock conditionally on the event that the price next year will be doubled.

The space $\mathbb{R}^d$ will be endowed with the usual Euclidean structure. The inner product of two vectors x and y in $\mathbb{R}^d$, written as (x, y), is to be interpreted as $(x, y) = x^1 y^1 + \cdots + x^d y^d$. In some cases we will simply write xy. We do not put a dot since this is reserved for stochastic integration. The norm of a vector x in $\mathbb{R}^d$ is written as $|x|$, reserving the notation $\| \, . \, \|$ for norms in L^p-spaces etc.

7.2 Introductory on Stochastic Processes

The following notation, coming from probability theory, will be used. We write $\overline{\mathbb{T}} = \mathbb{T} \cup +\infty$. A map $T : \Omega \to \overline{\mathbb{T}}$ is called a *stopping time* if for all $t \in \mathbb{T}$ the set $\{T \leq t\} \in \mathcal{F}_t$. For $T_1 \leq T_2$, two stopping times, we denote by $[\![T_1, T_2]\!]$ the set $\{(t, \omega) \mid t \in \mathbb{T}, \ \omega \in \Omega \text{ and } T_1(\omega) \leq t \leq T_2(\omega)\}$. Other stochastic intervals are denoted in an analogous way: $]\!] T_1, T_2]\!]$, $[\![T_1, T_2 [\![$, $]\!] T_1, T_2 [\![$. Remark that for $\mathbb{T} = \mathbb{R}_+$ the interval $[\![0, \infty]\!]$ denotes $\mathbb{R}_+ \times \Omega$ and not $[0, \infty] \times \Omega$. The symbol π denotes the projection $\pi : \mathbb{R}_+ \times \Omega \to \Omega$.

To avoid technical complications we suppose that in continuous time, the filtration $(\mathcal{F}_t)_t$ satisfies the *usual conditions*:

(i)   for all t we have: $\mathcal{F}_t = \bigcap_{s > t} \mathcal{F}_s$ (right continuity),

(ii) $\mathcal{F}_0$ contains all null sets of $\mathcal{F}_\infty$. This means: $A \subset B \in \mathcal{F}_\infty$ and $\mathbf{P}[B] = 0$ imply $A \in \mathcal{F}_0$ ($\mathcal{F}_0$ is saturated).

The natural filtration generated by a process X is defined as follows: we give the description for $\mathbb{T} = \mathbb{R}_+$:

(i)   for all t we define $\mathcal{H}_t = \sigma(X_u; u \leq t)$ and $\mathcal{H}_\infty = \bigvee_{t=0}^{\infty} \mathcal{H}_t = \sigma(X_t, 0 \leq t)$.
(ii) $\mathcal{G}_t = \bigcap_{s>t} \mathcal{H}_s$ and $\mathcal{G}_\infty = \mathcal{H}_\infty$.
(iii) $\mathcal{N} = \{A \mid \exists B \in \mathcal{G}_\infty, A \subset B \text{ and } \mathbf{P}[B] = 0\}$.
(iv) $\mathcal{F}_t = \sigma(\mathcal{G}_t, \mathcal{N})$.

The filtration $(\mathcal{F}_t)_{t \geq 0}$ is right continuous and satisfies the usual conditions. The filtration $(\mathcal{F}_t)_{t \geq 0}$ is called the natural filtration of X. The filtration $(\mathcal{H}_t)_{t \geq 0}$ is sometimes called the internal history of X.

In the case $\mathbb{T} = [0, 1]$ or $\mathbb{R}_+$ we suppose that the process S is càdlàg, i.e., for almost every $\omega \in \Omega$ the map $\mathbb{T} \to \mathbb{R}^d$, $t \mapsto S_t(\omega)$ is right continuous and has left limits (where meaningful).

If $X : \mathbb{T} \to \mathbb{R}^d$ is càdlàg we define $\Delta X_t(\omega) = X_t(\omega) - \lim_{s \nearrow t} X_s(\omega) = X_t(\omega) - X_{t-}(\omega)$. The process X is called continuous if almost surely, $\mathbb{T} \to \mathbb{R}^d$, $t \mapsto X_t(\omega)$ is continuous. Although the problems for $\mathbb{T} = \{0, \ldots, n\}$, $\mathbb{N}$, $[0, 1]$, or $\mathbb{R}_+$ are different, there is a possibility to treat many aspects in the same way. This is done through an embedding of $\mathbb{T}$ in $\mathbb{R}_+$. The finite discrete time case $\mathbb{T} = \{0, \ldots, n\}$ is treated in the following way. For $m \in \mathbb{N}$, $m \geq n$ we put $S_m = S_n$ and $\mathcal{F}_m = \mathcal{F}_n$, thus embedding the case $\mathbb{T} = \{1, \ldots, n\}$ into the case $\mathbb{T} = \mathbb{N}$. The case $[0, 1]$ is embedded in $\mathbb{R}_+$ in a similar way $S_u = S_1$ and $\mathcal{F}_u = \mathcal{F}_1$ for $u \geq 1$. To embed $\mathbb{N}$ in $\mathbb{R}_+$ we put for $n \leq t < n+1$, $S_t = S_n$ and $\mathcal{F}_t = \mathcal{F}_n$.

In view of this possibility to embed every time set into $\mathbb{R}_+$ we will only work with $\mathbb{T} = \mathbb{R}_+$.

On $\mathbb{R}_+ \times \Omega$ we consider different σ-algebras. They are the basis to do stochastic analysis. The σ-algebra consisting of Borel sets on $\mathbb{R}_+$ is denoted by $\mathcal{B}(\mathbb{R}_+)$. The σ-algebra $\mathcal{B}(\mathbb{R}_+) \otimes \mathcal{F}_\infty$ denotes the σ-algebra on $\mathbb{R}_+ \times \Omega$ of all measurable subsets. A process $X : \mathbb{R}_+ \times \Omega \to \mathbb{R}$, which is measurable for $\mathcal{B}(\mathbb{R}_+) \otimes \mathcal{F}_\infty$ is simply called measurable. The σ-algebra generated by all stochastic intervals of the form $[\![0, T[\![$ where T is a stopping time, is called the *optional* σ-algebra. It is denoted by $\mathcal{O}$. Under the usual conditions, right continuous adapted processes are measurable with respect to $\mathcal{O}$. Conversely $\mathcal{O}$ is generated by the set of all adapted right continuous real-valued processes (see Dellacherie [D 72]).

The σ-algebra generated by all stochastic intervals of the form $[\![0, T]\!]$ where T is a stopping time, is called the *predictable* σ-algebra. To be precise, when $\mathcal{F}_0$ is not trivial, we also have to include the sets of the form $\{0\} \times A$, where A runs through $\mathcal{F}_0$. The predictable σ-algebra is denoted by $\mathcal{P}$. It is generated by the set of all left continuous adapted real-valued processes. One can even show that $\mathcal{P}$ is already generated by the set of all continuous adapted real-valued processes. This implies in particular that $\mathcal{P} \subset \mathcal{O}$ (see [D 72]).

If $T : \Omega \to \mathbb{R}_+ \cup \{\infty\}$ is a stopping time, $\mathcal{F}_T$ is the σ-algebra of events prior to T i.e. $\mathcal{F}_T = \{A \mid A \in \mathcal{F}_\infty$ and for all $t \in \mathbb{R}_+$ we have $A \cap \{T \le t\} \in \mathcal{F}_t\}$. We also need the σ-algebra of events "strictly" prior to T. This σ-algebra, denoted by $\mathcal{F}_{T-}$, is not defined using a description of its elements. It is defined as the σ-algebra generated by $\mathcal{F}_0$ and by elements of the form $A \cap \{t < T\}$ where $A \in \mathcal{F}_t$. Clearly $\mathcal{F}_{T-} \subset \mathcal{F}_T$. It is easy to see that a stopping time $T : \Omega \to \mathbb{T}$ is $\mathcal{F}_{T-}$-measurable.

Remark 7.2.1. The difference between $\mathcal{F}_{T-}$ and $\mathcal{F}_T$ will turn out to be crucial. Typically the following happens. If T is a stopping time where an optional càdlàg process jumps, then $\mathcal{F}_{T-}$ does not provide any information on the jump size, whereas $\mathcal{F}_T$ does also contain this information. In insurance terms we could say that if T is the stopping time given by the arrival of a letter announcing a new claim, then $\mathcal{F}_{T-}$ gives all the information prior to this arrival, including the fact that a letter has arrived. The σ-algebra $\mathcal{F}_T$ also contains the information on the claim size.

Definition 7.2.2. *A stopping time T is called predictable if there is an increasing sequence of stopping times $(T_n)_{n=1}^\infty$ such that $T_n \nearrow T$ almost surely and $T_n < T$ on $\{0 < T\}$.*

Under the usual conditions T is predictable if and only if the set $[\![T]\!] = [\![T, T]\!] = \{(T(\omega), \omega) \mid T(\omega) < \infty\}$ is in the predictable σ-algebra $\mathcal{P}$. One can show that $\mathcal{P}$ is generated by the stochastic intervals $[\![0, T[\![$ where T is a predictable stopping time.

Definition 7.2.3. *The stopping time T is called totally inaccessible if for each predictable stopping time τ, we have $\mathbf{P}[\tau = T < \infty] = 0$.*

The following description is proved in the theory of stochastic processes (see [D 72]). Recall that $\pi : \Omega \times \mathbb{R}_+ \to \Omega$ denotes the canonical projection.

Proposition 7.2.4. *Let T be a stopping time.*

(i) $\mathcal{F}_T = \{\pi(A \cap [\![T]\!]) \mid A \in \mathcal{O}\}$,
(ii) $\mathcal{F}_{T-} = \{\pi(A \cap [\![T]\!]) \mid A \in \mathcal{P}\}$.

As a consequence, a function $f : \Omega \to \mathbb{R}$, that is $\mathcal{F}_\infty$-measurable is $\mathcal{F}_T$-measurable for a given stopping time T if and only if there is an optional process X, i.e., a process $X : \Omega \times \mathbb{R}_+ \to \mathbb{R}$ which is measurable with respect to $\mathcal{O}$, such that on $\{T < \infty\}$ we have $X_T = f$. The map f is $\mathcal{F}_{T-}$-measurable if and only if there is a predictable process Y such that on $\{T < \infty\}$ we have $Y_T = f$.

For a stopping time T, we define the process X^T as $(X^T)_t = X_{t \wedge T}$. We call X^T the *process X stopped at time T*.

Definition 7.2.5. *If (P) is a property of stochastic processes, then a stochastic process X satisfies (P) locally if there is an increasing sequence of stopping times $(T_n)_{n=1}^\infty$ such that $T_n \nearrow \infty$ almost surely, and for each n the process X^{T_n} satisfies (P).*

Sequences $T_n \nearrow \infty$, such that each X^{T_n} satisfies (P), are called localising sequences. In particular we have the following definition (see [D 72]).

Definition 7.2.6. (i) *A process* $S : \mathbb{R}_+ \times \Omega \to \mathbb{R}^d$ *is locally bounded if there is an increasing sequence of stopping times* $(T_n)_{n=1}^{\infty}$ *tending to* ∞ *a.s. and a sequence* $(K_n)_{n=1}^{\infty}$ *in* $\mathbb{R}_+$ *such that* $|S\mathbf{1}_{[0,T_n]}| \le K_n$.
(ii) *A process* $X : \mathbb{R}_+ \times \Omega \to \mathbb{R}$ *is a local martingale if there is an increasing sequence of stopping times* $(T_n)_{n=1}^{\infty}$ *tending to* ∞ *a.s. so that, for each* n, *the process* X^{T_n} *is a uniformly integrable martingale.*

One can show that X is a local martingale if and only if there is an increasing sequences of stopping times $T_n \nearrow \infty$ such that each X^{T_n} is a martingale or, equivalently, is a martingale bounded in $H^1(\mathbf{P})$ i.e. $\sup_{0 \le t \le T_n} |X_t| \in L^1(\mathbf{P})$ (compare Proposition 14.2.6 below).

Recall that, for a martingale M, the $H^p(\Omega, \mathcal{F}, \mathbf{P})$-norm is defined by

$$\|M\|_{H^p} = \left(\mathbf{E} \left[\left(\sup_t |M_t| \right)^p \right] \right)^{\frac{1}{p}}, \quad \text{for } 1 \le p < \infty. \tag{7.1}$$

The following proposition is almost obvious, but it has important consequences in mathematical finance.

Proposition 7.2.7. *If* $L : \mathbb{R}_+ \times \Omega \to \mathbb{R}$ *is a local martingale such that* $L \ge -1$, *then* L *is a super-martingale.*

Proof. Let $T_n \nearrow \infty$ be a localising sequence for L and let $U \le V$ be finite stopping times. For each n the process L^{T_n} is a uniformly integrable martingale with respect to the filtration $(\mathcal{F}_t)_{t \ge 0}$. For each $A \in \mathcal{F}_U$ and each $n \le m$ we therefore have $\int_{A \cap \{U \le T_n\}} L_{U \wedge T_n} = \int_{A \cap \{U \le T_n\}} L_{V \wedge T_m}$. Hence $\int_{A \cap \{U \le T_n\}} L_U = \int_{A \cap \{U \le T_n\}} L_{V \wedge T_m}$. If we let $m \to \infty$, observe that $L_{V \cap T_m} \ge -1$ and use Fatou's lemma to obtain that $\int_{A \cap \{U \le T_n\}} L_U \ge \int_{A \cap \{U \le T_n\}} L_V = \int_{A \cap \{U \le T_n\}} \mathbf{E}[L_V \mid \mathcal{F}_U]$. Hence on $\{U \le T_n\}$ we have $L_U \ge \mathbf{E}[L_V \mid \mathcal{F}_U]$. We now let n tend to ∞ to conclude. $\square$

Example 7.2.8. The archetype example of a local martingale which fails to be a martingale is the inverse Bessel (3) process. If we take $X : \mathbb{R}_+ \times \Omega \to \mathbb{R}^3$ to be a three dimensional Brownian motion, starting at $X_0 = (1, 0, 0)$, then with respect to the natural filtration of X, $L = \frac{1}{|X|}$ is a strictly positive local martingale that is not a martingale. The family $(L_t)_{t \ge 0}$ is uniformly integrable, but the family $\{L_T \mid T \text{ finite stopping time }\}$ is *not* uniformly integrable! One can show that the natural filtration of L is the filtration generated by a one-dimensional Brownian motion. Bessel processes are thoroughly studied by Pitman and Yor [PY 82]. See also [DS 95c] for applications of this theory to finance.

Example 7.2.9. If L is a local martingale, it is tempting to use the following sequence of stopping times as a "localising" sequence

$$\tau_n = \inf \{t \mid |L_t| \geq n\}$$

It is rather obvious that τ_n indeed defines a localising sequence in the case of *continuous* local martingales L. But if L fails to be continuous this is not a good choice in general. In fact there is a martingale $(M_n, \mathcal{F}_n)_{n=1}^{\infty}$ (indexed by the time set $\mathbb{T} = \mathbb{N}$ for convenience) such that for $T = \inf\{t \mid M_t \neq 0\}$, the stopped process M^T is not uniformly integrable. To construct such an example we start with a sequence of Bernoulli variables, this is a sequence of independent and identically distributed (i.i.d.) variables r_n such that $\mathbf{P}[r_n = 1] = \mathbf{P}[r_n = -1] = \frac{1}{2}$. We also need a variable X defined on Ω, independent of the sequence r_n and such that also for X we have $\mathbf{P}[X = 1] = \mathbf{P}[X = -1] = \frac{1}{2}$. We define the random time T as $T = \inf\{n \mid r_n = +1\}$. Hence we have $\mathbf{P}[T = n] = 2^{-n}$.

We now define a process $(M_n)_{n=0}^{\infty}$, indeed by $\mathbb{N}$. For $n < T$, we put $M_n = 0$ and at time T we put $M_T = X2^n$. After time T the process does not move anymore, meaning $M_n = M_T$ for $n \geq T$. The filtration $(\mathcal{F}_n)_{n=0}^{\infty}$ is defined as $\mathcal{F}_n = \sigma(M_1 \ldots M_n)$ so that T is a stopping time for $(\mathcal{F}_n)_{n=0}^{\infty}$. Clearly the process M is a martingale, hence a local martingale. But $T = \inf\{n \mid M_n \neq 0\}$ and we have that M_T is not integrable! Therefore the stopped process M^T cannot be a uniformly integrable martingale.

For positive local martingales one can do better as Remark 7.2.11 below shows. We first need a preparatory result.

Proposition 7.2.10. *If $L = (L_t)_{t \geq 0}$ is a local martingale such that*

$$\sup \{\mathbf{E}[|L_T|] \mid T \text{ finite stopping time } \} < \infty \tag{7.2}$$

then

$$T_n = \inf \{t \mid |L_t| \geq n\} \tag{7.3}$$

is a localising sequence. More precisely

(i) $\mathbf{P}[T_n = \infty] \nearrow 1$, *i.e., T_n increases in a stationary way to ∞.*
(ii) $L^{T_n} \in H^1$, *i.e.* $\mathbf{E}\left[L_{T_n}^*\right] = \mathbf{E}\left[\sup_{0 \leq t \leq T_n} |L_t|\right] < \infty$, *for each $n \in \mathbb{N}$.*

Proof. (i): Denote by K the sup appearing in (7.2). By Fatou's lemma we have $\mathbf{E}[|L_{T_n}|\mathbf{1}_{\{T_n < \infty\}}] \leq K$. Hence $\mathbf{P}[T_n < \infty] \leq \frac{K}{n}$ which gives (i).

(ii): Clearly $|L_{T_n}^*| \leq \max(n, |L_{T_n}|) \in L^1$ and hence $L^{T_n} \in H^1$. $\square$

Remark 7.2.11. (useful but often forgotten!) An $\mathbb{R}$-valued local martingale which is uniformly bounded from below certainly satisfies the hypothesis of Proposition 7.2.10. Indeed, for a stopping time τ we have by the super-martingale property of L (Proposition 7.2.7) that $\mathbf{E}[L_\tau] \leq \mathbf{E}[L_0]$.

The seemingly unimportant fact that $\mathbf{P}[T_n = \infty] \nearrow 1$ for the sequence of stopping times $(T_n)_{n=1}^{\infty}$ defined by (7.3), will be used when we deal with boundedness properties in the space $L^0(\Omega, \mathcal{F}, \mathbf{P})$.

7.3 Strategies, Semi-martingales and Stochastic Integration

The simplest strategy an agent can follow, is to buy at a deterministic time $T_1 \in \mathbb{R}$ and sell at a later time $T_2 \geq T_1$, $T_2 \in \mathbb{R}$. This situation was already encountered in Chap. 2 and further developed in Chaps. 5 and 6. So let us discuss this elementary strategy of buying and selling. To make decisions possible the random times T_1 and T_2 can only depend on past information and therefore need to be *stopping times*. Since we can give "limits" to our broker, the decision to buy/sell at time T_1 can depend on information available at time T_1. Therefore the number of assets we buy at time T_1 should be measurable with respect to $\mathcal{F}_{T_1}$. By acting in such a way the agent holds $f : \Omega \to \mathbb{R}^d$ assets from time T_1 to time T_2, where f is an $\mathcal{F}_{T_1}$-measurable $\mathbb{R}^d$-valued function. This action results in a gain (or loss) equal to $(f, S_{T_2} - S_{T_1})$.

Definition 7.3.1. *A predictable process* $H : \mathbb{R}_+ \times \Omega \to \mathbb{R}^d$ *with* $H_0 = 0$ *is said to be*

(i)   *a simple strategy if there are stopping times* $0 \leq T_0 \leq T_1 \cdots \leq T_n < \infty$, *as well as random variables* $f_0, \cdots, f_{n-1}$, *where each* f_k *is* $\mathcal{F}_{T_k}$-*measurable, such that* $H = \sum_{k=0}^{n-1} f_k \mathbf{1}_{]\!]T_k, T_{k+1}]\!]}$,
(ii) *a bounded simple strategy if in addition,* $f_0, \cdots, f_{n-1}$ *are in* L^∞,
(iii) *of bounded support if there is a real number* $t \in \mathbb{R}_+$ *such that* $H = H\mathbf{1}_{[\![0,t]\!]}$.

If H is a simple strategy then the ultimate gain equals

$$(H \cdot S)_\infty := \sum_{k=0}^{n-1} \left(f_k, S_{T_{k+1}} - S_{T_k} \right) \tag{7.4}$$

and at time t the portfolio has a gain equal to

$$(H \cdot S)_t := \sum_{k=0}^{n-1} \left(f_k, S_{T_{k+1} \wedge t} - S_{T_k \wedge t} \right). \tag{7.5}$$

The process $H \cdot S$ is called the stochastic integral of H with respect to S. It has to be seen as a process. The ultimate gain is described by the random variable $(H \cdot S)_\infty = \lim_{t \to \infty} (H \cdot S)_t$ (where the limit trivially exists). Another notation is

$$H \cdot S = \int H_u dS_u \tag{7.6}$$

Summing up, the definition of a stochastic integral for *simple* integrands goes exactly along the lines of the setting of finite discrete time as encountered in Chaps. 2, 5 and 6. There is no limiting procedure involved so far: the integrals (7.4), (7.5) and (7.6) reduce to finite sums.

The crucial step now consists in extending this notion from simple integrands to more general ones by an appropriate limiting procedure. This is quite a delicate task. The difficulties appear already in the case of Brownian motion; we concentrate for the moment on the case $S_t = W_t$, where $(W_t)_{t \geq 0}$ is a standard real-valued Brownian motion.

It was K. Itô's fundamental insight [I 44] that the good idea is *not to proceed in a pathwise way*, i.e., not to consider each $\omega \in \Omega$ separately. Instead one should take a functional-analytic point of view applying a basic isometry of Hilbert spaces. Consider the simple strategies $H = \sum_{k=0}^{n-1} f_k \mathbf{1}_{]\!]T_k, T_{k+1}]\!]}$ which are bounded and of bounded support as elements of $L^2(\Omega \times \mathbb{R}_+, \mathcal{P}, \mathbf{P} \otimes \lambda)$, where $\mathcal{P}$ denotes the predictable σ-algebra on $\Omega \times \mathbb{R}_+$ and $\mathbf{P} \otimes \lambda$ the product measure of $\mathbf{P}$ with Lebesgue measure λ on $\mathbb{R}_+$. This gives rise to the norm

$$\|H\|_{L^2(\mathbf{P} \otimes \lambda)} = \left(\mathbf{E} \left[\int_0^\infty H_s^2 ds \right] \right)^{\frac{1}{2}}. \tag{7.7}$$

The crucial isometry is that this $L^2(\mathbf{P} \otimes \lambda)$-norm of the *integrand* H equals the $L^2(\Omega, \mathcal{F}_\infty, \mathbf{P})$-norm of the *stochastic integral* $(H \cdot S)_\infty$, i.e.,

$$\|H\|_{L^2(\Omega \times \mathbb{R}_+, \mathcal{P}, \mathbf{P} \otimes \lambda)} = \|(H \cdot W)_\infty\|_{L^2(\Omega, \mathcal{F}, \mathbf{P})} = \left(\mathbf{E} \left[(H \cdot W)_\infty^2 \right] \right)^{\frac{1}{2}}, \tag{7.8}$$

where $\mathcal{F}_\infty$ denotes the σ-algebra generated by the Brownian motion $(W_t)_{t \geq 0}$.

In fact, this isometry is essentially a formality: for simple integrands H it is a straightforward consequence of the definition of Brownian motion (see, e.g., the beautiful introductory chapter of [RW 00]).

Having established (7.8) for the set of bounded simple integrands it now is one more formal step to extend this isometry to the closures in the respective Hilbert spaces $L^2(\mathbf{P} \otimes \lambda)$ and $L^2(\mathbf{P})$ respectively. For the former it follows from the definition of the predictable σ-algebra $\mathcal{P}$ in Sect. 7.2 above that this closure equals the entire space $L^2(\Omega \times \mathbb{R}_+, \mathcal{P}, \mathbf{P} \otimes \lambda)$, i.e., the predictable process H such that (7.7) remains finite. For the latter closure of the stochastic integrals $(H \cdot S)_\infty$ in $L^2(\mathbf{P})$, it turns out that this is the hyperplane in $L^2(\Omega, \mathcal{F}_\infty, \mathbf{P})$ formed by the random variables f with $\mathbf{E}[f] = 0$. This amounts to the martingale representation theorem 4.2.1 above.

We now define the process $H \cdot W = ((H \cdot W)_t)_{t \geq 0}$, where we have to be careful as this involves uncountably many $t \in \mathbb{R}_+$. This is done in the following way. For general $H \in L^2(\Omega \times \mathbb{R}_+, \mathcal{P}, \mathbf{P} \otimes \lambda)$ we take a sequence H^n of simple integrands, $H^n \in L^2(\Omega \times \mathbb{R}_+, \mathcal{P}, \mathbf{P} \otimes \lambda)$ so that $\|H - H^n\|_{L^2(\Omega \times \mathbb{R}_+, \mathcal{P}, \mathbf{P} \otimes \lambda)} \leq 4^{-n-1}$. By Doob's maximal integrability [RW 00, Chap. 5, Theorem 70.2] we get, for each $m, n \in \mathbb{N}$:

$$\left\| \sup_{t \geq 0}((H^m - H^n) \cdot W)_t \right\|_{L^2(\Omega, \mathcal{F}, \mathbf{P})} \leq 2 \left\| ((H^m - H^n) \cdot W)_\infty \right\|_{L^2(\Omega, \mathcal{F}, \mathbf{P})}$$

$$= 2 \left\| H^m - H^n \right\|_{L^2(\Omega \times \mathbb{R}_+, \mathcal{P}, \mathbf{P} \otimes \lambda)}.$$

Therefore $\mathbf{P} \left[\sup_{t \geq 0}((H^n - H^{n+1}) \cdot W)_t > 2^{-n} \right] \leq 2^{-n}$. The Borel-Cantelli lemma then implies that almost surely the sequence $(H^n \cdot W)_t(\omega)$ converges uniformly as a sequence of continuous functions on $\mathbb{R}_+$. Let us denote this limit process by $(H \cdot W)$. Of course $\left\| \sup_{t \geq 0} ((H - H^n) \cdot W)_t \right\|_2 \to 0$. The identity (7.8) then shows, by passing to the limit as $n \to \infty$:

$$\|H\mathbf{1}_{\rrbracket 0, t \rrbracket}\|_{L^2(\mathbf{P} \otimes \lambda)} = \|(H \cdot W)_t\|_{L^2(\mathbf{P})}.$$

At the same time $(H \cdot W)$, being the limit of the L^2-martingales $(H^n \cdot W)_{t \geq 0}$, also is a martingale bounded in $L^2(\mathbf{P})$.

We have taken some space to sketch these basic facts on stochastic integration, which can be found in much more detail in many beautiful textbooks (e.g., [P 90], [RY 91], [RW 00]), as we believe that the isometric identity (7.8) is the heart of the matter. Having clarified things for the case of Brownian motion W it is essentially a matter of routine techniques to extend the degree of generality.

To start we still restrict to the case of Brownian motion W but now consider predictable processes H such that H is only locally in $L^2(\mathbf{P} \otimes \lambda)$. This latter requirement is equivalent to the hypothesis $\int_0^t H_u^2 du < \infty$ a.s., for each $t < \infty$. In this case one can argue locally to define the stochastic integral $((H \cdot S)_t)_{t \geq 0}$ which then is a local martingale.

Passing to more general integrators than Brownian motion W consider a real-valued martingale $S = (S_t)_{t \geq 0}$ which we first assume to be L^2-bounded, i.e., $\sup_t \|S_t\|_{L^2(\Omega, \mathcal{F}_t, \mathbf{P})} < \infty$.

We then may define the *quadratic variation measure* $d[S]$ on the predictable σ-algebra $\mathcal{P}$ by

$$d[S]\left(\rrbracket \tau, \sigma \rrbracket\right) := \mathbf{E}\left[|S_\tau - S_\sigma|^2\right] \tag{7.9}$$

for all pairs of finite stopping times $\tau \leq \sigma$ and then extend this measure to $\mathcal{P}$ by sigma-additivity. The measure $d[S]$ is the analogue of the measure $\mathbf{P} \otimes \lambda$ in the case of Brownian motion $S = W$, and we again obtain the isometric identity

$$\|H\|_{L^2(d[S])} = \|(H \cdot S)_\infty\|_{L^2(\mathbf{P})}, \tag{7.10}$$

for each bounded simple integrand H such that the left hand side is finite. In fact, the identity (7.10) now simply is a reformulation of the definition (7.9). As in the case of Brownian motion, identity (7.10) allows to extend the stochastic integral from simple bounded integrands to general predictable processes H with finite $L^2(d[S])$-norm. By localisation this notion can be extended to the case of martingales S which are locally L^2-bounded as well as to integrands H, which are locally in $L^2(d[S])$. For the case of continuous local martingales S this is already the natural degree of generality as a *continuous* local martingale is automatically locally L^2-bounded. Finally we indicate that the theory may also be extended to the case of $\mathbb{R}^d$-valued local martingales by equipping $\mathbb{R}^d$ with its Euclidean norm $|.|$ and using the above Hilbert space techniques.

To develop the natural degree of generality also for processes with jumps we have to extend the theory of stochastic integration with respect to a local martingale S to the case, where S is not necessarily locally L^2-bounded (but only locally L^1-bounded). One has to replace the (easy) L^2-theory by some more refined functional analysis replacing $L^2(\mathbf{P})$ by $H^1(\mathbf{P})$ defined in (7.1). Similarly the (easy) maximal inequality for $L^2(\mathbf{P})$-bounded martingales has to be replaced by the more subtle Burkholder-Davis-Gundy maximal inequality pertaining to the norm of $H^1(\mathbf{P})$. We don't elaborate on these issues here and refer, e.g., to [P 90], [RW 00].

Rather we now extend the theory to the case of (càdlàg, adapted, $\mathbb{R}^d$-valued) processes S which are not necessarily local martingales. In the case when S is locally of bounded variation, i.e.

$$|S|_t = \sup_{0 \leq t_0 < \ldots < t_n \leq t} \sum_{i=1}^{n} |S_{t_i} - S_{t_{i-1}}| < \infty \quad \text{a.s., for each } t < \infty,$$

the integration theory is, in fact, rather simple as we now can indeed argue pathwise by considering each $\omega \in \Omega$ separately. For almost each $\omega \in \Omega$ the càdlàg function $(S_t(\omega))_{0 \leq t < \infty}$, which is of bounded variation on compact subsets of $\mathbb{R}_+$, defines a sigma-finite $\mathbb{R}^d$-valued Borel-measure $dS(\omega)$ on $\mathbb{R}_+$; it is defined on the intervals $]a, b]$, for $0 \leq a < b < \infty$, by

$$dS(\omega)(]a, b]) = S_b(\omega) - S_a(\omega).$$

Hence, for each bounded measurable $\mathbb{R}^d$-valued process H, the stochastic integral

$$(H \cdot S)_t(\omega) := \int_0^t (H_u(\omega), dS_u(\omega)) \tag{7.11}$$

is well-defined, for almost each $\omega \in \Omega$ and each $t \in \mathbb{R}_+$, as a classical Lebesgue-Stieltjes integral on the real positive line $\mathbb{R}_+$. One can still extend the stochastic integral (7.11) to the case, where the process H is not necessarily bounded, but only such that for almost every $\omega \in \Omega$ and each $t > 0$, $(H_u(\omega))_{0 \leq u \leq t}$ is $dS(\omega)$-integrable. This is an L^1-theory as opposed to the L^2-theory encountered in the setting of Brownian motion above.

We have thus briefly recapitulated the achievements of stochastic integration theory which were developed starting from the pioneering work of K. Itô [I 44] until the late seventies, notably by the Japanese school and the Strasbourg school of probability around Paul André Meyer. The notion of stochastic integral was pushed to increasingly more general classes: if the (càdlàg, adapted, $\mathbb{R}^d$-valued) process S can be written as $S = M + A$, where M is a local martingale and A is of locally bounded variation, then there is a good integration theory for S. For every locally bounded, predictable $\mathbb{R}^d$-valued process H the stochastic integral

$$(H \cdot S)_t := (H \cdot M)_t + (H \cdot A)_t, \quad t \in \mathbb{R}_+, \qquad (7.12)$$

is well-defined. One can even pass to not necessarily locally bounded predictable processes, provided the two terms on the right hand side make sense.

At this stage, around 1980, the pushing for greater and greater generality came to an end. Through the work of Bichteler [B 81] and Dellacherie [DM 80] it became clear that the class of *semi-martingales* defined in Definition 7.3.2 below is the largest class of processes for which the integration theory can be generalized from simple integrands to more general ones by continuous extension. The Bichteler-Dellacherie theorem (see, e.g., [P 90]) tells us that the semi-martingales S are precisely those processes which may be decomposed as $S = M + A$, where M is a local martingale and A is of locally bounded variation. We shall briefly recall this theorem.

The space $\mathcal{S}$ of bounded simple strategies is equipped with the topology of uniform convergence, which is given by the norm

$$\|H\|_\infty = \sup \left\{ \|H_t\|_{L^\infty(\Omega, \mathcal{F}_t, \mathbf{P})} \mid t \in \mathbb{R}_+ \right\}.$$

Definition 7.3.2. (i) S *is a strict semi-martingale if the operator*

$$I : \mathcal{S} \to L^0(\Omega, \mathcal{F}_\infty, \mathbf{P}) \quad and \quad I(H) = (H \cdot S)_\infty$$

is continuous for the topologies defined respectively by $\| \cdot \|_\infty$ *and by the convergence in probability,*
(ii) S *is a semi-martingale if it is locally a strict semi-martingale.*

It is an easy exercise to show that S is a semi-martingale if $\|H^n\|_\infty \to 0$ implies $(H^n \cdot S)_t \to 0$ in L^0 for all $t \geq 0$. It is easy to check that, for a process S of the form $S = M + A$ where M is a local martingale and A is a càdlàg process of finite variation i.e. for all t we have $\int_0^t |dA_u| < \infty$ a.s., this continuity property is satisfied. (Here and in the sequel we follow the usual terminology to call a process of bounded variation if it is locally of bounded variation). The Bichteler-Dellacherie theorem asserts that also the converse is true: a semi-martingale S in the sense of Definition 7.3.2 can be decomposed as $S = M + A$ in the above way.

We say that S is a *special* semi-martingale if $S = M + A$ where M is a local martingale, A is of finite variation and *predictable*. In this case the decomposition of S as a sum of a local martingale and a predictable process of finite variation is unique. We refer to it as the canonical decomposition (see [DM 80] and [P 90]). One can show that a semi-martingale S is special if and only if S is locally integrable, i.e. if there is a sequence of stopping times $T_n \nearrow \infty$ such that $\mathbf{E} \left[\sup_{0 \leq t \leq T_n} |S_t| \right] < \infty$.

We emphasize that being a semi-martingale does not depend on $\mathbf{P}$ but only on the null sets. In other words, if S is a semi-martingale under $\mathbf{P}$ and $\mathbf{Q} \sim \mathbf{P}$, then S is also a semi-martingale under $\mathbf{Q}$. However, if S is special for $\mathbf{P}$ and $\mathbf{Q} \sim \mathbf{P}$ is another probability equivalent to $\mathbf{P}$, then S does not need to be special for $\mathbf{Q}$. For more details on how to define stochastic integrals for bounded strategies and for general predictable strategies we refer to the literature (e.g. [DM 80], [P 90], [J 79], [B 81]).

If S is a special $\mathbb{R}^d$-valued semi-martingale and if K is a one-dimensional, predictable and bounded process, then the $\mathbb{R}^d$-valued stochastic integral $K \cdot S$ which is defined coordinatewise is still special. Indeed, suppose that $S = M + A$ where M is a local martingale and A is a predictable process of finite variation. Therefore there is a sequence of stopping times T_n tending to ∞, such that, for each $n \in \mathbb{N}$, M^{T_n} is in $H^1(\mathbf{P})$ and A^{T_n} is of integrable variation and predictable. The Burkholder-Davis-Gundy ([DM 80], [J 79]) inequalities show that $K \cdot M^{T_n}$ is still in H^1 and ordinary integration theory shows that $K \cdot A^{T_n}$ is still of integrable variation. As a result we see that $K \cdot S$ is a special semi-martingale. Moreover the canonical decomposition of $K \cdot S$ is $K \cdot M + K \cdot A$.

On the space of one-dimensional semi-martingales we put a vector space topology, the so-called *semi-martingale topology*, induced by the quasi-norm or distance function [E 79]

$$D[S] = \sum_{n=1}^{\infty} 2^{-n} \sup \left\{ \mathbf{E}\left[|(K \cdot S)_n| \wedge 1\right] \mid |K| \leq 1 \right\}, \tag{7.13}$$

where the processes K are assumed to be real-valued and predictable.

In this topology we have, for a sequence $(S^k)_{k=1}^{\infty}$ of semi-martingales, that $S^k \to 0$ if and only if, for each t, we have that $(K \cdot S^k)_t \xrightarrow{\mathbf{P}} 0$ uniformly in K, $|K| \leq 1$, K real-valued, predictable. An equivalent metric also inducing the semi-martingale topology is

$$D^*[S] = \sum 2^{-n} \sup \left\{ \mathbf{E}[(K \cdot S)_n^* \wedge 1] \mid |K| \leq 1 \right\}. \tag{7.14}$$

As usual Y^* denotes the maximal function defined as $Y_t^* = \sup_{0 \leq u \leq t} |Y_u|$. For càdlàg processes Y, the process Y^* is again càdlàg.

We can also define a stronger distance function, D_∞, inducing the semi-martingale topology on $\mathbb{T} = [0, \infty]$ as opposed to the time index set $\mathbb{T} = \mathbb{R}_+$. This distance is defined as

$$D_\infty[S] = \sup \left\{ \mathbf{E}\left[(K \cdot S)_\infty^* \wedge 1\right] \mid |K| \leq 1 \right\}.$$

For integration theory this topology is typically too strong but in Chap. 9 this notion will turn out to be useful.

We now extend the class of integrands for a given semi-martingale S from *locally bounded* predictable $\mathbb{R}^d$-valued processes H to processes H, which are not necessarily locally bounded. We say that a predictable $\mathbb{R}^d$-valued process H is S-integrable if $(H\mathbf{1}_{\{|H| \leq n\}} \cdot S)_{n=1}^{\infty}$ forms a Cauchy sequence in the space of one-dimensional semi-martingales with respect to the semi-martingale topology induced by (7.13) (or, equivalently, by (7.14)). The limit process is denoted by $H \cdot S$.

The reader should note the subtle, but for our later applications crucial difference to the sentence following (7.12) above: we shall see in Example 7.3.4 below that it may happen that, for a special semi-martingale $S = M + A$, a predictable process H is S-integrable in the sense introduced above, while the stochastic integral $H \cdot M$ does not exist as an integral in the sense of a local martingale. The next theorem clarifies the situation. It will play a crucial role in later sections.

Theorem 7.3.3. *If S is a special semi-martingale with canonical decomposition $S = M + A$ and if H is S-integrable, then $H \cdot S$ is special if and only if*

(i) *the process $H \cdot M$ is defined as an integral of local martingales and*
(ii) *the process $H \cdot A$ is defined as a Lebesgue-Stieltjes integral $\int H_u dA_u$.*

In this case the canonical decomposition of $H \cdot S$ is given by $H \cdot S = H \cdot M + H \cdot A$.

Before giving the proof we present an enlightening example due to M. Émery. It will be of central importance for Chap. 14 below.

Example 7.3.4 ([E 80], see also Example 14.2.2 below). Let $(\Omega, \mathcal{F}, \mathbf{P})$ be a probability space on which the following objects are defined: an exponentially distributed random variable T with parameter 2 (the 2 is only to keep in line with the notation in Example 14.2.2), i.e., $\mathbf{P}[T > \alpha] = e^{-2\alpha}$, and a Bernoulli variable B, i.e., $\mathbf{P}[B = 1] = \mathbf{P}[B = -1] = \frac{1}{2}$, which is independent of T.

The process $M = (M_t)_{t \geq 0}$ is defined as follows

$$M_t = \begin{cases} 0, & \text{for } t < T, \\ B, & \text{for } t \geq T. \end{cases}$$

Denoting by $(\mathcal{F}_t)_{t \geq 0}$ the natural filtration generated by the process M, it is straightforward to check (and intuitively rather obvious) that M is a martingale with respect to $(\mathcal{F}_t)_{t \geq 0}$. The process jumps at time T where it has a $50 : 50$ chance to either jump up to 1 or down to -1.

Define the process H by $H_u = \frac{1}{u}$, for $u > 0$. This (deterministic) process is M-integrable: indeed, the processes $\left(H \, \mathbf{1}_{\{|H| \leq n\}} \cdot S \right)_{n=1}^{\infty}$ converge in the semi-martingale topology to the process $X = H \cdot M$ which is given by

$$X_t = \begin{cases} 0, & \text{for } t < T, \\ \frac{B}{T}, & \text{for } t \geq T. \end{cases} \tag{7.15}$$

Morally speaking, one is tempted to believe that X should still be a martingale: the process X has the same chance of $50 : 50$ to jump upwards or downwards by $\frac{1}{T}$ at time T. But, mathematically speaking, X fails to be a martingale as we encounter integrability problems: for $t > 0$ we have

$$\mathbf{E}[|X_t|] = \int_0^t \left|\frac{B}{u}\right| d\mathbf{P}[T = u] = \int_0^t \frac{1}{u} 2e^{-2u} du = \infty.$$

Hence X is not a martingale as $\mathbf{E}[X_t]$ does not make sense. In fact, also stopping does not help to remedy the integrability problem! It is not hard to check that, for every stopping time τ w.r. to the filtration $(\mathcal{F}_t)_{t\geq 0}$ such that $\mathbf{P}[\tau > 0] > 0$, we still have (see [E 80] for the details)

$$\mathbf{E}[|X_\tau|] = \infty.$$

Hence X even fails to be a local martingale. In particular, $H \cdot M$ is not defined as a stochastic integral in the sense of integration with respect to a local martingale as developed above, as in this theory the integral of a local martingale is necessarily a local martingale.

In Sect. 8.3 below we shall define the notion of a sigma-martingale which will yield the proper framework enabling us to also interpret processes such as $X = H \cdot M$ above still as a "fair game".

This example of Émery is very simple, but it shows that one has to be careful when dealing with stochastic integrals. For a martingale M the integral $H \cdot M$ might exist as a semi-martingale but not as a local martingale! We conclude that the local martingales do not form a closed subspace (w.r. to the semi-martingale topology) of the space of semi-martingales and by the same example we see that the space of special semi-martingales is not closed in the space of semi-martingales. See Émery and Mémin for details [E 79], [M 80].

The proof of Theorem 7.3.3 will be based on two results, the first stating that under the assumptions of Theorem 7.3.3 the stochastic integral $H \cdot A$ necessarily exists as a Lebesgue-Stieltjes integral. The second is a necessary and sufficient condition for a stochastic integral of a local martingale to be a local martingale.

Lemma 7.3.5. *If S is a special $\mathbb{R}^d$-valued semi-martingale with canonical decomposition $S = M + A$, if H is an $\mathbb{R}^d$-valued predictable process, if the stochastic integral $H \cdot S$ is special, then the process $H \cdot A$ exists as a Lebesgue-Stieltjes integral.*

Proof. We start by localising which allows us to assume that the special semi-martingales S and $H \cdot S$ are of the form $S = M + A$ and $(H \cdot S) = N + B$ such that M and N are martingales bounded in $H^1(\mathbf{P})$ and A and B are predictable processes of integrable variation. We also may represent the $\mathbb{R}^d$-valued process A as $dA = \beta d|A|$, where $(|A|_t)_{t\geq 0}$ is the total variation process of $(A_t)_{t\geq 0}$ and β is an $\mathbb{R}^d$-valued predictable process taking values in the unit sphere of $\mathbb{R}^d$.

We now define the $\{-1, 0, 1\}$-valued predictable process

$$K_t = \text{sign}(H_t, \beta_t).$$

Clearly the Lebesgue-Stieltjes integrability of H with respect to A is tantamount to that of KH. The latter is easier to check as $KH \cdot A$ is an increasing process. Note that $KH \cdot S = K \cdot N + K \cdot B$ is still the sum of an $H^1(\mathbf{P})$-bounded martingale and a process of integrable variation so that, for each $t < \infty$,

$$\mathbf{E}\left[\sup_{0 \leq u \leq t} (KH \cdot S)_u\right] < \infty.$$

Fix $t < \infty$. Let $H^n = H\mathbf{1}_{\{|H| \leq n\}}$, for $n \in \mathbb{N}$, and define the stopping times τ_n by

$$\tau_n = \inf \left\{u \leq t \mid |(KH \cdot S)_u - (KH^n \cdot S)_u| > 1\right\} \wedge t.$$

As KH is S-integrable we have that $(\tau_n)_{n=1}^{\infty}$ increases stationary to t. We may estimate

$$|(KH^n \cdot S)_{\tau_n}| \leq |(KH \cdot S)_{\tau_n}| + 1 + |H_{\tau_n} \Delta S_{\tau_n}|$$
$$\leq 3 \sup_{0 \leq u \leq t} |((KH) \cdot S)_u| + 1.$$

As KH^n is bounded and M is bounded in $H^1(\mathbf{P})$, the process $(KH^n) \cdot M$ is an $H^1(\mathbf{P})$-bounded martingale too (starting at $((KH^n) \cdot M)_0 = 0$) so that

$$\mathbf{E}\left[((KH^n) \cdot A)_{\tau_n}\right] = \mathbf{E}\left[((KH^n) \cdot S)_{\tau_n}\right] - \mathbf{E}\left[((KH^n) \cdot M)_{\tau_n}\right]$$
$$\leq 3\mathbf{E}\left[\sup_{0 \leq u \leq t} ((KH) \cdot S)_u\right] + 1 < \infty.$$

Letting $n \to \infty$ we obtain by the monotone convergence theorem that

$$\mathbf{E}\left[((KH) \cdot A)_t\right] < \infty,$$

which proves the lemma. □

Example 7.3.6. Let S be a special semi-martingale with canonical decomposition $S = M + A$. For an S-integrable predictable process H, the integral $H \cdot S$ may exist, but $H \cdot A$ may not. Such an example can be made up as follows. Take a random variable T which is exponentially distributed with parameter 1 and let $S_t = \mathbf{1}_{\{t \geq T\}}$ with its natural filtration. Clearly the canonical decomposition of S is given by $S = M + A$ where

$$A_u = \int_0^{u \wedge T} ds = (u \wedge T)$$

is the compensator of the process S.

Let H be defined by $H_t = \frac{1}{1-t}$, for $0 \leq t < 1$, and $H_t = 0$ for $t \geq 1$. We find that

$$(H \cdot S)_t = \begin{cases} 0, & \text{for } t < T, \\ \frac{1}{1-T}, & \text{for } T \leq t \text{ and } T < 1, \\ 0, & \text{otherwise.} \end{cases}$$

On the other hand we have on $\{T > 1\}$ and for $t \geq 1$

$$(H \cdot A)_t = \int_0^{t \wedge T} \frac{du}{1 - u} = +\infty.$$

The next theorem gives a necessary and sufficient condition for a stochastic integral of a local martingale to be again a local martingale. The result in this form is due to Ansel and Stricker [AS 94]. An earlier form was given by Émery [E 79].

Theorem 7.3.7 (Ansel and Stricker). *Let M be an $\mathbb{R}^d$-valued local martingale, let H be $\mathbb{R}^d$-valued and predictable. Let H be M-integrable in the sense of semi-martingales. Then $H \cdot M$ is a local martingale if and only if there is an increasing sequence of stopping times $T_n \nearrow \infty$ as well as a sequence of integrable functions $\vartheta_n \leq 0$, such that $(H, \Delta M)^{T_n} \geq \vartheta_n$.*

Some explanation on the notation seems in order: the process $\Delta M = (\Delta M_t)_{t \geq 0}$ is the process formed by the jumps of M; it is different from zero only at the jumps of the càdlàg process M where the formula $(\Delta M)_t(\omega) = M_t(\omega) - M_{t-}(\omega)$ holds. Hence the process $(H, \Delta M) = (H_t, (\Delta M)_t)_{t \geq 0}$ is the process of jumps of $H \cdot M$. The condition $(H, \Delta M)^{T_n} \geq \vartheta_n$ means that, for a.e. $\omega \in \Omega$, the jumps of the process $((H \cdot M)_t)_{t \geq 0}$ such that $0 \leq t \leq T_n(\omega)$ all are bounded from below by $\vartheta_n(\omega)$ almost surely.

Proof of Theorem 7.3.7. We first prove necessity. If $H \cdot M$ is a local martingale then it is locally in H^1. Hence there is an increasing sequence of stopping times $T_n \nearrow \infty$ such that $\sup_{t \leq T_n} |(H \cdot M)_t| \in L^1$. Hence $|(H, \Delta M)^{T_n}| \leq 2 \sup_{t \leq T_n} |(H \cdot M)_t| \in L^1$.

The sufficiency of the hypothesis is less trivial. We will show that, for each $n \in \mathbb{N}$, the process $(H \cdot M)^{T_n}$ is a local martingale and we may therefore drop the stopping times T_n and replace ϑ_n by ϑ. Let

$$U_t = \sum_{s \leq t} \mathbf{1}_{\{|\Delta M_s| \geq 1 \text{ or } (H_s, \Delta M_s) \geq 1\}} \Delta M_s$$

Because M and $H \cdot M$ are semi-martingales, their jumps of high magnitude (here ≥ 1) form a discrete set, i.e., such that its intersection with each compact subset of $[0, \infty[$ is finite for a.e. $\omega \in \Omega$. The process U is therefore a càdlàg, adapted process of finite variation and hence a semi-martingale. Also H is U-integrable. Indeed, every ($\mathbb{R}^d$-valued, predictable) process is U-integrable. Let now $Y = M - U$. Since H is M- and U-integrable it is also Y-integrable. The semi-martingale Y has jumps of magnitude ≤ 1 and hence is a special semi-martingale. Its canonical decomposition is denoted as $Y = N + B$. Let $V = B + U = M - N$, the process V is the difference of two local martingales and is therefore a local martingale, moreover it has paths of locally bounded variation as it is the sum of the process B and U. Because $H \cdot Y = H \cdot M - H \cdot U$ has bounded jumps it is also special and by Lemma 7.3.5 $H \cdot B$ exists as an ordinary integral. Therefore $H \cdot V$ exists as an ordinary Lebesgue-Stieltjes integral too. We have to show two things

(i) $H \cdot V$ is a local martingale and

(ii) $H \cdot N$ is a local martingale.

The second one is standard. The jumps of Y and $H \cdot Y$ are bounded by 1 and hence for a predictable stopping time T we have $\Delta B_T = \mathbf{E}[\Delta Y_T \mid \mathcal{F}_{T-}]$ is bounded by 1. Also $|H_T \Delta Y_T| \leq 1$ and hence $\mathbf{E}[H_T \Delta Y_T \mid \mathcal{F}_{T-}] = H_T \Delta B_T$ is bounded by 1. For T totally inaccessible we have $\Delta B_T = 0$ a.s. as B is predictable and hence $H_T \Delta N_T = H_T \Delta Y_T$ as well as $\Delta N_T = \Delta Y_T$ are bounded by 1. It follows that $|H \Delta N|$ and $|\Delta N|$ are bounded by 2. The local martingale N is locally L^2 and the increasing process $\int_0^t H_u' d[N, N]_u H_u$ is also locally L^2. The L^2 theory of martingales shows that $(H \cdot N)$ is a local martingale (even locally L^2).

The first part is more tricky. For each p we define $R_p = \inf\{t \mid \int_0^t |H_u dB_u| \geq p\}$. This makes sense since $H \cdot B$ exists as an ordinary integral. The sequence $(R_p)_{p=1}^\infty$ increases to infinity. As $H \cdot B$ has jumps bounded by 1 we have $\int_0^{R_p} |H_u dB_u| \leq p + 1$.

We also define $S_p = \inf\{t \mid \int_0^t |H_u dU_u| \geq p\}$. Because $H \cdot U$ exists as an ordinary integral this makes sense again and also $(S_p)_{p=1}^\infty$ increases to infinity. For each n we now put $H^n = H 1_{\{|H| \leq n\}}$. Clearly $\int_0^{R_p} |H_u^n dB_u| \leq p + 1$ and because of the hypothesis $(H, \Delta M) \geq \vartheta$ we have $((H^n \cdot U)^{S_p})^- \leq p + |\vartheta|$. We now take one more sequence $(\tau_p)_{p=1}^\infty$ of stopping times increasing to infinity so that $V^{\tau_p} \in H^1(\mathbf{P})$. Because H^n is bounded, $(H^n \cdot V)^{\tau_p}$ is an $H^1(\mathbf{P})$-martingale and hence $\mathbf{E}\left[(H^n \cdot V)_{\tau_p \wedge S_p \wedge R_p}\right] = 0$ for all n and p. However, for the negative part we find $(H^n \cdot V)_{\tau_p \wedge S_p \wedge R_p}^- \leq (H^n \cdot B)_{\tau_p \wedge S_p \wedge R_p}^- + (H^n \cdot U)_{\tau_p \wedge S_p \wedge R_p}^- \leq p + 1 + p + |\vartheta| = 2p + 1 + |\vartheta|$. Therefore $\mathbf{E}\left[(H^n \cdot V)_{\tau_p \wedge S_p \wedge R_p}^-\right] \leq 2p + 1 + \mathbf{E}[|\vartheta|]$ and the same holds for $(H^n \cdot V)_{\tau_p \wedge S_p \wedge R_p}^+$. This in turn implies $\mathbf{E}\left[|(H^n \cdot V)_{\tau_p \wedge S_p \wedge R_p}|\right] \leq 4p + 2 + 2\mathbf{E}[|\vartheta|]$. If $n \to \infty$, we have $((H^n - H) \cdot V)^* = \sup_t |((H^n - H) \cdot V)_t|$. This implies that $(H^n \cdot V)_{\tau_p \wedge S_p \wedge R_p} \overset{\mathbf{P}}{\to} 0$ and an application of Fatou's lemma yields that also $\mathbf{E}\left[|(H \cdot V)_{\tau_p \wedge S_p \wedge R_p}|\right] \leq 4p + 2 + 2\mathbf{E}[|\vartheta|]$. Now the definition of S_p and R_p imply that $|(H \cdot V)_t| \leq 2p$ for $t < \tau_p \wedge S_p \wedge R_p$. We finally deduce that $\mathbf{E}\left[\sup_{0 \leq t \leq \tau_p \wedge S_p \wedge R_p} |(H \cdot V)_t|\right] \leq 2p + 2(2(p+1) + 2\mathbf{E}[|\vartheta|]) < \infty$.

The proof is almost complete now. The process $W = V^{\tau_p \wedge S_p \wedge R_p}$ is an H^1-martingale, the integral $(H \cdot W)$ has a maximal function $f = (H \cdot W)^*$ that is integrable. This is sufficient to prove that $H \cdot W$ is a martingale and hence an H^1-martingale. Again we use approximations. For each n let ν_n be defined as $\nu_n = \inf\{t \mid |(H^n \cdot W)_t - (H \cdot W)_t| > 1\}$. It is clear that $\nu_n \to \infty$. Each $(H^n \cdot W)^{\nu_n}$ is a martingale and its maximal function is bounded by $(H^n \cdot W)_{\nu_n}^* \leq (H \cdot W)_{\nu_n}^* + 1 + 2f$. The last term coming from a possible jump at ν_n. It follows that for all n we have the inequality $(H^n \cdot W)_{\nu_n}^* \leq 3f + 1$. A simple application of Lebesgue's dominated convergence theorem allows us to conclude that $H \cdot W$ is a martingale. The proof is now complete. $\square$

Corollary 7.3.8. *If M is a local martingale, if H is M-integrable and if $(H \cdot M)^-$ is locally integrable, i.e., there is a sequence $(\tau_n)_{n=1}^{\infty}$ of stopping times increasing to infinity such that $\mathbf{E}[-\inf_{0 \leq t \leq T_n}(H \cdot M)_t] < \infty$; then $H \cdot M$ is a local martingale.*

Proof. We only have to verify the hypothesis of theorem 7.3.7. For each n let R_n be defined as $R_n = \inf\{t \mid (H \cdot M)_t \geq n\}$. Let also τ_n be chosen in such a way that $H \cdot M^{\tau_n} \geq \vartheta_n$, where each ϑ_n is integrable and $\tau_n \nearrow \infty$. Let $T_n = \min(R_n, \tau_n)$. Clearly $T_n \nearrow \infty$ and the jumps $H\Delta M^{T_n} \geq -n + \vartheta_n$. Theorem 7.3.7 gives the desired result. $\square$

Proof of Theorem 7.3.3. Let H be S-integrable. If $H \cdot S$ is special then by Lemma 7.3.5 the process $H \cdot A$ exists as an ordinary Lebesgue-Stieltjes integral which yields (ii). As $H \cdot S$ is special it is locally integrable, i.e., there is a sequence of integrable functions $\vartheta_n \in L^1$ and an increasing sequence of stopping times $T_n \nearrow \infty$ such that $(H \cdot S)^{T_n} \geq \vartheta_n$. Take $R^n \nearrow \infty$ such that $(H \cdot A)^{R_n}$ has integrable variation. This is possible since the integral $H \cdot A$ is an ordinary integral and the result therefore is a predictable process. For each n we find $(H \cdot M)^{R_n \wedge T_n} \geq \vartheta_n - \int_0^{R_n} |H_u dA_u|$. We now apply the Ansel-Stricker theorem to show that (i) holds true.

Conversely, if (i) and (ii) hold true then $H \cdot S$ is the sum of the local martingale $H \cdot M$ and the predictable finite variation process $H \cdot A$ and therefore special. $\square$

8

Arbitrage Theory in Continuous Time: an Overview

8.1 Notation and Preliminaries

After all this preliminary work we are finally in a position to tackle the theme of no-arbitrage in full generality, i.e., for general models S of financial markets in continuous time, and for general (i.e., not necessarily simple) trading strategies H. The choice of the proper class of trading strategies will turn out to be rather subtle. In fact, for different applications (e.g., portfolio optimisation with respect to exponential utility to give a concrete example; see [DGRSSS 02] and [S 03a]) it will sometimes be necessary to consider different classes of appropriate trading strategies. But for the present purpose the concept of *admissible* strategies developed below will serve very well.

When defining an appropriate class of trading strategies, then, first of all, one has to restrict the choice of the integrands H to make sure that the process $H \cdot S$ exists. Besides the qualitative restrictions coming from the theory of stochastic integration considered in the previous chapter, one has to avoid problems coming from so-called doubling strategies. This was already noted in the paper by Harrison and Pliska [HP 81]. To explain this remark, let us consider the classical doubling strategy. We take the framework of a fair coin tossing game. We toss a coin, and when heads comes up, the player is paid 2 times his bet. If tails comes up, the player loses his bet.

The so-called "doubling strategy" is known for centuries and in French it is still referred to as "la martingale" (compare, e.g., [B 14, p. 77][1]). The player

---

[1] We cannot resist citing from Bachelier's book where he discusses the "suicide strategy" (see after Theorem 8.2.1 below), which is a close relative to the doubling strategy.

"La martingale est la cause unique des grosses fortunes, ... Pour devenir très riche, il faut être favorisé par des concours de circonstances extraordinaires et par des hasards constamment heureux. Jamais un homme n' est devenu très riche par sa valeur."

The martingale is the unique cause for big fortunes, ... To become very rich, you have to be favoured by extraordinary circumstances and by constantly lucky bets. Never a man became very rich by his value.

doubles his bet until the first time he wins. If he starts with $1 \, €$, his final gain (last payout minus the total sum of the preceding losses) is $1 \, €$ almost surely. He has an almost sure win. The probability that heads will eventually show up is indeed one (even if the coin is not fair). However, his accumulated losses are not bounded below. Everybody, especially a casino boss, knows that this is a very risky way of winning $1 \, €$. This type of strategy has to be ruled out: there should be a lower bound on the player's loss.

Here is the definition of the class of integrands which turns out to be appropriate for our purposes.

Definition 8.1.1. *Fix an $\mathbb{R}^d$-valued stochastic process $S = (S_t)_{t \geq 0}$ as defined in Chap. 5, which we now also assume to be a semi-martingale. An $\mathbb{R}^d$-valued predictable process $H = (H_t)_{t \geq 0}$ is called an admissible integrand for the semi-martingale S, if*

(i) *H is S-integrable, i.e., the stochastic integral $H \cdot S = ((H \cdot S)_t)_{t \geq 0}$ is well-defined in the sense of stochastic integration theory for semi-martingales,*
(ii) *there is a constant M such that*

$$(H \cdot S)_t \geq -M, \quad a.s., \text{ for all } t \geq 0.$$

Let us comment on this definition: we place ourselves into the "théorie générale" of integration with respect to semi-martingales: here we are on safe grounds as the theory, developed in particular by P.-A. Meyer and his school, tells us precisely what it means that a predictable process H is S-integrable (see Chap. 7 above). But in order to be able to apply this theory we have to make sure that S is a semi-martingale: this is precisely the class of processes allowing for a satisfactory integration theory, as we know from the theorem of Bichteler and Dellacherie ([B 81], [DM 80]; see also [P 90]).

How natural is the assumption that S is a semi-martingale from an economic point of view? In fact, it fits very nicely into the present no-arbitrage framework: it is shown in Theorem 9.7.2 below that, for a locally bounded, càdlàg process S, the assumption, that the closure of C^{simple} *with respect to the norm topology of $L^\infty(\mathbf{P})$ intersects $L^\infty(\mathbf{P})_+$ only in $\{0\}$*, implies already that S is a semi-martingale. The semi-martingale property therefore is implied by a very mild strengthening of the no-arbitrage condition for simple, admissible integrands. Loosely speaking, the message of this theorem is that a no-arbitrage theory for a stochastic process S modelling a financial market, only makes sense if we start with the assumption that S is a semi-martingale. For example, this rules out fractional Brownian motion (except for Brownian motion itself, of course). There is no reasonable no-arbitrage theory for these processes in the present setting of frictionless trading in continuous time. However, if one introduces transaction costs, then for fractional Brownian motion the picture changes completely and the arbitrage opportunities disappear (see [G 05]).

Regarding condition (ii) in the above definition: this is a strong and economically convincing requirement to rule out the above discussed doubling strategy, as well as similar schemes, which try to make a final gain at the cost of possibly going very deep into the red. Condition (ii) goes back to the work of Harrison and Pliska [HP 81]: the interpretation is that there is a finite credit line M obliging the investor to finance her trading in such a way that this credit line is respected at all times $t \geq 0$.

Definition 8.1.2. *Let S be an $\mathbb{R}^d$-valued semi-martingale and let*

$$K = \left\{ (H \cdot S)_\infty \ \middle| \ H \text{ admissible and } (H \cdot S)_\infty = \lim_{t \to \infty} (H \cdot S)_t \text{ exists a.s.} \right\},$$
(8.1)

which forms a convex cone of functions in $L^0(\Omega, \mathcal{F}, \mathbf{P})$, and

$$C = \{ g \in L^\infty(\mathbf{P}) \mid g \leq f \text{ for some } f \in K \}.$$
(8.2)

We say that S satisfies the condition of no free lunch with vanishing risk *(NFLVR), if*

$$\overline{C} \cap L^\infty_+(\mathbf{P}) = \{0\},$$

where $\overline{C}$ now denotes the closure of C with respect to the norm topology of $L^\infty(\mathbf{P})$.

Comparing the present definition with the notion of "no free lunch" *(NFL)*, the weak-star topology has been replaced by the topology of uniform convergence. Taking up again the discussion following the Kreps-Yan Theorem 5.2.2, we now find a better economic interpretation: S allows for a *free lunch with vanishing risk*, if there is $f \in L^\infty_+(\mathbf{P}) \setminus \{0\}$ and sequences $(f_n)_{n=0}^\infty = ((H^n \cdot S)_\infty)_{n=0}^\infty \in K$, where $(H^n)_{n=0}^\infty$ is a sequence of admissible integrands and $(g_n)_{n=0}^\infty$ satisfying $g_n \leq f_n$, such that

$$\lim_{n \to \infty} \|f - g_n\|_\infty = 0.$$

In particular the negative parts $((f_n)_-)_{n=0}^\infty$ and $((g_n)_-)_{n=0}^\infty$ tend to zero *uniformly*, which explains the term *"vanishing risk"*.

8.2 The Crucial Lemma

We now come back to the formulation of a *general version of the fundamental theorem of asset pricing*. We first restrict to the case of *locally bounded* processes S, as treated in Chap. 9. This technical assumption makes life much easier.

Theorem 8.2.1. *(Corollary 9.1.2) The following assertions are equivalent for an $\mathbb{R}^d$-valued locally bounded semi-martingale model $S = (S_t)_{t \geq 0}$ of a financial market:*

(i) *(EMM)*, *i.e., there is a probability measure* **Q**, *equivalent to* **P**, *such that S is a local martingale under* **Q**.

(ii) *(NFLVR)*, *i.e., S satisfies the condition of no free lunch with vanishing risk.*

The present theorem is a sharpening of the Kreps-Yan Theorem 5.2.2, as it replaces the weak-star convergence in the definition of "no free lunch" by the economically more convincing notion of uniform convergence. The price to be paid for this improvement is, that now we have to place ourselves into the context of *general admissible*, instead of *simple admissible* integrands.

The proof of Theorem 8.2.1 as given in Chap. 9 and its extension in Chap. 14 is surprisingly long and technical; despite of several attempts, no essential simplification of this proof has been achieved so far. We are not able to go in detail through this proof in this "guided tour", but we shall try to motivate and help the interested reader to find her way through the arguments in Chap. 9 below.

We start by observing that the implication (i) $\Rightarrow$ (ii) is still the easy one: supposing that S is a local martingale under **Q** and H is an admissible trading strategy, we may deduce from the Ansel-Stricker Theorem 7.3.7 and the fact that $H \cdot S$ is bounded from below, that $H \cdot S$ is a local martingale under **Q**, too. Using again the boundedness from below of $H \cdot S$, we also conclude that $H \cdot S$ is a super-martingale under **Q**, so that

$$\mathbf{E_Q}[(H \cdot S)_\infty] \leq 0. \tag{8.3}$$

Hence $\mathbf{E_Q}[g] \leq 0$, for all $g \in C$, and this equality extends to the norm closure $\overline{C}$ of C (in fact, it also extends to the weak-star-closure of C, but we don't need this stronger result here). Summing up, we have proved that *(EMM)* implies *(NFLVR)*.

Before passing to the reverse implication let us still have a closer look at the crucial inequality (8.3): its message is that the notion of equivalent local martingale measures **Q** and admissible integrands H has been designed in such a way, that the basic intuition behind the notion of a martingale holds true: *you cannot win in average by betting on a martingale*. Note, however, that the notion of admissible integrands does not rule out the possibility *to lose in average by betting on S*. An example, already noted in [HP 81], is the so-called "suicide strategy H" which is just the doubling strategy considered at the beginning of this chapter with opposite signs. Consider, similarly as above, a simplified roulette, where red and black both have probability $\frac{1}{2}$ and, as usual, when winning, your bet is doubled. The strategy consists in placing one € on red and then walking to the bar of the casino and regarding the roulette from a distance: if it happens that consecutively only red turns up in the next couple of games, you may watch a huddle of chips piling up with exponential growth (assuming, of course, that there is no limit to the size of the bets). But inevitably, i.e., with probability one, black will eventually turn up, which will cause the huddle — including your original $\in$ — to disappear. Translating this story into the language of stochastic integration, we have a martingale S (in fact, a random walk) and an admissible trading strategy H such that $(H \cdot S)_0 = 0$ while $(H \cdot S)_\infty = -1$ so that we have a strict inequality in (8.3).

A continuous analogue of the suicide strategy is given by the process $S_t = \exp\left(W_t - \frac{1}{2}t\right)$, where W is a standard Brownian motion. This process starts at 1 and moves up and down in $]0, \infty[$ according to a fair game (it is a martingale). But, as t tends to infinity, S_t tends to 0 almost surely. The reader can see that the process S can assume quite high values but eventually the player loses the initial bet $S_0 = 1$.

We now discuss the difficult implication *(NFLVR)* $\Rightarrow$ *(EMM)* of Theorem 8.2.1. It is reduced to the subsequent theorem which may be viewed as the "abstract" version of Theorem 8.2.1:

Theorem 8.2.2 (Theorem 9.4.2). *In the setting of Theorem 8.2.1 assume that (ii) holds true, i.e., that S satisfies (NFLVR).*
Then the cone $C \subseteq L^\infty(\mathbf{P})$ is weak-star-closed.

The fact that Theorem 8.2.2 implies Theorem 8.2.1 follows immediately from the Kreps-Yan Theorem 5.2.2, i.e., we find a probability measure $\mathbf{Q} \sim \mathbf{P}$ such that S is a local $\mathbf{Q}$-martingale. Theorem 8.2.2 tells us that we don't have to bother about passing to the weak-star-closure of C any more, as assumption (ii) of Theorem 8.2.1 implies that C *already is weak-star-closed*. In other words, our program of choosing the *"right"* class of admissible integrands has been successful: the passage to the limit which was necessary in the context of the Kreps-Yan theorem, i.e., the passage from C^{simple} to its weak-star-closure, is already taken care of by the passages to the limit in the stochastic integration theory from simple to general admissible integrands.

In fact, Theorem 8.2.2 tells us that — under the assumption of *(NFLVR)* — C equals precisely the weak-star-closure of C^{simple}. The fact that C^{simple} is weak-star dense in C follows from Chap. 7 where the general theory of stochastic integration is based on the idea of approximating a general integrand by simple integrands.

By rephrasing Theorem 8.2.1 in the form of Theorem 8.2.2, we did not come closer to a proof yet. But we see more clearly, what the heart of the matter is: for a net $(H^\alpha)_{\alpha \in I}$ of admissible integrands, $f_\alpha = (H^\alpha \cdot S)_\infty$ and $g_\alpha \le f_\alpha$ such that $(g_\alpha)_{\alpha \in I}$ weak-star converges in $L^\infty(\mathbf{P})$ to some f, we have to show that we can find an admissible integrand H such that $f \le (H \cdot S)_\infty$. This will prove Theorem 8.2.2 and therefore 8.2.1. Loosely speaking, we have to be able to pass from a net $(H^\alpha)_{\alpha \in I}$ of admissible trading strategies to a limiting admissible trading strategy H.

The first good news on our way to prove this result is that in the present context we may reduce from the case of a general net $(H^\alpha)_{\alpha \in I}$ to the case of a sequence $(H^n)_{n=0}^\infty$ and therefore to a sequence $f_n = (H^n \cdot S)_\infty$. This follows from a good old friend from functional analysis, the theorem of Krein-Smulian as worked out in Proposition 5.2.4 above.

Once we have reduced the problem to the case of sequences $(H^n)_{n=0}^\infty$ we may apply another good friend from functional analysis, the theorem of Banach-Steinhaus (also called principle of uniform boundedness): if *a sequence* $(g_n)_{n=0}^\infty$ *in a dual Banach space* X^* *is weak-star convergent, the norms* $(\|g_n\|)_{n=0}^\infty$ *remain bounded.* This result implies (see Chap. 9 below) that we may reduce to the case where the sequence $(H^n)_{n=0}^\infty$ admits a uniform admissibility bound M such that $H^n \cdot S \geq -M$, for all $n \in \mathbb{N}$.

Putting together these results from general functional analysis, it will suffice to prove the following result to complete the proof of Theorem 8.2.2.

Crucial Lemma 8.2.3. *Under the hypotheses of Theorem 8.2.2, let* $(H^n)_{n=0}^\infty$ *be a sequence of admissible integrands such that*

$$(H^n \cdot S)_t \geq -1, \quad a.s., \text{ for } t \geq 0 \text{ and } n \in \mathbb{N}. \tag{8.4}$$

Assume also that $f_n = (H^n \cdot S)_\infty$ *converges almost surely to* f. *Then there is an admissible integrand* H *such that*

$$(H \cdot S)_\infty \geq f. \tag{8.5}$$

The admissible strategy H *can be chosen in such a way that* $(H \cdot S)_\infty$ *is a maximal element in the set* K.

To convince ourselves that Lemma 8.2.3 indeed implies Theorem 8.2.2, we still have to justify one more reduction step which is contained in the statement of Lemma 8.2.3: we may reduce to the case, when $(f_n)_{n=0}^\infty$ converges almost surely. This is done by an elementary lemma in the spirit of Komlos' theorem (Lemma 9.8.1, compare also Lemma 6.6.1 above). In its simplest form it states the following: Let $(f_n)_{n=0}^\infty$ be an arbitrary sequence of random variables uniformly bounded from below. Then we may find convex combinations $h_n \in \text{conv}\{f_n, f_{n+1,\dots}\}$ converging almost surely to an $\mathbb{R} \cup \{+\infty\}$-valued random variable f. For more refined variations on this theme see Chap. 15.

Note that the passage to convex combinations does not cost anything in the present context, where our aim is to find a limit to a given sequence in a locally convex vector space; hence the above argument allows us to reduce to the case where we may assume, in addition to (8.4), that $(f_n)_{n=0}^\infty = ((H^n \cdot S)_\infty)_{n=0}^\infty$ converges almost surely to a function $f : \Omega \to \mathbb{R} \cup \{+\infty\}$. Using the assumption *(NFLVR)* we can quickly show in the present context that f must be a.s. finitely valued.

Summing up, Lemma 8.2.3 is a statement about the possibility of passing to a limit H, for a given sequence $(H^n)_{n=0}^\infty$ of admissible integrands. The crucial hypothesis is the uniform one-sided boundedness (8.4); apart from this strong assumption, we only have an information on the a.s. convergence of *the terminal values* $((H^n \cdot S)_\infty)_{n=0}^\infty$, but we do not have any a priori information on the convergence of *the processes* $((H^n \cdot S)_{t \geq 0})_{n=0}^\infty$.

Let us compare Lemma 8.2.3 with the previous literature. An important theorem of J. Mémin [M 80] states the following: if a sequence of stochastic integrals $((H^n \cdot S)_{t\geq0})_{n=0}^{\infty}$ on a given semi-martingale S is Cauchy with respect to the semi-martingale topology, then the limit exists (as a semi-martingale) and is of the form $(H \cdot S)_{t\geq0}$ for some S-integrable predictable process H.

This theorem will finally play an important role in proving Lemma 8.2.3; but we still have a long way to go before we can apply it, as the assumptions of Lemma 8.2.3 a priori do not tell us anything about the convergence of the sequence of processes $((H^n \cdot S)_{t\geq0})_{n=0}^{\infty}$.

Another line of results in the spirit of Lemma 8.2.3 assumes that the process S is a (local) martingale. The arch-example is the theorem of Kunita-Watanabe ([KW 67]; see also [P 90] or [Y 78a] or Chap. 7 above): suppose that S is a locally L^2-bounded martingale, that each $(H^n \cdot S)_{t\geq0}$ is an L^2-bounded martingale, and that the sequence $((H^n \cdot S)_{t\geq0})_{n=0}^{\infty}$ is Cauchy in the Hilbert space of square-integrable martingales (equivalently: that the sequence of terminal values $((H^n \cdot S)_{\infty})_{n=0}^{\infty}$ is Cauchy in the Hilbert space $L^2(\Omega, \mathcal{F}, \mathbf{P})$). Then the limit exists (as a square-integrable martingale) and it is of the form $(H \cdot S)_{t\geq0}$.

As the proof of this theorem is very simple and allows some insight into the present theme, we sketch it (assuming, for notational simplicity, that S is $\mathbb{R}$-valued): denote by $d[S]$ the quadratic variation measure of the locally L^2-bounded martingale S as in (7.9), which defines a sigma-finite measure on the σ-algebra $\mathcal{P}$ of predictable subsets of $\mathbb{R}_+ \times \Omega$. Denoting by $L^2(\mathbb{R}_+ \times \Omega, \mathcal{P}, d[S])$ the corresponding Hilbert space, the stochastic integration theory is designed in such a way that we have the isometric identity

$$\|H\|_{L^2(\mathbb{R}_+ \times \Omega, \mathcal{P}, d[S])} = \|(H \cdot S)_{\infty}\|_{L^2(\Omega, \mathcal{F}, \mathbf{P})}, \qquad (8.6)$$

for each predictable process H, for which the left hand side of (8.6) is finite (see Chap. 7 above).

Hence the assumption that $((H^n \cdot S)_{t\geq0})_{n=0}^{\infty}$ is Cauchy in the Hilbert space of square-integrable martingales is tantamount to the assumption that $(H^n)_{n=0}^{\infty}$ is Cauchy in $L^2(\mathbb{R}_+ \times \Omega, \mathcal{P}, d[S])$. Now, once more, the stochastic integration theory is designed in a way that $L^2(\mathbb{R}_+ \times \Omega, \mathcal{P}, d[S])$ consists precisely of the S-integrable, predictable processes H such that $H \cdot S$ is an L^2-bounded martingale. Hence by the completeness of the Hilbert space $L^2(\mathbb{R}_+ \times \Omega, \mathcal{P}, d[S])$ we can pass from the Cauchy-sequence $(H^n)_{n=0}^{\infty}$ to its limit $H \in L^2(\mathbb{R}_+ \times \Omega, \mathcal{P}, d[S])$. This finishes the sketch of the proof of the Kunita-Watanabe theorem.

The above argument shows in a nice and transparent way how to deduce from a completeness property of the *space of predictable integrands* H a completeness property of the corresponding *space of terminal results* $(H \cdot S)_{\infty}$ *of stochastic integrals*. In the context of the theorem of Kunita-Watanabe, the functional analytic background for this argument is reduced to the — almost trivial — isometric identification of the two corresponding Hilbert spaces in (8.6).

Using substantially more refined arguments, M. Yor [Y 78a] was able to extend this result to the case of Cauchy sequences $(H^n \cdot S)_{n=0}^\infty$ of martingales bounded in $L^p(\mathbf{P})$ or $H^p(\mathbf{P})$, for arbitrary $1 \le p \le \infty$, the most delicate and interesting case being $p = 1$ (compare also Chap. 15 below).

After this review of some of the previous literature on the topic of completeness of the space of stochastic integrals, let us turn back to Lemma 8.2.3.

Unfortunately the theorems of Kunita-Watanabe and Yor do not apply to its proof, as we don't assume that S is a local martingale. It is precisely the point, that we finally want to *prove* that S is a local martingale with respect to some measure $\mathbf{Q}$ equivalent to $\mathbf{P}$.

But in our attempt to build up some motivation for the proof of Lemma 8.2.3, let us cheat for a moment and suppose that we know already that S is a local martingale under some equivalent measure $\mathbf{Q}$ and let $(H^n)_{n=0}^\infty$ be a sequence of S-integrable predictable processes satisfying (8.4). Using again the theorem of Ansel-Stricker (Theorem 7.3.7 above) we conclude that $(H^n \cdot S)_{n=0}^\infty$ is a sequence of local martingales; inequality (8.4) quickly implies that this sequence is bounded in $L^1(\mathbf{Q})$-norm:

$$\|H^n \cdot S\|_{L^1(\mathbf{Q})} := \sup \left\{ \mathbf{E}_{\mathbf{Q}} \left[|(H^n \cdot S)_\tau| \right] \mid \tau \text{ stopping time} \right\} \le 2, \quad \text{for } n \ge 0.$$

Let us cheat once more and assume that each $H^n \cdot S$ is in fact a uniformly integrable $\mathbf{Q}$-martingale (instead of only being a local $\mathbf{Q}$-martingale) and that $((H^n \cdot S)_\infty)_{n=0}^\infty$ is Cauchy with respect to the $L^1(\mathbf{Q})$-norm defined above (instead of only being bounded with respect to this norm).

Admitting the above "cheating steps" we are in a position to apply Yor's theorem to find a limiting process H to the sequence $(H^n)_{n=0}^\infty$ for which (8.5) holds true, where we even may replace the inequality by an equality. But, of course, this is only motivation, why Lemma 8.2.3 should hold true, and we now have to find a mathematical proof, preferably without cheating.

We have taken some time for the above heuristic considerations to develop an intuition for the statement of Lemma 8.2.3 and to motivate the general philosophy underlying its proof: *we want to prove results which are — at least more or less — known for (local) martingales S, but replacing the martingale assumption on S by the assumption that S satisfies (NFLVR).*

As a starter we sketch the proof of a result which shows that, under the assumption of *(NFLVR)*, the technical condition imposed on the admissible integrand H in (8.1) is, in fact, automatically satisfied. The lemma is taken from Theorem 9.3.3 where it is stated for locally bounded semi-martingales, but it remains valid for general semi-martingales S.

Lemma 8.2.4 (9.3.3). *Let S be a semi-martingale satisfying (NFLVR) and let H be an admissible integrand.*
   Then

$$(H \cdot S)_\infty := \lim_{t \to \infty} (H \cdot S)_t$$

exists and is finite, almost surely.

This result is a good illustration for our philosophy: suppose *we know already* that the assumption of 8.2.4 implies that S is a local martingale under some $\mathbf{Q}$ equivalent to $\mathbf{P}$. Then the conclusion follows immediately from known results: from Ansel-Stricker (Theorem 7.3.7) we know that $H \cdot S$ is a super-martingale. As $H \cdot S$ is bounded from below, Doob's super-martingale convergence theorem (see, e.g., [W 91]) implies the almost sure convergence of $(H \cdot S)_t$ as $t \to \infty$, to an a.s. finite random variable.

Our goal is to replace these martingale arguments by some arguments relying only on *(NFLVR)*. The nice feature is that these arguments also allow for an economic interpretation.

Proof of Lemma 8.2.4. As in the usual proof of Doob's super-martingale convergence theorem we consider the number of up-crossings: to show almost sure convergence of $(H \cdot S)_t$, for $t \to \infty$, we consider, for any $\beta < \gamma$, the $\mathbf{P}$-measure of the set $\{\omega \mid (H \cdot S)_t(\omega) \text{ upcrosses }]\beta, \gamma[\text{ infinitely often}\}$. We shall show that it equals zero.

So suppose to the contrary that there is $\beta < \gamma$ such that the set

$$A = \{\omega \mid (H \cdot S)_t \text{ upcrosses }]\beta, \gamma[\text{ infinitely often}\}$$

satisfies $\mathbf{P}[A] > 0$. The economic interpretation of this situation is the following: an investor knows at time zero that, when applying the trading strategy H, with probability $\mathbf{P}[A] > 0$ her wealth will infinitely often be less than or equal to β and infinitely often be more than or equal to γ. A smart investor will realise that this offers a free lunch with vanishing risk, as she can modify H to obtain a very rewarding trading strategy K.

Indeed, define inductively the sequence of stopping times $(\sigma_n)_{n=0}^{\infty}$ and $(\tau_n)_{n=0}^{\infty}$ by $\sigma_0 = \tau_0 = 0$ and, for $n \geq 1$,

$$\sigma_n = \inf\{t \geq \tau_{n-1} \mid (H \cdot S)_t \leq \beta\},$$
$$\tau_n = \inf\{t \geq \sigma_n \mid (H \cdot S)_t \geq \gamma\}.$$

The set A then equals the set where, σ_n and τ_n are finite, for each $n \in \mathbb{N}$ (as usual, the inf over the empty set is taken to be $+\infty$).

What every investor wants to do is to "buy low and sell high". The above stopping times allow her to do that in a systematic way: define $K = H\mathbf{1}_{\{\cup_{n=1}^{\infty}]\sigma_n, \tau_n]\}}$, which is clearly a predictable S-integrable process. A more verbal description of K goes as follows: the investor starts by doing nothing (i.e., making a zero-investment into the risky assets $S^1, \ldots, S^d$) until the time σ_1 when the process $(H \cdot S)_t$ has dropped below β (If $\beta \geq 0$, we have $\sigma_1 = 0$)). At this time she starts to invest according to the rule prescribed by the trading strategy H; she continues to do so until time τ_1 when $(H \cdot S)_t$ first has passed beyond γ. Note that, if $\tau_1(\omega)$ is finite, our investor following the strategy K has at least gained the amount $\gamma - \beta$. At time τ_1 (if it happens to be finite) the investor clears all her positions and does not invest into the risky assets until time σ_2, when she repeats the above scheme.

One easily verifies (arguing either "mathematically" or "economically") that the process $K \cdot S$ satisfies

$$(K \cdot S)_t \geq -M \quad \text{a.s., for all } t,$$

where M is the uniform lower bound for $H \cdot S$, and

$$\lim_{t \to \infty} (K \cdot S)_t = \infty \quad \text{a.s. on } A.$$

Hence K describes a trading scheme, where the investor can lose at most a fixed amount of money, while, with strictly positive probability, she ultimately becomes infinitely rich. Intuitively speaking, this is "something like an arbitrage", and it is an easy task to formally deduce from these properties of K a "free lunch with vanishing risk": for example, it suffices to define $K^n = \frac{1}{n} K \mathbf{1}_{]\!]0, \tau_n \wedge T_n]\!]}$, for a sequence of (deterministic) times $(T_n)_{n=0}^\infty$, to let $f_n = (K^n \cdot S)_\infty = (K^n \cdot S)_{\tau_n \wedge T_n}$ and to define $g_n = f_n \wedge (\gamma - \beta) \mathbf{1}_B$ where $B = \bigcap_{n=0}^\infty \{\tau_n \leq T_n\}$. If $(T_n)_{n=1}^\infty$ tends to infinity sufficiently fast, we have $\mathbf{P}[B] > 0$, and one readily verifies that $(g_n)_{n=1}^\infty$ converges uniformly to $(\gamma - \beta) \mathbf{1}_B$.

Summing up, we have shown that *(NFLVR)* implies that, for $\beta < \gamma$, the process $H \cdot S$ almost surely upcrosses the interval $]\beta, \gamma[$ only finitely many times. Whence $(H \cdot S)_t$ converges almost surely to a random variable $(H \cdot S)_\infty$ with values in $\mathbb{R} \cup \{\infty\}$. The fact that $(H \cdot S)_\infty$ is a.s. finitely valued follows from another application (similar to but simpler than the above) of the assumption of *(NFLVR)*, which we leave to the reader. □

We now start to sketch the main arguments underlying the proof of Lemma 8.2.3. The strategy is to obtain from assumption (8.4) and from suitable modifications of the original sequence $(H^n)_{n=0}^\infty$ more information on the convergence of the sequence of *processes* $(H^n \cdot S)_{n=0}^\infty$. Eventually we shall be able to reduce the problem to the case where $(H^n \cdot S)_{n=0}^\infty$ converges in the semimartingale topology; at this stage Mémin's theorem [M 80] will give us the desired limiting trading strategy H.

So, what can we deduce from assumption (8.4) and the a.s. convergence of $(f_n)_{n=0}^\infty = ((H^n \cdot S)_\infty)_{n=0}^\infty$ for the convergence of the sequence of processes $(H^n \cdot S)_{n=0}^\infty$? The unpleasant answer is: a priori, we cannot deduce anything. To see this, recall the "suicide" strategy H which we have discussed in the context of inequality (8.3) above: it designs an admissible way to lose one €. Speaking mathematically, the corresponding stochastic integral $H \cdot S$ starts at $(H \cdot S)_0 = 0$, satisfies $(H \cdot S)_t \geq -1$ almost surely, for all $t \geq 0$, as well as $(H \cdot S)_\infty = -1$. But clearly this is not the only admissible way to lose one € and there are many other trading strategies K on the process S having the same properties. Taking up again the example discussed after (8.3), a trivial example is, to first wait without playing for a fixed number of games of the roulette, and to start the suicide strategy only after this waiting period; of course, this is a different way of losing one €.

Speaking mathematically, this means that — even when S is a martingale, as is the case in the example of the suicide strategy — the condition $(H \cdot S)_t \geq -1$ a.s., for all $t \geq 0$, and the final outcome $(H \cdot S)_\infty$ do not determine the process $H \cdot S$ uniquely. In particular there is no hope to derive from (8.4) and the a.s. convergence of the sequence of random variables $((H^n \cdot S)_\infty)_{n=0}^\infty$, a convergence property of the sequence of processes $(H^n \cdot S)$.

The idea to remedy the situation is to notice the following fact: the suicide strategy is a silly investment and obviously there are better trading strategies, e.g., not to gamble at all. By discarding such "silly investments", we hopefully will be able to improve the situation.

Here is the way to formalise the idea of discarding "silly investments": denote by D the set of all random variables h such that there is a random variable $f \geq h$ and a sequence $(H^n)_{n=0}^\infty$ of admissible trading strategies satisfying (8.4), and such that $(H^n \cdot S)_\infty$ converges a.s. to f. We call f_0 a maximal element of D if the conditions $h \geq f_0$ and $h \in D$ imply that $h = f_0$.

For example, in the context of the random walk $S = (S_t)_{t=0}^\infty$ on which we constructed the above "suicide strategy", $h \equiv -1$ is an element of D, but not a maximal element. A maximal element dominating h is, for example, $f_0 \equiv 0$.

More generally, it is not hard to prove under the assumptions of Lemma 8.2.3 that, for a given $f = (H \cdot S)_\infty \geq -1$, where H is an admissible integrand, there is a maximal element $f_0 \in D$ dominating f (see Lemma 9.4.4).

The point of the above concept is that, in the proof of Lemma 8.2.3, we may assume without loss of generality that f is a maximal element of D. Under this additional assumption it is indeed possible to derive from the a.s. convergence of the sequence of random variables $((H^n \cdot S)_\infty)_{n=0}^\infty$ some information on the convergence of the sequence of processes $(H^n \cdot S)_{n=0}^\infty$.

As the proof of this result is another nice illustration of our general approach of replacing "martingale arguments" by "economically motivated arguments" relying on the assumption *(NFLVR)*, we sketch the argument. Again we observe that the proof does not make use of the local boundedness of S.

Lemma 8.2.5 (9.4.6). *Let S be an $\mathbb{R}^d$-valued semi-martingale and let f be a maximal element of D. Let $(H^n)_{n=1}^\infty$ be a sequence of admissible integrands as in Lemma 8.2.3.*

Then the sequence of random variables

$$F_{n,m} = \sup_{0 \leq t < \infty} |(H^n \cdot S)_t - (H^m \cdot S)_t| \tag{8.7}$$

tends to zero in probability, as $n, m \to \infty$.

Proof. Suppose to the contrary that there is $\alpha > 0$, and sequences $(n_k, m_k)_{k=1}^\infty$ tending to infinity s.t. $\mathbf{P}[\sup_{0 \leq t}((H^{n_k} \cdot S)_t - (H^{m_k} \cdot S)_t) > \alpha] \geq \alpha$, for each $k \in \mathbb{N}$.

Define the stopping times τ_k as

$$\tau_k = \inf \left\{ t \mid (H^{n_k} \cdot S)_t - (H^{m_k} \cdot S)_t \geq \alpha \right\},$$

so that we have $\mathbf{P}[\tau_k < \infty] \geq \alpha$.

Define L^k as $L^k = H^{n_k} \mathbf{1}_{]\!]0,\tau_k]\!]} + H^{m_k} \mathbf{1}_{]\!]\tau_k,\infty[\![}$. Clearly the process L^k is predictable and $L^k \cdot S \geq -1$.

Translating the formal definition into prose: the trading strategy L^k consists of following the trading strategy H^{n_k} up to time τ_k, and then switching to H^{m_k}. The idea is that L^k produces a sensibly better final result $(L^k \cdot S)_\infty$ than either $(H^{n_k} \cdot S)_\infty$ or $(H^{m_k} \cdot S)_\infty$, which will finally lead to a contradiction to the maximality assumption on f.

Why is L^k "sensibly better" than H^{n_k} or H^{m_k}? For large k, the random variables $(H^{n_k} \cdot S)_\infty$ as well as $(H^{m_k} \cdot S)_\infty$ will both be close to f in probability; for the sake of the argument, assume that both are in fact equal to f (keeping in mind that the difference is "small with respect to convergence in probability"). A moment's reflection reveals that this implies that the random variable $(L^k \cdot S)_\infty$ equals f plus the random variable $((H^{n_k} \cdot S)_{\tau_k} - (H^{m_k} \cdot S)_{\tau_k}) \mathbf{1}_{\{\tau_k < \infty\}}$. The latter random variable is nonnegative and with probability α greater than or equal to α; this means that this difference between f and $(L^k \cdot S)_\infty$ is not "small with respect to convergence in probability"; this is, what we had in mind when saying that L^k is a "sensible" improvement as compared to H^{n_k} or H^{m_k}.

Modulo some technicalities, which are worked out in Lemma 9.4.6 below, this gives the desired contradiction to the maximality assumption on f, thus finishing the (sketch of the) proof of Lemma 8.2.5.   □

Lemma 8.2.5 is our first step towards a proof of Lemma 8.2.3: it gives some information on the convergence of the sequence of processes $(H^n \cdot S)_{n=0}^\infty$ in terms of the maximal functions defined in (8.7). But the assertion that these maximal functions tend to zero in probability is still much weaker than the convergence of $(H^n \cdot S)_{n=0}^\infty$ with respect to the semi-martingale topology, which we finally need in order to be able to apply Mémin's theorem. There is still a long way to go!

But it is time to finish this "guided tour" towards a proof of Theorem 8.2.2 and to advise the interested reader to find the remaining part of the proof in Chap. 9 below. We hope that we have succeeded to give some motivation for the proof and for the "economically motivated" arguments underlying it.

8.3 Sigma-martingales and the Non-locally Bounded Case

To finish this chapter we return to the basic assumption in Sect. 8.2 that the process S is *locally bounded*. What happens if we drop this — technically very convenient — assumption?

Before starting to answer this question, we remark that this question is not only of "academic" interest. It is also important from the point of view of applications: if one goes beyond the framework of continuous processes S — and there are good empirical reasons to do so — it is quite natural to allow for the jumps of the processes to be unbounded. As a concrete example we mention Lévy processes or the family of ARCH (Auto-Regressive Conditional Heteroskedastic) processes and their relatives (GARCH, EGARCH etc.). The former find increasing applications in financial engineering. The latter are very popular in the econometric literature: these are processes in discrete time where the conditional distribution of the jumps is Gaussian. In particular, these processes are not locally bounded (compare Example 8.3.3 below). There are many other examples of processes which fail to be locally bounded, used in the modelling of financial markets.

The answer to the question, whether Theorem 8.2.1 can be extended to this setting, is as we expect it to be: *mutatis mutandis* the fundamental theorem of asset pricing as well as the related theorems carry over to the case of not necessarily locally bounded $\mathbb{R}^d$-valued semi-martingales S. Not coming as a surprise, the techniques of the proofs have to be refined: in particular, we cannot entirely reduce to the study of the space $L^\infty(\Omega, \mathcal{F}, \mathbf{P})$, and the weak-star and norm topology of this space: there is no possibility anymore to reduce to the case of (one-sided) bounded stochastic integrals and we therefore have to use larger spaces than $L^\infty(\Omega, \mathcal{F}, \mathbf{P})$. Yet it turns out — and this is slightly surprising — that the duality between $L^\infty(\mathbf{P})$ and $L^1(\mathbf{P})$ still remains the central issue of the proof.

Here is the statement of the extension of the fundamental theorem of asset pricing as obtained in Chap. 14.

Theorem 8.3.1 (Main Theorem 14.1.1). *The following assertions are equivalent for an $\mathbb{R}^d$-valued semi-martingale model $S = (S_t)_{t \geq 0}$ of a financial market:*

(i) *(ESMM), i.e., there is a probability measure $\mathbf{Q}$ equivalent to $\mathbf{P}$ such that S is a sigma-martingale under $\mathbf{Q}$.*

(ii) *(NFLVR), i.e., S satisfies the condition of no free lunch with vanishing risk.*

There is a slight change in the statement (i) as compared to the statement of Theorem 8.2.1 above: the term "local martingale" in the definition of *(EMM)* was replaced by the term "sigma-martingale" thus replacing the acronym *(EMM)* by *(ESMM)*. On the other hand, condition (ii) remained completely unchanged.

The notion of a sigma-martingale is a generalisation of the notion of a local martingale:

Definition 8.3.2 (Chap. 14). *An $\mathbb{R}^d$-valued semi-martingale $S = (S_t)_{t \geq 0}$ is called a sigma-martingale if there is a predictable process $\varphi = (\varphi_t)_{t \geq 0}$,* *taking its values in* $]0, \infty[$, *such that the* $\mathbb{R}^d$-*valued stochastic integral* $\varphi \cdot S$ *is a martingale.*

To motivate this definition we recall Émery's example 7.3.4: we have seen there that the process $X = H \cdot M$ defined in (7.15) fails to be a local martingale. Nevertheless, the above definition gives us a tool to interpret the process X as "something which has the essential features of a martingale": defining $\varphi_t = t$ we find a (deterministic and therefore predictable) process φ such that $\varphi \cdot X = (\varphi H) \cdot M = M$ is a martingale. Hence X is a sigma-martingale.

The notion of sigma-martingales was introduced (using slightly different notation) by Chou [C 77] and further analyzed by Émery [E 80]. It is tailor-made for our present purposes for the following two reasons:

Fact 1: In the setting of Theorem 8.3.1 (i) it is unavoidable to pass to a concept going beyond the notion of a local martingale.

Fact 2: For the purposes of hedging contingent claims the notion of a sigma-martingale is just as useful as the notion of a local martingale (or even that of a martingale).

To justify these two facts we start with the second one: assume that $S = (S_t)_{t \geq 0}$ is a sigma-martingale so that there is a $]0, \infty[$-valued predictable process φ such that $\widetilde{S} = \varphi \cdot S$ is a martingale. Let H be any $\mathbb{R}^d$-valued predictable process. Then H is S-integrable, iff $\widetilde{H} := \frac{H}{\varphi}$ is $\widetilde{S}$-integrable and in this case the processes $H \cdot S$ and $\widetilde{H} \cdot \widetilde{S}$ are identical. This follows from the rather trivial formula

$$\widetilde{H} \cdot \widetilde{S} = \left(\tfrac{H}{\varphi}\right) \cdot (\varphi \cdot S) = H \cdot S.$$

As a consequence, the class of processes $\{H \cdot S \mid H \text{ is } S\text{-integrable}\}$ and $\{\widetilde{H} \cdot \widetilde{S} \mid \widetilde{H} \text{ is } \widetilde{S}\text{-integrable}\}$ coincide. Every statement pertaining only to this class (such as Theorem 8.3.1 (ii)) remains unaffected by the passage from the sigma-martingale S to the martingale $\widetilde{S}$.

As regards Fact 1 above, we construct in Example 14.2.3 a slight variant of Émery's Example 7.3.4 with the following property: the process S is a sigma-martingale (under $\mathbf{P}$) but, for each $\mathbf{Q} \ll \mathbf{P}$, S fails to be a local $\mathbf{Q}$-martingale. Hence the process S satisfies *(NFLVR)*, but there is no probability measure $\mathbf{Q} \ll \mathbf{P}$ such that S is a local $\mathbf{Q}$-martingale.

We now try to give a sketch of the strategy for the proof of Theorem 8.3.1, where S is a general (not necessarily locally bounded) semi-martingale. As usual the implication (i) $\Rightarrow$ (ii) is the easy one: it follows from the discussion of Fact 2 above and the Ansel-Stricker Theorem 7.3.7 that, if S is a sigma-martingale under $\mathbf{Q}$, and H an admissible integrand for S, the process $H \cdot S$ is a super-martingale under $\mathbf{Q}$. Hence $\mathbf{E}_{\mathbf{Q}}[f] \leq 0$, for all $f \in C$, which implies Theorem 8.3.1 (ii) by the usual arguments.

The subtle issue is the implication (ii) $\Rightarrow$ (i) of Theorem 8.3.1. The first good news is that for the validity of Theorem 8.2.2, which asserts that *(NFLVR)* implies the weak-star-closureness of the cone C defined in (8.2), the assumption of local boundedness is not needed: The proof of Theorem 8.2.2 does not use the local boundedness of S and works in full generality. Hence we may apply the Kreps-Yan Theorem 5.2.2 also under the assumptions of Theorem 8.3.1 (ii) to find a probability measure $\mathbf{Q} \sim \mathbf{P}$ such that $\mathbf{E_Q}[f] \leq 0$, for each $f \in C$. The set of these probability measures, i.e., the probability measures $\mathbf{Q} \ll \mathbf{P}$ such that $\mathbf{Q}|_C \leq 0$ is denoted by $\mathcal{M}_s^e$ in Proposition 14.4.5. Y.M. Kabanov [K 97] proposed the name "separating measures" for this set.

In the case of S being (locally) bounded we have seen that, for a separating measure $\mathbf{Q}$, the semi-martingale S is a (local) $\mathbf{Q}$-martingale. We have used this rather obvious fact (Lemma 5.1.3) to deduce Theorem 8.2.1 from Theorem 8.2.2 above. But in the present case of a general semi-martingale S this implication breaks down. The subsequent easy example illustrates the situation.

Example 8.3.3. Let X be a normally distributed real random variable with mean $\mu \in \mathbb{R}$ and variance $\sigma^2 > 0$. Define the process $S = (S_t)_{t \geq 0}$ by

$$S_t = \begin{cases} 0, & \text{for } 0 \leq t < 1, \\ X, & \text{for } t \geq 1. \end{cases}$$

which we consider under its natural filtration $(\mathcal{F}_t)_{t \geq 0}$.

Observe that, for every admissible integrand H, we have $H \cdot S \equiv 0$. Expressing this property in prose: the only *admissible* way, i.e., with uniformly bounded risk, to bet on the random variable X, is the zero bet. This follows from the fact that X is unbounded from below as well as from above.

Hence the cone K defined in (8.1) is reduced to zero, and the cone C defined in (8.2) equals the negative orthant $L^\infty(\Omega, \mathcal{F}_\infty, \mathbf{P})$. It follows that *every* probability measure $\mathbf{Q} \ll \mathbf{P}$ is a separating measure. But S is a sigma-martingale w.r. to $\mathbf{Q}$, iff it is a martingale w.r. to $\mathbf{Q}$, iff $\mathbf{E_Q}[X] = 0$.

The example shows that, if we allow S to have unbounded jumps, the separating measures $\mathbf{Q}$ are not necessarily sigma-martingale measures any more. The important observation which will eventually prove Theorem 8.3.1 is the following: the set $\mathcal{M}_\sigma^e$ of measures $\mathbf{Q} \sim \mathbf{P}$ such that S is a sigma-martingale under $\mathbf{Q}$ is dense in the set of separating measures $\mathcal{M}_s^e$ (see Proposition 14.4.5) which we restate here for convenience). Of course, this density assertion is a more precise information than the assertion of Theorem 8.3.1 that $\mathcal{M}_\sigma^e$ is not empty.

Proposition 8.3.4. *Denote by $\mathcal{M}_s^e$ the set of probability measures*

$$\mathcal{M}_s^e = \{\mathbf{Q} \mid \mathbf{Q} \sim \mathbf{P} \text{ and for each } f \in C : \mathbf{E_Q}[f] \leq 0\}.$$

If S satisfies (NFLVR), then

$$\mathcal{M}_\sigma^e = \{\mathbf{Q} \mid S \text{ is a } \mathbf{Q} \text{ sigma-martingale}\},$$

is dense in $\mathcal{M}_s^e$ with respect to the norm of $L^1(\Omega, \mathcal{F}_\infty, \mathbf{P})$.

Let us illustrate this fact for the easy Example 8.3.3 above: fix $\mu \in \mathbb{R}$, $\sigma^2 > 0$, $\varepsilon > 0$ and a probability measure $\mathbf{Q}_1 \sim \mathbf{P}$. We want to find a measure $\mathbf{Q} \sim \mathbf{P}$, $\|\mathbf{Q}_1 - \mathbf{Q}\| < \varepsilon$ such that $\mathbf{E}_\mathbf{Q}[X] = 0$. Suppose w.l.g. that $\mathbf{E}_{\mathbf{Q}_1}[X] < 0$. (If $\mathbf{E}_{\mathbf{Q}_1}[X] = 0$ there is nothing to prove; if $\mathbf{E}_{\mathbf{Q}_1}[X] > 0$ it suffices to reverse the inequalities in (8.8).)

Define, for $\xi \in \mathbb{R}$, $\alpha > 1$, $0 < \beta < 1$, the measure $\mathbf{Q}(\xi, \alpha, \beta)$ by

$$\frac{d\mathbf{Q}(\xi, \alpha, \beta)}{d\mathbf{Q}_1} = \alpha \mathbf{1}_{\{X > \xi\}} + \beta \mathbf{1}_{\{X \le \xi\}}. \tag{8.8}$$

As X is unbounded under the measure $\mathbf{Q}_1$, one easily verifies that one may find (ξ, α, β) such that $\mathbf{Q} := \mathbf{Q}(\xi, \alpha, \beta)$ is a probability measure and such that $\|\mathbf{Q} - \mathbf{Q}_1\| < \varepsilon$ as well as $\mathbf{E}_\mathbf{Q}[X] = 0$ (compare Lemma 14.3.4).

Summing up, we have shown the validity of Proposition 8.3.4 in the very special case of Example 8.3.3. This strategy of proof also applies to the proof of Proposition 8.3.4 in full generality, modulo some delicate technicalities as we now shall try to explain.

To sketch the idea of the proof of Proposition 8.3.4 suppose that $S = (S_t)_{t \ge 0}$ is a semi-martingale satisfying *(NFLVR)*, so that $\mathcal{M}_s^e$ is non-empty. Fix $\mathbf{Q}_1 \in \mathcal{M}_s^e$.

The problem is that $\mathbf{Q}_1$ may fail to be a sigma-martingale measure; this is due to the fact that the semi-martingale S may have "big jumps". So let us deal with the jumps in a systematic way. We know from the general theory [D 72, DM 80] that the jumps of a càdlàg process S can be exhausted by countably many stopping times; in addition, these stopping times can be classified into the predictable ones and the totally inaccessible ones (Definition 7.2.2 and 7.2.3). More precisely, for a given càdlàg process S we may find sequences $(T_n^p)_{n=1}^\infty$ and $(T_n^i)_{n=1}^\infty$ such that T_n^p (resp. T_n^i) are predictable (resp. totally inaccessible) stopping times and such that

$$\{(\omega, t) \in \Omega \times \mathbb{R}_+ \mid \Delta S_t(\omega) \ne 0\} \subseteq \bigcup_{n=1}^\infty [\![T_n^p]\!] \cup \bigcup_{n=1}^\infty [\![T_n^i]\!]. \tag{8.9}$$

In addition we may assume that the sets $([\![T_n^p]\!])_{n=1}^\infty$ and $([\![T_n^i]\!])_{n=1}^\infty$ are mutually disjoint.

To sketch the idea of the proof of Proposition 8.3.4 we start by considering the case where there is only one predictable jump in (8.9), i.e.

$$\{(\omega, t) \in \Omega \times \mathbb{R}_+ \mid \Delta S_t(\omega) \ne 0\} \subseteq [\![T^p]\!], \tag{8.10}$$

for some predictable stopping time T^p. This is the case, e.g., in Example 8.3.3, where $T^p \equiv 1$. In order to show Proposition 8.3.4 we proceed similarly as in this example (supposing for the moment that S is real-valued and that $\mathcal{F}_{T^p_-}$ is trivial to simplify even further). If the jump ΔS_{T^p} is unbounded from above as well as from below we may proceed just as in (8.8) above to find a probalility measure $\mathbf{Q} \in \mathcal{M}^e_s$ with $\|\mathbf{Q} - \mathbf{Q}_1\|_1 < \varepsilon$ such that $\mathbf{E_Q}[\Delta S_{T^p} \mid \mathcal{F}_{T^p_-}] = \mathbf{E_Q}[\Delta S_{T^p}] = 0$. Hence $\mathbf{Q} \in \mathcal{M}^e_\sigma$ as we now have that the (conditional) expectation at the jump time T^p assumes the correct value, namely zero.

Now suppose that the jump ΔS_{T^p} is only one-sided bounded, say bounded from below, but unbounded from above. This is a slight variation of the situation of Example 8.3.3. In this case it is important to note that, for $\mathbf{Q}_1 \in \mathcal{M}^e_s$, we must have that

$$\mathbf{E_{Q_1}}[\Delta S_{T^p} \mid \mathcal{F}_{T^p_-}] = \mathbf{E_{Q_1}}[\Delta S_{T^p}] \le 0. \tag{8.11}$$

Indeed, the predictable process $H = \mathbf{1}_{[\![T^p]\!]}$ then is admissible and we have $(H \cdot S)_\infty = \Delta S_{T^p}$ which implies (8.11).

In the case when we have strict inequality in (8.11) (otherwise there is nothing to prove) we can again change the measure $\mathbf{Q}_1$ to a probability measure $\mathbf{Q}$ similarly as in (8.8) above to *increase* the value $\mathbf{E_Q}[\Delta S_{T^p}]$ by putting more mass of the probability on the upper tail of the distribution of ΔS_{T^p} (recall that ΔS_{T^p} is assumed to be unbounded from above). Hence we may increase this value until we have equality in (8.11).

Summing up, we have managed to pass from a given $\mathbf{Q}_1 \in \mathcal{M}^e_s$ to $\mathbf{Q} \in \mathcal{M}^e_\sigma$ such that $\|\mathbf{Q}-\mathbf{Q}_1\|_1 < \varepsilon$; but we have used some very restrictive assumptions. Now let us get rid of them. The most obvious step is to drop the assumption that $\mathcal{F}_{T^p_-}$ is trivial: it suffices to apply the above arguments conditionally on $\mathcal{F}_{T^p_-}$. More delicate is the passage from $\mathbb{R}$-valued processes S to $\mathbb{R}^d$-valued ones. In the $\mathbb{R}$-valued case there are only two possibilities of one-sided boundedness of ΔX_{T^p}, i.e., either from above or from below. For $d \ge 2$ we have to consider general cones of directions in $\mathbb{R}^d$ into which the jump ΔX_{T^p} is bounded. The corresponding arguments are worked out in Chap. 14 below. Finally, we can generalise the assumption (8.10) to the case when the jumps of S are exhausted not by one stopping time T^p but by a sequence $(T^p_n)^\infty_{n=1}$ of predictable stopping times: repeat inductively the above argument and apply an $\frac{\varepsilon}{2^n}$-argument. This program takes care of the predictable stopping times in (8.9).

We still have to deal with the totally inaccessible stopping times $(T^i_n)^\infty_{n=1}$ in (8.9). They form a different league as in this case we cannot argue conditionally on $\mathcal{F}_{(T^i_n)_-}$ as in the predictable case above. Instead we have to consider the compensators of the jumps of S at the totally inaccessible stopping times T^i_n. This requires some machinery from semi-martingale theory as developed, e.g., in [JS 87]. The technicalities are more complicated than in the case of predictable stopping times; nevertheless it is possible to proceed in a similar spirit and to argue inductively on $(T^i_n)^\infty_{n=1}$ to make sure that all the relevant conditional expectations are well-defined and have the desired value, namely zero. This program is worked out in Chap. 14 below in full detail and eventually yields a proof of Proposition 8.3.4 and therefore of Theorem 8.3.1.

After proving the *Fundamental Theorem of Asset Pricing* in the version of Theorem 8.3.1 in the first part of Chap. 14 we also extend its twin, the "superreplication theorem" (see Theorem 2.4.2 for the most elementary version) to its natural degree of generality in the setting of general semi-martingale models (see Theorem 14.5.9).

We now arrived at the point where we finish this "guided tour" which should help to develop the intuition by some informal arguments. Now we have to refer the reader to the original papers for a rigorous mathematical treatment.

Part II

The Original Papers

A General Version of the Fundamental Theorem of Asset Pricing (1994)

9.1 Introduction

A basic result in mathematical finance, sometimes called the *fundamental theorem of asset pricing* (see [DR 87]), is that for a stochastic process $(S_t)_{t \in \mathbb{R}_+}$, the existence of an equivalent martingale measure is *essentially* equivalent to the absence of arbitrage opportunities. In finance the process $(S_t)_{t \in \mathbb{R}_+}$ describes the random evolution of the discounted price of one or several financial assets. The equivalence of no-arbitrage with the existence of an equivalent probability martingale measure is at the basis of the entire theory of "pricing by arbitrage". Starting from the economically meaningful assumption that S does not allow arbitrage profits (different variants of this concept will be defined below), the theorem allows the probability $\mathbf{P}$ on the underlying probability space $(\Omega, \mathcal{F}, \mathbf{P})$ to be replaced by an equivalent measure $\mathbf{Q}$ such that the process S becomes a martingale under the new measure. This makes it possible to use the rich machinery of martingale theory. In particular the problem of fair pricing of contingent claims is reduced to taking expected values with respect to the measure $\mathbf{Q}$. This method of pricing contingent claims is known to actuaries since the introduction of actuarial skills, centuries ago and known by the name of "equivalence principle".

The theory of martingale representation allows to characterise those assets that can be reproduced by buying and selling the basic assets. One might get the impression that martingale theory and the general theory of stochastic processes were tailor-made for finance (see [HP 81]).

The change of measure from $\mathbf{P}$ to $\mathbf{Q}$ can also be seen as a result of risk aversion. By changing the physical probability measure from $\mathbf{P}$ to $\mathbf{Q}$, one can attribute more weight to unfavourable events and less weight to more favourable ones.

[DS 94] A General Version of the Fundamental Theorem of Asset Pricing. *Mathematische Annalen*, vol. 300, pp. 463–520, Springer, Berlin, Heidelberg, New York (1994).

As an example that this technique has in fact a long history, we quote the use of mortality tables in insurance. The actual mortality table is replaced by a table reflecting more mortality if a life insurance premium is calculated but is replaced by a table reflecting a lower mortality rate if e.g. a lump sum buying a pension is calculated. Changing probabilities is common practice in actuarial sciences. It is therefore amazing to notice that today's actuaries are introducing these modern financial methods at such a slow pace.

The present paper focuses on the question: "What is the precise meaning of the word *essentially* in the first paragraph of the paper?" The question has a twofold interest. From an economic point of view one wants to understand the precise relation between concepts of no-arbitrage type and the existence of an equivalent martingale measure in order to understand the exact limitations up to which the above sketched approach may be extended. From a purely mathematical point of view it is also of natural interest to get a better understanding of the question which stochastic processes are martingales after an appropriate change to an equivalent probability measure. We refer to the well-known fact that a semi-martingale becomes a quasi-martingale under a well-chosen equivalent law (see [P 90]); from here to the question whether we can obtain a martingale, or more generally a local martingale, is natural.

We believe that the main theorem (Theorem 9.1.1 below) of this paper contributes to both theories, mathematics as well as economics. In economic terms the theorem contains essentially two messages. First that it is possible to characterise the existence of an equivalent martingale measure for a general class of processes in terms of the concept of no free lunch with vanishing risk, a concept to be defined below. In this notion the aspect of vanishing risk bears economic relevance. The second message is that — in a general setting — there is no way to avoid general stochastic integration theory. If the model builder accepts the possibility that the price process has jumps at all possible times, he needs a sophisticated integration theory, going beyond the theory for "simple integrands". In particular the integral of unbounded predictable processes of general nature has to be used. From a purely mathematical point of view we remark that the proof of the Main Theorem 9.1.1 below, turns out to be surprisingly hard and requires heavy machinery from the theory of stochastic processes, from functional analysis and also requires some very technical estimates.

The process S, sometimes denoted $(S_t)_{t \in \mathbb{R}_+}$ is supposed to be $\mathbb{R}$-valued, although all proofs work with a d-dimensional process as well. However, we prefer to avoid vector notation in d dimensions. If the reader is willing to accept the 1-dimensional notation for the d-dimensional case as well, nothing has to be changed. The theory of d-dimensional stochastic integration is a little more subtle than the one-dimensional theory but no difficulties arise.

The general idea underlying the concept of no-arbitrage and its weakenings, stated in several variants of "no free lunch" conditions, is that there should be no trading strategy H for the process S, such that the final pay-off described by the stochastic integral $(H \cdot S)_\infty$, is a non-negative function, strictly positive with positive probability. The economic interpretation is that by betting on the process S and without bearing any risk, it should not be possible to make something out of nothing. If one wants to make this intuitive idea precise, several problems arise. First of all one has to restrict the choice of the integrands H to make sure that $(H \cdot S)_\infty$ exists. Besides the qualitative restrictions coming from the theory of stochastic integration, one has to avoid problems coming from so-called doubling strategies. This was already noted in the papers [HK 79] and [HP 81]. To explain this remark let us consider the classical doubling strategy. We draw a coin and when heads comes out the player is paid 2 times his bet. If tails comes up, the player loses his bet. The strategy is well-known: the player doubles his bet until the first time he wins. If he starts with 1 €, his final gain (= last pay out − total sum of the preceding bets) is almost surely 1 €. He has an almost sure win. The probability that heads will eventually show up is indeed one, even if the coin is not fair. However, his accumulated losses are not bounded below. Everybody, especially the casino boss, knows that this is a very risky way of winning 1 €. This type of strategy has to be ruled out: there should be a lower bound on the player's loss. The described doubling strategy is known for centuries and in French it is still referred to as "la martingale".

One possible way to avoid these difficulties is to restrict oneself to simple predictable integrands. These are defined as linear combinations of buy and hold strategies. Mathematically such a buy and hold strategy is described as an integrand of the form $H = f\mathbf{1}_{]\!]T_1,T_2]\!]}$, where $T_1 \leq T_2$ are finite stopping times and f is $\mathcal{F}_{T_1}$, measurable. The advantage of using such integrands is that they have a clear interpretation: when time $T_1(\omega)$ comes up, buy $f(\omega)$ units of the financial asset, keep them until time $T_2(\omega)$ and sell. A linear combination of such integrands is called a simple integrand. An elementary integrand is a linear combination of buy and hold strategies with stopping times that are deterministic. This terminology agrees with standard terminology of stochastic integration (see [P 90, DM 80, CMS 80]). Even if the process S is not a semimartingale the stochastic integral $(H \cdot S)$ for $H = f\mathbf{1}_{]\!]T_1,T_2]\!]}$ can be defined as the process $(H \cdot S)_t = f \cdot (S_{min(t,T_2)} - S_{min(t,T_1)})$. Also the definition of the limit $(H \cdot S)_\infty = \lim_{t \to \infty}(H \cdot S)_t = f \cdot (S_{T_2} - S_{T_1})$ poses no problem. The net profit of the strategy is precisely $(H \cdot S)_\infty$. The use of stopping times is interpreted as the use of signals coming from available, observable information. This explains why in financial theories the filtration and the derived concepts such as predictable processes, are important. It is clear that the use of simple integrands rules out the introduction of doubling strategies. This led [HK 79, K 81, HP 81] to define no-arbitrage and no free lunch in terms of simple integrands and to obtain theorems relating these notions to the existence of an equivalent martingale measure. In various directions these results were extended in [DH 86, Str 90, DMW 90, AS 93, MB 91, L 92, D 92, S 94, K 93].

To relate our work to earlier results, let us summarise the present state of the art. The case when the time set is finite is completely settled in [DMW 90] and the use of simple or even elementary integrands is no restriction at all (see [S 92, KK 94, R 94] for elementary proofs). For the case of discrete but infinite time sets, the problem is solved in [S 94]. The case of continuous and bounded processes in continuous time, is solved in [D 92]. In these two cases the theorems are stated in terms of simple integrands and limits of sequences and by using the concept of no free lunch with bounded risk. We shall review these issues in Sect. 9.6.

In the general case, i.e. a time set of the form $[0, \infty[$ or $[0, 1]$ and with a possibility of random jumps, the situation is much more delicate. The existence of an equivalent martingale measure can be characterised in terms of "no free lunch" involving the convergence of nets or generalised sequences, see e.g. [K 81, L 92]. S. Kusuoka [K 93] used convergence in Orlicz spaces and Fuffie, Huang and Stricker [DH 86, Str 90] used L^p convergence for $1 \leq p < \infty$. In the latter case the restrictions posed on S were such that the new measure has a density in L^q where $q = \frac{p}{p-1}$. Contrary to the case of continuous processes or to the case of discrete time sets, no general solution was known in terms of "no free lunch" involving convergent sequences. Hence there remained the natural question whether for a general adapted process S, the existence of an equivalent martingale measure could be characterised in such terms.

The answer turns out to be no if one only uses simple integrands. In Sect. 9.7, we give an example of a process $S = M + A$ where M is a uniformly bounded martingale, A is a predictable process of finite variation, S admits no equivalent martingale measure but there is "no free lunch with bounded risk" if one only uses simple integrands. A closer look at the example shows that if one allows strategies of the form: "sell before each rational number and buy back after it", then there is even a "free lunch with vanishing risk". Of course such a trading strategy is difficult to realise in practice but if we allow discontinuities for the price process at arbitrary times, then we should also allow strategies involving the same kind of pathology. The example shows that we should go beyond the simple integration theory to cover these cases as well. To back this assertion let us recall that the basis of the whole theory of asset pricing by arbitrage is, of course, the celebrated Black-Scholes formula (see [BS 73, M 73]), widely used today by practitioners in option trading. Also in this case the trading strategy H, which perfectly replicates the payoff of the given option, is not a simple integrand. It is described as a smooth function of time and the underlying stock price. Being a smooth function of the stock price, its trajectories are in fact of unbounded variation. One can argue that in practice already this strategy is difficult to realise. In this case, however, one shows that the integrand can be approximated by simple integrands in a reasonable way; for details we refer the reader to books an stochastic integration theory with special emphasis an Brownian motion, e.g. [KS 88]. In the case of the example of Sect. 9.7, this reduction is not possible and as already advocated, general integrands are really needed.

Summing up we are forced to leave the framework of simple integrands. However, we immediately face new problems. First the process S should be restricted in order to allow the definition of integrals $H \cdot S$ for more general trading strategies. S has to be a semi-martingale to realise this. This is precisely the content of the Bichteler-Dellacherie theorem (see [P 90]). It turns out that this is not really a restriction. From the work of [FS 91, AS 93], we know that no free lunch conditions stated with simple integrands, imply that a càdlàg adapted process is a special semi-martingale. (A process is called càdlàg, if almost every trajectory admits left limits and is right continuous). We refer to Sect. 9.7 of this paper for a general version of this result, adapted to our framework. The second difficulty arises from the fact that doubling-like strategies have to be excluded. This may be done by using the concept of *admissible integrands* H, requiring that the process $H \cdot S$ is uniformly bounded from below, a concept going back to [HP 81] and developed in [D 92, MB 91, S 94]. The concept of admissible integrand is a mathematical formulation of the requirement that an economic agent's position cannot become too negative, a practice sometimes referred to as "your friendly broker calls for extra margin". The third problem is to make sure that $(H \cdot S)_\infty = \lim_{t \to \infty} (H \cdot S)_t$ has a meaning. We shall see that this problem has a very satisfactory solution if one restricts to admissible integrands.

The condition of *no free lunch with vanishing risk (NFLVR)* can now be described as follows. There should be no sequence of final payoffs of admissible integrands, $f_n = (H^n \cdot S)_\infty$ such that the negative parts f_n^- tend to 0 *uniformly* und such that f_n tends almost surely to a $[0, \infty]$-valued function f_0 satisfying $\mathbf{P}[f_0 > 0] > 0$. We will give a detailed discussion of this property below in Sect. 9.3. For the time being let us remark that the property *(NFLVR)* is different from the previously considered concept of no free lunch with bounded risk in the sense that we require that the risk taken, the lower bounds on the processes $(H^n \cdot S)$, tend to zero uniformly. In the property *(NFLBR)* one only requires that this risk is uniformly bounded below und that the variables f_n^- tend to zero in probability. The main theorem of the paper can now be stated as:

Theorem 9.1.1. *Let S be a bounded real-valued semi-martingale. There is an equivalent martingale measure for S if und only if S satisfies (NFLVR).*

One implication in the above theorem is almost trivial: if there is an equivalent martingale measure for S then it is easy to see that S satisfies *(NFLVR)*, see the first part of the proof in the beginning of Sect. 9.4. The interesting aspect of Theorem 9.1.1 lies in the reverse implication: the (economically meaningful) assumption *(NFLVR)* guarantees the existence of an equivalent martingale measure for S und thus opens the way to the wide range of applications from martingale theory.

If the process S is only a locally bounded semi-martingale we still obtain the following partial result:

Corollary 9.1.2. *Let S be a locally bounded real-valued semi-martingale. There is an equivalent local martingale measure for S if and only if S satisfies (NFLVR).*

In [DS 94a] counter-examples are given which show that in the above corollary one can only assert the existence of a measure **Q** under which S is a *local* martingale. Even if the variables S_t, are uniformly bounded in L^p for some $p > 1$, this does not imply that S is a martingale. On the other hand we do not know whether the hypothesis of local boundedness is essential for the corollary to hold. There is some hope that the condition is superfluous but at present this remains an open question.[†] In the discrete time case the local boundedness assumption is not needed as shown in [S 94].

The proof of Theorem 9.1.1 is quite technical and will be the subject of Sect. 9.4. The rest of the paper is organised as follows. Sect. 9.2 deals with definitions, notation and results of general nature. In Sect. 9.3 we examine the property *(NFLVR)* and we prove that under this condition, the limit $(H \cdot S)_\infty = \lim_{t\to\infty} (H \cdot S)_t$ exists almost surely for admissible integrands. The fifth section is devoted to the study of the set of local martingale measures. Here we give a new characterisation of a complete market. It turns out that if each local martingale measure that is absolutely continuous with respect to the original measure, is already equivalent to the original measure, then the market is complete and there is only one equivalent (absolutely continuous) local martingale measure. These results are related to results in [AS 94, J 92]. We also show that the framework of admissible integrands allows to formulate a general duality theorem (Theorem 9.5.8). In Sect. 9.6 we investigate the relation between the no free lunch with vanishing risk *(NFLVR)* property and the no free lunch with bounded risk *(NFLBR)* property. In the case of an infinite horizon the latter property permits to restrict to strategies that are of bounded support. They have a more intuitive interpretation since they only require 'planning' up to a bounded time. In Sect. 9.7 we introduce the no free lunch properties *(NFLVR)*, *(NFLBR)* and *(NFL)* stated in terms of simple strategies. It is shown that in the case of continuous price processes one can avoid the use of general integrands and restrict oneself to simple integrands. The result generalises the main theorem of [D 92] in the case of a finite dimensional price process. The relation between the no free lunch with vanishing risk property for simple integrands and the semi-martingale property is also investigated in Sect. 9.7. We also give examples that show that the use of simple integrands is not enough to obtain a general theorem and relate the present results to previous ones, in particular to [D 92, S 94]. Appendix 9.8 contains some technical lemmas already used in [S 94]. We state versions which are more general and provide somewhat easier proofs.

[†] Note added in this reprint: The answer to this question is given in Chap. 14 below. In fact the notion of a local martingale measure has to be replaced by the notion of a sigma-martingale measure. This is precisely the theme of Chap. 14 below.

9.2 Definitions and Preliminary Results

Throughout the paper we will work with random variables and stochastic processes which are defined on a fixed probability space $(\Omega, \mathcal{F}, \mathbf{P})$. We will without further notice identify variables that are equal almost everywhere. The space $L^0(\Omega, \mathcal{F}, \mathbf{P})$, sometimes written as L^0, is the space of equivalence classes of measurable functions, defined up to equality almost everywhere. The space L^0 is equipped with the topology of convergence in measure. It is a complete metrisable topological vector space, a Fréchet space, but it is not locally convex. The space $L^1(\Omega, \mathcal{F}, \mathbf{P})$ is the Banach space of all integrable $\mathcal{F}$-measurable functions. The dual space is identified with $L^\infty(\Omega, \mathcal{F}, \mathbf{P})$ the space of bounded measurable functions. The weak-star topology on L^∞ is the topology $\sigma(L^\infty, L^1)$.

The existence of an equivalent martingale measure is proved using Hahn-Banach type theorems. Central in this approach is the construction of a convex weak-star-closed subset of L^∞. To prove that a set is weak-star-closed we will use the following result. The proof essentially consists of a combination of the classical Krein-Smulian theorem and the fact that the unit ball of L^∞ under the weak-star topology is an Eberlein compact. (see [D 75] or [G 54, Exercise 1, p. 321].

Theorem 9.2.1. *If C is a convex cone of L^∞ then C is weak-star-closed if and only if for each sequence $(f_n)_{n \geq 1}$ in C that is uniformly bounded by 1 and converges in probability to a function f_0, we have that $f_0 \in C$.*

The properties of stochastic processes are always defined relative to a fixed filtration $(\mathcal{F}_t)_{t \in \mathbb{R}_+}$. This filtration is supposed to satisfy the usual conditions i.e. the filtration is right continuous and contains all negligible sets: if $B \subset A \in \mathcal{F}$ and $\mathbf{P}[A] = 0$ then $B \in \mathcal{F}_0$. We also suppose that the σ-algebra $\mathcal{F}$ is generated by $\bigcup_{t \geq 0} \mathcal{F}_t$. Stochastic intervals are denoted as $[\![T, S]\!]$ where $S \leq T$ are stopping times and $[\![T, S]\!] = \{(t, \omega) \mid t \in \mathbb{R}_+, \; \omega \in \Omega, \; T(\omega) \leq t \leq S(\omega)\}$. Stochastic intervals of the form $]\!]T, S]\!]$ etc. are defined in the same way. The interval $[\![T, T]\!]$ is denoted by $[\![T]\!]$ and it is the graph of the stopping time $T, \{(T(\omega), \omega) \mid T(\omega) < \infty\}$. We note that according to this definition the set $[\![0, \infty]\!]$ equals $\mathbb{R}_+ \times \Omega$. Stochastic processes are indexed by a time set. In this paper the time set will be $\mathbb{R}_+$. This will cover the case of infinite horizon and indeed represents the general case since bounded time sets $[0, t]$ can of course be imbedded by requiring the processes to be constant after time t. It also contains the case of discrete time sets, by requiring the processes and the filtration to be constant between two consecutive natural numbers. A mapping $X \colon \mathbb{R}_+ \times \Omega \to \mathbb{R}$ is called an adapted stochastic process if for each $t \in \mathbb{R}_+$ the mapping $\omega \mapsto X(t, \omega) = X_t(\omega)$ is $\mathcal{F}_t$-measurable. X is called continuous (right continuous, left continuous), if for almost all $\omega \in \Omega$, the mapping $t \mapsto X_t(\omega)$ is continuous (right continuous, left continuous). Stochastic processes that are indistinguishable are always identified. Other concepts such as optional and predictable processes are also used in this paper and we refer the reader to [P 90] for the details. The predictable σ-algebra $\mathcal{P}$ on $\mathbb{R}_+ \times \Omega$ is the σ-algebra generated by the stochastic intervals $[\![0, T]\!]$, where T runs through all the stopping times. A predictable process H is a process that is measurable for the σ-algebra $\mathcal{P}$. For the theory of stochastic integration we refer to [P 90] and to [CMS 80]. If X is a real-valued stochastic process the variable X^* is defined as $X^* = \sup_{t \geq 0} |X_t(\omega)|$. This variable is measurable if X is right or left continuous. Sometimes we will use X_t^* which is defined as $\sup_{t \geq u \geq 0} |X_u(\omega)|$. X^* is called the maximum function and it plays a central role in martingale theory. If X is a càdlàg process, i.e. a right continuous process possessing left limits for each $t > 0$, then ΔX denotes the process that describes the jumps of X. More precisely $(\Delta X)_t = X_t - X_{t-}$ and $(\Delta X)_0 = X_0$.

If X is a semi-martingale then X defines a continuous operator on the space of bounded predictable processes of bounded support into the space L^0. The space of semi-martingales can therefore be considered as a space of linear operators. The semi-martingale topology is precisely induced by the topology of linear operators. It is therefore metrisable by a translation invariant metric given by the distance of X to the zero semi-martingale:

$$\mathbf{D}(X) = \sup \left\{ \sum_{n \geq 1} 2^{-n} \mathbf{E} \left[\min \left(|(H \cdot X)_n|, 1 \right) \right] \, \middle| \, H \text{ predictable}, \, |H| \leq 1 \right\}.$$

For this metric, the space of semi-martingales is complete, see [E 79]. A semi-martingale X is called special if it can be decomposed as $X = M + A$ where M is a local martingale and A is a *predictable* process of finite variation. In this case such a decomposition is unique and it is called the canonical decomposition. It is well-known (see [CMS 80]) that a semi-martingale is special if and only if X is locally integrable, i.e. there is an increasing sequence of stopping times T_n, tending to ∞ such that $X_{T_n}^*$ is integrable. The following theorem on special semi-martingales will be used on several occasions, for a proof we refer to [CMS 80].

Theorem 9.2.2. *If X is a special semi-martingale with canonical decomposition $X = M + A$ and if H is X-integrable then the semi-martingale $H \cdot X$ is special if and only if*

(1) *H is M-integrable in the sense of stochastic integrals of local martingales and*
(2) *H is A-integrable in the usual sense of Stieltjes-Lebesgue integrals.*

In this case the canonical decomposition of $H \cdot X$ is given by $H \cdot X = H \cdot M + H \cdot A$.

The following theorem seems to be folklore. Essentially it may be deduced from (the proof of) an inequality of Stein ([St 70]), see also [L 78, Y 78b]. For a survey of these results and related inequalities see [DS 95d]. For convenience of the reader we include the easy proof, suggested by Stricker, of Theorem 9.2.3.

The theorem and more precisely its Corollary 9.2.4, will be used in Sect. 9.4. It allows to control the jumps of the martingale part in the canonical decomposition of a special semi-martingale.

Theorem 9.2.3. *If X is a semi-martingale satisfying $\|(\Delta X)^*\|_p < \infty$, where $1 < p \le \infty$, then*

(a) *X is special and has a canonical decomposition $X = M + A$*
(b) *A satisfies $\|(\Delta A)^*\|_p \le \frac{p}{p-1}\|(\Delta X)^*\|_p$;*
(c) *M satisfies $\|(\Delta M)^*\|_p \le \frac{2p-1}{p-1}\|(\Delta X)^*\|_p$.*

Proof. Since X is locally p-integrable it is certainly locally integrable and hence is special. (a) is therefore proved. Let $X = M + A$ be the canonical decomposition where A is the predictable process of finite variation and M is the local martingale part. Let Y be the càdlàg martingale defined as $Y_t = \mathbf{E}[(\Delta X)^* \mid \mathcal{F}_t]$.

Since A is predictable the set $\{\Delta A \ne 0\}$ is the union of a sequence of sets of the form $[\![T_n]\!]$ where T_n are predictable stopping times. For each predictable stopping time T we have that $\Delta A_T = \mathbf{E}[\Delta X_T \mid \mathcal{F}_{T-}]$ and hence

$$|\Delta A_T| \le \mathbf{E}[|\Delta X_T| \mid \mathcal{F}_{T-}] \le \mathbf{E}[(\Delta X)^* \mid \mathcal{F}_{T-}] = Y_{T-} \le Y^*.$$

This implies that $(\Delta A)^* \le Y^*$. From Doob's maximal inequality, see [DM 80], it now follows that

$$\|Y^*\|_p \le \frac{p}{p-1}\|(\Delta X)^*\|_p \text{ and therefore}$$

$$\|(\Delta A)^*\|_p \le \frac{p}{p-1}\|(\Delta X)^*\|_p \text{ and } \|(\Delta M)^*\|_p \le \frac{2p-1}{p-1}\|(\Delta X)^*\|_p. \qquad \square$$

Corollary 9.2.4. *If T is a stopping time then:*

$$\|(\Delta A)_T\|_p \le \frac{p}{p-1}\|(\Delta X)^*\|_p;$$

$$\|(\Delta M)_T\|_p \le \frac{2p-1}{p-1}\|(\Delta X)^*\|_p.$$

Corollary 9.2.5. *If $1 < p \le \infty$ and the semi-martingale X satisfies $\sup\{\|(\Delta X)_T\|_p \mid T \text{ stopping time }\} = N < \infty$, then for $p' < p$ there is constant $k(p, p')$ depending only an p and p' such that*

$$\|(\Delta A)^*\|_{p'} \le k(p, p')N.$$

Proof. Let the stopping time T be defined as $T = \inf\{t \mid |(\Delta X)_t| \ge c\}$. From the hypothesis we deduce that $\int |\Delta X_T| \le N^p$ and this implies, by the Markov-Tchebycheff inequality, that $c^p \mathbf{P}[(\Delta X)^* > c] \le N^p$. The rest follows easily. $\qquad \square$

Remark 9.2.6. In Corollary 9.2.4 we cannot replace $(\Delta X)^*$ by $(\Delta X)_T$. The following example illustrates this. We construct a bounded semi-martingale X such that for each $\varepsilon > 0$ there is a stopping time T with $|\Delta A_T| = 1$ and $|\Delta X_T| \leq \varepsilon$. This clearly shows that there is no constant K such that $\|(\Delta A)_T\|_p \leq K\|(\Delta X)_T\|_p$. The construction is as follows: For $0 \leq t < 1$ put $X_t = 0$. We now proceed by recursion. For n a natural number we suppose the process X is already constructed for $t < n$. The filtration $\mathcal{F}_s$ is defined as $\mathcal{F}_s = \sigma(X_u; u \leq s)$ and $\mathcal{F}_{s-} = \sigma(X_u; u < s)$. At $t = n$ we put a jump $(\Delta X)_n$ such that $|(\Delta X)_n|$ is uniformly distributed over the interval $[0, 2]$ and is independent of the past $\mathcal{F}_{n-}$ of the process. This means that $|(\Delta X)_n|$ is independent of the variables $(\Delta X)_1, \ldots, (\Delta X)_{n-1}$. If $(X)_{n-} \geq 0$ then $(\Delta X)_n$ is uniformly distributed over the interval $[-2, 0]$, otherwise if $(X)_{n-} < 0$ then $(\Delta X)_n$ is uniformly distributed over $[0, 2]$. For $n \leq t < n+1$ we put $X_t = X_n$. The filtration $\mathcal{F}_s$ is clearly right continuous and if we augment it with the null sets we obtain that the natural filtration of X satisfies the usual conditions. For $\varepsilon > 0$ we now define $T = \inf\{t \mid |(\Delta X)_t| \leq \varepsilon\}$. Clearly $T < \infty$ almost surely and satisfies the desired properties.

If A is a predictable process of finite variation with $A_0 = 0$, we can associate with it a (random) measure on $\mathbb{R}_+$. The variation of A, a process denoted by V, is given by

$$V_t = \sup\left\{ \sum_{k=1}^n |A_{s_k} - A_{s_{k-1}}| \,\middle|\, 0 = s_0 < s_1 < \ldots < s_n = t \right\}.$$

The process V is predictable and it also defines a (random) measure on $\mathbb{R}_+$. The process V defines a σ-finite measure μ_V on the predictable σ-algebra on $\mathbb{R}_+ \times \Omega$. The definition of μ_V is, for K a predictable subset of $\mathbb{R}_+ \times \Omega$:

$$\mu_V(K) = \mathbf{E}\left[\int_0^\infty (\mathbf{1}_K)_u \, dV_u \right].$$

The measure μ_A is defined in a similar way, but its definition is restricted to a σ-ring to avoid expressions like $\infty - \infty$. It is well-known, see [M 76, Chap. I]), that the measure μ_V is precisely the variation measure of μ_A. From the Hahn decomposition theorem we deduce that there is a partition of $\mathbb{R}_+ \times \Omega$, in two sets, B_+ and B_-, both predictable, such that $(\mathbf{1}_{B_+} \cdot A)$ and $(-\mathbf{1}_{B_-} \cdot A)$ are increasing. Moreover $V = ((\mathbf{1}_{B_+} - \mathbf{1}_{B_-}) \cdot A)$. For almost all ω the measure dA on $\mathbb{R}_+$ is absolutely continuous with respect to dV and the Radon-Nikodým derivative is precisely $\mathbf{1}_{F_+} - \mathbf{1}_{F_-}$ where $F_\pm = \{t \mid (t, \omega) \in B_\pm\}$. We will refer to this decomposition as the **Hahn decomposition** of A. Note that the difficulty in the definition of the pathwise decomposition of the measures $dA(\omega)$ comes from the fact that the sets F_+ and F_- have to be glued together in order to form the predictable sets B_+ and B_-. See [M 76, Chap. I] for the details of this result which is due to Cathérine Doléans-Dade.

Throughout the paper, with the exception of Sect. 9.7, S will be a fixed semi-martingale. As mentioned in the introduction S represents the discounted price of a financial asset.

Definition 9.2.7. *Let a be a positive real number. An S-integrable predictable process H is called **a-admissible** if $H_0 = 0$ and $(H \cdot S) \geq -a$ (i.e. for all $t \geq 0 : (H \cdot S)_t \geq -a$ almost everywhere). H is called admissible if it is admissible for some $a \in \mathbb{R}_+$.*

Given the semi-martingale S we denote, in a similar way as in [Str 90], by K_0 the convex cone in L^0, formed by the functions

$$K_0 = \left\{ (H \cdot S)_\infty \;\middle|\; H \text{ admissible and } (H \cdot S)_\infty = \lim_{t \to \infty} (H \cdot S)_t \text{ exists a.s.} \right\}.$$

By C_0 we denote the cone of functions dominated by elements of K_0 i.e. $C_0 = K_0 - L_+^0$. With C and K we denote the corresponding intersections with the space L^∞ of bounded functions $K = K_0 \cap L^\infty$ and $C = C_0 \cap L^\infty$. By $\overline{C}$ we denote the closure of C with respect to the norm topology of L^∞ and by $\overline{C}^*$ we denote the weak-star-closure of C.

Definition 9.2.8. *We say that the semi-martingale S satisfies the condition*

(i) ***no-arbitrage** (NA) if $C \cap L_+^\infty = \{0\}$*
(ii) ***no free lunch with vanishing risk** (NFLVR) if $\overline{C} \cap L_+^\infty = \{0\}$.*

It is clear that *(ii)* implies *(i)*. The no-arbitrage property *(NA)* is equivalent to $K_0 \cap L_+^0 = \{0\}$ and has an obvious interpretation: there should be no possibility of obtaining a positive profit by trading alone (according to an admissible strategy): it is impossible to make something out of nothing without risk. It is well-known that in general the notion *(NA)* is too restrictive to imply the existence of an equivalent martingale measure for S, see Sect. 9.7. Compare also to the results in [DMW 90] and [S 94, Remark 4.11].

The notion *(NFLVR)* is a slight generalisation of *(NA)*. If *(NFLVR)* is not satisfied then there is a f_0 in L_+^∞ not identically 0, as well as a sequence $(f_n)_{n \geq 1}$ of elements in C, tending almost surely to f_0 such that for all n we have that $f_n \geq f_0 - \frac{1}{n}$. In particular we have $f_n \geq -\frac{1}{n}$. In economic terms this amounts to almost the same thing as *(NA)*, as the risk of the trading strategies becomes arbitrarily small. See also Proposition 9.3.7 below.

We emphasize that the set C and hence the properties *(NA)* and *(NFLVR)* are defined using *general admissible predictable processes* H. This is a more general definition than the one usually taken in the literature and used by the authors in previous papers (see [S 94, D 92]). These classical concepts were defined using simple integrands or/and integrands with bounded support. In these cases we will say that S satisfies *(NA) for simple integrands*, *(NFLVR) for integrands with bounded support*, etc. These notions will reappear in Sect. 9.7, where we will emphasize on the differences between these notions.

We close this section by quoting a result due to Émery and Ansel and Stricker. The result states that under suitable conditions the stochastic integral of a local martingale is again a local martingale. A counter-example due to [E 80] shows that in general a stochastic integral of a local martingale need not be a local martingale. From Theorem 9.2.2 it follows that if M is a local martingale with respect to a measure $\mathbf{P}$, then $H \cdot M$ is a local martingale if and only if it is a special semi-martingale, i.e. if it is locally integrable. The next theorem gives us a criterion that is related to admissibility of H.

Theorem 9.2.9. *If M is a local martingale and if H is an admissible integrand for M, then $H \cdot M$ is a local martingale. Consequently $H \cdot M$ is a super-martingale.*

Proof. We refer to [E 80] and [AS 94, Corollaire 3.5]. It is an easy consequence of Fatou's lemma that if $H \cdot M$ is a local martingale uniformly bounded from below, then it is a super-martingale.   □

9.3 No Free Lunch with Vanishing Risk

The main result of this section states that for a semi-martingale S, under the condition of no free lunch with vanishing risk *(NFLVR)*, the limit $(H \cdot S)_\infty = \lim_{t \to \infty}(H \cdot S)_t$ exists and is finite whenever the integrand H is admissible. To get a motivation for this result, consider the case where we already know that there is an equivalent local martingale measure $\mathbf{Q}$. In this case, by Theorem 9.2.9, the stochastic integral $H \cdot S$ is a $\mathbf{Q}$-local martingale if H is admissible. This implies that it is a super-martingale and the classical convergence theorem shows that the limit $(H \cdot S)_\infty = \lim_{t \to \infty}(H \cdot S)_t$, exists and is finite almost everywhere. But of course we do not know yet that there is an equivalent martingale measure $\mathbf{Q}$ and the art of the game is to derive the convergence result simply from the property *(NFLVR)*. We start with two preparatory results.

Proposition 9.3.1. *If S is a semi-martingale with the property (NFLVR), then the set*

$$\{(H \cdot S)_\infty \mid H \text{ is } 1\text{-admissible and of bounded support}\}$$

is bounded in L^0.

Proof. H 1-admissible means that H is S-integrable and $(H \cdot S)_t \geq -1$. Being of bounded support means that H is 0 outside $[\![0, T]\!]$ where T is a positive real number. The limit $(H \cdot S)_\infty = \lim_{t \to \infty}(H \cdot S)_t$ exists without difficulty because $(H \cdot S)_t$ becomes eventually constant. Suppose that the set $\{(H \cdot S)_\infty \mid H$ is 1-admissible and of bounded support$\}$ is not bounded in L^0. This implies the existence of a sequence H^n of 1-admissible integrands of bounded support and the existence of $\alpha > 0$ such that $\mathbf{P}[(H^n \cdot S)_\infty \geq n] > \alpha > 0$. The sequence $f_n = \min\left(\frac{1}{n}(H^n \cdot S)_\infty, 1\right)$ is in C, $\mathbf{P}[f_n = 1] > \alpha > 0$ and $\|f_n^-\|_\infty \leq \frac{1}{n}$. By taking convex combinations we may take $g_n \in \text{conv}\{(f_n, f_{n+1}, \ldots\}$ that converge a.s. to $g : \Omega \to [0, 1]$. (We can use Lemma 9.8.1, but a simpler argument in L^∞ can do the job, compare [S 94, Remark 3.4]). Clearly $\mathbf{E}[g] \geq \alpha$ and therefore $\mathbf{P}[g > 0] = \beta \geq \alpha > 0$. By Egorov's theorem $g_n \to g$ uniformly on a set Ω' of measure at least $1 - \frac{\beta}{2}$. The functions $h_n = \min(g_n, \mathbf{1}_{\Omega'})$ are still in the set C and $h_n \to g\mathbf{1}_{\Omega'}$ in the norm topology of L^∞. Since $\mathbf{P}[g\mathbf{1}_{\Omega'} > 0] \geq \frac{\beta}{2} > 0$ we obtain a contradiction to *(NFLVR)*. $\qquad\square$

Proposition 9.3.2. *If S is a semi-martingale satisfying (NFLVR), then for each admissible H the function $(H \cdot S)^* = \sup_{0 \leq t} |(H \cdot S)_t|$ is finite almost everywhere and the set $\{(H \cdot S)^* \mid H \text{ 1-admissible}\}$ is bounded in L^0.*

Proof. If the set is not bounded, we can find a sequence of 1-admissible integrands H^n, stopping times T_n and $\alpha > 0$ such that $\mathbf{P}[T_n < \infty] > \alpha > 0$ and $(H^n \cdot S)_{T_n} > n$ on $\{T_n < \infty\}$. For each natural number n take t_n large enough so that $\alpha < \mathbf{P}[T_n \leq t_n]$ and observe that for $K^n = H^n \mathbf{1}_{[\![0,\min(T_n,t_n)]\!]}$ we have that K^n is of bounded support and $\mathbf{P}[(K^n \cdot S)_\infty > n] > \alpha > 0$, a contradiction to Proposition 9.3.1. $\qquad\square$

We now prove the main result of this section. It extends from [S 94, Proposition 4.2] to the present case of a general semi-martingale S.

Theorem 9.3.3. *If S is a semi-martingale satisfying (NFLVR), then for H admissible the limit $(H \cdot S)_\infty = \lim_{t\to\infty}(H \cdot S)_t$ exists and is finite almost everywhere.*

Proof. We will mimic the proof of the martingale convergence theorem of Doob. The classical idea of considering upcrossings through an interval $[\beta, \gamma]$ may in mathematical finance be interpreted as the well-known procedure: "Buy low, sell high". We may suppose that H is 1-admissible and hence $(H \cdot S)^* = \sup_{0 \leq t} |(H \cdot S)_t| < \infty$ almost surely by Proposition 9.3.2. We therefore only have to show that $\liminf_{t\to\infty}(H \cdot S)_t = \limsup_{t\to\infty}(H \cdot S)_t$ a.s.. Suppose this were not the case and that $\mathbf{P}[\liminf_{t\to\infty}(H \cdot S)_t < \limsup_{t\to\infty}(H \cdot S)_t] > 0$. Take $\beta < \gamma$ and $\alpha > 0$ so that $\mathbf{P}[\liminf_{t\to\infty}(H \cdot S)_t < \beta < \gamma < \limsup_{t\to\infty}(H \cdot S)_t] > \alpha$. We will construct finite stopping times $(U_n, V_n)_{n \geq 1}$, such that

(1) $U_1 \leq V_1 \leq U_2 \leq V_2 \leq \ldots \leq U_n \leq V_n \leq U_{n+1} \leq \ldots$
(2) $L^n = \sum_{k=1}^n H\mathbf{1}_{]\!]U_k,V_k]\!]}$ is $(1 + \beta)$-admissible
(3) $\mathbf{P}[(L^n \cdot S)_\infty > n(\gamma - \beta)] > \frac{\alpha}{2}$.

The existence of such a sequence clearly violates the conclusion of Proposition 9.3.2 and this will prove the Theorem.

The stopping times are constructed by induction. Take $(\varepsilon_n)_{n \geq 1}$ strictly positive and such that the sum $\sum_{n \geq 1} \varepsilon_n < \frac{\alpha}{100}$. Let A be the set defined as

$A = \{\liminf_{t\to\infty}(H \cdot S)_t < \beta < \gamma < \limsup_{t\to\infty}(H \cdot S)_t\}$. Since the Boolean algebra $\bigcup_{0 \le t} \mathcal{F}_t$ is dense in the σ-algebra $\mathcal{F}$ we have that there is t_1 and $A_1 \in \mathcal{F}_{t_1}$ such that $\mathbf{P}[A \triangle A_1] < \varepsilon_1$. For $\omega \ne A_1$ we put $U_1 = V_1 = t_1$ and we concentrate on $\omega \in A_1$.

First define

$$U_1' = \inf\{t \mid t \ge t_1 \text{ and } (H \cdot S)_t < \beta\} \text{ for } \omega \text{ in } A_1$$
$$V_1' = \inf\{t \mid t \ge U_1' \text{ and } (H \cdot S)_t > \gamma\} \text{ for } \omega \text{ in } A_1\,.$$

The variables U_1' and V_1' are clearly stopping times and take values in $[0, \infty]$. By construction of A_1 we have that

$$\mathbf{P}[V_1' < \infty] \ge \mathbf{P}[A \cap A_1] > \alpha - \varepsilon_1\,.$$

Take $s_1 > t_1$ so that $\mathbf{P}[V_1' \le s_1] > \alpha - \varepsilon_1$ and define

$$U_1 = \min(U_1', s_1)\,,$$
$$V_1 = \min(V_1', s_1)\,.$$

The set $B_1 = \{(H \cdot S)_{U_1} \le \beta < \gamma \le (H \cdot S)_{V_1}\}$ is in $\mathcal{F}_{s_1}$ and $\mathbf{P}[B_1 \cap A] > \alpha - \varepsilon_1$. Put $K^1 = H\mathbf{1}_{\rrbracket U_1, V_1 \rrbracket}$. We claim that K^1 is $(1 + \beta)$-admissible. Indeed on A_1^c clearly $(K^1 \cdot S)_t = 0$ for all t. For $\omega \in A_1$ and $t \le U_1$ we also have $(K^1 \cdot S)_t(\omega) = 0$. For $\omega \in A_1$ and $U_1 < t \le V_1$ we have

$$(K^1 \cdot S)_t = (H \cdot S)_t - (H \cdot S)_{U_1} \ge -1 - \beta = -(1 + \beta)\,.$$

Let us put $L^1 = K^1$. We now apply the same reasoning on the set $(B_1 \cap A)$ i.e. we take $t_2 \ge s_1$, $A_2 \in \mathcal{F}_{t_2}$ such that $A_2 \subset B_1$, $\mathbf{P}[A_2 \triangle (B_1 \cap A)] > \alpha - \varepsilon_1 - \varepsilon_2$. On the set A_2 we define

$$U_2' = \inf\{t \mid t \ge t_2 \text{ and } (H \cdot S)_t < \beta\}$$
$$V_2' = \inf\{t \mid t \ge U_2' \text{ and } (H \cdot S)_t > \gamma\}\,.$$

$\mathbf{P}[V_2' < \infty] > \alpha - \varepsilon_1 - \varepsilon_2$ and we select $s_2 > t_2$ so that $\mathbf{P}[V_2' \le s_2] > \alpha - \varepsilon_1 - \varepsilon_2$. Take

$$U_2 = \min(U_2', s_2)$$
$$V_2 = \min(V_2', s_2)$$
$$K^2 = H\mathbf{1}_{\rrbracket U_2, V_2 \rrbracket}\,.$$

The integrand is $(1 + \beta)$-admissible, but outside the set B_1 the process $(K^2 \cdot S)$ is zero. On the set B_1, however, $(L^1 \cdot S)_{t_2} = (L^1 \cdot S)_{s_2} \ge \gamma - \beta > 0$. The integrand $L^2 = L^1 + K^2$ remains therefore $(1 + \beta)$-admissible. Furthermore $\mathbf{P}[(L^2 \cdot S)_{t_2} \ge 2(\gamma - \beta)] > \alpha - \varepsilon_1 - \varepsilon_2$. This permits us to continue the construction and to define L^n by induction. $\square$

The rest of this section is devoted to some results giving a better understanding of the property *(NFLVR)* of no free lunch with vanishing risk and relating this property to previous results of [D 92, S 94].

Corollary 9.3.4. *If the semi-martingale S satisfies (NFLVR) then the set*

$$\{(H \cdot S)_\infty \mid H \text{ is } 1\text{-admissible}\}$$

is bounded in L^0.

Proof. This follows immediately from the existence of the limit $(H \cdot S)_\infty$ and from Proposition 9.3.1. □

Remark 9.3.5. The convergence theorem shows in particular that in the definition of K_0 the requirement that the limit exists is superfluous. We also want to point out that to derive the above results 9.3.1 to 9.3.4, we only used the condition *(NFLVR)* for integrands with bounded support, i.e. for integrands that are zero outside a stochastic interval $[\![0, k]\!]$ for some real number k.

The next result only uses the (very weak) assumption of no-arbitrage. We emphasize that the property *(NA)*, as we defined it, refers to general integrands.

Proposition 9.3.6. (compare [S 94, Proposition 4.2]) *If the semi-martingale S satisfies (NA) then for every admissible integrand H, such that $(H \cdot S)_\infty = \lim_{t \to \infty}(H \cdot S)_t$ exists, we have for each $t \in \mathbb{R}_+$:*

$$\|(H \cdot S)_t^-\|_\infty \leq \|(H \cdot S)_\infty^-\|_\infty .$$

Proof. If $\|(H \cdot S)_t^-\|_\infty > \|(H \cdot S)_\infty^-\|_\infty$ then we define the set $A \in \mathcal{F}_t$ as

$$A = \{(H \cdot S)_t < -\|(H \cdot S)_\infty^-\|_\infty\} .$$

The integrand $K = \mathbf{1}_A \mathbf{1}_{]\!]t, \infty[\![}$ is admissible, the random variable $(K \cdot S)_\infty$ exists, is non-negative and $\mathbf{P}[(K \cdot S)_\infty > 0] > 0$. This violates *(NA)*. □

The next result may be seen as a sharpening of [S 94, Proposition 1.5]. It combines the property *(NA)* with the conclusion of Proposition 9.3.1.

Proposition 9.3.7. *If the semi-martingale S fails the property (NFLVR) then either S fails (NA) or there exists $f_0 : \Omega \to [0, \infty]$ not identically 0, a sequence of variables $(f_n)_{n \geq 1} = ((H^n \cdot S)_\infty)_{n \geq 1}$ in K_0 with H^n a $\frac{1}{n}$-admissible integrand and such that $\lim_{n \to \infty} f_n = f_0$ in probability.*

Proof. It is clear that the existence of such sequences violates *(NFLVR)*. Indeed the set $\{n(H^n \cdot S)_\infty; n \geq 1\}$ is unbounded in L^0, whereas the integrands $(nH^n)_{n \geq 1}$ are 1-admissible. This contradicts Proposition 9.3.1.

The converse is less obvious. Suppose that S satisfies *(NA)* and suppose that $(g_n)_{n \geq 1}$ is a sequence in C such that $g_0 = \lim_{n \to \infty} g_n$ in L^∞, $g_0 \geq 0$, $\mathbf{P}[g_0 > \alpha] > \alpha > 0$. From the hypothesis on the sequence $(g_n)_{n \geq 1}$ we deduce that $\|g_n^-\|_\infty$ tends to 0. By passing to a subsequence, if necessary, we may suppose that $\|g_n^-\|_\infty \leq \frac{1}{n}$. For each n we take a function h_n in K_0 such that

$h_n \geq g_n$. If $h_n = (L^n \cdot S)_\infty$ then $\|h_n^-\|_\infty \leq \frac{1}{n}$ and hence L^n is $\frac{1}{n}$-admissible by Proposition 9.3.6 and the property *(NA)* of S. Lemma 9.8.1 allows us to replace h_n by $f_n \in \text{conv}\{h_n, h_{n+1}, \ldots\}$ such that f_n converges to $f_0 : \Omega \to [0, \infty]$ in probability. Let H^n be the corresponding convex combination of the integrands $(L^k)_{k \geq n}$. Obviously H^n is still $\frac{1}{n}$-admissible and f_n^- tends to 0 in L^∞. For n large enough we have $\|g_n - g_0\|_\infty \leq \frac{\alpha}{2}$ and hence $\mathbf{P}[h_n > 0] \geq \mathbf{P}[g_n > \frac{\alpha}{2}] > \frac{\alpha}{2}$. Lemma 9.8.1 now shows that $\mathbf{P}[f_0 > 0] > 0$. □

The following corollary relates the condition *(NFLVR)* with the condition (d) in [D 92] (which in turn is just reformulating the concept of *(NFLBR)* to be defined in Sect. 9.6 below).

Corollary 9.3.8. *The semi-martingale S satisfies the condition (NFLVR) if and only if for a sequence $(g_n)_{n \geq 1}$ in K_0, the condition $\|g_n^-\|_\infty \to 0$ implies that g_n tends to 0 in probability.*

Proof. We first observe that the condition stated in the corollary implies *(NA)*. The corollary is now a direct consequence of the Proposition 9.3.7 and the Lemma 9.8.1. □

Corollary 9.3.9. *Under the assumption (NA), the semi-martingale S satisfies the condition (NFLVR) if and only if the set*

$$\{(H \cdot S)_\infty \mid H \text{ 1-admissible and of bounded support}\}$$

is bounded in L^0.

Proof. From the proof of Proposition 9.3.2, it follows that the set $\{\sup_{0 \leq t}(H \cdot S)_t \mid H \text{ 1-admissible}\}$ is also bounded in L^0. If the sequence $(g_n)_{n \geq 1}$ in K_0, satisfies $\|g_n^-\|_\infty \to 0$, then by the *(NA)* property and Proposition 9.3.6, $g_n = (H^n \cdot S)_\infty$ where H^n is ε_n-admissible with $\varepsilon_n = \|g_n^-\|_\infty$. The sequence $\frac{1}{\varepsilon_n} g_n$ has to be bounded which is only possible when g_n tends to 0 in probability. The conclusion now follows from the preceding corollary. □

9.4 Proof of the Main Theorem

In this section we prove the main theorem of the paper. The proof follows the following plan: prove that the set C, introduced in Sect. 9.2, is weak-star-closed in L^∞ and apply the separation theorem of Kreps and Yan (see [S 94]), which in turn is a consequence of the Hahn-Banach theorem. We use similar arguments as in [D 92] and [S 94]. The technicalities are, however, different and more complicated.

Definition 9.4.1 (compare [MB 91] and [S 94], Definition 3.4). *A subset D of L^0 is Fatou closed if for every sequence $(f_n)_{n \geq 1}$ uniformly bounded from below and such that $f_n \to f$ almost surely, we have $f \in D$.*

We remark that if D is a cone then D is *Fatou closed* if for every sequence $(f_n)_{n \geq 1}$ in D with $f_n \geq -1$ and $f_n \to f$ almost surely, we have $f \in D$.
The next result is the technical version of the main theorem.

Theorem 9.4.2. *If S is a bounded semi-martingale satisfying (NFLVR), then*

(1) C_0 *is Fatou closed and hence*
(2) $C = C_0 \cap L^\infty$ *is $\sigma(L^\infty, L^1)$-closed.*

Proof. We will not prove the first part of Theorem 9.4.2 immediately, its proof is quite complicated and will fill the rest of this section.

The second assertion is proved using Theorem 9.2.1. If C_0 is Fatou closed then we have to prove that $C = C_0 \cap L^\infty$ is closed for the topology $\sigma(L^\infty, L^1)$. Take a sequence $(f_n)_{n \geq 1}$ in C, uniformly bounded in absolute value by 1 and such that $f_n \to f$ almost surely. Since C_0 is Fatou closed the element f belongs to C_0 and hence also $f \in C$. □

We now show how Theorem 9.4.2 implies the main theorem of the paper. For convenience of the reader we restate the main Theorem 9.1.1.

Theorem 9.1.1 (Main Theorem). *Let S be a bounded real-valued semi-martingale. There is an equivalent martingale measure $\mathbf{Q}$ for S if and only if S satisfies (NFLVR).*

Proof. We proceed on a well-known path ([D 92, MB 91, S 92, Str 90, L 92, S 94]). Since S satisfies *(NA)* we have $C \cap L^\infty_+ = \{0\}$. Because C is weak-star-closed in L^∞ we know that there is an equivalent probability measure $\mathbf{Q}$ such that $\mathbf{E_Q}[f] \leq 0$ for each f in C. This is precisely the Kreps-Yan separation theorem, for a proof of which we refer to [S 94, Theorem 3.1]. For each $s < t$, $B \in \mathcal{F}_s$, $\alpha \in \mathbb{R}$ we have $\alpha(S_t - S_s)\mathbf{1}_B \in C$ (S is bounded!). Therefore $\mathbf{E_Q}[(S_t - S_s)\mathbf{1}_B] = 0$ and $\mathbf{Q}$ is a martingale measure for S.

The condition *(NFLVR)* is not altered if we replace the original probability measure by an equivalent one. In the proof that condition *(NFLVR)* is also necessary, we may therefore suppose that $\mathbf{P}$ is already a martingale measure for the bounded semi-martingale S. If H is an admissible integrand then by Theorem 9.2.9 we know that the process $(H \cdot S)$ is a super-martingale. Therefore $\mathbf{E}[(H \cdot S)_\infty] \leq \mathbf{E}[(H \cdot S)_0] = 0$. Every function f in C therefore satisfies $\mathbf{E}[f] \leq 0$. The same applies for elements in the norm closure $\overline{C}$ of C. Therefore $\overline{C} \cap L^\infty_+ = \{0\}$. □

We now show how the main theorem implies Corollary 9.1.2 pertaining to the locally bounded case. We refer to [DS 94a] for examples that show that we can only obtain an equivalent local martingale measure for the process S. The proof of Corollary 9.1.2 is similar to [S 94, Theorem 5.1].

Corollary 9.1.2. *Let S be a locally bounded real-valued semi-martingale. There is an equivalent local martingale measure $\mathbf{Q}$ for S if and only if S satisfies (NFLVR).*

Proof. Since S is locally bounded, there is a sequence $\alpha_n \to +\infty$ and an increasing sequence of stopping times $T_n \to \infty$ so that on $[\![0, T_n]\!]$ the process S is bounded by α_n. We replace S by

$$\widetilde{S} = S\mathbf{1}_{[\![0,T_1]\!]} + \sum_{n \geq 1} 2^{-n} \frac{1}{\alpha_n + \alpha_{n+1}} (\mathbf{1}_{]\!]T_n, T_{n+1}]\!]} \cdot S),$$

$\widetilde{S}$ is bounded and satisfies *(NFLVR)* since the outcomes of admissible integrands are the same for S and $\widetilde{S}$. A martingale measure for $\widetilde{S}$ is a local martingale measure for S and therefore the corollary follows from the main theorem. The proof of the necessity of the condition *(NFLVR)* is proved in the same way as in the Theorem 9.1.1. □

Remark 9.4.3. The necessity of the condition *(NFLVR)* and Theorem 9.4.2 show that if S is a locally bounded local martingale then the set C_0 is Fatou closed.

We now proceed with the proof of Theorem 9.4.2. The bounded semi-martingale S will be assumed to satisfy the property *(NFLVR)*. We take a sequence $h_n \in C_0$, $h_n \geq -1$ and $h_n \to h$ a.s.; we have to show $h \in C_0$. This is the same as showing that there is a $f_0 \in K_0$ with $f_0 \geq h$. For each n we take $g_n \in K_0$ such that $g_n \geq h_n$. The sequence g_n is not necessarily convergent and even if it were, this does not give good information about the sequence of integrands used to construct g_n. To overcome this difficulty we introduce a maximal element (compare Remark 9.4.5 below). Define $\mathfrak{D}$ as the set $\mathfrak{D} = \{f \mid \text{there is a sequence } K^n \text{ of 1-admissible integrands such that } (K^n \cdot S)_\infty \to f \text{ a.s. and } f \geq h\}$.

Lemma 9.4.4. *The set $\mathfrak{D}$ is not empty and contains a maximal element f_0.*

Proof. $\mathfrak{D}$ is not empty. Indeed $\mathfrak{D}$ contains an element g that dominates h. To see this we take g_n as above and apply Lemma 9.8.1. Next observe that the set $\mathfrak{D}$ is bounded in L^0 since it is contained in the closure of the set $\{(H \cdot S)_\infty \mid H \text{ 1-admissible}\}$ which is bounded by Corollary 9.3.4. The set $\mathfrak{D}$ is clearly closed for the convergence in probability. We now apply the well-known fact that a bounded closed set of L^0 contains a maximal element. For completeness we give a proof. We will use transfinite induction. For $\alpha = 1$ take an arbitrary element f_1 of $\mathfrak{D}$. If α is of the form $\alpha = \beta + 1$ and if f_β is not maximal then choose $f_\alpha \geq f_\beta$; $\mathbf{P}[f_\alpha > f_\beta] > 0$ and $f_\alpha \in \mathfrak{D}$. If α is a countable limit ordinal then $\alpha = \lim \beta_n$ where β_n is increasing to α. The sequence f_{β_n}, is increasing and converges to a function f_α finite a.s. ($\mathfrak{D}$ is bounded!). In this way we construct for each countable ordinal the variable f_α. Since $\mathbf{E}[\exp(-f_\alpha)]$ is well-defined and form a decreasing "long sequence", this sequence has to become eventually stationary, say at a countable ordinal α_0. By construction $f_0 = f_{\alpha_0}$ is maximal. □

Remark 9.4.5. Let us motivate why we introduced the maximal element f_0 in the above lemma. As already observed the sequence g_n introduced before Lemma 9.4.4 is not of immediate use. Our goal is, of course, to find a 1-admissible integrand H_0 which is, in some sense, a limit of the sequence H_n of the 1-admissible integrands used to construct the sequence g_n. But the convergence of $(g_n)_{n \geq 1}$ (which we may assume by Lemma 9.8.1) does not imply the convergence of the sequence $(H_n)_n$ in any reasonable sense. We illustrate this with the following example in discrete time. Let $(r_m)_{m \geq 1}$ be a sequence of Rademacher functions i.e. a sequence of independent identically distributed variables with $\mathbf{P}[r_m = +1] = \mathbf{P}[r_m = -1] = \frac{1}{2}$. Let $S_m = \sum_{k=1}^{m} r_k$ and $S_0 = 0$. For each n, an odd natural number, we take for the strategy H^n the so called doubling strategy. This strategy is defined as

$$H_t^n = \begin{cases} 2^{t-1} & \text{if } r_1 = \ldots = r_{t-1} = 1 \\ 0 & \text{elsewhere.} \end{cases}$$

Clearly $(H^n \cdot S)_t = H_1^n r_1 + \cdots + H_{t-1}^n r_t$ hence we obtain

$$(H^n \cdot S)_t = \begin{cases} 2^t - 1 & \text{with probability } 2^{-t} \\ -1 & \text{with probability } 1 - 2^{-t} \end{cases}$$

For odd n the final outcome g_n satisfies $g_n = \lim_{t \to \infty} (H^n \cdot S)_t = -1$ almost surely.

For each n, an even natural number, we introduce a "doubling strategy" H^n starting at time n. More precisely

$$H_t^n = \begin{cases} 0 & \text{for } t \leq n \\ 2^{m-1} & \text{if } t = n + m \text{ and } r_{n+1} = \ldots = r_{n+m-1} = 1 \\ 0 & \text{elsewhere.} \end{cases}$$

Clearly for $t \leq n$: $(H^n \cdot S)_t = 0$ and for $t > n$: $(H^n \cdot S)_t = H_n^n(r_{n+1}) + \cdots + H_{t-1}^n(r_t)$ hence for $t > n$:

$$(H^n \cdot S)_t = \begin{cases} 2^{(t-n)} - 1 & \text{with probability } 2^{-(t-n)} \\ -1 & \text{with probability } 1 - 2^{-(t-n)}. \end{cases}$$

Again, for each even number n, the final outcome g_n satisfies $g_n = \lim_{t \to \infty} (H^n \cdot S)_t = -1$ almost surely. Hence all the variables g_n, for odd as well as for even n, are equal to -1 almost surely and hence trivially $g = \lim g_n = -1$ a.s.. On the other hand the sequence H^n, along the even numbers, tends to zero on $\mathbb{R}_+ \times \Omega$. Along the odd numbers the sequence H^n is constant and equal to the same doubling strategy. The sequence H^n is therefore not converging. Note, however, that the limit function g is not maximal in the sense of Lemma 9.4.4. If we take limits along the even numbers then the pointwise limit H of H^n is zero and hence $(H \cdot S)_\infty = 0$. The example suggests that the outcome 0, which is larger than g, can be obtained by looking at limits of the strategies H^n. So the remedy is to replace the function g

by the larger outcome 0. Replacing g by a maximal element is in this sense a "best try".

Of course, this is only a very simple example and the reader may construct examples where even more pathological phenomena occur. But the present example shows in a convincing way, that the convergence of the final outcome g_n does not imply any kind of convergence of the corresponding integrands H^n.

The difficulties arising from the above introduced "suicide strategies" H^n were already addressed in [HP 81].

We finish this remark by giving an example of a process $(S_t)_{t\geq 0}$ such that $K = K_0 \cap L^\infty$ is not $\sigma(L^\infty, L^1)$-closed. This underlines again the importance of considering the cone C_0 of elements dominated by elements of K_0, a phenomenon already encountered in the Kreps-Yan theorem (see [S 94, Theorem 3.1]). The example is in discrete time. We consider a sequence Y_n of independent variables taking 3 possible values $\{a, b, c\}$. The probability is defined as $\mathbf{P}[Y_n = a] = \frac{1}{2}$; $\mathbf{P}[Y_n = b] = \frac{1}{2} - 4^{-n}$; $\mathbf{P}[Y_n = c] = 4^{-n}$. We again use the sequence of Rademacher functions defined this time as $r_n = 1$ if $Y_n = a$, and $r_n = -1$ if $Y_n = b$ or c. Let T be defined as the first n so that $Y_n = c$. It is clear that $\mathbf{P}[\text{there is } n \text{ such that } Y_n = c] \leq \frac{1}{3}$. We define the process S as $S_m = \sum_{n=1}^{\min(m,T)} r_n$. More precisely we take the sum of the first m Rademacher functions but we stop the process at T. The original measure is clearly a martingale measure for S. Let us now define B_n as the set $\{T > n\}$ and let H^n be the doubling strategy starting at time n. From the definition of T it follows that the final outcome $g_n = (H^n \cdot S)_\infty = -\mathbf{1}_{B_n}$. The sequence g_n tends weak-star to $g = -\mathbf{1}_{\{T=\infty\}}$. This random variable g, however, is not in the set K. Suppose on the contrary that H is a predictable integrand such that $(H \cdot S)_\infty = -\mathbf{1}_{\{T=\infty\}}$. On the set $\{T \leq n-1\}$ we can without disturbing the final outcome, replace $H_1, \ldots, H_n$ by 0. This new integrand is still denoted by H. Let now n be the first integer such that H_n is not identically 0. On the set $\{T = n\}$ the product $H_n r_n$ is also the final outcome. Since this set is disjoint from the set $\{T = \infty\}$ we find that $H_n = 0$ on the set $\{T = n\}$. The variable H_n is $\mathcal{F}_{n-1}$-measurable and by independence of $\mathcal{F}_{n-1}$ and Y_n we therefore have $H_n = 0$ on the set $\{T > n-1\}$. This contradicts the assumption on n. $\qquad\square$

For the rest of the proof of Theorem 9.4.2 we will denote by f_0 a maximal element of $\mathfrak{D}$, $(f_n)_{n\geq 1}$ is a sequence of elements, obtained as $f_n = (H^n \cdot S)_\infty$, where H^n are 1-admissible strategies H^n, and the sequence f_n converges to f_0 almost surely. Remark that if we can prove that $f_0 \in K_0$, we finish the proof of Theorem 9.4.2.

Lemma 9.4.6. *With the notation introduced above we have that the random variables*

$$F_{n,m} = \left((H^n - H^m) \cdot S\right)^* = \sup_{t \in \mathbb{R}_+} |(H^n \cdot S)_t - (H^m \cdot S)_t|$$

tend to zero in probability as $n, m \to \infty$.

Proof. Suppose to the contrary that there is $\alpha > 0$, sequences $(n_k, m_k)_{k \geq 1}$ tending to ∞ and for each k: $\mathbf{P}\left[\sup_{t \geq 0}\left((H^{n_k} \cdot S)_t - (H^{m_k} \cdot S)_t\right) > \alpha\right] \geq \alpha$.

Define the stopping times T_k as

$$T_k = \inf\left\{t \mid (H^{n_k} \cdot S)_t - (H^{m_k} \cdot S)_t \geq \alpha\right\}$$

so that we have $\mathbf{P}[T_k < \infty] \geq \alpha$.

Define L^k as $L^k = H^{n_k}\mathbf{1}_{[\![0, T_k]\!]} + H^{m_k}\mathbf{1}_{]\!]T_k, \infty[\![}$. The process L^k is predictable and it is 1-admissible. Indeed for $t \leq T_k$ we have $(L^k \cdot S)_t = (H^{n_k} \cdot s)_t \geq -1$ since H^{n_k} is 1-admissible. For $t \geq T_k$ we have

$$(L^k \cdot S)_t = (H^{n_k} \cdot S)_{T_k} + (H^{m_k} \cdot S)_t - (H^{m_k} \cdot S)_{T_k}$$
$$\geq (H^{m_k} \cdot S)_t + \alpha \geq -1 + \alpha.$$

Denote $\lim_{t \to \infty}(L^k \cdot S)_t$ by ρ_k. From the preceding inequalities we deduce that ρ_k can be written as $\rho_k = \varphi_k + \psi_k$ where

$$\varphi_k = f_{n_k}\mathbf{1}_{\{T_k = \infty\}} + f_{m_k}\mathbf{1}_{\{T_k < \infty\}} \quad \text{and} \quad \mathbf{P}[\psi_k \geq \alpha] \geq \alpha.$$

By assumption $\varphi_k \to f_0$ and by taking convex combination as in Lemma 9.8.1 we may suppose that $\psi_k \to \psi_0$ where $\mathbf{P}[\psi_0 > 0] > 0$. Therefore convex combinations of ρ_k converge almost surely to an element $f_0 + \psi_0$, a contradiction to the maximality of f_0. $\qquad\square$

Remark 9.4.7. Let us give an economic interpretation of the argument of the proof. At time T_k we know that the trading strategy H^{n_k} has obtained the result $(H^{n_k} \cdot S)_{T_k}$, which is at least α better than $(H^{m_k} \cdot S)_{T_k}$ on a set of measure bigger than α. On the other hand we know that, for k big enough, both strategies yield at time ∞ a result close to f_0. Having this information the economic agent will switch from the strategy H^{n_k} to H^{m_k} since, starting from a lower level, H^{m_k} yields almost the same final result, i.e. the gain on the interval $]\!]T_k, \infty[\![$ is better for H^{m_k} than for H^{n_k}. The strategy L^k precisely describes this attitude.

The proof used convergence in probability. In the rest of the proof we will make use of decomposition theorems, estimation of maximal functions etc. These methods are easier when applied in an "L^2-environment". We therefore replace the original measure $\mathbf{P}$ by a new equivalent measure $\mathbf{Q}$ we will now construct.

First we observe that $(H^n \cdot S)_t$ converges uniformly in t. The variable $q = \sup_n \sup_t |(H^n \cdot S)_t|$ is therefore finite almost surely. For $\mathbf{Q}$ we now take a probability measure equivalent with $\mathbf{P}$ and such that $q \in L^2(\mathbf{Q})$ e.g. we can take $\mathbf{Q}$ with density $\frac{d\mathbf{Q}}{d\mathbf{P}} = \frac{\exp(-q)}{\mathbf{E}_{\mathbf{P}}[\exp(-q)]}$. From the dominated convergence theorem we then easily deduce that

$$\lim_{n, m \to \infty}\left\|\sup |(H^n \cdot S)_t - (H^m \cdot S)_t|\right\|_{L^2(\mathbf{Q})} = 0.$$

From now on $\mathbf{Q}$ will be fixed. Since S is bounded it is a special semi-martingale and its canonical decomposition (with respect to $\mathbf{Q}$) will be denoted as $S = M + A$, where M is the local martingale part and A is of finite variation and predictable. The symbols M and A are from now on reserved for this decomposition.

The next lemma is crucial in the proof of the main theorem. It is used to obtain bounds on $H^n \cdot M$. Because we shall need such an estimate also for other integrands we state it in a more abstract way. For $\lambda > 0$, let $\mathcal{H}_\lambda$ be the convex set of 1-admissible integrands H with the extra property $\|(H \cdot S)^*\|_{L^2(\mathbf{Q})} \leq \lambda$.

Lemma 9.4.8. *For $\lambda > 0$ the set of maximal functions $\{(H \cdot M)^* \mid H \in \mathcal{H}_\lambda\}$ is bounded in $L^0(\mathbf{Q})$.*

Proof. Fix $\lambda > 0$ and abbreviate the set $\mathcal{H}_\lambda$ by $\mathcal{H}$. The semi-martingales $H \cdot S$ where H is in $\mathcal{H}$, are special (with respect to $\mathbf{Q}$) because their maximal functions are in $L^2(\mathbf{Q})$. Therefore, by Theorem 9.2.2, the canonical decomposition of $H \cdot S$ comes from the decomposition $S = M + A$ i.e., $H \cdot M$ is the local martingale part of $H \cdot S$ and $H \cdot A$ is the predictable part of finite variation.

Because the proof of the lemma is rather lengthy let us roughly sketch the idea, which is quite simple. If K^n is a sequence in $\mathcal{H}$ such that $(K^n \cdot M)^*$ is unbounded in probability, then $K^n \cdot A$ is also unbounded and — keeping in mind that $K^n \cdot A$ is predictable — using good strategies we might take advantage of positive gains. This turns out to be possible as the calculations will show that the gains coming from the predictable part A in the long run overwhelm the possible losses coming from the martingale part M. This will contradict the property *(NFLVR)*. Very roughly speaking, the gains coming from the predictable part A add up proportionally in time, whereas the expected losses from the martingale part only add up proportionally to $\sqrt{\text{time}}$. These phenomena are due to the orthogonality of martingale differences, whereas the variation of the predictable part over the union of two intervals is the sum of the variations over each interval.

Let us now turn to the technicalities. If $\{(H \cdot M)^* \mid H \in \mathcal{H}\}$ in not bounded in L^0, there is a sequence $(K^n)_{n \geq 1}$ in $\mathcal{H}$, as well as $\alpha > 0$, such that for all $n \geq 1$ we have $\mathbf{Q}[(K^n \cdot M)^* > n^3] > 8\alpha$. From the L^2 bound on $(H \cdot S)^*$ and Tchebycheff's inequality we deduce that $\mathbf{Q}[\sup_t |(K^n \cdot S)_t| > n] \leq \frac{\lambda^2}{n^2}$ and for n large enough (say $n \geq N$) this expression is smaller than $\frac{\alpha}{3}$. For each n we now define T_n as

$$T_n = \inf \left\{ t \mid |(K^n \cdot M)_t| \geq n^3 \text{ or } |(K^n \cdot S)_t| \geq n \right\}.$$

If we now define the integrand $L^n = \frac{1}{n^2} K^n \mathbf{1}_{[\![0, T_n]\!]}$ we obtain that

(i) $L^n \cdot M$ are local martingales

(ii) $\mathbf{Q}[(L^n \cdot M)^* \geq n] \geq \mathbf{Q}[(K^n \cdot M)^* \geq n^3] - \mathbf{Q}[(K^n \cdot S)^* \geq n] \geq 8\alpha - \frac{\lambda^2}{n^2} \geq 7\alpha$ for all $n \geq N$.

(iii) $L^n \cdot M$ is constant after T_n.

(iv) The jumps of $L^n \cdot S$ are bounded from below by $-\frac{n+1}{n^2}$. Indeed the process $(K^n \cdot S)^{T_n}$ is bounded above by n on $[\![0, T_n[\![$. Its value is always bigger than -1 and hence jumps of $(K^n \cdot S)^{T_n}$ are bounded from below by $-(n+1)$.

(v) $\|(L^n \cdot M)^*\|_{L^2(\mathbf{Q})} \leq n + \|\Delta(L^n \cdot M)_{T_n}\|_{L^2(\mathbf{Q})} \leq n + \frac{3\lambda}{n^2}$. The last inequality follows from Corollary 9.2.4 and the inequality $\|(L^n \cdot S)^*\|_{L^2(\mathbf{Q})} \leq \frac{\lambda}{n^2}$.

The local martingale $L^n \cdot M$ is therefore an $L^2(\mathbf{Q})$-martingale. For each n we define a sequence of stopping times $(T_{n,i})_{i \geq 0}$. We start with $T_{n,0} = 0$ and put (eventually the value is $+\infty$)

$$T_{n,i} = \inf \left\{ t \mid t \geq T_{n,i-1} \text{ and } |(L^n \cdot M)_t - (L^n \cdot M)_{T_{n,i-1}}| \geq 1 \right\}.$$

We then may estimate

$$\left\|(L^n \cdot M)_{T_{n,i}} - (L^n \cdot M)_{T_{n,i-1}}\right\|_{L^2(\mathbf{Q})} \leq 1 + \left\|\Delta(L^n \cdot M)_{T_{n,i}}\right\|_{L^2(\mathbf{Q})}$$

$$\leq 1 + \frac{3\lambda}{n^2} \leq 1 + \alpha \leq 2 \quad \text{for all } n \geq N.$$

Let k_n be the integer part of $\frac{n\alpha}{4}$. We claim that for $i = 1, \ldots, k_n$ and all $n \geq N$, we have $\mathbf{Q}[T_{n,i} < \infty] > 6\alpha$. An inequality of this type is suggested by the fact that the variables $f_{n,i} = (L^n \cdot M)_{T_{n,i}} - (L^n \cdot M)_{T_{n,i-1}}$ are bounded by 2 in $L^2(\mathbf{Q})$ but their sum has to be large, so we need many of them. To prove that for each $i \leq k_n$ we have $\mathbf{Q}[T_{n,i} < \infty] > 6\alpha$, it is of course sufficient to prove that

$$\mathbf{Q}[T_{n,k_n} < \infty] = \mathbf{Q}[|(L^n \cdot M)_{T_{n,k_n}} - (L^n \cdot M)_{T_{n,k_n-1}}| \geq 1] > 6\alpha.$$

Put $B = \{T_{n,k_n} < \infty\}$ and estimate, for $n \geq N$, the $L^2(\mathbf{Q})$-norm of $(L^n \cdot M)^* \mathbf{1}_{B^c}$:

$$\|(L^n \cdot M)^* \mathbf{1}_{B^c}\|_{L^2(\mathbf{Q})}$$

$$\leq \left\| \sum_{i=1}^{k_n} (L^n \mathbf{1}_{]\!]T_{n,i-1}, T_{n,i}]\!]} \cdot M)^* \mathbf{1}_{B^c} \right\|_{L^2(\mathbf{Q})}$$

$$\leq \sum_{i=1}^{k_n} \left\| (L^n \mathbf{1}_{]\!]T_{n,i-1}, T_{n,i}]\!]} \cdot M)^* \mathbf{1}_{B^c} \right\|_{L^2(\mathbf{Q})}$$

$$\leq \sum_{i=1}^{k_n} \left\| (L^n \mathbf{1}_{]\!]T_{n,i-1}, T_{n,i}]\!]} \cdot M)^* \right\|_{L^2(\mathbf{Q})}$$

$$\leq 2 \sum_{i=1}^{k_n} \left\| (L^n \mathbf{1}_{]\!]T_{n,i-1}, T_{n,i}]\!]} \cdot M)_\infty \right\|_{L^2(\mathbf{Q})} \quad \text{(by Doob's inequality)}$$

$$\leq 4 k_n$$

$$\leq n\alpha.$$

Tchebycheff's inequality now yields $\mathbf{Q}[(L^n \cdot M)^* \mathbf{1}_{B^c} \geq n] \leq \alpha^2$ which implies $\mathbf{Q}[B^c \cap \{(L^n \cdot M)^* \geq n\}] \leq \alpha^2 \leq \alpha$ and hence

$$\mathbf{Q}[B] \geq \mathbf{Q}[(L^n \cdot M)^* \geq n] - \mathbf{Q}[B^c \cap \{(L^n \cdot M)^* \geq n\}] > 7\alpha - \alpha = 6\alpha.$$

For $n \geq N$ and $i = 1, \ldots, k_n$, the random variables $f_{n,i}$ are bounded in $L^2(\mathbf{Q})$-norm by 2 but in $L^0(\mathbf{Q})$ they satisfy the lower bound $\mathbf{Q}[|f_{n,i}| \geq 1] > 6\alpha$. This will allow us to obtain a lower $L^0(\mathbf{Q})$ estimate for $f_{n,i}^-$. Let $\beta = \alpha^2$ and $B_{n,i} = \{f_{n,i}^- \geq \alpha\}$. We will show that $\mathbf{Q}[B_{n,i}] > \beta$.

The martingale property implies that

$$\mathbf{E}_{\mathbf{Q}}[f_{n,i}^-] = \mathbf{E}_{\mathbf{Q}}[f_{n,i}^+] = \frac{\mathbf{E}_{\mathbf{Q}}[|f_{n,i}|]}{2} > 3\alpha.$$

Therefore as $f_{n,i}^-$ is bounded by α outside $B_{n,i}$:

$$\mathbf{E}_{\mathbf{Q}}[f_{n,i}^- \mathbf{1}_{B_{n,i}}] \geq \mathbf{E}_{\mathbf{Q}}[f_{n,i}^-] - \alpha > 2\alpha.$$

On the other hand the Cauchy-Schwarz inequality gives

$$\mathbf{E}_{\mathbf{Q}}[f_{n,i}^- \mathbf{1}_{B_{n,i}}] \leq \|f_{n,i}\|_{L^2(\mathbf{Q})} \mathbf{Q}[B_{n,i}]^{\frac{1}{2}} \leq 2\mathbf{Q}[B_{n,i}]^{\frac{1}{2}}.$$

Both inequalities show that $\mathbf{Q}[B_{n,i}] > \alpha^2 = \beta$.

We now turn to $L^n \cdot A$. Because $L^n \cdot S = L^n \cdot M + L^n \cdot A$ and we know that $L^n \cdot S$ is small and the negative parts of $L^n \cdot M$ are big, we can deduce that positive parts in $L^n \cdot A$ are also big. Let us formalise this idea: from the definition of λ we infer that for all i

$$\|(L^n \cdot S)_{T_{n,i}} - (L^n \cdot S)_{T_{n,i-1}}\|_{L^2(\mathbf{Q})} \leq \frac{2\lambda}{n^2}.$$

Tchebycheff's inequality implies

$$\mathbf{Q}\left[|(L^n \cdot S)_{T_{n,i}} - (L^n \cdot S)_{T_{n,i-1}}| \geq \frac{2\lambda}{n}\right] \leq \left(\frac{2\lambda}{n^2}\right)^2 \frac{n^2}{4\lambda^2} = n^{-2}.$$

Because $\mathbf{Q}[((L^n \cdot M)_{T_{n,i}} - (L^n \cdot S)_{T_{n,i-1}})^- \geq \alpha] > \beta$ we necessarily have $\mathbf{Q}\left[(L^n \cdot A)_{T_{n,i}} - (L^n \cdot A)_{T_{n,i-1}} \geq \alpha - \frac{2\lambda}{n}\right] > \beta - n^{-2}$ and this holds for all $i \leq k_n$ and $n \geq N$.

We will now construct a strategy that allows us to take profit of these k_n positive differences. The process $L^n \cdot A$ is of bounded variation. The Hahn decomposition of this measure, see the discussion preceding Definition 9.2.7, produces a partition of $\mathbb{R}_+ \times \Omega$ in two predictable sets B_+^n and B_-^n on which this measure is respectively positive and negative. The processes $(L^n \mathbf{1}_{B_+^n} \cdot A)$ and $(-L^n \mathbf{1}_{B_-^n} \cdot A)$ are therefore increasing. Let R^n be the process $L^n \mathbf{1}_{B_+^n \cap [0, T_{n,k_n}]}$.

The process $(R^n \cdot A) = (L^n \mathbf{1}_{B_+^n \cap [0, T_{n,k_n}]} \cdot A)$ satisfies

$$(R^n \cdot A)_{T_{n,i}} - (R^n \cdot A)_{T_{n,i-1}} \geq (L^n \cdot A)_{T_{n,i}} - (L^n \cdot A)_{T_{n,i-1}}$$

and we therefore obtain

$$\mathbf{Q}\left[(R^n \cdot A)_{T_{n,i}} - (R^n \cdot A)_{T_{n,i-1}} \geq \alpha - \frac{2\lambda}{n}\right] > \beta - n^{-2}$$

for $i = 1, \ldots, k_n$ and all $n \geq N$.

Unfortunately we do not know that R^n is 1-admissible or even admissible. A final stopping time argument and some estimates will allow us to control the "admissibility" of R^n. The jumps of $R^n \cdot S$ are part of the jumps of $L^n \cdot S$ and hence

$$\Delta(R^n \cdot S) \geq \Delta(L^n \cdot S) \geq -\frac{n+1}{n^2} \geq -\frac{2}{n}.$$

An upper bound for $(R^n \cdot M)$ is obtained by

$$\left\|(R^n \cdot M)_{T_{n,k_n}}\right\|^2_{L^2(\mathbf{Q})} \leq \left\|(L^n \cdot M)_{T_{n,k_n}}\right\|^2_{L^2(\mathbf{Q})}$$

$$\leq \sum_{i=1}^{k_n} \|f_{n,i}\|^2_{L^2(\mathbf{Q})}.$$

For $n \geq N$ this is smaller than $4k_n$. Doob's maximal inequality applied on the $L^2(\mathbf{Q})$-martingale $(R^n \cdot M)^{T_{n,k_n}}$ yields

$$\left\|\sup_{t\geq 0} |(R^n \cdot M)_t|\right\|_{L^2(\mathbf{Q})} \leq 4\sqrt{k_n}.$$

This inequality will show that $R^n \cdot S$ will not become too negative on big sets. First note that we may estimate $(R^n \cdot S)$ from below by $R^n \cdot M$. Indeed, $R^n \cdot S = R^n \cdot M + R^n \cdot A \geq R^n \cdot M$ since $R^n \cdot A$ is increasing and hence positive. The following estimates hold

$$\mathbf{Q}\left[\inf_{t\geq 0}(R^n \cdot S)_t \leq -k_n n^{-\frac{1}{4}}\right]$$

$$\leq \mathbf{Q}\left[\sup_{t\geq 0} |(R^n \cdot M)_t| \geq k_n n^{-\frac{1}{4}}\right]$$

$$\leq 16\frac{\sqrt{n}}{k_n} \text{ by Tchebycheff's inequality and the above estimate}$$

$$\leq 64\alpha\frac{1}{\sqrt{n}}.$$

Let now $U_n = \inf\{t \mid (R^n \cdot S)_t < -k_n n^{-\frac{1}{4}}\}$. The preceding inequality says that $\mathbf{Q}[U_n < \infty] \leq 64\alpha\frac{1}{\sqrt{n}}$. We define yet another integrand: let $V^n = \frac{1}{k_n}R^n \mathbf{1}_{[\![0, U_n]\!]}$. The jumps of $V^n \cdot S$ are then bounded from below by $\frac{-2}{nk_n}$ and the process $(V^n \cdot S)$ is therefore bounded below by $-n^{-\frac{1}{4}} - \frac{2}{nk_n}$. The integrands V^n are therefore admissible and their uniform lower bound tends to zero. We now claim that $(V^n \cdot S)_\infty$ is positive with high probability.

From $\mathbf{Q}\left[(R^n \cdot A)_{T_{n,i}} - (R^n \cdot A)_{T_{n,i-1}} \geq \alpha - \frac{2\lambda}{n}\right] > \beta - n^{-2}$ and from Corollary 9.8.7 we deduce that

$$\mathbf{Q}\left[(R^n \cdot A)_{T_{n,k_n}} \geq \frac{k_n}{2}\left(\alpha - \frac{2\lambda}{n}\right)\left(\beta - n^{-2}\right)\right] > \frac{\beta - n^{-2}}{2}.$$

It follows that

$$\mathbf{Q}\left[(V^n \cdot A)_{T_{n,k_n}} \geq \frac{1}{2}\left(\alpha - \frac{2\lambda}{n}\right)\left(\beta - n^{-2}\right)\right] > \frac{\beta - n^{-2}}{2} - \mathbf{Q}[U_n < \infty]$$

or

$$\mathbf{Q}\left[(V^n \cdot A)_\infty \geq \left(\frac{\alpha}{2} - \frac{\lambda}{n}\right)\left(\beta - n^{-2}\right)\right] > \frac{\beta - n^{-2}}{2} - 64\alpha\frac{1}{\sqrt{n}}.$$

Since $\left(\frac{\alpha}{2} - \frac{\lambda}{n}\right)\left(\beta - n^{-2}\right)$ tends to $\gamma = \frac{\alpha\beta}{2}$ we obtain that for n large enough, say $n \geq N'$

$$\mathbf{Q}\left[(V^n \cdot A)_\infty \geq \frac{\gamma}{2}\right] > \frac{\beta}{4}.$$

Let us now look at $(V^n \cdot S)_\infty = (V^n \cdot M)_\infty + (V^n \cdot A)_\infty$. The first term $(V^n \cdot M)_\infty$ tends to zero in $L^2(\mathbf{Q})$. Indeed

$$\|(V^n \cdot M)_\infty\|_{L^2(\mathbf{Q})} \leq \frac{1}{k_n}\|(R^n \cdot M)_{T_{n,k_n}}\|_{L^2(\mathbf{Q})} \leq 2\frac{1}{\sqrt{k_n}} \to 0.$$

The second term satisfies $\mathbf{Q}\left[(V^n \cdot A)_\infty > \frac{\gamma}{2}\right] > \frac{\beta}{4}$.

Tchebycheff's inequality therefore implies that for n large enough, say $n \geq N''$ we have

$$\mathbf{Q}\left[(V^n \cdot S)_\infty > \frac{\gamma}{4}\right] \geq \frac{\beta}{4} - \mathbf{Q}\left[(R^n \cdot M)_{T_{n,k_n}} > \frac{\gamma}{4}\right] \geq \frac{\beta}{8}.$$

The functions $g_n = (V^n \cdot S)_\infty$ have their negative parts going to zero in the norm of L^∞. This is a contradiction to Corollary 9.3.8. □

The next step in the proof is to obtain convex combinations $L^n \in \text{conv}\{H^n; n \geq 1\}$ so that the local martingales $L^n \cdot M$ converge in the semi-martingale topology. If we knew that the elements $H^n \cdot M$ were bounded in $L^2(\mathbf{Q})$ then we could proceed as follows: by taking convex combinations the elements H^n can be replaced by elements L^n such that $L^n \cdot M$ converge in the $L^2(\mathbf{Q})$-topology, whence in the semi-martingale topology. Afterwards we then should concentrate on the processes $L^n \cdot A$. Unfortunately we do not dispose of such an $L^2(\mathbf{Q})$-bound but only a L^0-bound and a slightly more precise information given by the preceding lemma. It suggests that we should stop the local martingales $H^n \cdot M$ when they cross the level $c > 0$, apply Corollary 9.2.4 to control the final jumps in $L^2(\mathbf{Q})$ and apply some L^2-argument on the so obtained L^2-bounded martingales. Afterwards we should take care of the remaining parts and let c tend to ∞. Again the idea is simpler than the technique. Let us introduce the following sequence of stopping times (c is supposed to be > 0).

$T_c^n = \inf\{t \mid |(H^n \cdot M)_t| \geq c\}$. The local martingales $(H^n \cdot M)$ will be stopped at T_c^n, causing an error $K_c^n \cdot M$ where $K_c^n = H^n 1_{]\!]T_c^n, \infty[\![}$.

Lemma 9.4.9. *For all $\varepsilon > 0$, there is $c_0 > 0$ such that for arbitrary n, for all convex weights $(\lambda_1, \ldots, \lambda_n)$ and all $c \geq c_0$, we have*

$$\mathbf{Q}\left[\left(\sum_{i=1}^n \lambda_i K_c^i \cdot M\right)^* > \varepsilon\right] < \varepsilon.$$

Proof. Suppose on the contrary that there is $\alpha > 0$ such that for all c_0 there are convex weights $(\lambda_1, \ldots, \lambda_n)$ and $c \geq c_0$, such that

$$\mathbf{Q}\left[\left(\sum_{i=1}^n \lambda_i K_c^i \cdot M\right)^* > \alpha\right] > \alpha.$$

From this we will deduce the existence of a sequence of 1-admissible integrands L^n such that $\sup_n \|(L^n \cdot S)^*\|_{L^2(\mathbf{Q})}$ is bounded and such that $(L^n \cdot M)^*$ is unbounded in $L^0(\mathbf{Q})$. This will contradict Lemma 9.4.8.

Let N be large enough so that $\mathbf{Q}[q > N] < \frac{\alpha}{4}$ (remember $q = \sup_n \sup_t |(H^n \cdot S)_t|$). This is easy since q is finite a.s.. If we define τ as the stopping time

$$\tau = \inf\{t \mid \text{ for some } n \geq 1 : |(H^n \cdot S)_t| > N\}$$

we trivially have $\mathbf{Q}[\tau < \infty] < \frac{\alpha}{4}$. From Lemma 9.4.8, applied with $\lambda = \sup \|(H^n \cdot S)^*\|_{L^2(\mathbf{Q})}$, we deduce that $\lim_{c \to \infty} \sup_n \mathbf{Q}[T_c^n < \infty] \leq \lim_{c \to \infty} \sup_n \mathbf{Q}[(H^n \cdot M)^* \geq c] = 0$. For $0 < \delta < \frac{\alpha}{4}$, let c_1 be chosen so that for all n and all $c \geq c_1$ we have $\mathbf{Q}[T_c^n < \infty] < \delta^2$. For each n we have

$$\|(K_c^n \cdot S)^*\|_{L^2(\mathbf{Q})} \leq \|2(H^n \cdot S)^* 1_{\{T_c^n < \infty\}}\|_{L^2(\mathbf{Q})}$$
$$\leq 2\|q\|_{L^2(\mathbf{Q})}\mathbf{Q}[T_c^n < \infty]^{\frac{1}{2}}.$$

If follows that there is c_2 so that for all n and all $c \geq c_2$

$$\|(K_n^c \cdot S)^*\|_{L^2(\mathbf{Q})} \leq \delta.$$

For $c \geq \max(c_1, c_2)$ take $\lambda_1 \ldots \lambda_n$ a convex combination that guarantees $\mathbf{Q}\left[\left(\sum_{i=1}^n \lambda_i K_c^i \cdot M\right)^* > \alpha\right] > \alpha$ and let $\sigma = \inf\left\{t \mid |(\sum_{i=1}^n \lambda_i K_c^i \cdot M)_t| \geq \alpha\right\}$. Put $K = (\sum_{i=1}^n \lambda_i K_c^i) 1_{[\![0, \min(\tau, \sigma)]\!]}$.

Clearly $\mathbf{Q}[(K \cdot M)^* \geq \alpha] > \alpha - \mathbf{Q}[\tau < \infty] = \frac{3\alpha}{4}$ and the inequality $(K \cdot S)^* \leq \sum_{i=1}^n \lambda_i (K_c^i \cdot S)^*$ implies $\|(K \cdot S)^*\|_{L^2(\mathbf{Q})} \leq \delta$. Let us now investigate whether K is admissible.

$$(K \cdot S)_t = \sum_{i=1}^{n} \lambda_i \mathbf{1}_{\{t > T_c^i\}} \left((H^i \cdot S)_{\min(t,\tau,\sigma)} - (H^i \cdot S)_{\min(T_c^i,\tau,\sigma)} \right)$$

$$\geq \sum_{i=1}^{n} \lambda_i \mathbf{1}_{\{t > T_c^i\}} (-1 - N)$$

$$\geq -(N+1) \sum_{i=1}^{n} \lambda_i \mathbf{1}_{\{t > T_c^i\}}$$

$$\geq -(N+1) F_t$$

where F is the process $F = \sum_{i=1}^{n} \lambda_i \mathbf{1}_{]\!]T_c^i, \infty[\![}$. F is an increasing adapted left continuous process, it is therefore predictable. By construction $\mathbf{E}_{\mathbf{Q}}[F_\infty] \leq \delta^2$ and therefore $\mathbf{Q}[F_\infty > \delta] \leq \delta$. This implies that the stopping time ν, defined as $\nu = \inf\{t \mid F_t > \delta\}$, satisfies $\mathbf{Q}[\nu < \infty] < \delta < \frac{\alpha}{4}$.

This implies that $K' = K \mathbf{1}_{[\![0,\nu]\!]}$, satisfies

$$\|(K' \cdot S)^*\|_{L^2(\mathbf{Q})} \leq \delta$$

and

$$\mathbf{Q}[(K' \cdot M)^* > \alpha] > \alpha - \mathbf{Q}[\tau < \infty] - \mathbf{Q}(\nu < \infty) \geq \frac{\alpha}{2}$$

as well as

$$(K' \cdot S) \geq -(N+1)\delta.$$

The integrand $L^\delta = \frac{K'}{(N+1)\delta}$ therefore is 1-admissible and

$$\|(L^\delta \cdot S)^*\|_{L^2(\mathbf{Q})} \leq \left(\frac{1}{N+1} \right).$$

Furthermore $\mathbf{Q} \left[(L^\delta \cdot M)^* > \frac{\alpha}{(N+1)\delta} \right] > \frac{\alpha}{2}$.

For δ tending to zero this produces a contradiction to Lemma 9.4.8. $\square$

The following lemma relates, in the L^0-topology, the maximal function of a local martingale with the maximal function of a stochastic integral for an integrand that is bounded by 1. The proof uses the fact that the sequence $(H^n \cdot M)_{n \geq 1}$ is a sequence of local L^2-martingales with uniform L^2-control of the jumps.

Lemma 9.4.10. *With the same notation as in Lemma 9.4.9, for all $\varepsilon > 0$ there is $c_0 > 0$ such that for all h predictable $|h| \leq 1$, all convex weights $(\lambda_1 \dots \lambda_n)$ and all $c \geq c_0$*

$$\mathbf{Q} \left\{ \left[\left(h \sum_{i=1}^{n} \lambda_i K_c^i \right) \cdot M \right]^* > \varepsilon \right\} < \varepsilon.$$

In particular $\mathbf{D} \left(\sum \lambda_i K_c^i \cdot M \right)^* < 2\varepsilon$ where $\mathbf{D}$ is the quasi-norm introduced in Sect. 9.2 and inducing the semi-martingale topology.

Proof. Let $\varepsilon > 0$ and take c_0 as in Lemma 9.4.9 i.e.

$$\mathbf{Q}\left[\left(\sum \lambda_i K_c^i \cdot M\right)^* > \varepsilon\right] < \varepsilon$$

for all $(\lambda_1 \ldots \lambda_n)$ convex combination and all $c \geq c_0$. By enlarging c_0 we also may suppose that $\sup_n \|(K_c^n \cdot S)^*\|_{L^2(\mathbf{Q})} \leq \frac{\varepsilon}{3}$ (see the proof of the Lemma 9.4.9). Corollary 9.2.4 now implies that for all n and every stopping time σ

$$\|\Delta(K_c^n \cdot M)_\sigma\|_{L^2(\mathbf{Q})} \leq \varepsilon.$$

Take now h predictable and bounded by 1, take $c \geq c_0, \lambda_1 \ldots \lambda_n$ a convex combination. Define σ as

$$\sigma = \inf\left\{t \,\middle|\, \left|\left(\sum_{i=1}^n \lambda_i (K_c^i \cdot M)_t\right)\right| > \varepsilon\right\}.$$

The following estimate holds:

$$\sup_{t \leq \sigma} \left|\left(\sum_{i=1}^n \lambda_i K_c^i\right) \cdot M\right|_t \leq \varepsilon + \sum \lambda_i \left|\Delta(K_c^i \cdot M)_\sigma\right|.$$

The L^2-norm of the left hand side is therefore smaller than 2ε and we have an L^2-martingale. This implies that the martingale $\left(h \sum \lambda_i K_c^i\right) \mathbf{1}_{[0,\sigma]} \cdot M$ is in L^2 and its norm is smaller than 2ε. Hence

$$\mathbf{Q}\left[\left(\left(h \sum \lambda_i K_c^i\right) \cdot M\right)^* > \sqrt{\varepsilon}\right]$$
$$\leq \mathbf{Q}\left[\left(\left(h \sum \lambda_i K_c^i\right) \mathbf{1}_{[0,\sigma]} \cdot M\right)^* > \sqrt{\varepsilon}\right] + \mathbf{Q}[\sigma < \infty]$$
$$\leq \frac{4\varepsilon^2}{\varepsilon} + \varepsilon = 5\varepsilon. \qquad \square$$

Lemma 9.4.11. *There is a sequence of convex combinations $L^n \in \text{conv}\{H^k, k \geq n\}$ such that $(L^n \cdot M)$ converges in the semi-martingale topology.*

Proof. We use the notation introduced before Lemma 9.4.9. For $\varepsilon = \frac{1}{n}$ we apply Lemma 9.4.10 to find c_n such that

$$\mathbf{D}\left(\left(\sum_{i=1}^m \lambda_i K_{c_n}^i\right) \cdot M\right) \leq \frac{1}{n} \text{ for all convex weights } \lambda_1 \ldots \lambda_m.$$

For each n and each k we have $(H^k \mathbf{1}_{[0,T_{c_n}^k]} \cdot M)^* \leq c_n + |\Delta(H^k \cdot M)_{T_{c_n}^k}|$ and an application of Corollary 9.2.4 yields that each $H^k \mathbf{1}_{[0,T_{c_n}^k]} \cdot M$ is an $L^2(\mathbf{Q})$-martingale with bound $c_n + 3\|q\|_{L^2(\mathbf{Q})}$. A standard diagonalisation argument shows the existence of convex weights $\lambda_0^k, \lambda_1^k, \ldots, \lambda_{N_k}^k$, such that

$$Y_n^k = \sum_{j=0}^{N_k} \lambda_j^k H^{k+j} \mathbf{1}_{[0,T_{cn}^k]} \cdot M .$$

is, for each n, converging in the space of $L^2(\mathbf{Q})$-martingales. An easy way to prove this assertion, is via the following reasoning in Hilbert spaces.

Let $\mathcal{M}^2$ be the Hilbert space of $L^2(\mathbf{Q})$-martingales and let $\mathfrak{H} = \left(\sum \oplus \mathcal{M}^2\right)_{\ell^2}$ be its ℓ^2-sum (see [D 75]). An element of this space is a sequence $X = (X_n)_n$ where each X_n is in $\mathcal{M}^2$. This space is also a Hilbert space when equipped with the norm $\|X\|^2 = \sum_{n\geq 1} \|X_n\|_2^2$. The sequence X^k, defined by the co-ordinates

$$X_n^k = \frac{1}{2^n \left(c_n + 3\|q\|_{L^2(\mathbf{Q})}\right)} \left(H^k \mathbf{1}_{[0,T_{cn}^k]} \cdot M\right)$$

is bounded in the Hilbert space $\mathfrak{H}$ and hence there are convex combinations $Y^k \in \text{conv}\{X^k, X^{k+1}, \ldots\}$ that converge with respect to the norm of $\mathfrak{H}$. It follows that each "co-ordinate" converges in $\mathcal{M}^2$. This implies the existence of convex weights $\lambda_0^k, \lambda_1^k, \ldots, \lambda_{N_k}^k$ such that

$$Y_n^k = \sum_{j=0}^{N_k} \lambda_j^k H^{k+j} \mathbf{1}_{[0,T_{cn}^k]} \cdot M$$

is, for each n, converging in the space of $L^2(\mathbf{Q})$-martingales.

The sequence $L^k = \sum_{j=0}^{N_k} \lambda_j^k H^{k+j} \cdot M$ is now a Cauchy sequence in the space of semi-martingales. Indeed for given $\varepsilon > 0$ take N such that $\frac{1}{N} < \varepsilon$. We find that for k, l:

$$\mathbf{D}\big((L^k - L^l) \cdot M\big)$$

$$\leq \mathbf{D}(Y_N^k - Y_N^l) + \mathbf{D}\left(\sum_{j=1}^{n} \lambda_j^k K_{c_N}^{k+j} \cdot M\right) + \mathbf{D}\left(\sum_{j=1}^{n} \lambda_j^l K_{c_N}^{l+j} \cdot M\right)$$

$$\leq \mathbf{D}(Y_N^k - Y_N^l) + 2\varepsilon .$$

For k and l large enough this is smaller than 3ε.   □

Lemma 9.4.12. *The sequence $(L^k)_{k\geq 1}$ of Lemma 9.4.11 is such that $(L^k \cdot A)$ converges in the semi-martingale topology.*

Proof. We know that $L^k \cdot S \geq -1$ and that $(L^k \cdot M)$ converges in the semi-martingale topology. To show that $(L^k \cdot A)$ converges in the semi-martingale topology we have to prove that for each $t \geq 0$ the total variation $\int_0^t |d((L^k - L^m) \cdot A)|$ converges to 0 in probability as k and m tend to ∞. We will show the stronger statement that $\int_0^\infty |d((L^k - L^m) \cdot A)|$ tend to 0 in probability as k and m tend to ∞. If this were not the case then by the Hahn decomposition, described in Sect. 9.2, we could find h^k predictable with values in $\{+1, -1\}, \alpha > 0$ and two increasing sequences $(i_k, j_k)_{k \geq 1}$ such that $\mathbf{Q}[\varphi_k > \alpha] > \alpha$ where

$$\varphi_k = \int_{[0,\infty[} h_u^k d\big((L^{i_k} - L^{j_k}) \cdot A\big)_u$$

$$= \int_{[0,\infty[} h_u^k (L_u^{i_k} - L_u^{j_k}) dA_u$$

$$= \int_{[0,\infty[} |L_u^{i_k} - L_u^{j_k}| \, |dA_u|.$$

We now define the integrand R^k as

$$R^k = \big(L^{j_k} + \tfrac{1}{2}(1 + h^k)(L^{i_k} - L^{j_k})\big)$$

$$= \tfrac{1}{2}\Big(L^{i_k} + L^{j_k} + h^k(L^{i_k} - L^{j_k})\Big).$$

The idea is simple if $h^k = 1$ i.e. if $(L^{i_k} - L^{j_k}) \cdot dA \geq 0$ we take L^{i_k}, if $h^k = -1$ i.e if $(L^{i_k} - L^{j_k})dA \leq 0$ we take L^{j_k}. In some sense R^k takes the best of both. The processes $(R^k - L^{i_k})$ and $(R^k - L^{j_k}) \cdot A$ define positive measures and are therefore increasing. Indeed

$$(R^k - L^{i_k}) \cdot A = \big((L^{j_k} - L^{i_k}) + \tfrac{1}{2}(1 + h^k)(L^{i_k} - L^{j_k})\big) \cdot A$$

$$= \tfrac{1}{2}\Big((h^k - 1)(L^{i_k} - L^{j_k})\Big) \cdot A \text{ and}$$

$$(R^k - L^{j_k}) \cdot A = \tfrac{1}{2}\Big((h^k - 1)(L^{i_k} - L^{j_k})\Big) \cdot A.$$

Both measures are positive by the construction of h^k. Also

$$\varphi_k = \big((R^k - L^{i_k}) \cdot A\big)_\infty + \big((R^k - L^{j_k}) \cdot A\big)_\infty.$$

We may therefore suppose that $\mathbf{Q}\big[\big((R^k - L^{i_k}) \cdot A\big)_\infty > \tfrac{\alpha}{2}\big] > \tfrac{\alpha}{2}$ (if necessary we interchange i_k and j_k and take subsequences to keep them increasing). Because $(R^k - L^{i_k}) \cdot M = \tfrac{1}{2}((h^k - 1)(L^{i_k} - L^{j_k}) \cdot M)$ and because $(L^{i_k} - L^{j_k}) \cdot M$ tend to zero in the semi-martingale topology on $[0, \infty[$ we deduce that the maximal functions $((R^k - L^{i_k}) \cdot M)^*$ tend to zero in probability. The same holds for $((R^k - L^{j_k}) \cdot M)^*$. Let now $(\delta_k)_{k \geq 1}$ be a sequence of strictly positive numbers tending to 0. By taking subsequences and by the above observation we may suppose that $\mathbf{Q}[((R^k - L^{i_k}) \cdot M)^* > \delta_k$ or $((R^k - L^{j_k}) \cdot M)^* > \delta_k] < \delta_k$ holds for all k. This implies that the stopping time τ_k defined as $\tau_k = \inf\{t \mid (R^k \cdot M)_t \leq \max((L^{i_k} \cdot M)_t, (L^{j_k} \cdot M)_t) - \delta_k\}$ satisfies $\mathbf{Q}[\tau_k < \infty] < \delta_k$. Define now $\widetilde{R}^k = R^k \mathbf{1}_{[0,\tau_k]}$. We claim that the integrands $\widetilde{R}^k$ are $(1 + \delta_k)$-admissible!

For $t < \tau_k$ we have

$$(\widetilde{R}^k \cdot S)_t = (R^k \cdot S)_t$$
$$= (R^k \cdot A)_t + (R^k \cdot M)_t$$
$$\geq \max\left((L^{i_k} \cdot A)_t, (L^{j_k} \cdot A)_t\right) + (R^k \cdot M)_t$$
$$\geq \max\left((L^{i_k} \cdot A)_t, (L^{j_k} \cdot A)_t\right) + \max\left((L^{i_k} \cdot M)_t, (L^{j_k} \cdot M)_t\right) - \delta_k$$
$$\geq \max\left((L^{i_k} \cdot S)_t, (L^{j_k} \cdot S)_t\right) - \delta_k$$
$$\geq -1 - \delta_k.$$

At time τ_k the jump $\Delta(\widetilde{R}^k \cdot S)$ is either $\Delta(L^{i_k} \cdot S)$ or $\Delta(L^{j_k} \cdot S)$ and hence $(R^k \cdot S)_{\tau_k} \geq -1 - \delta_k$ because the left limit of $(\widetilde{R}^k \cdot S)$ at τ_k is at least $\max((L^{i_k} \cdot S)_{\tau_{k-}}, (L^{j_k} \cdot S)_{\tau_{k-}}) - \delta_k$.

The integrands $(1 + \delta_k)^{-1}\widetilde{R}^k$ are 1-admissible. We will use them to construct a contradiction to the maximal property of $f_0 = \lim_{k\to\infty}(L^{j_k} \cdot S)_\infty = \lim_{m\to\infty}(H^m \cdot S)_\infty$.

$$\left(\frac{\widetilde{R}^k}{1 + \delta_k} \cdot S - L^{i_k} \cdot S\right)_\infty = \frac{1}{1 + \delta_k}\left((\widetilde{R}^k - L^{i_k}) \cdot S\right)_\infty - \frac{\delta_k}{1 + \delta_k}(L^{i_k} \cdot S)_\infty$$
$$= \left(\frac{1}{1 + \delta_k}\right)\left((\widetilde{R}^k - L^{i_k}) \cdot A\right)_\infty$$
$$+ \frac{1}{1 + \delta_k}\left((\widetilde{R}^k - L^{i_k}) \cdot M\right)_\infty - \frac{\delta_k}{1 + \delta_k}(L^{i_k} \cdot S)_\infty.$$

This first term is estimated from below

$$\mathbf{Q}\left[\left((\widetilde{R}^k - L^{i_k}) \cdot A\right)_\infty > \frac{\alpha}{2}\right] > \frac{\alpha}{2} \text{ and } \left((\widetilde{R}^k - L^{i_k}) \cdot A\right)_\infty \geq 0.$$

The second term is estimated from above

$$\mathbf{Q}\left[\left((\widetilde{R}^k - L^{i_k}) \cdot M\right)_\infty \leq -\delta_k\right] < \delta_k \text{ and } \left((\widetilde{R}^k - L^{i_k}) \cdot M\right)_\infty \to 0.$$

The third term tends to zero since $\delta_k \to 0$. From Lemma 9.8.1 we know that there are convex combinations $V^k \in \text{conv}\{\widetilde{R}^k, \widetilde{R}^{k+1}, \ldots\}$ such that $(V^k \cdot S)_\infty$ will converge to a function g. Because $(L^{i_k} \cdot S)_\infty \to f_0$ and because $\mathbf{Q}\left[((\widetilde{R}^k - L^{i_k}) \cdot S)_\infty > \frac{\alpha}{2} - \delta_k\right] > \frac{\alpha}{2} - \delta_k$ we deduce from Lemma 9.8.6 that $\mathbf{Q}[g > f_0] > 0$. Also $g \geq f_0$, a contradiction to the construction of f_0. □

Final part of the proof of Theorem 9.4.2. From Lemmas 9.4.12 and 9.4.11 we deduce the existence of 1-admissible integrands $L^k \in \text{conv}\{H^k, H^{k+1}, \ldots\}$ such that $L^k \cdot M$ and $L^k \cdot A$ both converge in the semi-martingale topology. The sequence $(L^k \cdot S)_{k\geq 1}$ is therefore convergent in the semi-martingale topology. Mémin's theorem (see [M 80]) now implies the existence of a predictable process L such that $L^k \cdot S \to L \cdot S$ in the semi-martingale topology. In particular L is 1-admissible and the final value satisfies

$$(L \cdot S)_\infty = \lim_{t \to \infty} (L \cdot S)_t = \lim_{t \to \infty} \lim_{n \to \infty} (L^n \cdot S)_t$$
$$= \lim_{n \to \infty} \lim_{t \to \infty} (L^n \cdot S)_t = \lim_{n \to \infty} (L^n \cdot S)_\infty = f_0 \,.$$

The interchange of the limits is allowed because almost surely $(L^n \cdot S)_t \to (L \cdot S)_t$ uniformly in t, by Lemma 9.4.6. Indeed $(H^n \cdot S)_t$ converge uniformly on $\mathbb{R}_+$ and the convex combinations $L^k \in \text{conv}\{H^k, H^{k+1}, \ldots\}$ preserve this uniform convergence. This shows that $f_0 \in K_0$ and as remarked before Lemma 9.4.6 this implies Theorem 9.4.2.   □

Remark 9.4.13. The topology of semi-martingales was defined in Sect. 9.2. It was defined using the open end interval $[0, \infty[$. A similar but stronger topology could have been defined using the time interval $[0, \infty]$. This amounts to using the distance function:

$$\mathbf{D}(X) = \sup\{\mathbf{E}[\min(|(H \cdot X)_\infty|, 1)] \mid H \text{ predictable}, |H| \le 1\} \,.$$

The difference between the two topologies is comparable to the difference between uniform convergence on compact sets of $[0, \infty[$ and uniform convergence on $[0, \infty]$. A careful inspection of the proofs, mainly devoted to checking the existence of the limits at ∞, shows that the semi-martingales $(L^n \cdot S)$ tend to $(L \cdot S)$ in the semi-martingale topology on $[0, \infty]$ and not only on $[0, \infty[$. We preferred not to use this approach in order to keep the proofs easier.

9.5 The Set of Representing Measures

In this section we use the results obtained in Ansel and Stricker [AS 94] "Couverture des actifs contingents" and we give a new criterion under which the market is complete. Throughout this paragraph the process S is supposed to be locally bounded and to be a local martingale under the measure $\mathbf{P}$. This will facilitate the notation. We will study the following sets of "representing measures" defined on the σ-algebra $\mathcal{F}$ (see e.g. [D 92] for an explanation concerning the name "representing measures"):

$$\mathcal{M}(\mathbf{P}) = \{\mathbf{Q} \mid \mathbf{Q} \ll \mathbf{P}, \mathbf{Q} \text{ is } \sigma\text{-additive and } S \text{ is a } \mathbf{Q}\text{-local martingale}\}$$
$$\mathcal{M}^e(\mathbf{P}) = \{\mathbf{Q} \mid \mathbf{Q} \sim \mathbf{P}, \mathbf{Q} \text{ is } \sigma\text{-additive and } S \text{ is a } \mathbf{Q}\text{-local martingale}\} \,.$$

The space $\mathcal{M}(\mathbf{P})$ consists of all absolutely continuous local martingale measures and it can happen that some of the elements will give a measure zero to events that under the original measure are supposed to have a strictly positive probability to occur. This phenomenon was studied in detail in [D 92]. We will show that $\mathcal{M}^e(\mathbf{P}) = \mathcal{M}(\mathbf{P})$ implies $\mathcal{M}(\mathbf{P}) = \{\mathbf{P}\}$.

We will need the following set of attainable assets:

$$W^0 = \{f \mid \text{there is an } S\text{-integrable } H, H \cdot S \text{ bounded and } (H \cdot S)_\infty = f\} \,.$$

The set W^0 is a subspace of L^∞. There is no problem in this notation since if $H \cdot S$ is bounded, then H as well as $(-H)$ is admissible and therefore $f = (H \cdot S)_\infty$ exists and is a bounded random variable. From Proposition 9.3.6 it follows that $W^0 = K \cap (-K)$. The same notation for a space related to W^0 is already used in [D 92]. The set W is simply $\{\alpha + f \mid \alpha \in \mathbb{R}$ and $f \in W^0\}$. Because S is supposed to be locally bounded these vector spaces are quite big. The following lemma seems to be obvious but, because unbounded S-integrable processes are used, it is not so trivial as one might suspect. The proof we give uses rather heavy material but it saves place.

Lemma 9.5.1. *If H is S-integrable and $H \cdot S$ is bounded, then $H \cdot S$ is a $\mathbf{Q}$-martingale for all $\mathbf{Q} \in \mathcal{M}(\mathbf{P})$.*

Proof. Take $\mathbf{Q} \in \mathcal{M}(\mathbf{P})$. Clearly S is a special semi-martingale under the measure $\mathbf{Q}$. Since it is a local martingale it decomposes as $S = S + 0$. The stochastic integral $H \cdot S$ is bounded and hence is a special martingale under $\mathbf{Q}$. Its decomposition is, according to Theorem 9.2.2, $H \cdot S = H \cdot S + H \cdot 0$, i.e. $H \cdot S$ is a $\mathbf{Q}$-local martingale. Being bounded it is a martingale under $\mathbf{Q}$. $\square$

It follows from the martingale property that if H and G are two S-integrable processes such that $H \cdot S$ and $G \cdot S$ are bounded and such that $(H \cdot S)_\infty = (G \cdot S)_\infty$ then necessarily $(H \cdot S) = (G \cdot S)$. (This also follows from arbitrage considerations.)

The following theorem is due to [AS 94] and [J 92] (see also Chap. 11). Earlier versions can be found in [KLSX 91]. The theorem is particularly important in the setting of incomplete markets (e.g. semi-martingales with more than one equivalent martingale measure). It shows exactly what elements can be constructed or hedged, using admissible strategies.

Theorem 9.5.2. *If $f \in L^0(\Omega, \mathcal{F}, \mathbf{P})$ with $f^- \in L^\infty(\Omega, \mathcal{F}, \mathbf{P})$ then the following are equivalent*

(i) *there is H predictable, S-integrable, $\mathbf{Q} \in \mathcal{M}^e(\mathbf{P})$ and $\alpha \in \mathbb{R}$ such that $H \cdot S$ is a $\mathbf{Q}$-uniformly integrable martingale with $f = \alpha + (H \cdot S)_\infty$*
(ii) *there is $\mathbf{Q} \in \mathcal{M}^e(\mathbf{P})$ such that $\mathbf{E}_{\mathbf{R}}[f] \leq \mathbf{E}_{\mathbf{Q}}[f]$ for all $\mathbf{R} \in \mathcal{M}^e(\mathbf{P})$.*

For f bounded these two properties are also equivalent to

(iii) *$\mathbf{E}_{\mathbf{R}}[f]$ is constant as a function of $\mathbf{R} \in \mathcal{M}(\mathbf{P})$.*

Proof. We refer to [AS 94, Theorem 3.2]. For (iii) we remark that $\mathcal{M}^e(\mathbf{P})$ is $L^1(\mathbf{P})$-dense in $\mathcal{M}(\mathbf{P})$ and hence $\mathbf{E}_{\mathbf{R}}[f]$ is constant on $\mathcal{M}^e(\mathbf{P})$ if and only if it is constant on $\mathcal{M}(\mathbf{P})$. $\square$

Corollary 9.5.3. *W is $\sigma(L^\infty, L^1)$-closed in L^∞.*

Proof. This follows immediately from (iii) of the theorem. W is the subspace of these elements in L^∞ that are constant on a subset of L^1. $\square$

Remark 9.5.4. The corollary was known long before Theorem 9.5.2 was known. The earliest versions of it are due to Yor [Y 78a]. Contrary to intuition, the boundedness condition needed in (iii) of Theorem 9.5.2 cannot be relaxed to f being a member of $L^1_+(\mathbf{R})$ for each $\mathbf{R}$ in $\mathcal{M}^e(\mathbf{P})$. A counter-example can be found in [S 93].

The next theorem is a new criterion for the completeness of the market.

Theorem 9.5.5. *If S is locally bounded and $\mathbf{P}$ is a local martingale measure for S, then*

(i) $\mathcal{M}(\mathbf{P})$ *is a closed convex bounded set of $L^1(\Omega, \mathcal{F}, \mathbf{P})$*
(ii) $\mathcal{M}(\mathbf{P}) = \mathcal{M}^e(\mathbf{P})$ *implies that $\mathcal{M}(\mathbf{P}) = \{\mathbf{P}\}$.*

Proof. (i): We only have to show that $\mathcal{M}(\mathbf{P})$ is closed. Take $\mathbf{Q}_n$ a sequence in $\mathcal{M}(\mathbf{P})$ and suppose that $\mathbf{Q}_n$ converges to $\mathbf{Q}$. Take T a stopping time such that S^T is bounded. If $t < s$ and $A \in \mathcal{F}_t$ then we can see that: $\mathbf{E_Q}[S_t^T \mathbf{1}_A]$ $= \lim \mathbf{E_{Q_n}}[S_t^T \mathbf{1}_A] = \lim \mathbf{E_{Q_n}}[S_s^T \mathbf{1}_A] = \mathbf{E_{Q_n}}[S_s^T \mathbf{1}_A]$. This proves $\mathbf{Q} \in \mathcal{M}(\mathbf{P})$.

(ii): If $\mathcal{M}(\mathbf{P}) = \mathcal{M}^e(\mathbf{P})$ then $\mathcal{M}^e(\mathbf{P})$ is a closed, bounded, convex set. The Bishop-Phelps theorem, see [D 75], states that the set G of elements f of $L^\infty(\Omega, \mathcal{F}, \mathbf{P})$ that attains their supremum on $\mathcal{M}^e(\mathbf{P})$, is a norm dense set in $L^\infty(\Omega, \mathcal{F}, \mathbf{P})$. The preceding theorem, part (ii), states that G is a subset of W. Since W is weak-star-closed it is certainly norm closed. Since W is closed and G is dense for the norm topology we obtain $W = L^\infty(\Omega, \mathcal{F}, \mathbf{P})$. By the Hahn-Banach theorem, two distinct elements of L^1 can be separated by an element of L^∞ i.e. by an element of W. However, elements of W are constant on $\mathcal{M}(\mathbf{P})$. This implies that $\mathcal{M}(\mathbf{P}) = \{\mathbf{P}\}$. □

As we remarked in the introduction our results remain true for $\mathbb{R}^d$-valued processes. The same holds for Theorem 9.5.5. As the example of [AH 95] shows, Theorem 9.5.5 is no longer true for an infinite number of assets. The example uses the set $\{0, 1\}$ as time set, but as easily seen and stated in [AH 95] it is easy to transform the example into a setting with continuous time.

In [D 92] the following identity was proved for a continuous process S. For every $f \in L^\infty$:

$$\sup_{\mathbf{Q} \in \mathcal{M}(\mathbf{P})} \mathbf{E_Q}[f] = \inf\{x \mid \text{ there is } h \in W^0 \text{ with } x + h \geq f\}.$$

In the general case this equality becomes false as the following example in discrete time shows. The left hand side of the equality is always dominated by the right hand side. The example shows that a "gap" is possible. Some further properties displayed by Example 9.5.6 are: W^0 is weak-star-closed but the set $W^0 - L^\infty_+$ is not even norm closed. We will also see that the norm closure and the weak-star-closure of $W^0 - L^\infty_+$ are different.

Example 9.5.6. The set Ω is the set $\mathbb{N} = \{1, 2, 3, \ldots\}$ of natural numbers. The σ-algebra $\mathcal{F}_n$ is the σ-algebra generated by the atoms $\{k\}$ for $k \leq 3n$ and the atom $\{3n + 1, 3n + 2, \ldots\}$. $S_0 = 0$ and $S_n - S_{n-1}$ is defined as the variable $g_n(3(n - 1) + 1) = n; g_n(3(n - 1) + 2) = 1; g_n(3n) = -1$. The process S is not bounded but a normalisation of the functions g_n allows us to replace S by a bounded process. To keep the notation simple we prefer to continue with the locally bounded process S given above. For the measure $\mathbf{P}$ we choose any measure that gives a strictly positive mass to all natural numbers and such that for all n we have $\mathbf{E}_{\mathbf{P}}[g_n] = 0$. The space W^0 is precisely the set:

$$\left\{ \sum_{n \geq 1} a_n g_n \mid (n a_n)_{n \geq 1} \text{ is bounded} \right\}.$$

Take now for f the function defined as

for all $n \geq 1: \ f(3(n - 1) + 1) = 0 \, ; \ f(3(n - 1) + 2) = 1 \text{ and } f(3n) = 0 \, .$

From the description of W^0 it follows that for h in W^0 and $x \in \mathbb{R}$ the random variable $x + h$ can only dominate f if $x \geq 1$. The constant function 1 clearly dominates f. This shows that

$$\inf\{x \mid \text{ there is } h \in W^0 \text{ with } x + h \geq f\} = 1 \, .$$

On the other hand if $\mathbf{Q}$ is a local martingale measure for S then $n\mathbf{Q}[3(n - 1) + 1] + \mathbf{Q}[3(n-1) + 2] = \mathbf{Q}[3n]$ implies that $\mathbf{Q}[3(n-1) + 2] \leq \frac{1}{2}\mathbf{Q}[\{3(n-1) + 1, 3(n - 1) + 2, 3n\}]$, with strict inequality if $\mathbf{Q}$ is equivalent to $\mathbf{P}$. Therefore $\mathbf{E}_{\mathbf{Q}}[f] \leq \frac{1}{2}$ with strict inequality for $\mathbf{Q}$ in $\mathcal{M}^e(\mathbf{P})$. If we take any measure $\mathbf{Q}$ such that $\mathbf{Q}[3(n-1) + 1] = 0$ and $\mathbf{Q}[3(n-1) + 2] = \mathbf{Q}[3n]$ then $\mathbf{Q}$ is in $\mathcal{M}(\mathbf{P})$ and $\mathbf{E}_{\mathbf{Q}}[f] = \frac{1}{2}$. It is now clear that $\max_{\mathbf{Q} \in \mathcal{M}(\mathbf{P})} \mathbf{E}_{\mathbf{Q}}[f] = \frac{1}{2}$.

This example also shows that in Theorem 9.5.2 (ii), the condition $\mathbf{Q} \in \mathcal{M}^e(\mathbf{P})$ may not be replaced by the condition $\mathbf{Q} \in \mathcal{M}(\mathbf{P})$. Referring to the proof of [D 92, Lemma 5.7], we remark that in this example the function f is not in $\frac{1}{2} + W^0 - L_+^\infty$ but it is in the weak-star-closure of it. To see this let f_n be the function defined as $f_n(3(k - 1) + 2) = 1$ for all $k \leq n$ and 0 elsewhere. The functions f_n are smaller than $\frac{1}{2} + \sum_{k=1}^{n}(\frac{1}{2}g_k)$ and therefore are in $\frac{1}{2} + W^0 - L_+^\infty$, they converge weak-star to f. The set $W^0 - L_+^\infty$ is not even norm closed as the following reasoning shows. An element h in $W^0 - L_+^\infty$ is of the form $\sum_{n \geq 1} a_n g_n - k$ where k is in L_+^∞ and $|n\, a_n|$ is bounded, say by m. If a_n is positive then $h(3(n-1)+2) \leq a_n g_n \leq \frac{m}{n}$ and if a_n is negative then $h(3(n - 1) + 2) \leq 0$. In any case $h(3(n - 1) + 2) \leq \frac{m}{n}$. Take now the function p defined as $p(3(n - 1) + 1) = 0$, $p(3(n - 1) + 2) = \frac{1}{\sqrt{n}}$ and $p(3n) = \frac{-1}{\sqrt{n}}$. It is easy to see that p is in the norm closure of $W^0 - L_+^\infty$ but it cannot be in $W^0 - L_+^\infty$ since the converge of $p(3(n-1)+2)$ to 0 is too slow. This reasoning also shows that the element f, described above, cannot be in the norm closure of the set $x + W^0 - L_+^\infty$ for any $x < 1$. $\qquad\square$

To remedy this "gap" phenomenon, well-known in *infinite* dimensional linear programming, we will use another set to calculate the infimum. The set we will use is precisely the set C introduced in Sections 9.2, 9.3 and 9.4. In Sect. 9.4, Theorem 9.4.2, it is proved that C is weak-star-closed in L^∞. In the case of processes which are not necessarily continuous, C is the exact substitute for the set $W^0 - L^\infty_+$, so useful in the continuous case. The polar C° of the cone C is by definition

$$C^\circ = \{g \mid g \in L^1, \ E[gh] \le 0 \text{ for all } h \text{ in } C\}.$$

Theorem 9.5.7.

$$\mathcal{M}(\mathbf{P}) = \{\mathbf{Q} \mid \mathbf{Q} \in L^1, \ \mathbf{Q}[\Omega] = 1 \text{ and } \mathbf{Q} \in C^\circ\}.$$

Proof. If $\mathbf{Q}$ is in $\mathcal{M}(\mathbf{P})$ then for H admissible we know by Theorem 9.2.9 that $H \cdot S$ is a $\mathbf{Q}$-super-martingale. Therefore $\mathbf{E_Q}[h] \le 0$ for every h in C. Conversely let $\mathbf{Q}$ be in L^1, of norm 1 and $\mathbf{Q} \in C^\circ$. The set $-L^\infty_+$ is a subset of C and hence every element of C° is in L^1_+. Therefore $\mathbf{Q}$ is a probability measure. If T is a stopping time and S^T is bounded then the random variables $\alpha(S^T_u - S^T_t)\mathbf{1}_A$ for $u \ge t$, α real and A in $\mathcal{F}_t$, are in C and hence $\mathbf{Q}$ is a local martingale measure for S. □

The following theorem is the precise form of the duality equality stated above. We will prove it for bounded functions, referring to [AS 94] for the case of measurable functions with bounded negative parts.

Theorem 9.5.8. *For every f in L^∞ we have*

$$\sup_{\mathbf{Q} \in \mathcal{M}^e(\mathbf{P})} \mathbf{E_Q}[f] = \sup_{\mathbf{Q} \in \mathcal{M}(\mathbf{P})} \mathbf{E_Q}[f]$$
$$= \inf\{x \mid \text{ there is } h \in C \text{ with } x + h \ge f\}$$
$$= \inf\{x \mid \text{ there is } h \in C \text{ with } x + h = f\}.$$

Proof. From the definition of C it follows that $x + h \ge f$ for h in C if and only if there is h in C with $f = x + h$. The second equality is therefore obvious. From the preceding theorem it follows that

$$\sup_{\mathbf{Q} \in \mathcal{M}(\mathbf{P})} \mathbf{E_Q}[f] \le \inf\{x \mid \text{ there is } h \in C \text{ with } x + h \ge f\}.$$

If $z < \inf\{x \mid \text{ there is } h \in C \text{ with } x + h \ge f\}$ then $f - z$ is not an element of the weak-star-closed cone C. By the Hahn-Banach theorem there is a signed measure $\mathbf{Q} \in L^1$, $\mathbf{E_Q}[h] \le 0$ for all h in C and $\mathbf{E_Q}[f - z] > 0$. The preceding theorem shows that $\mathbf{Q}$ can normalise as $\mathbf{Q}[\Omega] = 1$ and then it is in $\mathcal{M}(\mathbf{P})$. It follows that $z < \mathbf{E_Q}[f] \le \sup_{\mathbf{R} \in \mathcal{M}(\mathbf{P})} \mathbf{E_R}[f]$. This shows that

$$\sup_{\mathbf{Q} \in \mathcal{M}(\mathbf{P})} \mathbf{E_Q}[f] \ge \inf\{x \mid \text{ there is } h \in C \text{ with } x + h = f\}. □$$

Remark 9.5.9. The infimum is a minimum since C is weak-star and hence norm closed.

Remark 9.5.10. Let us recall that the dual of L^∞ is $ba(\Omega, \mathcal{F}, \mathbf{P})$, the space of all bounded, finitely additive measures on the σ-algebra $\mathcal{F}$, absolutely continuous with respect to $\mathbf{P}$. We can try to define the set of all finitely additive measures that can be considered as local martingale measures for S. It is not immediately clear how this can be done in a canonical way. But, if we define $\mathcal{M}_{ba}(\mathbf{P}) = \{\mathbf{Q} \mid \mathbf{Q} \in ba, \mathbf{Q}[\Omega] = 1, \mathbf{E}_{\mathbf{Q}}[h] \leq 0 \text{ for all } h \text{ in } C\}$, then it is easy to see, via the equality in Theorem 9.5.8, that $\mathcal{M}(\mathbf{P})$ is $\sigma(ba, L^\infty)$-dense in $\mathcal{M}_{ba}(\mathbf{P})$. In other words $\mathcal{M}_{ba}(\mathbf{P})$ is the $\sigma(L^\infty, L^1)$-closure of $\mathcal{M}(\mathbf{P})$ in the space $ba(\Omega, \mathcal{F}, \mathbf{P})$, the dual of L^∞. This is of course the good definition of $\mathcal{M}_{ba}(\mathbf{P})$. We remark that the set C has to be used and not just the set W^0. Indeed Example 9.5.6 shows that the set $\mathcal{M}(\mathbf{P})$ is not necessarily $\sigma(ba, L^\infty)$-dense in the set $\{\mathbf{Q} \mid \mathbf{Q} \text{ finitely additive, positive}, \mathbf{Q}[\Omega] = 1 \text{ and } \mathbf{E}_{\mathbf{Q}}[h] = 0 \text{ for all } h \text{ in } W^0\}$. To see this, we observe that the function f defined in Example 9.5.6 is not in the norm closure of $x + W^0 - L_+^\infty$ for any $x < 1$. By the Hahn-Banach theorem there is a finitely additive positive probability $\mathbf{Q}$ such that $\mathbf{E}_{\mathbf{Q}}[f] = 1$ and $\mathbf{E}_{\mathbf{Q}}[h] = 0$ for all h in W^0. Because $\sup_{\mathbf{Q} \in \mathcal{M}(\mathbf{P})} \mathbf{E}_{\mathbf{Q}}[f] = \frac{1}{2}$ this element $\mathbf{Q}$ cannot be in the $\sigma(ba, L^\infty)$-closure of the set $\mathcal{M}(\mathbf{P})$. This suggests that the "good" definition of such finitely additive measures should use the inequality $\mathbf{E}_{\mathbf{Q}}[h] \leq 0$ for all h in C and not only for all h in $W^0 - L_+^\infty$.

9.6 No Free Lunch with Bounded Risk

In this section we will compare the property of no free lunch with vanishing risk *(NFLVR)* with the previously used property of no free lunch with bounded risk *(NFLBR)*. This property was used in a series of papers: [MB 91, D 92, S 94]. The property *(NFLBR)* is a generalisation of the property *(NFLVR)*. To define this property we need some more notation. By $\overline{C}$ we denoted the closure of C with respect to the norm topology of L^∞, by $\overline{C}^*$ we will denote the weak-star-closure of C. The set $\widetilde{C}$ is the set of all limits of weak-star converging *sequences* of elements of C. Although the fact that a convex set in L^∞ is weak-star-closed if and only if it is sequentially closed for the weak-star topology, the closure of a convex set cannot necessarily be obtained by taking all limits of sequences. (In [B 32, Annexe théorème 1] one can find for each k, examples of convex sets such that after k iterations of taking weak-star limits of sequences, the weak-star-closure is not obtained, but after $k+1$ iterations the closure is found.) Therefore in general, there is a difference between $\overline{C}^*$ and $\widetilde{C}$ and the use of nets is essential to find the weak-star-closure of C.

Definition 9.6.1. *If S is a semi-martingale then we say that S satisfies the property*

(i) *no free lunch with bounded risk (NFLBR) if $\widetilde{C} \cap L_+^\infty = \{0\}$,*
(ii) *no free lunch (NFL) if $\overline{C}^* \cap L_+^\infty = \{0\}$.*

From the definitions and the results of Sect. 9.3 it follows that *(NFL)* implies *(NFLBR)* implies *(NFLVR)* implies *(NA)*. As regards the notion of no free lunch *(NFL)*, this was introduced by [K 81] and is at the basis of subsequent work on the topic. It requires that there should not exist f_0 in L^∞_+, not identically 0, as well as a net $(f_\alpha)_\alpha$ of elements in C such that f_α, converges to f_0 in the weak-star topology of L^∞. Because nets are used, there is no bound on the negative part f_α^- of f_α. It is not excluded that e.g. $\|f_\alpha^-\|_\infty$ tends to ∞, reflecting the enormous amount of risk taken by the agent. It is well-known that for bounded càdlàg adapted processes S, *(NFL)* (even when defined by simple strategies) is equivalent to the existence of an equivalent martingale measure. See [S 94] for a proof of this theorem which is essentially due to [K 81, Y 80]. The drawback of this theorem is twofold. First it is stated in terms of nets, a highly non-intuitive concept. Second it involves the use of very risky positions. The main theorem of the present paper remedies this drawback. We therefore focus attention on variants of the properties *(NFLVR)*. The following characterisation, the proof of which is almost the same as the proof of Proposition 9.3.7 and Corollary 9.3.8, was proved in [S 94]. The proof makes essential use of the Banach-Steinhaus theorem on the boundedness of weak-star convergent *sequences*.

Proposition 9.6.2. *The semi-martingale S satisfies the condition (NFLBR) if and only if for a sequence of 1-admissible integrands $(H^n)_{n \geq 1}$ with final values $g_n = (H^n \cdot S)_\infty$ the condition $g_n^- \to 0$ in probability implies that g_n tends to 0 in probability.*

Proof. Suppose that S satisfies the property *(NFLBR)* and let $(H^n)_{n \geq 1}$ be a sequence of 1-admissible integrands $(H^n)_{n \geq 1}$ with final values $g_n = (H^n \cdot S)_\infty$ such that $g_n^- \to 0$ in probability. Suppose that this sequence does not tend to 0 in probability. By selecting a subsequence, still denoted by $(g_n)_{n \geq 1}$ we may suppose that there is $\alpha > 0$ such that $\mathbf{P}[g_n > \alpha] > \alpha$ for all n. By Lemma 9.8.1 we may take convex combinations $f_n \in \text{conv}\{g_k; k \geq n\}$ that converge in probability to a function f. The negative parts f_n^- still tend to 0 in probability and hence $f : \Omega \to [0, \infty]$. The function f satisfies $\mathbf{P}[f > 0] > 0$. The functions $h_n = \min(f_n, 1)$ are in the convex set C and converge in probability to $h = \min(f, 1)$. The functions h_n are uniformly bounded by 1 and therefore the convergence in probability implies the convergence in the weak-star topology of L^∞. The function h is therefore in $\widetilde{C}$ and the property *(NFLBR)* now implies that $h = 0$ almost everywhere. This, however, is a contradiction to $\mathbf{P}[f > 0] > 0$.

Suppose conversely that S satisfies the announced property. It is clear that S satisfies the no-arbitrage property *(NA)*. Suppose now that h_n is a sequence in C that converges weak-star to h. We have to prove that $h = 0$ almost everywhere. By the Banach-Steinhaus property on weak-star bounded sets, the sequence h_n is uniformly bounded. Without loss of generality we may suppose that it is uniformly bounded by 1 and hence $h_n \geq -1$ almost surely. Since the sequence h_n tends to h weak-star in L^∞ it certainly converges weakly to h in

L^2. Therefore there is a sequence of convex combinations $g_n \in \text{conv}\{h_k; k \geq n\}$ that converges to h in L^2 and therefore in probability. The sequence g_n is bigger than -1 and by the no-arbitrage property g_n is the final value of 1-admissible integrands H_n (see Proposition 9.3.6). The property of S now says that $h = 0$.   □

The difference between *(NFLVR)* and *(NFLBR)* is now clear. In the no free lunch with vanishing risk property we deal with sequences such that the negative parts tend to 0 uniformly. In the no free lunch with bounded risk property we only require these negative parts to tend to 0 in probability and remain uniformly bounded!

If the case of an infinite time horizon the set K_0 was defined using general admissible integrands. The infinite time horizon and especially strategies that require action until the very end, are not easy to interpret. It would be more acceptable if we could limit the properties *(NFLBR)* and *(NFLVR)* to be defined with integrands having bounded support. The following proposition remedies this. (We recall as already stated in the remark following Corollary 9.3.4 that an integrand H is of bounded support if H is zero outside a stochastic interval $[\![0, k]\!]$ for some real number k.)

Proposition 9.6.3.

(1) *If the semi-martingale S satisfies (NFLBR) for integrands with bounded support, then it satisfies (NA) for general admissible integrands.*

(2) *If the semi-martingale S satisfies (NFLVR) for integrands with bounded support and (NA) for general integrands, then it satisfies (NFLVR) for general integrands.*

Proof. We start with the remark that if S satisfies *(NFLVR)* for integrands with bounded support then from Theorem 9.3.3 it follows that for each admissible H, the limit $(H \cdot S)_\infty = \lim_{t \to \infty} (H \cdot S)_t$ exists and is finite almost everywhere. We now show (1) of the proposition. Let $g = (H \cdot S)_\infty$ for H 1-admissible and suppose that $g \geq 0$ almost everywhere. Let $g_n = (H \cdot S)_n$. Clearly g_n^- tends to 0 in probability and each g_n is the result of a 1-admissible integrand with bounded support. The property *(NFLBR)* for integrands with bounded support shows that $g \geq 0$ implies that $g = 0$. The semi-martingale therefore satisfies *(NA)* for admissible integrands.

We now turn to (2) of the proposition. Let $g_n = (H^n \cdot S)_\infty$ with H^n admissible, be a sequence such that the sequence g_n^- tends to 0 in L^∞-norm. Because the process S satisfies *(NA)* it follows from Proposition 9.3.6 that each H^n is $\|q_n^-\|_\infty$-admissible. For each n we take t_n big enough so that $h_n = (H^n \cdot S)_{t_n}$ is close to g_n in probability, e.g. such that $\mathbf{E}[\min(|h_n - g_n|, 1)] \leq \frac{1}{n}$. Since each h_n is the result of a $\|g_n^-\|_\infty$-admissible integrand with bounded support, the property *(NFLVR)* for integrands with bounded support implies that h_n tends to 0 in probability. As a result we obtain that also g_n tends to 0 in probability.   □

The Proposition 9.6.3 allows us to obtain a sharpening of the main theorem of [S 94, Theorem 1.6]. We leave the economic interpretation to the reader.

Proposition 9.6.4. *Let $(S_n)_n$ be a locally bounded adapted stochastic process for the discrete time filtration $(\mathcal{F}_n)_n$. If there does not exist an equivalent local martingale measure for S then at least one of the following two conditions must hold:*

(1) *S fails (NA) for general admissible integrands, i.e. there is an admissible integrand H such that $(H \cdot S)_\infty \geq 0$ a.s. and $\mathbf{P}[(H \cdot S)_\infty > 0] > 0$.*
(2) *S fails (NFLVR) for elementary integrals, i.e. there is a sequence $(H_n)_n$ of elementary integrals such that $(H_n \cdot S) \geq -n^{-1}$ and $(H_n \cdot S)_\infty$ tends almost surely to a function $f : W \to [0, \infty]$ with $\mathbf{P}[f > 0] > 0$.*

Proof. For discrete time processes, elementary integrands and general integrands with bounded support are the same. Therefore if S satisfies both conditions (1) and (2), then by Proposition 9.6.3, S also satisfies *(NFLVR)* for general integrands. The main Theorem 9.1.1 now asserts that S admits an equivalent local martingale measure. The proposition is the contraposition of this statement. $\qquad \square$

The following example shows that in general the no free lunch with vanishing risk property for admissible integrands with bounded support does not imply the no free lunch with vanishing risk property for general admissible integrands! As Proposition 9.6.3 indicates there should be arbitrage for general integrands.

Example 9.6.5. We give the example in discrete time. The extension to continuous time processes is obvious. The set Ω is the compact space of all sequences of -1 or $+1$: $\{-1, +1\}^{\mathbb{N}}$. The σ-algebras $\mathcal{G}_n$ of the filtration are defined as the smallest σ-algebras making the first n co-ordinates measurable. On Ω we put two measures $\mathbf{P}$ and $\mathbf{Q}$. The measure $\mathbf{P}$ is defined as the Haar measure, this is the only measure such that the co-ordinates r_n are a sequence of independent, identically distributed variables with $\mathbf{P}[r_n = \pm 1] = \frac{1}{2}$. The measure $\mathbf{Q}$ is defined as $\frac{1}{2}(\mathbf{P} + \delta_a)$, where δ_a is the Dirac measure giving all its mass to the element a, the sequence identically 1. Define f as the variable $f = -\mathbf{1}_{\{a\}} + \mathbf{1}_{\Omega \setminus \{a\}}$. Clearly $\mathbf{E}_{\mathbf{Q}}[f] = 0$. Define now the process S_n by $S_n = \mathbf{E}_{\mathbf{Q}}[f \mid \mathcal{G}_n]$. The σ-algebras $\mathcal{F}_n$ of the filtration are defined as the smallest σ-algebras making the $S_1, \ldots, S_n$ measurable, i.e. the natural filtration of S. The σ-algebra $\mathcal{F}$ is generated by the sequence $(S_n)_n$. It is easy to see that on $\mathcal{F}$, S admits only one equivalent martingale measure, namely $\mathbf{Q}$. We will now consider the process S under the measure $\mathbf{P}$. On each σ-algebra $\mathcal{F}_n$ the two measures, $\mathbf{P}$ and $\mathbf{Q}$, are equivalent. Suppose now that H^n is a sequence of boundedly supported predictable integrands such that $g_n = (H^n \cdot S)_\infty \geq \frac{-1}{n}$ almost everywhere for the measure $\mathbf{P}$. For each n there is k_n big enough such that g_n is measurable for $\mathcal{F}_n$. Therefore also $\mathbf{Q}\left[g_n \geq \frac{-1}{n}\right] = 1$ for each n. Since $\mathbf{Q}$ is a martingale measure for S it follows that $\mathbf{E}_{\mathbf{Q}}[g_n] = 0$ and that the sequence $(g_n)_{n\geq 1}$ tends to 0 in $L^1(\mathbf{Q})$. Therefore the sequence $(g_n)_{n\geq 1}$ tends to 0 in probability for the measure $\mathbf{Q}$. Because $\mathbf{P}$ is absolutely continuous with respect to $\mathbf{Q}$ we deduce that g_n tends to 0 in probability for the probability $\mathbf{P}$. This implies that S satisfies the *(NFLVR)* property for integrands with bounded support. Because $\mathbf{Q}$ is the only martingale measure for S and because $\mathbf{Q}$ is not absolutely continuous with respect to $\mathbf{P}$, the process S cannot satisfy the no free lunch with vanishing risk property for general integrands (Theorem 9.1.1). In fact, precisely as predicted in Proposition 9.6.4, there is already arbitrage if general integrands are allowed! Take e.g. H the predictable process identically one. Because $S_0 = 0$, we have $H \cdot S = S$ and H is therefore admissible. Now S_n tends to f for the probability $\mathbf{Q}$ and hence S_n tends to $\mathbf{1}_{\Omega \setminus \{a\}}$ for the measure $\mathbf{P}$, i.e. tends to the constant function 1 for the probability $\mathbf{P}$. The process S does not satisfy *(NA)* for general integrands.

9.7 Simple Integrands

In this section we investigate the consequences of the no-free-lunch like properties when defined with simple integrands. It turns out that there is a relation between the semi-martingale property and the no free lunch with vanishing risk *(NFLVR)* property for simple integrands. For continuous processes we are able to strengthen Theorem 9.1.1 and the main theorem of [D 92].

Definition 9.7.1. *A **simple predictable integrand** is a linear combination of processes of the form $H = f\mathbf{1}_{]\!]T_1,T_2]\!]}$ where f is $\mathcal{F}_{T_1}$-measurable and $T_1 \leq T_2$ are finite stopping times with respect to the filtration $(\mathcal{F}_t)_{t\in\mathbb{R}_+}$ (see also [P 90]). The expression "elementary predictable integrand" is reserved for processes of the same kind but with the restriction that the stopping times are deterministic times.*

Simple predictable integrands seem to be the easiest strategies an investor can use. The integrand $H = f\mathbf{1}_{]\!]T_1,T_2]\!]}$ corresponds to buying f units at time T_1 and selling them at time T_2. The requirement that only stopping times and predictable integrands are used reflects the fact that only information available from the past can be used. The interpretation of simple integrands is therefore straightforward. The use of general integrands, however, seems more difficult to interpret and their use can be questioned in economic models. It is therefore reasonable to investigate how far one can go in requiring the integrands to be simple.

As pointed out in Sect. 9.2, we can define the concepts such as no-arbitrage, ... with the extra restriction that the integrands are simple. In the case of simple integrands, stochastic integrals are defined for adapted processes. In this section we therefore suppose that S is a càdlàg adapted process. The following theorem shows that the condition of no free lunch with vanishing risk for simple integrands, already implies that S is a semi-martingale. In

particular this theorem shows that in the Main Theorem 9.1.1, the hypothesis
that the price process is a semi-martingale is not a big restriction. The theorem
is a version of [AS 93, Theorem 8]. The proof follows the same lines but control
in L^2-norm is replaced by other means. The theorem only uses conditions that
are invariant under the equivalent changes of measure. The context of the
following theorem is therefore more natural than the same theorem stated in
an L^2-environment. We, however, pay a price by requiring the process S to
be locally bounded. A counter-example will show that the local boundedness
cannot be dropped.

Theorem 9.7.2. *Let $S : \mathbb{R}_+ \times \Omega :\to \mathbb{R}$ be an adapted càdlàg process. If S is
locally bounded and satisfies the no free lunch with vanishing risk property for
simple integrands, then S is a semi-martingale.*

The proof requires some intermediate results that have their own merit.
Since S is locally bounded there is an increasing sequence of stopping times
$(\tau_n)_{n\geq 1}$ such that each stopped process S^{τ_n} is bounded and $\tau_n \to \infty$ a.s..
To prove that S is a semi-martingale it is sufficient to prove that each S^{τ_n}
is a semi-martingale. We therefore may and do suppose that S is bounded. To
simplify notation we suppose that $|S| \leq 1$. In the following lemmas it is
always assumed that S satisfies *(NFLVR)* for simple integrands.

Lemma 9.7.3. *Under the assumptions of Theorem 9.7.2, let $\mathcal{H}$ be a family
of simple predictable integrands each bounded by 1, i.e. $|H_t(\omega)| \leq 1$ for all t
and $\omega \in \Omega$. If*

$$\left\{ \sup_{0\leq t}(H \cdot S)_t^- \mid H \in \mathcal{H} \right\} \text{ is bounded in } L^0, \text{ then}$$
$$\left\{ \sup_{0\leq t}(H \cdot S)_t^+ \mid H \in \mathcal{H} \right\} \text{ is also bounded in } L^0.$$

Proof. Suppose that the set $\{\sup_{0\leq t}(H \cdot S)_t^+ \mid H \in \mathcal{H}\}$ is not bounded in L^0.
This implies the existence of a sequence $c_n \to \infty$, $H^n \in \mathcal{H}$ and $\varepsilon > 0$ such that
$\mathbf{P}[\sup_{0\leq t}(H \cdot S)_t^+ > c_n] > \varepsilon$. Take K such that $\mathbf{P}[\sup_{0\leq t}(H \cdot S)_t^- < -K] < \frac{\varepsilon}{2}$
for all n and all $H \in \mathcal{H}$ and define the stopping times

$$T'_n = \inf\{t \mid (H^n \cdot S)_t \geq c_n\},$$
$$U_n = \inf\{t \mid (H^n \cdot S)_t < -K\}.$$

Clearly $(H^n \cdot S)_t \geq -K - 2$ on $[\![0, U_n]\!]$ since each H^n is bounded by 1 and
$| S | \leq 1$. Take $T_n = \min(T'_n, U_n)$ and observe that

$$\mathbf{P}[(H^n \cdot S)_{T_n} \geq c_n, \sup_{0\leq t\leq T_n} (H^n \cdot S)_t \leq K + 2] \geq \frac{\varepsilon}{2}.$$

Take now $\delta_n \to 0$ so that $\delta_n c_n \to \infty$ and remark that

(a) $(\delta_n H^n \mathbf{1}_{[\![0,T_n]\!]} \cdot S)_\infty^- \leq \delta_n(K + 2)$
(b) $f_n = (\delta_n H^n \mathbf{1}_{[\![0,T_n]\!]} \cdot S)_\infty$ satisfies $\mathbf{P}[f_n \geq \delta_n c_n] \geq \frac{\varepsilon}{2}$.

By Lemma 9.8.1 there are $g_n \in \text{conv}\{f_n, f_{n+1} \dots\}$ such that $g_n \to g$ a.e. where $g : \Omega \to [0, \infty]$. Also $\mathbf{P}[g > 0] > 0$. If $g_n = \lambda_0^n f_n + \cdots + \lambda_k^n f_{n+k}$ is the convex combination, let us put $K^n = \lambda_0^n H^n + \cdots + \lambda_k^n H^{n+k}$. Clearly

(a) $\|(K^n \cdot S)_\infty^-\| \to 0$ and
(b) $(K^n \cdot S)_\infty \to g : \Omega \to [0, \infty]$.

Since $\mathbf{P}[g > 0] > 0$, this is a contradiction to *(NFLVR)* with simple integrands. $\qquad\square$

Lemma 9.7.4. *The set*

$$\mathcal{G} = \left\{ \sum_{k=0}^{n} (S_{T_{k+1}} - S_{T_k})^2 \ \middle| \ 0 \leq T_0 \leq T_1 \leq \ldots \leq T_{n+1} < \infty \ \text{stopping times} \right\}$$

is bounded in L^0.

Proof. For $0 \leq T_0 \leq T_1 \leq \ldots \leq T_{n+1} < \infty$ stopping times put:

$$H = -2 \sum_{k=0}^{n} S_{T_k} \mathbf{1}_{]\!] T_k, T_{k+1}]\!]}.$$

Because $|S| \leq 1$ we have that H is bounded by 2. Also

$$(H \cdot S)_\infty = \sum_{k=0}^{n} (S_{T_{k+1}} - S_{T_k})^2 - S_{T_{n+1}}^2 + S_{T_0}^2$$

and hence $(H \cdot S)_\infty \geq -1$. The same calculation applied to the sequence of stopping times $\min(T_1, t), \ldots, \min(T_n, t)$ yields $(H \cdot S)_t \geq -1$ and therefore $\sup_{0 \leq t}(H \cdot S)_t^- \leq 1$. The preceding lemma now implies that $\mathcal{G}$ is bounded in L^0. $\qquad\square$

Proof of Theorem 9.7.2. We have to show that if H^n is a sequence of simple predictable processes such that $H^n \to 0$ uniformly over $\mathbb{R}_+ \times \Omega$, then $(H^n \cdot S)_\infty \to 0$ in probability. By the Bichteler-Dellacherie theorem this implies the classical definition of a semi-martingale. (In [P 90] this property is used as the definition of a semi-martingale). It is of course, sufficient to show that the sequence $(H_n \cdot S)_\infty$ is bounded in L^0. If this were not true then there would exist a subsequence of simple integrands, still denoted by $(H^n)_{n \geq 1}$, such that

(a) $H^n \to 0$ uniformly over $\mathbb{R}_+ \times \Omega$;
(b) $\mathbf{P}[(H^n \cdot S)_\infty \geq n] \geq \varepsilon > 0$.
(c) Each H^n can be written as

$$H^n = \sum_{k=0}^{N_n} f_k^n \mathbf{1}_{]\!] T_k^n, T_{k+1}^n]\!]}$$

where $0 \le T_0 \le \ldots \le T_{N_n+1} < \infty$ are stopping times and the functions f_k^n are $\mathcal{F}_{T_k^n}$-measurable functions, bounded by 1.

For each n we put ζ_t^n the process defined as

$$\zeta_t^n = \sum (S_{T_{k+1}^n} - S_{T_k^n})^2 \, ,$$

where the summation is done over the set of indices $k = 0, \ldots, N_n$ such that $T_{k+1}^n \le t$.

Since by the preceding lemma $\mathcal{G}$ is bounded in L^0, there is $c > 0$ such that $\mathbf{P}[\zeta_\infty^n \ge c] \le \frac{\varepsilon}{2}$. Let for each n the stopping time T_n' be defined as $T_n' = \inf\{t \mid \zeta_t^n \ge c\}$. This definition implies that T_n' takes values in the set $\{T_0, \ldots, T_{n+1}, \infty\}$ and is a stopping time with respect to the discrete time filtration $(\mathcal{F}_{T_k^n})_{k=0,\ldots,N_n+1}$. The bound $\zeta_{T_n'}^n \le c + 4$ (since $|S| \le 1$) and $\mathbf{P}[T_n' < \infty] \le \frac{\varepsilon}{2}$ are straightforward. Take now $K^n = H^n \mathbf{1}_{[0,T_n']}$ and observe that $\mathbf{P}[(K^n \cdot S)_\infty \ge n] \ge \frac{\varepsilon}{2}$.

Each discrete time, stopped, process $\left(S_{\min(T_k^n, T_n')}\right)_{k=0,\ldots,N_n+1}$ is now decomposed according to the discrete time Doob decomposition:

$$A_{T_{k+1}^n}^n - A_{T_k^n}^n = \mathbf{E}\left[S_{\min(T_{k+1}^n, T_n')} - S_{\min(T_k^n, T_n')} \,\Big|\, \mathcal{F}_{T_k^n}\right]$$

$$M_{T_{k+1}^n}^n - M_{T_k^n}^n = \left(S_{\min(T_{k+1}^n, T_n')} - S_{\min(T_k^n, T_n')}\right)$$
$$- \mathbf{E}\left[S_{\min(T_{k+1}^n, T_n')} - S_{\min(T_k^n, T_n')} \,\Big|\, \mathcal{F}_{T_k^n}\right] .$$

$\left(M_{T_k^n}^n\right)_{k=0,\ldots,N_n+1}$ is now a martingale bounded in L^2. Indeed

$$\mathbf{E}\left[\left(M_{T_{N_n+1}^n}^n\right)^2\right] = \sum_{k=0}^{N_n} \mathbf{E}\left[\left(M_{T_{k+1}^n}^n - M_{T_k^n}^n\right)^2\right] + \mathbf{E}\left[\left(M_{T_0^n}^n\right)^2\right]$$

$$\le \sum_{k=0}^{N_n} \mathbf{E}\left[\left(S_{\min(T_{k+1}^n, T_n')} - S_{\min(T_k^n, T_n')}\right)^2\right] + \mathbf{E}\left[\left(S_{\min(T_0^n, T_n')}\right)^2\right]$$

$$\le \left[(\zeta_{T_n'}^n)^2\right] + 1 \le c + 5 \, .$$

For each t we put $M_t^n = \mathbf{E}\left[M_{T_{N_n+1}^n}^n \,\Big|\, \mathcal{F}_t\right]$ and we take a càdlàg version of this martingale. Because of the optional sampling theorem this definition coincides with the previously given construction of M_t^n for times $t = T_{N_k}^n$. In the definition of H^n we now replace each f_k^n by $\widetilde{f}_k^n = f_k^n \, \mathrm{sign}\left(A_{T_{k+1}^n}^n - A_{T_k^n}^n\right)$. The functions $\widetilde{f}_k^n$ are still measurable with respect to the σ-algebra $\mathcal{F}_{T_k^n}$. The resulting process is denoted by $\widetilde{K}^n$ i.e.

$$\widetilde{K}^n = \sum_{k=0}^{N_n} \widetilde{f}_k^n \mathbf{1}_{]\!]T_k^n, T_{k+1}^n]\!]} \, .$$

Since $\mid \widetilde{K}^n \mid \le 1$ we still have

$$\mathbf{E}\left[\left((\widetilde{K}^n \cdot M^n)_{T_{N_n+1}^n}\right)^2\right] \le c+5.$$

On the other hand

$$(\widetilde{K}^n \cdot S)_\infty = (\widetilde{K}^n \cdot S)_{T_{N_n+1}^n}$$

$$= (\widetilde{K}^n \cdot M)_{T_{N_n+1}^n} + \sum_{k=0}^{N_n} |f_k^n| \, |A_{T_{k+1}^n}^n - A_{T_k^n}^n|$$

$$\ge (\widetilde{K}^n \cdot M)_{T_{N_n+1}^n} + \sum_{k=0}^{N_n} (f_k^n)\left(A_{T_{k+1}^n}^n - A_{T_k^n}^n\right)$$

$$\ge (\widetilde{K}^n \cdot M)_{T_{N_n+1}^n} + (K^n \cdot S)_\infty - (K^n \cdot M)_{T_{N_n+1}^n}.$$

Because the sequences $(\widetilde{K}^n \cdot M)_{T_{N_n+1}^n}$ and $(K^n \cdot M)_{T_{N_n+1}^n}$ are bounded in L^2 and the sequence $(K^n \cdot S)_\infty^+$ is unbounded in L^0, the sequence $(\widetilde{K}^n \cdot S)_\infty$ is necessarily unbounded in L^0. On the other hand $\sup_{0 \le t}(\widetilde{K}^n \cdot S)_t^-$ is a bounded sequence in L^0. Indeed for $t = T_k^n$ we have

$$(\widetilde{K}^n \cdot S)_{T_k^n}^- \le (\widetilde{K}^n \cdot M)_{T_k^n} \le \sup_{0 \le t} \mid (\widetilde{K}^n \cdot M)_t \mid.$$

And for $T_k^n \le t \le T_{k+1}^n$ we find:

$$(\widetilde{K}^n \cdot S)_t^- \le (\widetilde{K}^n \cdot S)_{T_k^n}^- + |f_k^n| \, |S_t - S_{T_k^n}| \le 2 + \sup_{0 \le t} |(\widetilde{K}^n \cdot M)_t|$$

and hence

$$\left\|\sup_{0 \le t}(\widetilde{K}^n \cdot S)_t^-\right\|_2 \le 2 + \left\|\sup_{0 \le t}(\widetilde{K}^n \cdot M)_t\right\|_2$$

$$\le 2 + 2\|(\widetilde{K}^n \cdot M)_t\|_2$$

$$\le 2 + 2\sqrt{c+5} \quad \text{(by Doob's maximum inequality).}$$

This proves the boundedness in L^0. From Lemma 9.7.3 it now follows that $(\widetilde{K}^n \cdot S)_t^+$ is bounded in L^0. This contradicts the choice of the sequence $\widetilde{K}^n$. $\square$

The following example shows that the requirement that S is locally bounded cannot be dropped. The same notation will also be used in a later example.

Example 9.7.5. We suppose that on a probability space $(\Omega, \mathcal{F}, \mathbf{P})$ following sequences of variables are defined: a sequence $(\gamma_n)_{n \ge 1}$ of Gaussian normalised $\mathcal{N}(0, 1)$ variables, a sequence $(\varphi_n)_{n \ge 1}$ of random variables with distribution $\mathbf{P}[\varphi_n = 1] = 2^{-n}$ and $\mathbf{P}[\varphi_n = 0] = 1 - 2^{-n}$. All these variables are supposed to be independent. The countable set of rationals in the interval $]0,1[$ is enumerated as $(q_n)_{n\geq 1}$. Because $\sum \mathbf{P}[\varphi_n = 1] < \infty$ the Borel-Cantelli lemma tells us that for almost all $\omega \in \Omega$ there are only a finite number of natural numbers n such that $\varphi_n = 1$.

The stochastic process X defined as

$$X_t = \sum_{n;\, q_n \leq t} \varphi_n \gamma_n$$

is therefore right continuous, even piecewise constant (by the above Borel-Cantelli argument). The natural filtration generated by this process is therefore right continuous (see [P 90, Theorem 25] for a proof that can be adapted to this case) and so is the filtration augmented with the negligible sets. The filtration so constructed therefore satisfies the usual conditions.

Take now $F : [0,1] \rightarrow \mathbb{R}$ a continuous function of unbounded variation, e.g. $F(t) = t \sin\left(\frac{1}{t}\right)$. Let now $S_t = X_t + F(t)$. It is easy to verify that X is an L^2-martingale and hence it is a semi-martingale. If S were a semi-martingale then F would also be a semi-martingale. This, however, implies that F is of bounded variation. We conclude that S is not a semi-martingale. We will now show that S satisfies the *(NFL)* property for simple integrands. This certainly implies that S satisfies the *(NFLVR)* property for simple integrands and it shows that the local boundedness condition in Theorem 9.7.2 is not superfluous. To verify the *(NFL)* property with simple integrands, let us start with an integrand $H = f\mathbf{1}_{]\!]T,T']\!]}$ where $T \leq T'$ are two stopping times and f is $\mathcal{F}_T$-measurable. We will show that $H \cdot S$ is not uniformly bounded from below unless $H = 0$. Suppose on the contrary that $\mathbf{P}[\{T < T'\} \cap \{f > 0\}] > 0$ (the case $\{f > 0\}$ is similar). Take t real and q_n rational such that $t < q_n$ and $\mathbf{P}[\{f > 0\} \cap \{T \leq t\} \cap \{q_n \leq T'\}] > 0$. Because f is $\mathcal{F}_T$-measurable, $t < q_n$ and T' is a stopping time we obtain that $A = \{f > 0\} \cap \{T \leq t\} \cap \{q_n \leq T'\} \in \mathcal{F}_{q_{n-}}$ and hence is independent of $\varphi_n \gamma_n$. Because $\varphi_n \gamma_n$ is unbounded from below (and from above for the other similar case) we obtain that $\mathbf{P}[A \cap \{\varphi_n \gamma_n < -K\}] > 0$ for all $K > 0$. It is now easy to see that this implies that $H \cdot S$ is unbounded from below. It also follows that the only simple integrand H for which $H \cdot S$ is bounded from below is the zero integrand. Since there are no admissible simple integrands, the *(NFL)* property with simple integrands is trivially satisfied!   □

Theorem 9.7.2 and the Main Theorem 9.1.1 allow us to strengthen the main theorem in [D 92]. The theorem shows that in the case of continuous price processes and finite horizon, the condition (d) in [D 92], an equivalent form of the no free lunch with bounded risk for simple integrands, can be relaxed. The case of infinite horizon is already treated in Example 9.6.5. By using the techniques developed in [S 93] one can translate this example into an example where S is a continuous process.

Theorem 9.7.6.

(a) *If* $S : [0,1] \times \Omega \to \mathbb{R}$ *is a continuous semi-martingale having the no-arbitrage property (NA), if* S *satisfies (NFLVR) for simple integrands, then* S *has an equivalent local martingale measure.*[†]

(b) *If* $S : \mathbb{R}_+ \times \Omega \to \mathbb{R}$ *is a continuous semi-martingale having the no-arbitrage property (NA), if* S *satisfies the no free lunch with bounded risk property (NFLBR) for simple integrands, then* S *has an equivalent local martingale measure.*[†]

Proof. Let H^n be a sequence of general admissible integrands. Suppose that $g_n = (H^n \cdot S)_1 \geq -\varepsilon_n$ where ε_n tends to zero. By *(NA)* the integrand H^n is now ε_n-admissible. We have to prove that g_n tends to 0 in probability, which will prove part (a) in view of the Main Theorem 9.1.1. From the theory of stochastic integration (see [CMS 80]) we deduce that there are simple integrands L^n such that $\mathbf{P}[\sup_{0 \leq t \leq 1} |(L^n \cdot S)_t - (H^n \cdot S)_t| \geq \varepsilon_n] \leq \varepsilon_n$. For each n we define the stopping time T_n as $\inf\{t \mid (L^n \cdot S)_t < -2\varepsilon_n\}$. Clearly $\mathbf{P}[T_n < 1] \leq \varepsilon_n$. Since the process S is continuous, the random variables $h_n = (L^n \cdot S)_{T_n}$ are bounded below by $-2\varepsilon_n$ and are therefore results of $2\varepsilon_n$-admissible simple integrands. Because $\mathbf{P}[T_n < 1] \leq \varepsilon_n$ and $\mathbf{P}[\sup_{0 \leq t \leq 1} |(L^n \cdot S)_t - (H^n \cdot S)_t| \geq \varepsilon_n] \leq \varepsilon_n$ the sequence $h_n - g_n$ tends to 0 in probability. From the property no free lunch with vanishing risk *(NFLVR)* for simple integrands we deduce that h_n and hence g_n tend to 0 in probability. Therefore the property no free lunch with vanishing risk property *(NFLVR)* is satisfied and by the Main Theorem 9.1.1 there is an equivalent martingale measure.

For the second part we refer to [S 94, Sect. 5, Proposition 5.1].   □

The above theorem seems to indicate that for continuous processes simple integrands are sufficient to describe no-arbitrage conditions. This is not true in general. The Bes³(1)-process, $(R_t)_{0 \leq t \leq 1}$ gives a counter-example. This process can be seen as the Euclidean norm of a three dimensional Brownian motion starting at the point $(1,0,0)$ of $\mathbb{R}^3$. It plays a major role in the theory of continuous martingales and Brownian motion, see [RY 91] for details. The process R satisfies the no-arbitrage *(NA)* property for simple integrands but fails the no-arbitrage *(NA)* property for general integrands. We refer to [DS 95c] for the details. The inverse of this process, $L = R^{-1}$, a local martingale, has been used in [DS 94a].

As a general question one might ask whether for continuous processes the no-arbitrage *(NA)* property for general integrands is sufficient for the existence of an equivalent local martingale measure. The following example shows that this is not true.

[†] Note added in this reprint: In the original paper [DS 94] the hypothesis that S satisfies *(NA)* for general admissible integrands was forgotten but used in the proof! Several people (including ourselves) have noted this. A counter-example for a process satisfying *(NFLVR)* for simple integrands but not satisfying *(NA)* for general admissible integrands is the Bessel process in dimension 3 (see [DS 95c]).

Example 9.7.7. We take a standard Wiener process W with its natural filtration $(\mathcal{G}_t)_{0 \le t \le 1}$. Before we define the price process S, we first define a local martingale of exponential type by:

$$L_t = \begin{cases} \exp\left(-(f \cdot W)_t - \frac{1}{2}\left(\int_0^t f^2(u)du\right)\right), & \text{if } t < 1 \\ 0, & \text{if } t = 1. \end{cases}$$

where f is the deterministic function defined as $f(t) = \frac{1}{\sqrt{1-t}}$.

We define the stopping time T as $T = \inf\{t \mid L_t \ge 2\}$. The stopped process L^T is a bounded martingale starting at zero. Clearly $L_T = 2$ if $T < 1$ and equals 0 if $T = 0$. Therefore $\mathbf{P}[T < 2] = \frac{1}{2}$. We now define the price process by its differential

$$dS_t = \begin{cases} dW_t + \frac{1}{\sqrt{1-t}}dt, & \text{if } t \le T \\ 0, & \text{if } t \ge T. \end{cases}$$

The filtration is now defined as $(\mathcal{F}_t)_{0 \le t \le 1} = (\mathcal{G}_{\min(t,T)})_{0 \le t \le 1}$. Except for sets of measure zero, this is also the natural filtration of the process S and of the stopped Wiener process W^T. All local martingales with respect to this filtration are stochastic integrals with respect to the Wiener process (stopped at T) (see [RY 91, Theorem 4.2] and stop all the local martingales at the stopping time T). Girsanov's formula therefore implies that the only probability measure $\mathbf{Q}$, absolutely continuous with respect to $\mathbf{P}$ and for which S is a local martingale, is precisely the measure $\mathbf{Q}$ defined through its density on $\mathcal{F}_1$ as $d\mathbf{Q} = L_T d\mathbf{P}$. As we shall see, S satisfies the property of no-arbitrage *(NA)*. Important in the proof of this, is the fact that for $t < 1$, the measures $\mathbf{Q}$ and $\mathbf{P}$ are equivalent on $\mathcal{F}_t$, (the density L_t^T is strictly positive). Because the process S is continuous the proof that S satisfies *(NA)* reduces to verifying the statement that for H admissible, $(H \cdot S)_1$ cannot be almost everywhere positive without being zero a.s.. Take H admissible and suppose that $(H \cdot S)_1 \ge 0$, $\mathbf{P}$-a.s.. This certainly implies that $(H \cdot S)_1 \ge 0$, $\mathbf{Q}$-a.s.. Because S is a continuous $\mathbf{Q}$-local martingale, we know that $H \cdot S$ is a continuous $\mathbf{Q}$-local martingale and because H is admissible for $\mathbf{Q}$, $\mathbf{Q}$ being absolutely continuous with respect to $\mathbf{P}$, $H \cdot S$ is a $\mathbf{Q}$-super-martingale. From this it follows that $\mathbf{E}_{\mathbf{Q}}[(H \cdot S)_1] \le 0$ and by positivity of $(H \cdot S)_1$, this in turn implies that $(H \cdot S)_1 = 0$, $\mathbf{Q}$-a.s.. Under the probability $\mathbf{Q}$, the process S is a local martingale and hence satisfies *(NA)* with respect to $\mathbf{Q}$! For each $\varepsilon > 0$, let now V be the stopping time defined as $\inf\{t \mid (H \cdot S)_t \ge \varepsilon\}$. The integrand $K = (1_{[\![0,V]\!]}H)$ is clearly admissible and $(K \cdot S)_1 = 0$ on $\{V = 1\}$, whereas on $\{V < 1\}$ the outcome is ε, i.e. strictly positive. The *(NA)* property for S (under $\mathbf{Q}$!) implies that $\mathbf{Q}[V < 1] = 0$. In other words the process $H \cdot S$ never exceeds ε $\mathbf{Q}$-a.s.. This implies $(H \cdot S)_t \le 0$, $\mathbf{Q}$-a.s. for all $t < 1$. Because $\mathbf{Q}$ and $\mathbf{P}$ are equivalent on $\mathcal{F}_t$, for $t < 1$, this is the same as $(H \cdot S)_t \le 0$, $\mathbf{P}$-a.s. for all $t < 1$. From this and the continuity of the process $(H \cdot S)$ we deduce that $(H \cdot S)_1 \le 0$, $\mathbf{P}$-a.s.. This in turn implies that $(H \cdot S)_1 = 0$, $\mathbf{P}$-a.s.. The process S therefore satisfies *(NA)* under the probability $\mathbf{P}$. $\square$

We now give some more examples motivating the introduction of general integrands. As seen in the above theorems and examples, the case of continuous processes can essentially be reduced to simple integrands. The following examples show that for general semi-martingales the no free lunch with bounded risk *(NFLBR)* property for simple integrands is not sufficient to imply the existence of an equivalent local martingale measure.

The examples are very similar in nature; the problems arise from the fact that the jumps do not occur at an increasing sequence $(T_n)_{n \geq 1}$ of *predictable* stopping times (a case already solved in [S 94]). In our examples the jumps occur at an *increasing* sequence of *accessible* stopping times, similarly as in Example 9.7.5. The first example of this kind is an unbounded process but it contains all the ingredients and the general idea. The second example of this kind gives a bounded process. Of course the price to pay is the use of more technique.

Example 9.7.8. The first example uses the process X introduced in Example 9.7.5. The semi-martingale S we will need, is defined as $S_t = X_t + t$. The process S is now a special semi-martingale and again if H is simple predictable with $H \cdot S$ bounded from below then $H = 0$. Therefore S trivially satisfies the no free lunch *(NFL)* property with simple integrands. If, however, we put $H = \mathbf{1}_{([0,1] \setminus \mathbf{Q}) \times \Omega}$ (sell before each rational and buy back immediately after it) we have $(H \cdot S)_t = t$ (for $0 \leq t \leq 1$) and this violates *(NA)* for general integrands. If $\mathbf{Q}$ were an equivalent local martingale measure for the process S, then because $H = \mathbf{1}_{([0,1] \setminus \mathbf{Q}) \times \Omega}$ is bounded, $H \cdot S$ is also a local martingale (see [P 90, Theorem 2.9]). This is absurd.   □

The previous example has at least one disadvantage: the process S is unbounded. The next example overcomes this problem. This time we will work with a doubly indexed sequence of Rademacher variables $(r_{n,m})_{n \geq 1, m \geq 1}$, i.e. variables with distribution $\mathbf{P}[r_{n,m} = 1] = \mathbf{P}[r_{n,m} = -1] = \frac{1}{2}$, and with a doubly indexed sequence of variables $(\varphi_{n,m})_{n \geq 1, m \geq 1}$ with the property $\mathbf{P}[\varphi_{n,m} = 1] = 2^{-(n+m)}$ and $\mathbf{P}[\varphi_{n,m} = 0] = 1 - 2^{-(n+m)}$. We also need a sequence of Brownian motions W^n starting at 0. All these variables and processes are supposed to be independent. The rationals in $]0, 1[$ are again enumerated as $(q_n)_{n \geq 1}$. We first define the L^2-martingales Y^m as:

$$Y_t^m = \sum_{n;\, q_n \leq t} \varphi_{n,m} r_{n,m} \, .$$

The Borel-Cantelli implies, as in Example 9.7.5, that each Y^m is piecewise constant. We define the stopping time T_m as:

$$T_m = \min \left(\inf\{t \mid |W_t^m| = m \text{ or } Y_t^m \neq 0\}, 1 \right).$$

We make the crucial observation that

$$\mathbf{P}[T_m < 1] \le \mathbf{P}\left[\sup_{0 \le t \le 1} |W_t^m| > m\right] + \sum_{m \ge 1} 2^{-(n+m)}$$

$$\le \sqrt{\frac{2}{\pi}}\frac{1}{m}e^{-\frac{m^2}{2}} + 2^{-m}$$

and hence $\sum \mathbf{P}[T_m < 1] < \infty$. This implies, via the Borel-Cantelli lemma, that for almost all $\omega \in \Omega$, $T_m(\omega)$ becomes eventually 1.

The process Z^m is now defined as

$$Z_t^m := \begin{cases} Y_t^m + \alpha_m(W_t^m + m^2 t) & \text{for } t \le T_m, \\ Y_{T_m}^m + \alpha_m(W_{T_m}^m + m^2 T_m) & \text{for } T_m \le t \le 1. \end{cases}$$

The sequence α_m will be chosen later, but will satisfy $0 < \alpha_m \le 1$.

The process Z^m is clearly bounded by $1 + (m + m^2)\alpha_m \le 1 + m + m^2$. Finally we define

$$S_t := \begin{cases} \frac{1}{2}Z_t^1 & \text{for } 0 \le t \le 1, \\ S_{m-1} + 2^{-m}Z_{t-(m-1)}^m & \text{for } m - 1 \le t \le m. \end{cases}$$

The process S is càdlàg and $|S| \le \sum_{m \ge 1} 2^{-m}(1 + m + m^2)\alpha_m \le 24$. It is a semi-martingale with decomposition $S = M + A$, where A is given by the recurrence relations

$$A_{m-1+t} - A_{m-1} = \begin{cases} 2^{-m}\alpha_m m^2 t & \text{for } t \le T_m, \\ 2^{-m}\alpha_m m^2 T_m & \text{for } T_m \le t \le 1. \end{cases}$$

The martingale M is uniformly bounded on each interval $[\![0, m]\!]$.

With respect to its natural filtration, augmented with the zero sets, S is a special semi-martingale and the filtration satisfies the usual conditions. The last statement is not trivial to verify but it follows from the same property of the filtration of the Brownian motion.

Lemma 9.7.9. *For each sequence $(\alpha_m)_{m \ge 1}$ in $]0, 1]$, the process S fails the equivalent (local) martingale property.*

Proof. Consider the sequence $(H^m)_{m \ge 1}$ defined as

$$H^m = \alpha_m^{-1}m^{-2}2^m \mathbf{1}_{(]m-1,m]\setminus\mathbf{Q})\times\Omega}.$$

Each H^m is a deterministic process, hence predictable. The process $(H^m \cdot S)$ is uniformly bounded from below by -1 and $((H^m \cdot S)_\infty)_{m \ge 1}$ equals $\frac{1}{m^2}W_{T_m}^m + T_m \ge T_m - \frac{1}{m}$.

Because $T_m = 1$ for m big enough we see that $(H^m \cdot S)_\infty$ tends to 1 for m tending to ∞. This clearly violates *(NFLVR)*. Because of the Main Theorem 9.1.1 we see that S cannot have an equivalent martingale measure. $\square$

Lemma 9.7.10. *If $(\alpha_m)_{m\geq 1}$ is a sequence in $]0,1]$ such that $\alpha_m \to 0$ fast enough, then S satisfies (NFLBR) for simple integrands.*

(By fast enough we mean that for all m_0 we have:

$$\sum_{m>m_0} 2^{m+1}m^2\alpha_m < \frac{\beta_{m_0}}{2m_0} \text{ where } \beta_{m_0} = \exp(-3m_0^5)).$$

Proof. For each m natural number, we know that the process S^m, i.e. S stopped at m, admits an equivalent martingale measure $\mathbf{Q}_m$. Indeed we can use a Girsanow transformation to find an equivalent martingale measure such that for k fixed, the process $(W_t^k + k^2 t)_{0\leq t\leq 1}$ stopped at T_k is a martingale. The density of this measure is given by $\exp(\delta W_{T_k}^k - \frac{1}{2}\delta^2 T_k)$ where $\delta = -k^2$. This density is bounded above by $\exp(k^3)$ and below by $\exp(-k^3 - \frac{1}{2}k^4)$. The density of $\mathbf{Q}_m$ on $\mathcal{F}_m$ is therefore bounded below by $\exp(-\sum_{k=1}^m (k^3 + \frac{1}{2}k^4)) \geq \exp(-2m^5)$ and bounded above by $\exp(\sum_{k=1}^m k^3) \leq \exp(m^4)$. Under the measure $\mathbf{Q}_m$ the process S^m is a martingale and hence for each H that is 1-admissible, $H \cdot S^m$ is a $\mathbf{Q}_m$-super-martingale (by Theorem 9.2.9) and hence for each 1-admissible integrand we find

$$\mathbf{E}_{\mathbf{Q}_m}[(H \cdot S)_m^+] \leq \mathbf{E}_{\mathbf{Q}_m}[(H \cdot S)_m^-]$$

and hence

$$\exp(-2m^5)\mathbf{E}_{\mathbf{P}}[(H \cdot S)_m^+] \leq \exp(m^4)\mathbf{E}_{\mathbf{P}}[(H \cdot S)_m^-]$$

and

$$\mathbf{E}_{\mathbf{P}}[(H \cdot S)_m^-] \geq \beta_m \mathbf{E}_{\mathbf{P}}[(H \cdot S)_m^+] \quad \text{for} \quad \beta_m = \exp(-3m^5).$$

We will show that if $\alpha_m \to 0$ as announced, the process S satisfies *(NFLBR)* with simple integrands.

Suppose on the contrary that S does not satisfy the *(NFLBR)* property for simple integrands. We then choose H^j simple, predictable, 1-admissible such that $(H^j \cdot S)_\infty$ tends to $f_0 \geq 0$ where $\mathbf{P}[f_0 > 0] > 0$. Find m_0 so that $\mathbf{E}_{\mathbf{P}}[\min(f_0,1)] > \frac{2}{m_0}$. For each j we define the stopping time U_j as $\inf\{t \mid (H^j \cdot S)_t \geq 1\}$ and let $L^j = H^j \mathbf{1}_{[0,U_j]}$. For each j the simple predictable process L^j is still 1-admissible and $(L^j \cdot S)_\infty \geq \min((H^j \cdot S)_\infty, 1)$, therefore $\liminf_{j\to\infty}(L^j \cdot S)_\infty \geq \min(f_0, 1)$. Each L^j is of the form $\sum_{k=1}^n f_k \mathbf{1}_{]V_{k-1},V_k]}$ where f_k is $\mathcal{F}_{T_k}$-measurable and $V_0 \leq V_1 \leq \ldots \leq V_n < \infty$ are stopping times.

If $]V_{k-1},V_k] \cap]m-1,m-1+T_m]$ is not equivalent to the zero process, then the probability of a jump between V_{k-1} and V_k is strictly positive by the same arguments as in Example 9.7.5. Because the jumps of S are positive or negative with the same probability we conclude that the downward jump of $(L^j \cdot S)$ cannot be smaller than -2. (Indeed the process is always bigger than -1 and is stopped when it hits the level 1). We conclude that also the positive jump is bounded by 2. Therefore $|L_{T_m}^j \Delta S_{T_m}| \leq 2$. We conclude that $|L^j| \leq 2^{m+1}$ on $]m-1,m-1+T_m]$. Because we stopped the process $(L^j \cdot S)$ when it exceeds the level 1, we see that:

$$\min((L^j \cdot S)_m, 1) - \min((L^j \cdot S)_{m-1}, 1) \le (L^j \cdot S)_m - (L^j \cdot S)_{m-1}.$$

The process L^j is bounded in intervals $[0, m]$ and because S is also uniformly bounded with only one jump in each interval $[k, k+1]$, the semi-martingale $L^j \cdot S$ is locally bounded, therefore special and decomposed as $L^j \cdot S = L^j \cdot M + L^j \cdot A$. The local martingale part is a square integrable martingale and hence:

$$\mathbf{E_P}[(L^j \cdot M)_m - (L^j \cdot M)_{m-1}] = 0.$$

This yields the following estimates:

$$
\begin{aligned}
&\mathbf{E_P}[\min((L^j \cdot S)_m, 1) - \min((L^j \cdot S)_{m-1}, 1)] \\
&\le \mathbf{E_P}[(L^j \cdot S)_m - (L^j \cdot S)_{m-1}] \\
&\le \mathbf{E_P}[(L^j \cdot A)_m - (L^j \cdot A)_{m-1}] \\
&\le \mathbf{E_P}\left[\int_{]m-1, m]} L_u^j \alpha_m m^2 \, du\right] \\
&\le 2^{m+1} m^2 \alpha_m.
\end{aligned}
$$

This implies that

$$
\begin{aligned}
&\mathbf{E_P}[\min((L^j \cdot S)_{m_0}, 1)] \\
&\ge \mathbf{E_P}[\min((L^j \cdot S)_\infty, 1)] - \sum_{m > m_0} 2^{m+1} m^2 \alpha_m \\
&\ge \mathbf{E_P}[\min((L^j \cdot S)_\infty, 1)] - \frac{\beta_{m_0}}{2m_0} \quad \text{(by the choice of } \alpha_m).
\end{aligned}
$$

Because

$$\liminf_{j \to \infty} \mathbf{E_P}[\min((L^j \cdot S)_\infty, 1)] > \frac{2}{m_0} \quad \text{we can deduce that}$$

$$\liminf_{j \to \infty} \mathbf{E_P}[\min((L^j \cdot S)_{m_0}, 1)] > \frac{2}{m_0} - \frac{\beta_{m_0}}{2m_0} > \frac{1}{m_0}.$$

We may now suppose that $\mathbf{E_P}[\min((L^j \cdot S)_{m_0}, 1)] > \frac{1}{m_0}$ for all j. Because of the choice of β_m we also see that

$$\mathbf{E_P}[\min((L^j \cdot S)_{m_0}^-, 1)]$$

$$\ge \beta_m \mathbf{E_P}[\min((L^j \cdot S)_{m_0}^+, 1)] \ge \beta_m \mathbf{E_P}[\min((L^j \cdot S)_{m_0}, 1)] > \frac{\beta_m}{m_0}.$$

Let the set A_j be defined as $A_j = \{(L^j \cdot S)_{m_0} < 0\}$.

Because $\liminf_{j \to \infty} \min((L^j \cdot S)_\infty, 1) \ge \min(f_0, 1)$ we also have that $\liminf_{j \to \infty}(\mathbf{1}_{A_j} \min((L^j \cdot S)_\infty, 1)) \ge \liminf_{j \to \infty}(\mathbf{1}_{A_j} \min(f_0, 1))$.

An application of Fatou's lemma yields that

$$\mathbf{E_P} \left[\liminf_{j \to \infty} (\mathbf{1}_{A_j} \min(f_0, 1)) \right]$$

$$\leq \mathbf{E_P} \left[\liminf_{j \to \infty} \mathbf{1}_{A_j} \min((L^j \cdot S)_\infty, 1) \right]$$

$$\leq \liminf_{j \to \infty} \mathbf{E_P} \left[\mathbf{1}_{A_j} \min((L^j \cdot S)_\infty, 1) \right]$$

$$\leq \liminf_{j \to \infty} \mathbf{E_P} \left[\mathbf{1}_{A_j} \min((L^j \cdot S)_{m_0}, 1) \right]$$

$$+ \sum_{m > m_0} \mathbf{E_P} \left[\min((L^j \cdot S)_m, 1) - \min((L^j \cdot S)_{m-1}, 1) \right]$$

$$\leq -\frac{\beta_{m_0}}{m_0} + \sum_{m > m_0} 2^{m+1} m^2 \alpha_m$$

$$\leq -\frac{\beta_{m_0}}{2m_0}.$$

This is clearly a contradiction to $f_0 \geq 0$. $\square$

9.8 Appendix: Some Measure Theoretical Lemmas

In this appendix we prove two lemmas we used at several places. We assume that, especially regarding the second lemma, the results are known, but we could not find a reference. We therefore give full proofs and we also add some remarks that are of independent interest but are not used elsewhere in this paper. The first lemma was already proved in [S 92, Lemma 3.5]. We give a similar but simpler proof.

Lemma 9.8.1. *Let $(f_n)_{n \geq 1}$ be a sequence of $[0, \infty[$-valued measurable functions on a probability space $(\Omega, \mathcal{F}, \mathbf{P})$. There is $g_n \in \mathrm{conv}\{f_n, f_{n+1}, \ldots\}$, such that $(g_n)_{n \geq 1}$ converges almost surely to a $[0, \infty]$-valued function g.*

If $\mathrm{conv}\{f_n; n \geq 1\}$ is bounded in L^0, then g is finite almost surely. If there are $\alpha > 0$ and $\delta > 0$ such that for all n: $\mathbf{P}[f_n > \alpha] > \delta$, then $\mathbf{P}[g > 0] > 0$.

Proof. Let $u : \mathbb{R}_+ \cup \{+\infty\} \to [0, 1]$ be defined as $u(x) = 1 - e^{-x}$. Economists may see u as a utility function but there is no need to. Define s_n as

$$s_n = \sup\{\mathbf{E}[u(g)] \mid g \in \mathrm{conv}\{f_n, f_{n+1}, \ldots\}\}$$

and choose $g_n \in \mathrm{conv}\{f_n, f_{n+1}, \ldots\}$ so that

$$\mathbf{E}[u(g)] \geq s_n - \frac{1}{n}.$$

Clearly s_n decreases to $s_0 \geq 0$ and $\lim_{n \to \infty} E[u(g_n)] = s_0$. We shall show that the sequence $(g_n)_{n \geq 1}$ converges in probability to a function g. We will work with the compact (metrisable) space $[0, \infty]$. A sequence $(x_n)_{n \geq 1}$ of elements of $[0, \infty]$ is a Cauchy sequence in $[0, \infty]$ if and only if for each $\alpha > 0$ there is n_0 so that for all $n, m \geq n_0$ we have $|x_n - x_m| \leq \alpha$ or $\min(x_n, x_m) \geq \alpha^{-1}$. From the properties of u it also follows that for $\alpha > 0$ there is $\beta > 0$, so that $|x - y| > \alpha$ and $\min(x, y) \leq \alpha^{-1}$, implies $u\left(\frac{x+y}{2}\right) > \frac{1}{2}(u(x) + u(y)) + \beta$.

We can now easily proceed with the proof of the lemma. By the observation on the topology of $[0, \infty]$ we have to show that $\lim_{n,m \to \infty} \mathbf{P}[|g_n - g_m| > \alpha$ and $\min(g_n, g_m) \leq \alpha^{-1}] = 0$.

For given $\alpha > 0$ we take β as above and we obtain

$$\mathbf{E}\left[u\left(\frac{g_n + g_m}{2}\right)\right] \geq \frac{1}{2}\mathbf{E}[u(g_n)] + \frac{1}{2}\mathbf{E}[u(g_m)]$$

$$+\beta \mathbf{P}[|g_n + g_m| > \alpha \text{ and } \min(g_n, g_m) < \alpha^{-1}].$$

By construction $\mathbf{E}\left[u\left(\frac{g_n+g_m}{2}\right)\right] \leq s_n$, but by concavity of u we have

$$\mathbf{E}\left[u\left(\frac{g_n + g_m}{2}\right)\right] \geq \frac{1}{2}\left(\mathbf{E}[u(g_n)] + \mathbf{E}[u(g_m)]\right).$$

From this it follows

$$\beta \mathbf{P}\left[|g_n - g_m| > \alpha \text{ and } \min(g_n, g_m) < \alpha^{-1}\right]$$
$$\leq \mathbf{E}\left[u\left(\frac{g_n + g_m}{2}\right)\right] - \frac{1}{2}\left(\mathbf{E}[u(g_n)] + \mathbf{E}[u(g_m)]\right).$$

The choice of the sequence $(g_n)_{n \geq 1}$ implies that the right hand side tends to 0. We therefore proved that $(g_n)_{n \geq 1}$ is a Cauchy sequence in probability and hence there is a function $g : \Omega \to [0, \infty]$ so that g_n converges to g in probability. If one wants a sequence converging almost surely one can pass to a subsequence.

If $\text{conv}\{f_n; n \leq 1\}$ is bounded in L^0 then for each $\varepsilon > 0$ there is N so that $\mathbf{P}[h > N] < \varepsilon$ for all $h \in \text{conv}\{f_n; n \geq 1\}$. In particular this implies that $\mathbf{P}[g_n > N] < \varepsilon$ and hence $\mathbf{P}[g > N] \leq \varepsilon$. The function g so obtained is therefore finite almost surely.

If $\mathbf{P}[f_n > \alpha] > \delta > 0$ for each n and fixed $\alpha > 0$, we obtain that $\mathbf{E}[u(g_n)] \geq \delta u(\alpha) > 0$. Since g_n tends to g we find $u(g_n) \to u(g)$ and by the bounded convergence theorem we obtain $\mathbf{E}[u(g)] \geq \delta u(\alpha) > 0$ and therefore $\mathbf{P}[g > 0] > 0$. $\qquad\square$

Remark 9.8.2. If $(f_n)_{n \geq 1}$ is a sequence of $[0, \infty]$-valued measurable functions then the same conclusion can be obtained. The proof is the same up to minor changes in the notation. The reader can convince himself that there is almost no gain in generality.

Remark 9.8.3. If $(f_n)_{n \geq 1}$ is a sequence of $\mathbb{R}$-valued measurable functions such that $\text{conv}\{f_n^-; n \geq 1\}$ is bounded in L^0, then there are $g_n \in \text{conv}\{f_n; n \geq 1\}$ so that g_n converges almost surely to a $]-\infty, +\infty]$-valued measurable function g.

Proof. We first take convex combinations of $\{f_n^-; n \geq 1\}$ that converge almost surely. Since $\mathrm{conv}\{f_n^-; n \geq 1\}$ is bounded in L^0, the limit is finite almost surely. We now apply the lemma to the same convex combination of f_n^+. This procedure yields convex combinations of the original sequence $(f_n)_{n\geq 1}$, converging almost surely to a $]-\infty, +\infty]$-valued function. $\qquad\square$

Remark 9.8.4. If in Remark 9.8.3 we only require that $\{f_n^-; n \geq 1\}$ is bounded in L^0 then the conclusion breaks down. Indeed take $(f_n)_{n\geq 1}$ a sequence of 1-stable (see [Lo 78] for a definition) independent random variables. If there were convex combinations converging a.s. we could make the convex combinations so that $g_k \in \mathrm{conv}\{f_{n_k+1}, \ldots, f_{n_{k+1}}\}$ where $n_1 < n_2 \ldots$. This implies that $(g_k)_{k\geq 1}$ is an independent sequence. Since convex combinations of independent 1-stable variables are 1-stable this would produce an i.i.d. sequence converging almost surely, a contradiction.

Remark 9.8.5. If in the setting of Lemma 9.8.1 the sequence $\{f_n; n \geq 1\}$ is bounded in L^0, but $\mathrm{conv}\{f_n; n \geq 1\}$ not bounded in L^0, then the procedure used in the proof does not necessarily yield a function g that is finite almost surely. The next example shows that there is a sequence $\{f_n; n \geq 1\}$ bounded in L^0 and such that *every* g that is a limit of functions $g_n \in \mathrm{conv}\{f_n, f_{n+1}, \ldots\}$, is identically $+\infty$. Before we give the construction let us recall some results from the theory of Brownian motion (see [RY 91] for details). If $(B_t)_{0\leq t}$ is a standard 1-dimensional Brownian motion, let us denote by T_β the stopping time defined as $T_\beta = \inf\{t | B_t = \beta\}$. It is known (see [RY 91, p. 67]) that for $\beta > 0$, $T_\beta < \infty$ a.s. and for each $u \geq 0$: $\mathbf{E}[\exp(-uT_\beta)] = \exp(-\beta\sqrt{2u})$. It follows that if f has the same distribution as T_β, then for $\lambda > 0$, λf has the same distribution as $T_{(\lambda)^{\frac{1}{2}}\beta}$. If $f_1, \ldots, f_N$ are independent and have the same distribution as $T_{\beta_1}, \ldots, T_{\beta_N}$ then $f_1 + \cdots + f_N$ has the same distribution as $T_{\beta_1+\cdots+\beta_N}$, (this follows easily from the interpretation of f_n as the hitting time of β_n). Take now $(f_n)_{n\geq 1}$ a sequence of independent identically distributed variables, each having the same distribution as T_1. Suppose that $g_n \in \mathrm{conv}\{f_n, f_{n+1}, \ldots\}$ and $g_n \to g$ a.e. We will show $g = +\infty$ a.e. We can assume that the functions g_n are independent, eventually we take subsequences. Each g_n has a distribution of the form

$$\lambda_1^n f_1 + \cdots + \lambda_{N_n}^n f_{N_n}$$

where $(\lambda_1^n, \ldots, \lambda_{N_n}^n)$ is a convex combination. From preceding considerations it follows that the distribution of g_n is T_{α_n} where $\alpha_n = \sum_{i=1}^{N_n} \sqrt{\lambda_i^n} \geq 1$. The 0-1-law gives us that either $g = +\infty$ or that $\mathbf{P}[g < \infty] = 1$. In this case we conclude that there is a real number α such that $\alpha_n \to \alpha \geq 1$ and g has the same distribution as T_α. From the 0-1-law it follows again that the distribution of g is degenerate, impossible if $\alpha \geq 1$. Therefore $g = +\infty$ identically. $\qquad\square$

The following lemma is quite simple, it was used above in the proof of Lemma 9.4.8 above.

Lemma 9.8.6. *Let $(g_k)_{1 \le k \le n}$ be non-negative functions defined on the probability space $(\Omega, \mathcal{F}, \mathbf{P})$. Suppose that there are positive numbers $(a_k)_{1 \le k \le n}$ as well as $\delta > 0$ so that for every k: $\mathbf{P}[g_k \ge a_k] \ge \delta > 0$. If $g = \sum_{j=1}^n g_j$ then for all $0 < \eta < 1$ we have $\mathbf{P}[g \ge \eta(\sum_{j=1}^n a_j)\delta] \ge \frac{\delta(1-\eta)}{1-\eta\delta}$.*

Proof. Let $A = \{g \ge (\sum_{j=1}^n a_j)\delta\eta\}$. Clearly

$$\mathbf{E}[g\mathbf{1}_{A^c}] \le \left(\sum_{j=1}^n a_j\right)\delta\eta\mathbf{P}[A^c] = \left(\sum_{j=1}^n a_j\right)\delta\eta(1 - \mathbf{P}[A]).$$

On the other hand

$$\mathbf{E}[g\mathbf{1}_{A^c}] = \left(\sum_{j=1}^n \mathbf{E}[g_j\mathbf{1}_{A^c}]\right)$$

$$\ge \left(\sum_{j=1}^n a_j\mathbf{P}\Big[A^c \cap \{g_j \ge a_j\}\Big]\right)$$

$$\ge \left(\sum_{j=1}^n a_j\Big(\mathbf{P}[g_j \ge a_j] - \mathbf{P}[A]\Big)\right)$$

$$\ge \left(\sum_{j=1}^n a_j\right)\delta - \left(\sum_{j=1}^n a_j\right)\mathbf{P}[A].$$

Both inequalities imply

$$\left(\sum_{j=1}^n a_j\right)\mathbf{P}[A](1 - \delta\eta) \ge \left(\sum_{j=1}^n a_j\right)\delta(1 - \eta).$$

We may of course suppose that $\sum_{j=1}^n a_j > 0$ and this yields the desired result $\mathbf{P}[A] \ge \frac{\delta(1-\eta)}{1-\delta\eta}$. $\qquad\square$

Corollary 9.8.7. *If $(g_j)_{1 \le j \le n}$ are non-negative functions defined on the probability space $(\Omega, \mathcal{F}, \mathbf{P})$ and if for $j = 1, \ldots, n$ we have $\mathbf{P}[g_j \ge a] \ge b$ where $a, b > 0$, then for $g = \sum_{j=1}^n g_j$ we have $\mathbf{P}\left[g \ge \frac{nab}{2}\right] \ge \frac{b}{2}$.*

Acknowledgement

The authors want to thank P. Artzner, M. Émery, P. Müller and Ch. Stricker for fruitful discussions on this paper. Part of this research was supported by the European Community Stimulation Plan for Economic Science contract No SPES-CT91-0089.

10

A Simple Counter-Example to Several Problems in the Theory of Asset Pricing (1998)

Abstract. We give an easy example of two strictly positive local martingales which fail to be uniformly integrable, but such that their product is a uniformly integrable martingale. The example simplifies an earlier example given by the second author. We give applications in Mathematical Finance and we show that the phenomenon is present in many incomplete markets.

10.1 Introduction and Known Results

Let $S = M + A$ be a continuous semi-martingale, which we interpret as the discounted price process of some traded asset; the process M is a continuous local martingale and the continuous process A is of finite variation. It is obviously necessary that $dA \ll d\langle M, M \rangle$, for otherwise we would invest in the asset when A moves but M doesn't and make money risklessly. Thus we have for some predictable process α:

$$dS_t = dM_t + \alpha_t \, d\langle M, M \rangle_t . \tag{10.1}$$

It has long been recognised (see [HK 79]) that the absence of *arbitrage* (suitably defined) in this market is equivalent to the existence of some probability $\mathbf{Q}$, equivalent to the reference probability $\mathbf{P}$, under which S becomes a local martingale (see Chap. 9 for the definition of *arbitrage* and a precise statement and proof of this fundamental result). Such a measure $\mathbf{Q}$ is then called an equivalent local martingale measure or *(ELMM)*. The set of all such *(ELMM)* is then denoted by $\mathcal{M}^e(S)$, or $\mathcal{M}^e$ for short. The result of that paper

[DS 98a]   A Simple Counter-Example to Several Problems in the Theory of Asset Pricing. *Mathematical Finance*, vol. 8, pp. 1–12, (1998).

* Part of this research was supported by the European Community Stimulation Plan for Economic Science contract Number SPES-CT91-0089. We thank an anonymous referee for substantial advice and even rewriting the main part of the introduction. We also thank Ch. Stricker for helpful discussions.

applies to the more general situation of a locally bounded semi-martingale S, but in the situation of continuous S, perhaps the result can be proved more simply (see e.g. Chap. 12 for the case of continuous processes and its relation to no *arbitrage*). In particular, it is tempting to define $\mathbf{Q}$ by looking at the decomposition (10.1) of S and by setting

$$\left.\frac{d\mathbf{Q}}{d\mathbf{P}}\right|_{\mathcal{F}_t} = \mathcal{E}(-\alpha \cdot M)_t$$

provided the exponential local martingale $\mathcal{E}(-\alpha \cdot M)$ is a true martingale. Is it possible that there exist an equivalent local martingale measure for S, and yet the exponential local martingale $\mathcal{E}(-\alpha \cdot M)$ fails the martingale property? The answer is yes; our example shows it. In the terminology of [FS 91], this means that the *minimal local martingale measure* for the process S does not exist, although $\mathcal{M}^e(S)$ is non-empty.

A second question where our example finds interesting application is in hedging of contingent claims in incomplete markets. The positive contingent claim g, or more generally a function that is bounded below by a constant, can be hedged if g can be written as

$$g = c + (H \cdot S)_\infty, \tag{10.2}$$

where c is a positive constant and where H is some admissible integrand (i.e. for some constant $a \in \mathbb{R}, (H \cdot S) \geq -a$). In order to avoid *suicide strategies* we also have to impose that $(H \cdot S)_\infty$ is maximal in the set of outcomes of admissible integrands (see Chaps. 11 and 13) for information on maximal elements and [HP 81] for the notion of *suicide strategies*). We recall that an outcome $(H \cdot S)_\infty$ of an admissible strategy H, is called maximal if for an admissible strategy K the relation $(K \cdot S)_\infty \geq (H \cdot S)_\infty$ implies that $(K \cdot S)_\infty = (H \cdot S)_\infty$. S. Jacka [J 92], J.P. Ansel, Ch. Stricker [AS 94] and the authors showed that g can be hedged if and only if there is an equivalent local martingale measure $\mathbf{Q} \in \mathcal{M}^e$ such that

$$\mathbf{E}_\mathbf{Q}[g] = \sup\{\mathbf{E}_\mathbf{R}[g] \mid \mathbf{R} \in \mathcal{M}^e\}.$$

Looking at (10.2) we then can show that $H \cdot S$ is a $\mathbf{Q}$-uniformly integrable martingale and hence that $c = \mathbf{E}_\mathbf{Q}[g]$. Also the outcome $(H \cdot S)_\infty$ is then maximal. It is natural to conjecture that in fact for all $\mathbf{R} \in \mathcal{M}^e$, we might have $\mathbf{E}_\mathbf{R}[g] = c$, and the sup becomes unnecessary, which is the case for bounded functions g. However, our example shows that this too is false in general.

To describe our example, suppose that B and W are two independent Brownian motions and let $L_t = \exp(B_t - \frac{1}{2}t)$. Then L is a strict local martingale. For information on continuous martingales and especially martingales related to Brownian motion we refer to Revuz and Yor [RY 91]. Let us recall that a local martingale that is not a uniformly integrable martingale is called a strict local martingale. The terminology was introduced by Elworthy, Li, and Yor [ELY 99]. The stopped process L^τ where $\tau = \inf\{t \mid L_t = \frac{1}{2}\}$ is still a strict local martingale and $\tau < \infty$. If we stop L^τ at some independent random time σ, then $L^{\tau \wedge \sigma}$ will be uniformly integrable if $\sigma < \infty$ a.s. and otherwise it will not be. If we thus define $M_t = \exp(W_t - \frac{1}{2}t)$ and $\sigma = \inf\{t \mid M_t = 2\}$ then $L^{\tau \wedge \sigma}$ is not uniformly integrable since $\mathbf{P}[\sigma = \infty] = \frac{1}{2}$. However, if we change the measure using the uniformly integrable martingale $M^{\tau \wedge \sigma}$, then under the new measure we have that W becomes a Brownian motion with drift $+1$ and so $\sigma < \infty$ a.s.. The product $L^{\tau \wedge \sigma} M^{\tau \wedge \sigma}$ becomes a uniformly integrable martingale!

The problem whether the product of two strictly positive strict local martingales could be a uniformly integrable martingale goes back to Karatzas, Lehoczky, and Shreve [KLSX 91]. Lépingle [L 91] gave an example in discrete time. Independently, Karatzas, Lehoczky and Shreve also gave such an example but the problem remained open whether such a situation could occur for continuous local martingales. The first example on the continuous case was given in [S 93], but it is quite technical (although the underlying idea is rather simple).

In this note we simplify the example considerably. A previous version of this paper, only containing the example of Sect. 10.2, was informally distributed with the title *A Simple Example of Two Non Uniformly Integrable Continuous Martingales Whose Product is a Uniformly Integrable Martingale.* Our sincere thanks go to L.C.G. Rogers, who independently constructed an almost identical example and kindly supplied us with his manuscript, see [R 93].

We summarise the results of [S 93] translated into the present context. The basic properties of the counter-example are described by the following

Theorem 10.1.1. *There is a continuous process X that is strictly positive, $X_0 = 1$, $X_\infty > 0$ a.s. as well as a strictly positive process Y, $Y_0 = 1$, $Y_\infty > 0$ a.s. such that*

(1) *The process X is a strict local martingale under $\mathbf{P}$, i.e. $\mathbf{E_P}[X_\infty] < 1$.*
(2) *The process Y is a uniformly integrable martingale.*
(3) *The process XY is a uniformly integrable martingale.*

Depending on the interpretation of the process X we obtain the following results.

Theorem 10.1.2. *There is a continuous semi-martingale S such that*

(1) *The semi-martingale admits a Doob-Meyer decomposition of the form $dS = dM + d\langle M, M \rangle h$.*
(2) *The local martingale $\mathcal{E}(-h \cdot M)$ is strict.*
(3) *There is an equivalent local martingale measure for S.*

Proof. Take X as in the preceding theorem and define, through the stochastic logarithm, the process S as $dS = dM + d\langle M, M\rangle$ where $X = \mathcal{E}(-M)$. The measure $\mathbf{Q}$ defined as $d\mathbf{Q} = X_\infty Y_\infty \, d\mathbf{P}$ is an *(ELMM)* for S. Obviously the natural candidate for an *(ELMM)* suggested by the Girsanov-Maruyama-Meyer formula, i.e. the *density* X_∞, does not define a probability measure. □

Remark 10.1.3. If in the previous theorem we replace S by $\mathcal{E}(S)$, then we can even obtain a positive price system.

Theorem 10.1.4. *There is a process S that admits an (ELMM) as well as a hedgeable element g such that $\mathbf{E_R}[g]$ is not constant on the set $\mathcal{M}^e$.*

Proof. For the process S we take X from Theorem 10.1.1. The original measure $\mathbf{P}$ is an *(ELMM)* and since there is an *(ELMM)* $\mathbf{Q}$ for X such that X becomes a uniformly integrable martingale, we necessarily have that $\mathbf{E_Q}[X_\infty - X_0] = 0$ and that $X_\infty - X_0$ is maximal. However, $\mathbf{E_P}[X_\infty - X_0] < 0$. □

As for the economic interpretation, let us consider a contingent claim f that is maximal and such that $\mathbf{E_R}[f] < 0 = \sup\{\mathbf{E_Q}[f] \mid \mathbf{Q} \in \mathcal{M}^e\}$ for some $\mathbf{R} \in \mathcal{M}^e$. Suppose now that a new instrument T is added to the market and suppose that the instrument T has a price at time t equal to $\mathbf{E_R}[f \mid \mathcal{F}_t]$. The measure $\mathbf{R}$ is still a local martingale measure for the couple (S, T), hence the financial market described by (S, T) still is *arbitrage free* — more precisely it does not admit a free lunch with vanishing risk; but as easily seen the element f is no longer maximal in this expanded market. Indeed the element $T_\infty - T_0 = f - \mathbf{E_R}[f]$ dominates f by the quantity $-\mathbf{E_R}[f] > 0$! In other words, before the introduction of the instrument T the hedge of f as $(H \cdot S)_\infty$ may make sense economically, after the introduction of T it becomes a *suicide strategy* which only an idiot will apply.

Note that an economic agent cannot make arbitrage by going short on a strategy H that leads to $(H \cdot S)_\infty = f$ and by buying the financial instrument T. Indeed the process $-(H \cdot S) + T - T_0$ in not bounded below by a constant and therefore the integrand $(-H, 1)$ is not admissible!

On the other hand the maximal elements f such that $\mathbf{E_R}[f] = 0$ for all measures $\mathbf{R} \in \mathcal{M}^e$ have a stability property. Whatever new instrument T will be added to the market, as long as the couple (S, T) satisfies the *(NFLVR)* property, the element f will remain maximal for the new market described by the price process (S, T). The set of all such elements as well as the space generated by the maximal elements is the subject of Chap. 13.

Sect. 10.2 of this paper gives an easy example that satisfies the properties of Theorem 10.1.1. Sect. 10.3 shows that the construction of this example can be mimicked in most incomplete markets with continuous prices.

10.2 Construction of the Example

We will make use of two independent Brownian motions, B and W, defined on a filtered probability space $(\Omega, (\mathcal{F}_t)_{0 \leq t}, \mathbf{P})$, where the filtration $\mathcal{F}$ is the natural filtration of the couple (B, W) and is supposed to satisfy the usual conditions. This means that $\mathcal{F}_0$ contains all null sets of $\mathcal{F}_\infty$ and that the filtration is right continuous. The process L defined as

$$L_t = \exp\left(B_t - \tfrac{1}{2}t\right)$$

is known to be a strict local martingale with respect to the filtration $\mathcal{F}$. Indeed, the process L tends almost surely to 0 at infinity, hence it cannot be a uniformly integrable martingale. Let us define the stopping time τ as

$$\tau = \inf\left\{t \mid L_t = \tfrac{1}{2}\right\}.$$

Clearly $\tau < \infty$ a.s.. Using the Brownian motion W we similarly construct

$$M_t = \exp\left(W_t - \tfrac{1}{2}t\right).$$

The stopping time σ is defined as

$$\sigma = \inf\{t \mid M_t = 2\}.$$

In the case the process M does not hit the level 2 the stopping time σ equals ∞ and we therefore have that M_σ either equals 2 or equals 0, each with probability $\tfrac{1}{2}$. The stopped process M^σ defined by $M_t^\sigma = M_{t \wedge \sigma}$ is a uniformly integrable martingale. It follows that also the process $Y = M^{\tau \wedge \sigma}$ is uniformly integrable and because $\tau < \infty$ a.s. we have that Y is almost surely strictly positive on the interval $[0, \infty]$.

The process X is now defined as the process L stopped at the stopping time $\tau \wedge \sigma$. Note that the processes L and M are independent since they were constructed using independent Brownian motions. Stopping the processes using stopping times coming from the other Brownian motion destroys the independence and it is precisely this phenomenon that will allow us to make the counter-example.

Theorem 10.2.1. *The processes X and Y, as defined above, satisfy the properties listed in Theorem 10.1.1*

(1) *The process X is a strict local martingale under $\mathbf{P}$, i.e. $\mathbf{E}_\mathbf{P}[X_\infty] < 1$ and $X_\infty > 0$ a.s..*
(2) *The process Y is a uniformly bounded integrable martingale.*
(3) *The process XY is a uniformly integrable martingale.*

Proof. Let us first show that X is not uniformly integrable. For this it is sufficient to show that $\mathbf{E}[X_\infty] = \mathbf{E}[L_{\tau \wedge \sigma}] < 1$. This is quite easy. Indeed

$$\mathbf{E}[L_{\tau \wedge \sigma}] = \int_{\{\sigma = \infty\}} L_\tau + \int_{\{\sigma < \infty\}} L_{\sigma \wedge \tau}.$$

In the first term the variable L_τ equals $\tfrac{1}{2}$ and hence this term equals $\tfrac{1}{2}\mathbf{P}[\sigma = \infty] = \tfrac{1}{4}$. The second term is calculated using the martingale property of L and the optional stopping time theorem.

$$\int_{\{\sigma<\infty\}} L_{\sigma\wedge\tau} = \int_0^\infty \mathbf{P}[\sigma \in dt]\mathbf{E}[L_{\tau\wedge t}]$$
$$= \mathbf{P}[\sigma < \infty].$$

The first line follows from the independence of σ and the process L. Putting together both terms yields $\mathbf{E}[L_{\tau\wedge\sigma}] = \frac{1}{2}\mathbf{P}[\sigma = \infty] + \mathbf{P}[\sigma < \infty] = \frac{3}{4} < 1$.

On the other hand the product XY is a uniformly integrable martingale. To see this, it is sufficient to show that $\mathbf{E}[X_\infty Y_\infty] = 1$. The calculation is similar to the preceding calculation and uses the same arguments.

$$\mathbf{E}[X_\infty Y_\infty] = \mathbf{E}[L_{\tau\wedge\sigma} M_{\tau\wedge\sigma}]$$
$$= \mathbf{E}[L_{\tau\wedge\sigma} M_\sigma] \text{ because } M^\sigma \text{ is a uniformly integrable martingale}$$
$$= 2\,\mathbf{E}[L_{\tau\wedge\sigma}\mathbf{1}_{\{\sigma<\infty\}}]$$
$$= 2\,\mathbf{P}[\sigma < \infty] = 1. \qquad\qquad \Box$$

10.3 Incomplete Markets

All processes will be defined on a filtered probability space $(\Omega, (\mathcal{F}_t)_{t\geq 0}, \mathbf{P})$. For the sake of generality the time set is supposed to be the set $\mathbb{R}_+$ of all non-negative real numbers. The filtration is supposed to satisfy the usual hypothesis, i.e. it is right continuous and the σ-algebra $\mathcal{F}_0$ is saturated with all the negligible sets of $\mathcal{F}_\infty = \bigvee_{t\geq 0} \mathcal{F}_t$. The symbol S denotes a d-dimensional semi-martingale $S\colon \Omega \times \mathbb{R}_+ \to \mathbb{R}^d$. For vector stochastic integration we refer to [J 79]. If needed, we denote by x' the transpose of a vector x.

We assume that S has the (NFLVR) property and the set of (ELMM) is denoted by $\mathcal{M}^e$. The market is supposed to be incomplete in the following sense. We assume that there is a real-valued non-zero continuous local martingale W such that the bracket $\langle W, S\rangle = 0$ but such that the measure $d\langle W, W\rangle$ (defined on the predictable σ-algebra of $\Omega \times \mathbb{R}_+$) is not singular with respect to the measure $d\lambda$ where $\lambda = \text{trace}\langle S, S\rangle$.

Let us first try to give some economic interpretation to this hypothesis. The existence of a local martingale W such that $\langle S, W\rangle = 0$ implies that the process S is not sufficient to hedge all the contingent claims. The extra assumption that the measure $d\langle W, W\rangle$ is not singular to $d\,\text{trace}\langle S, S\rangle$ then means that at least part of the local martingale W moves at the same time as the process S. The incompleteness of the market, therefore, is not only due to the fact that S and W are varying in disjoint time sets but the incompleteness is also due to the fact that locally the process S does not span all of the random movements that are possible.

Theorem 10.3.1. *If S is a continuous d-dimensional semi-martingale with the (NFLVR) property, if there is a continuous local martingale W such that $\langle W, S\rangle = 0$ but $d\langle W, W\rangle$ is not singular to $d\,\text{trace}\langle S, S\rangle$, then for each $\mathbf{R}$ in $\mathcal{M}^e$, there is a maximal element f such that $\mathbf{E}_\mathbf{R}[f] < 0$.*

Proof. The proof is broken up in different lemmata. Let W' be the martingale component in the Doob-Meyer decomposition of W with respect to the measure $\mathbf{R}$. Clearly $\langle W', W' \rangle = \langle W, W \rangle$.   □

Lemma 10.3.2. *Under the hypothesis of the theorem, there is a real-valued $\mathbf{R}$-local martingale $U \neq 0$ such that*

(1) $\langle S, U \rangle = 0$
(2) *there is a bounded $\mathbb{R}^d$-valued predictable process H that is S-integrable and such that $d\langle U, U \rangle = H' \, d\langle S, S \rangle \, H$ so that the process $N = H \cdot S$ satisfies $\langle N, N \rangle = \langle U, U \rangle$.*

Proof of Lemma 10.3.2. Let $\lambda = \operatorname{trace}\langle S, S \rangle$. Since $d\langle W, W \rangle$ is not singular with respect to $d\lambda$ there is a predictable set A such that $\mathbf{1}_A \, d\langle W, W \rangle$ is not identically zero and absolutely continuous with respect to $d\lambda$. From the predictable Radon-Nikodým theorem, see Chap. 12, it follows that there is a predictable process h such that $\mathbf{1}_A \, d\langle W, W \rangle = h \, d\lambda$. For n big enough the process $h\mathbf{1}_{\{\|h\| \leq n\}}\mathbf{1}_{[0,n]}$ is λ-integrable and is such that $\mathbf{1}_A \mathbf{1}_{\{\|h\| \leq n\}}\mathbf{1}_{[0,n]}d\langle W, W \rangle$ is not zero a.s.. We take $U = \left(\mathbf{1}_A \mathbf{1}_{\{\|h\| \leq n\}}\mathbf{1}_{[0,n]} \right) \cdot W'$. To find H we first construct a strategy K such that $\frac{d\langle K \cdot S, K \cdot S \rangle}{d\lambda} \neq 0$ a.e.. This is easy. For each coordinate i, we take an investment $P_i = (0, 0, \ldots, 0, 1, 0, \ldots)$ in asset number i. On the predictable set $\frac{d\langle P_1 \cdot S, P_1 \cdot S \rangle}{d\lambda} \neq 0$ we take $K = P_1$, on the predictable set where $\frac{d\langle P_1 \cdot S, P_1 \cdot S \rangle}{d\lambda} = 0$ and $\frac{d\langle P_2 \cdot S, P_2 \cdot S \rangle}{d\lambda} \neq 0$ we take $K = P_2$, etc.. We now take $H = K\mathbf{1}_A \mathbf{1}_{\{\|h\| \leq n\}}\mathbf{1}_{[0,n]}h^{\frac{1}{2}}\left(\frac{d\lambda}{d\langle K \cdot S, K \cdot S \rangle} \right)$.   □

Remark 10.3.3. We define the stopping time ν_u as $\nu_u = \inf\{t \mid \langle N, N \rangle_t > u\}$, where N is defined as in Lemma 10.3.2 above. If we replace the process (N, U), the filtration $\mathcal{F}_t$ and the probability $\mathbf{R}$ by, respectively, the process $(N_{\nu_0 + t}, U_{\nu_0 + t})_{t \geq 0}$, $(\mathcal{F}_{\nu_0 + t})_{t \geq 0}$, and the conditional probability $\mathbf{R}[\,.\,\mid \nu_0 < \infty]$, we may without loss of generality suppose that $\mathbf{R}[\nu_0 = 0] = \mathbf{R}[\langle N, N \rangle_\infty > 0] = 1$. In this case we have that $\lim_{u \to 0} \mathbf{R}[\nu_u < \infty] = 1$ and $\lim_{u \to 0} \nu_u = 0$. We will do so without further notice.

Remark 10.3.4. The idea of the subsequent construction is to see the strongly orthogonal local martingales U and N as time-transformed independent Brownian motions and to use the construction of Sect. 10.2. The first step is to prove that there is a strict local martingale that is an exponential. The idea is to use the exponential $\mathcal{E}(B)$ where B is a time transform of a Brownian motion. However, the exponential only tends to zero on the set $\{\langle B, B \rangle = \infty\}$.

Lemma 10.3.5. *There is a predictable process K such that the local $\mathbf{R}$-martingale $\mathcal{E}(K \cdot N)$ is not uniformly $\mathbf{R}$-integrable.*

Proof of Lemma 10.3.5. Take a strictly decreasing sequence of strictly positive real numbers $(\varepsilon_n)_{n \geq 1}$ such that $\sum_{n \geq 1} \varepsilon_n 2^n < \frac{1}{8}$.

We take u_1 small enough so that $\mathbf{R}[\nu_{u_1} < \infty] > 1 - \varepsilon_1$. From the definition of ν_{u_1} it follows that $\langle N, N \rangle_\infty > u_1$ on the set $\{\nu_{u_1} < \infty\}$. For each k we look at the exponential $\mathcal{E}(k \cdot N)$ and we let $f_k = (\mathcal{E}(k \cdot N))_{\nu_{u_1}}$. Since $\langle N, N \rangle_{\nu_{u_1}} > 0$ we have that f_k tends to zero a.s. as k tends to ∞.

Take k_1 big enough to have $\mathbf{R}[f_{k_1} < \frac{1}{2}] > 1 - \varepsilon_1$. We now define

$$\tau_1 = \inf \left\{ t \mid (\mathcal{E}(k_1 \cdot N))_t > 2 \text{ or } < \tfrac{1}{2} \right\} \wedge \nu_{u_1}.$$

Clearly $\mathbf{R}[\tau_1 < \nu_{u_1}] > 1 - \varepsilon_1$ and hence

$$\mathbf{R}\left[(\mathcal{E}(k_1 \cdot N))_{\tau_1} \in \left\{ \tfrac{1}{2}, 2 \right\} \right] > 1 - \varepsilon_1.$$

For later use we define $X_1 = (\mathcal{E}(k_1 \cdot N))_{\tau_1}$ and we observe that $\mathbf{R}[X_1 = 2] > \frac{1}{3} - \varepsilon_1$ and $\mathbf{R}[X_1 = \frac{1}{2}] > \frac{2}{3} - \varepsilon_1$.

We now repeat the construction at time ν_{u_1}. Of course this can only be done on the set $\{\nu_{u_1} < \infty\} = \{\langle N, N \rangle_\infty > u_1\}$. Take $u_2 > u_1$ small enough so that

$$\mathbf{R}[\nu_{u_2} < \infty] > \mathbf{R}[\nu_{u_1} < \infty](1 - \varepsilon_2).$$

We define $f_k = (\mathcal{E}(k \cdot (N - N^{\nu_{u_1}})))_{\nu_{u_2}}$ and observe that f_k tends to zero on the set $\{\nu_{u_1} < \infty\}$ as k tends to infinity. Indeed this follows from the statement that $\langle N - N^{\nu_{u_1}}, N - N^{\nu_{u_1}} \rangle_\infty > 0$ on the set $\{\nu_{u_1} < \infty\}$.

So we take k_2 big enough to guarantee that $\mathbf{R}[f_{k_2} < \frac{1}{2}] > \mathbf{R}[\nu_{u_1} < \infty](1 - \varepsilon_2)$. We define $\tau_2 = \inf \left\{ t > \nu_{u_1} \mid (\mathcal{E}(k \cdot (N - N^{\nu_{u_1}})))_t > 2 \text{ or } < \tfrac{1}{2} \right\} \wedge \nu_{u_2}$. Clearly $\mathbf{R}[\tau_2 < \nu_{u_2}] > \mathbf{R}[\nu_{u_1} < \infty](1 - \varepsilon_2)$. We define $X_2 = (\mathcal{E}(k \cdot (N - N^{\nu_{u_1}})))_{\tau_2}$ and we observe that $\frac{1}{2} \le X_2 \le 2$, $\mathbf{R}[X_2 \in \{\frac{1}{2}, 2\} \mid \nu_{u_1} < \infty] > 1 - \varepsilon_2$.

Since $\mathbf{E_R}[X_2 \mid \nu_{u_1} < \infty] = 1$ we therefore have that $\mathbf{R}[X_2 = 2 \mid \nu_{u_1} < \infty] > \frac{1}{3} - \varepsilon_2$ and $\mathbf{R}[X_2 = \frac{1}{2} \mid \nu_{u_1} < \infty] > \frac{2}{3} - \varepsilon_2$.

Continuing this way we construct sequences of

(1) stopping times ν_{u_n} with $\mathbf{R}[\nu_{u_n} < \infty] > \mathbf{R}[\nu_{u_{n-1}} < \infty](1 - \varepsilon_n)$, $\nu_0 = 0$
(2) real numbers k_n
(3) stopping times τ_n with $\nu_{u_{n-1}} \le \tau_n \le \nu_{u_n}$
(4) $X_n = (\mathcal{E}(k_n \cdot (N - N^{\nu_{u_{n-1}}})))_{\tau_n}$,

so that

(1) $\frac{1}{2} \le X_n \le 2$
(2) $X_n = 1$ on the set $\{\nu_{u_{n-1}} = \infty\}$
(3) $\mathbf{R}[X_n = 2 \mid \nu_{u_{n-1}} < \infty] > \frac{1}{3} - \varepsilon_n$
(4) $\mathbf{R}[X_n = \frac{1}{2} \mid \nu_{u_{n-1}} < \infty] > \frac{2}{3} - \varepsilon_n$.

Let now $K = \sum_{n \ge 1} k_n \mathbf{1}_{[\nu_{u_{n-1}}, \tau_n]}$. Clearly $\mathcal{E}(K \cdot N)$ is defined and $(\mathcal{E}(K \cdot N))_{\tau_n} = \prod_{k=1}^n X_k$. We claim that $\mathbf{E_R}[(\mathcal{E}(K \cdot N))_\infty] < 1$, showing that $\mathcal{E}(K \cdot N)$ is not uniformly integrable.

Obviously $(\mathcal{E}(K \cdot N))_\infty = \prod_{k \ge 1} X_k$. From the strong law of large numbers for martingale differences we deduce that a.s.

$$\frac{1}{n} \sum_{k=1}^{n} \left(\log X_k - \mathbf{E}_{\mathbf{R}} \left[\log X_k \, \big| \, \mathcal{F}_{\nu_{u_{k-1}}} \right] \right) \to 0 \,.$$

On the set $\{\nu_{u_{k-1}} < \infty\}$ we have that $\mathbf{E}_{\mathbf{R}} \left[\log X_k \, | \, \mathcal{F}_{\nu_{u_{k-1}}} \right] \leq (\frac{2}{3} - \varepsilon_k) \log \frac{1}{2} + (\frac{1}{3} + \varepsilon_k) \log 2 \leq -\frac{1}{3} \log 2 + 2\varepsilon_k \log 2 \leq -\frac{1}{6} \log 2$, at least for k large enough. It follows that on the set $\bigcap_{n \geq 1} \{\nu_{u_n} < \infty\}$, we have that $\sum_{k=1}^{n} \log X_k \to -\infty$, and hence $(\mathcal{E}(K \cdot N))_\infty = 0$ on this set. On the complement, i.e. on $\bigcup_{n \geq 1} \{\nu_{u_n} = \infty\}$, we find that the maximal function $(\mathcal{E}(K \cdot N))_\infty^*$ is bounded by 2^n where n is the first natural number such that $\nu_{u_n} = \infty$. The probability of this event is bounded by ε_n and hence $\mathbf{E}_{\mathbf{R}}[(\mathcal{E}(K \cdot N))_\infty] \leq \eta = \sum_n \varepsilon_n 2^n \leq \frac{1}{8}$. □

Remark 10.3.6. By adjusting the ε_n we can actually obtain a predictable process K such that $(\mathcal{E}(K \cdot N))_\infty = 0$ on a set with measure arbitrarily close to 1.

Lemma 10.3.7. *If L is a continuous positive strict local martingale, starting at 1, then for $\alpha > 0$ small enough the process L stopped when it hits the level α is still a strict local martingale.*

Proof of Lemma 10.3.7. Simply let $\tau = \inf\{t \mid L_t < \alpha\}$. Clearly $\mathbf{E}_{\mathbf{R}}[L_\tau] < \alpha + \mathbf{E}_{\mathbf{R}}[L_\infty] < 1$ for $\alpha < 1 - \mathbf{E}_{\mathbf{R}}[L_\infty]$.

If we apply the previous lemma on the exponential martingale $L = \mathcal{E}(K \cdot N)$ and to $\alpha = \eta$, we obtain a stopping time τ and a strict local martingale $\mathcal{E}(K \cdot N)^\tau$ that is bounded away from zero.

We now use the same integrand K to construct $Z = \mathcal{E}(K \cdot U)$ and we define $\sigma = \inf\{t \mid Z_t = 2\}$.

We will show that $\mathbf{E}_{\mathbf{R}}[L_{\tau \wedge \sigma}] < 1$ and that $\mathbf{E}_{\mathbf{R}}[Z_{\tau \wedge \sigma} L_{\tau \wedge \sigma}] = 1$. This will complete the proof of the theorem since the measure $\mathbf{Q}$ defined by $d\mathbf{Q} = Z_{\tau \wedge \sigma} d\mathbf{R}$ is an equivalent martingale measure and the element $f = L_{\tau \wedge \sigma} - 1$ is therefore maximal. On the other hand $\mathbf{E}_{\mathbf{R}}[f] < 0$.

Both statements will be shown using a time transform argument. The fact that the processes $K \cdot N$ and $K \cdot U$ both have the same bracket will now turn out to be useful. The time transform can be used to transform both these processes into Brownian motions at the same time.

Following [RY 91, Chap. V, Sect. 1], we define

$$T_t = \inf \left\{ u \, \middle| \, \langle K \cdot N, K \cdot N \rangle_u = \int_0^u K_s^2 \, d\langle N, N \rangle_s > t \right\} .$$

As well-known [RY 91], there are

(1) a probability space $(\widetilde{\Omega}, \widetilde{\mathcal{F}}, \widetilde{\mathbf{R}})$,
(2) a map $\pi : \widetilde{\Omega} \to \Omega$,
(3) a filtration $(\widetilde{\mathcal{F}}_t)_{t \geq 0}$ on $\widetilde{\Omega}$,

(4) two processes β^1 and β^2 that are Brownian motions with respect to $(\widetilde{\mathcal{F}}_t)_{t\geq 0}$ and such that $\langle \beta^1, \beta^2 \rangle = 0$,

(5) the variable $\gamma = \int_0^\infty K_s^2 \, d\langle N, N \rangle_s \circ \pi$ is a stopping time with respect to $(\widetilde{\mathcal{F}}_t)_{t\geq 0}$,

(6) $\beta^1_{t\wedge\gamma} = (K \cdot N)_{T_t} \circ \pi$,

(7) $\beta^2_{t\wedge\gamma} = (K \cdot U)_{T_t} \circ \pi$,

(8) $\widetilde{L} = \mathcal{E}(\beta^1)$ satisfies $L_{T_t} \circ \pi = \widetilde{L}_{t\wedge\gamma}$,

(9) $\widetilde{Z} = \mathcal{E}(\beta^2)$ satisfies $Z_{T_t} \circ \pi = \widetilde{Z}_{t\wedge\gamma}$,

(10) $\widetilde{\tau} = \inf\{t \mid (\mathcal{E}(\beta^1))_t < \frac{1}{2}\}$ satisfies $\tau \circ \pi = T_{\widetilde{\tau}}$,

(11) $\widetilde{\sigma} = \inf\{t \mid (\mathcal{E}(\beta^2))_t > 2\}$ satisfies $\sigma \circ \pi = T_{\widetilde{\sigma}}$.

In this setting we have to show that $\mathbf{E}_{\mathbf{R}}[L_{\tau\wedge\sigma}] = \widetilde{\mathbf{E}}_{\widetilde{\mathbf{R}}}[\widetilde{L}_{\widetilde{\tau}\wedge\widetilde{\sigma}\wedge\gamma}] < 1$. But on the set $\{\int_0^\infty K_s^2 \, d\langle N, N \rangle_s < \infty\}$ we have, as shown above,

$$\mathbf{E}_{\mathbf{R}} \left[\mathbf{1}_{\{\int_0^\infty K_s^2 \, d\langle N,N \rangle_s < \infty\}} L_{\tau\wedge\sigma} \right]$$
$$\leq \mathbf{E}_{\mathbf{R}} \left[\mathbf{1}_{\{\int_0^\infty K_s^2 \, d\langle N,N \rangle_s < \infty\}} L^* \right]$$
$$\leq \eta.$$

In other words, $\widetilde{\mathbf{E}}_{\widetilde{\mathbf{R}}}[\mathbf{1}_{\{\gamma<\infty\}}\widetilde{L}^*_\gamma] \leq \eta$. So it remains to be shown that

$$\widetilde{\mathbf{E}}_{\widetilde{\mathbf{R}}} \left[\mathbf{1}_{\{\gamma=\infty\}}\widetilde{L}_{\widetilde{\tau}\wedge\widetilde{\sigma}} \right] < 1 - \eta.$$

Actually, we will show that

$$\widetilde{\mathbf{E}}_{\widetilde{\mathbf{R}}} \left[\widetilde{L}_{\widetilde{\tau}\wedge\widetilde{\sigma}} \right] < 1 - \eta.$$

This is easy and follows from the independence of β^1 and β^2, a consequence of $\langle \beta^1, \beta^2 \rangle = 0$! As in Sect. 10.2 we have

$$\widetilde{\mathbf{E}}_{\widetilde{\mathbf{R}}} \left[\widetilde{L}_{\widetilde{\tau}\wedge\widetilde{\sigma}} \right] = \eta\widetilde{\mathbf{R}}[\widetilde{\sigma} = \infty] + \widetilde{\mathbf{R}}[\widetilde{\sigma} < \infty] = \frac{1}{2}\eta + \frac{1}{2} < 1 - \eta$$

since $\eta \leq \frac{1}{8}$. To show that

$$\mathbf{E}_{\mathbf{R}} \left[L_{\tau\wedge\sigma} Z_{\tau\wedge\sigma} \right] = 1$$

we again use the extension and time transform. But $(\widetilde{L}\widetilde{Z})^{\widetilde{\tau}\wedge\widetilde{\sigma}}$ is a uniformly integrable martingale, as follows from the easy calculation in Sect. 10.2, and hence we obtain

$$\mathbf{E}_{\mathbf{R}}[L_{\tau\wedge\sigma} Z_{\tau\wedge\sigma}] = \widetilde{\mathbf{E}}_{\widetilde{\mathbf{R}}}[\widetilde{L}_{\widetilde{\tau}\wedge\widetilde{\sigma}\wedge\gamma} Z_{\widetilde{\tau}\wedge\widetilde{\sigma}\wedge\gamma}] = 1.$$

The proof of the theorem is complete now. $\square$

11

The No-Arbitrage Property
under a Change of Numéraire (1995)

Abstract. For a price process that has an equivalent risk neutral measure, we investigate if the same property holds when the numéraire is changed. We give necessary and sufficient conditions under which the price process of a particular asset — which should be thought of as a different currency — can be chosen as new numéraire. The result is related to the characterisation of attainable claims that can be hedged. Roughly speaking: the asset representing the new currency is a reasonable investment (in terms of the old currency) if and only if the market does not permit arbitrage opportunities in terms of the new currency as numéraire. This rough but economically meaningful idea is given a precise content in this paper. The main ingredients are a duality relation as well as a result on maximal elements. The paper also generalises results previously obtained by Jacka, Ansel-Stricker and the authors.

11.1 Introduction

In this paper we deal with the change of numéraire problem. Let us assume that a d-dimensional process S describes the price of d assets in a fixed chosen currency unit. If e.g. the currency unit is changed, the price process S will be multiplied by the exchange ratio describing the old currency in function of the new one. We shall give examples showing that the no-arbitrage property of the process S may depend on the choice of numéraire. Such an example was already given in [DS 94a]. The question now arises when the value of an asset or more generally of a portfolio, can be used as a new numéraire without destroying the no-arbitrage property. Of course this will depend on the kind of no-arbitrage we use. We will give precise definitions further in the paper but for the moment let us assume (oversimplifying things) that no-

[DS 95b] The No-Arbitrage Property under a Change of Numéraire. *Stochastics and Stochastic Reports*, vol. 53, pp. 213–226, (1995).

* Part of this research was supported by the European Community Stimulation Plan for Economic Science contract Number SPES-CT91-0089.

arbitrage stands for the existence of an equivalent risk neutral (i.e. for a local martingale) measure.

It turns out that the problem is related to the characterisation of those contingent claims that can be hedged. This topic was studied by Jacka [J 92] and Ansel-Stricker [AS 94]. These authors use the H^1-BMO duality. We will give a measure independent characterisation in terms of maximal elements of attainable claims. These elements were already used, as a technical device, in Chap. 9. The proofs of the theorems below use these results as well as an extension of a duality relation from Delbaen [D 92].

The technique of a change of numéraire together with the change of the risk neutral measure was used by El Karoui, Geman and Rochet [EGR 95] and Jamshidian [J 87] to facilitate calculations of prices of contingent claims[*].

The results of this paper can also be used to build consistent models of exchange rates of currencies. In this case the discounting procedure depends on the currency since the interest rate in different currencies will be different. We refer to Delbaen-Shirakawa [DSh 96] for details.

The rest of this section is devoted to the introduction of the basic notation. Sect. 11.2 recalls known facts from arbitrage theory. In Sect. 11.3 we extend the duality equality and relate it to properties of maximal elements. Sect. 11.4 finally contains the main theorem on the change of numéraire and the application to the theory of hedgeable elements.

The setup in this paper is the usual setup in mathematical finance. A probability space $(\Omega, \mathcal{F}, \mathbf{P})$ with a filtration $(\mathcal{F}_t)_{0 \leq t}$ is given. In order to cover the most general case, the time set is supposed to be $\mathbb{R}_+$. The filtration is assumed to satisfy the "usual conditions", i.e. it is right continuous and $\mathcal{F}_0$ contains all null sets of $\mathcal{F}$. A price process S, describing the evolution of the discounted price of d assets, is defined on $\mathbb{R}_+ \times \Omega$ and takes values in $\mathbb{R}^d$. In order to use the results of Chap. 9, we suppose that the process S is locally bounded. This assumption is fairly general, in particular it covers the case of continuous price processes. As shown under a wide range of hypotheses, the assumption that S is a semi-martingale follows from arbitrage considerations. We can therefore assume that the process S is a semi-martingale. Since it is also locally bounded it is a special semi-martingale. Stochastic integration is used to describe outcomes of investment strategies. When dealing with processes in dimension higher than 1 it is understood that vector stochastic integration is used. We refer to Protter [P 90] and Jacod [J 79] for details on these matters.

The authors want to thank Ch. Stricker and H. Shirakawa for helpful discussions on the topic. Part of the research was done while the first author was on visit in the University of Tokyo. Discussions with the colleagues and especially with S. Kusuoka, S. Kotani and N. Kunitomo contributed to the development of this paper.

---

[*] Note added in this reprint: The idea of changing the numéraire can be traced back to the work of Margrabe [M 78a], [M 78b]

11.2 Basic Theorems

Before proving the main results of the paper we need to recall some definitions and notations introduced in Chap. 9.

Definition 11.2.1. *An $\mathbb{R}^d$-valued predictable process H is called a-admissible if it is S-integrable, if $H_0 = 0$, if the stochastic integral satisfies $H \cdot S \geq -a$ and if $(H \cdot S)_\infty = \lim_{t\to\infty}(H \cdot S)_t$ exists a.s.. We say that H is admissible if it is a-admissible for some number a.*

The following notations will be used:

$$\mathcal{K} = \{(H \cdot S)_\infty \mid H \text{ is admissible}\}$$
$$\mathcal{K}_a = \{(H \cdot S)_\infty \mid H \text{ is } a\text{-admissible}\}$$
$$\mathcal{C}_0 = \mathcal{K} - L^0_+$$
$$\mathcal{C} = \mathcal{C}_0 \cap L^\infty.$$

The basic Theorem 9.1.1 uses the concept of no free lunch with vanishing risk. This is a rather weak form of no-arbitrage-type and it is stated in terms of L^∞ convergence. The *(NFLVR)* property is therefore independent of the choice of equivalent probability measure. Only the class of negligible sets comes into play.

Definition 11.2.2. *We say that the locally bounded semi-martingale S satisfies the no free lunch with vanishing risk or property (NFLVR), with respect to general admissible integrands, if*

$$\overline{\mathcal{C}} \cap L^\infty_+ = \{0\},$$

where the bar denotes the closure in the sup-norm topology of L^∞.

The locally bounded semi-martingale S satisfies the no-arbitrage or (NA) property with respect to general admissible integrands, if

$$\mathcal{C} \cap L^\infty_+ = \{0\}.$$

The fundamental theorem of asset pricing can now be formulated as follows:

Theorem 11.2.3. *The locally bounded semi-martingale S satisfies the property (NFLVR), with respect to general admissible integrands, if and only if there is an equivalent probability measure $\mathbf{Q}$ such that S is a $\mathbf{Q}$-local martingale. In this case the set $\mathcal{C}$ is already weak-star (i.e. $\sigma(L^\infty, L^1)$) closed in L^∞.*

Remark 11.2.4. If $\mathbf{Q}$ is an equivalent local martingale measure for S and if H satisfies $H \cdot S \geq -a$ then the result of Ansel-Stricker [AS 94] shows that $H \cdot S$ is still a local martingale and hence, being bounded from below, is a super-martingale. It follows that the limit $(H \cdot S)_\infty$ exists a.s. and that $\mathbf{E}_{\mathbf{Q}}[(H \cdot S)_\infty] \leq 0$.

The proof of the fundamental theorem is quite complicated and we cannot repeat it here. The basic idea in Chap. 9, see Lemma 9.4.4 and the remark following it, is the use of maximal elements in $\mathcal{K}_1$. For convenience we give a definition of what we mean by this.

Definition 11.2.5. *We say that an element $f \in \mathcal{K}_a$ is maximal in $\mathcal{K}_a$ if the properties $g \geq f$ a.s. and $g \in \mathcal{K}_a$ imply that $g = f$ a.s..*

It is easy to see that if S satisfies the no-arbitrage condition then the fact that f is maximal in $\mathcal{K}_a$ already implies that f is maximal in $\mathcal{K}_b$ for all $b \geq a$ and therefore with the obvious definition also in $\mathcal{K}$. Indeed suppose that $f \in \mathcal{K}_a$, $g = (H \cdot S)_\infty \in \mathcal{K}$ and $g \geq f$ a.s., then $g \geq -a$. From Proposition 9.3.6 it then follows that g is a-admissible and hence the maximality of f in $\mathcal{K}_a$ implies that $g = f$ a.s.. An example of an element in $\mathcal{K}_1$ that is not maximal will be given below. The *(NA)* property with respect to general admissible integrands is now equivalent to the fact that the zero function is maximal in the set $\mathcal{K}$.

In the proof of the fundamental theorem the following intermediate results are shown, again for the (complicated) proof we refer to Lemma 9.8.1, Lemma 9.4.4 and the proof of Theorem 9.4.2.

Theorem 11.2.6. *If the locally bounded semi-martingale S satisfies the property (NFLVR) with respect to general admissible integrands, if $(f_n)_{n \geq 1}$ is a sequence in $\mathcal{K}_1$, then*

(1) *there is a sequence of convex combinations $g_n \in \mathrm{conv}\{f_n, f_{n+1}, \ldots\}$ such that g_n tends in probability to a function g, taking finite values a.s.,*
(2) *there is a maximal element h in $\mathcal{K}_1$ such that $h \geq g$ a.s..*

Corollary 11.2.7. *If the locally bounded semi-martingale S satisfies the property (NFLVR) with respect to general admissible integrands, then the maximal elements of the closure of $\mathcal{K}_1$ in L^0, are in $\mathcal{K}_1$.*

Remark 11.2.8. The set $\mathcal{K}_1$ is not necessarily closed in the space L^0. However, under the *(NFLVR)* property with respect to general admissible integrands, the set $\mathcal{K}_1$ and hence its closure are convex and bounded in L^0. When we define maximal elements of this closure in the obvious way, these maximal elements are already in $\mathcal{K}_1$.

The following theorem, in the spirit of Chap. 9, gives another description of the *(NFLVR)* property.

Theorem 11.2.9. *The locally bounded semi-martingale S satisfies the (NFLVR) property with respect to general admissible integrands if and only if it satisfies the (NA) property with respect to general admissible integrands and if there exists a strictly positive local martingale L such that $L_\infty > 0$ a.s. with $L S$ a local martingale.*

Proof. The necessity is clear. If $\mathbf{Q}$ is an equivalent local martingale measure, then the Radon-Nikodým derivative $\frac{d\mathbf{Q}}{d\mathbf{P}}$ defines a strictly positive $\mathbf{P}$-martingale L such that LS is a $\mathbf{P}$-local martingale. Also the process S necessary satisfies the *(NA)* property with respect to general admissible integrands.

The converse is less obvious. We recall from Corollary 9.3.9, that it is sufficient to prove that S satisfies *(NA)* with respect to general admissible integrands and that the set $\mathcal{K}_1$ is bounded in L^0. If L is a strictly positive local martingale, then the sequence of stopping times defined as

$$T_n = \inf\{t \mid L_t \geq n\}$$

satisfies $\mathbf{P}[T_n = \infty] \to 1$ and L^{T_n} is a uniformly integrable martingale. These properties follow from the fact that L is a super-martingale and the fact that the jumps of L are necessarily integrable. Also we may and do suppose that $L_0 = 1$. For each n the measure $\mathbf{Q}_n$ defined by $\frac{d\mathbf{Q}_n}{d\mathbf{P}} = L_{T_n}$ is a local martingale measure for the stopped process S^{T_n}. It follows that the set $\mathcal{K}_1$ is bounded when restricted to the event $\{T_n = \infty\}$. Because $\mathbf{P}[T_n = \infty] \to 1$, this implies that $\mathcal{K}_1$ is bounded in L^0. □

The theorem yields the following result, see [DS 94a] and Chap. 10 for a different approach and for related results. For details on continuous martingales and Bessel processes we refer to Revuz-Yor [RY 91].

Corollary 11.2.10. *If R is the Bessel(3) process, stopped at time 1 and with its natural filtration then R allows arbitrage with respect to general admissible integrands.*

Proof. The process $L = \frac{1}{R}$ is a local martingale and from stochastic calculus it follows that it is the only local martingale X such that $X_0 = 1$ and such that XR is a local martingale. If now $\mathbf{Q}$ were a local martingale measure for R, then the martingale X defined as $\mathbf{E}_\mathbf{P}[\frac{d\mathbf{Q}}{d\mathbf{P}} \mid \mathcal{F}_t]$ satisfies that XR is a local martingale and hence $X = L$. Since L is only a local martingale and not a true martingale we arrive at a contradiction. It follows that R does not have an equivalent local martingale measure. Since it satisfies the second part of the preceding theorem, it cannot satisfy the *(NA)* property with respect to general admissible integrands. □

Remark 11.2.11. The element $L_1 - 1$ is not maximal in the set $\mathcal{K}_1$ constructed with the process L. To see this recall that $\mathbf{E}[L_1] < 1$ and that $L_1 - \mathbf{E}[L_1]$ is by the predictable representation property of L, the result of a uniformly integrable martingale of the form $K \cdot L$. It is clear that $(K \cdot L)_1 = (L_1 - \mathbf{E}[L_1]) > L_1 - 1$.

If a locally bounded semi-martingale S satisfies the *(NFLVR)* property with respect to general admissible integrands, then the following two non-empty sets will play a role in the theory:

$$\mathcal{M}^e(\mathbf{P}) = \left\{ \mathbf{Q} \;\middle|\; \begin{array}{l} \mathbf{Q} \text{ is equivalent to } \mathbf{P} \\ \text{and the process } S \text{ is a } \mathbf{Q}\text{-local martingale} \end{array} \right\}$$

$$\mathcal{M}(\mathbf{P}) = \left\{ \mathbf{Q} \;\middle|\; \begin{array}{l} \mathbf{Q} \text{ is absolutely continuous with respect to } \mathbf{P} \\ \text{and the process } S \text{ is a } \mathbf{Q}\text{-local martingale} \end{array} \right\}.$$

We identify absolutely continuous measures with their Radon-Nikodým derivatives. It is clear that the set $\mathcal{M}^e(\mathbf{P})$ is L^1-dense in $\mathcal{M}(\mathbf{P})$.

11.3 Duality Relation

In this section we extend the duality formula of Delbaen [D 92] and Chap. 9 to the case of unbounded functions. We denote by $\mathcal{C}^\circ$ the polar of the cone $\mathcal{C}$, i.e.

$$\mathcal{C}^\circ = \left\{ f \mid f \in L^1(\mathbf{P}) \text{ and for each } h \in \mathcal{C} \text{ we have } \mathbf{E}_{\mathbf{P}}[f\,h] \le 0 \right\}.$$

Theorem 11.3.1. *If S is a locally bounded semi-martingale that satisfies (NFLVR) with respect to general admissible integrands then*

$$\mathcal{M}(\mathbf{P}) = \mathcal{C}^\circ \cap \{\mathbf{Q} \mid \mathbf{Q} \text{ probability measure}, \mathbf{Q} \ll \mathbf{P}\}.$$

Proof. If $\mathbf{Q} \in \mathcal{M}(\mathbf{P})$ then for each admissible integrand H we have, by the Ansel-Stricker theorem, [AS 94], that $H \cdot S$ is a $\mathbf{Q}$-local martingale and hence it is a super-martingale. Therefore $\mathbf{E}_{\mathbf{Q}}[f] \le 0$ for each $f \in \mathcal{K}$. The same inequality pertains for elements of $\mathcal{C}$.

Conversely if $\mathbf{Q}$ is a probability measure in $\mathcal{C}^\circ$ then S will be a $\mathbf{Q}$-local martingale. Indeed take T_n an increasing sequence of stopping times, $T_n \nearrow \infty$, such that each S^{T_n} is bounded. For each $s < t$ and each $A \in \mathcal{F}_s$ we have that $\mathbf{1}_A (S_t^{T_n} - S_s^{T_n})$ is in $\mathcal{C}$ and hence we have $\mathbf{E}_{\mathbf{Q}}[\mathbf{1}_A (S_t^{T_n} - S_s^{T_n})] \le 0$. Replacing $\mathbf{1}_A$ by $-\mathbf{1}_A$ gives that $\mathbf{E}_{\mathbf{Q}}[\mathbf{1}_A (S_t^{T_n} - S_s^{T_n})] = 0$. These equalities show that S is a $\mathbf{Q}$-local martingale. $\qquad\square$

Corollary 11.3.2. *Suppose that the locally bounded semi-martingale S satisfies the (NFLVR) property with respect to general admissible integrands. The set $\mathcal{M}(\mathbf{P})$ is then closed in $L^1(\mathbf{P})$.*

We remark that this is essentially a consequence of the local boundedness of S. It is easy to give counter-examples in the general case.

Theorem 11.3.3. *If the locally bounded semi-martingale S satisfies the (NFLVR) property with respect to general admissible integrands, then for bounded elements f in L^∞ we have that*

$$\sup_{\mathbf{Q} \in \mathcal{M}^e(\mathbf{P})} \mathbf{E_Q}[f] = \sup_{\mathbf{Q} \in \mathcal{M}(\mathbf{P})} \mathbf{E_Q}[f]$$
$$= \inf\{x \mid \exists h \in \mathcal{C} \ \ x + h \geq f\}$$
$$= \inf\{x \mid \exists h \in \mathcal{C} \ \ x + h = f\}$$
$$= \inf\{x \mid (f - x) \in \mathcal{C}\}$$
$$= \inf\{x \mid \exists h \in \mathcal{K} \ \ x + h \geq f\}.$$

Furthermore all infima are minima.

Proof. The proof of this theorem is an application of the previous theorem and duality theory.

The first equality is almost trivial since $\mathcal{M}^e(\mathbf{P})$ is dense in $\mathcal{M}(\mathbf{P})$ for the norm topology of $L^1(\mathbf{P})$. Suppose that $f \leq x + h$ where $h \in \mathcal{C}$. It follows from the preceding theorem that for all $\mathbf{Q} \in \mathcal{M}(\mathbf{P})$ we have that $\mathbf{E_Q}[f] \leq x + \mathbf{E_Q}[h] \leq x$. It is therefore obvious that

$$\sup_{\mathbf{Q} \in \mathcal{M}(\mathbf{P})} \mathbf{E_Q}[f] \leq \inf\{x \mid \exists h \in \mathcal{C} \ \ x + h \geq f\}.$$

The converse inequality is proved using the Hahn-Banach theorem and the fact that the set $\mathcal{C}$ is weak-star-closed, see Theorem 11.2.3 above. Let z be a real number such that

$$z < \inf\{x \mid \exists h \in \mathcal{C} \ \ x + h \geq f\}.$$

We have that $f - z \notin \mathcal{C}$. By the Hahn-Banach theorem there is a weak-star continuous functional on L^∞, denoted by the corresponding measure $\mathbf{Q}$, such that for all $h \in \mathcal{C}$ we have

$$\int (f - z)\, d\mathbf{Q} > \int h\, d\mathbf{Q}.$$

Since $\mathcal{C}$ is a cone containing $-L^\infty_+$, this necessarily implies that for all $h \in \mathcal{C}$ we have

$$0 \geq \int h\, d\mathbf{Q} \quad \text{and that} \quad \int (f - z)\, d\mathbf{Q} > 0.$$

We deduce that $\mathbf{Q}$ is necessarily positive and we may therefore suppose that $\mathbf{Q}$ is normalised in such a way that $\mathbf{Q}(\Omega) = 1$. In that case $\mathbf{Q}$ is a probability measure, is an element of $\mathcal{C}^\circ$ and hence an element of $\mathcal{M}(\mathbf{P})$. But then the second inequality shows that $\mathbf{E_Q}[f] > z$. We obtain that

$$\sup_{\mathbf{Q} \in \mathcal{M}(\mathbf{P})} \mathbf{E_Q}[f] \geq \inf\{x \mid \exists h \in \mathcal{C} \ \ x + h \geq f\}$$

and this ends the proof of the equalities. The fact that all infima are minima is an easy consequence of the closedness of $\mathcal{C}$ for the norm topology of L^∞. Indeed, the set $\{x \mid (f - x) \in \mathcal{C}\}$ is closed. □

We will now generalise the preceding equalities to arbitrary positive functions. The proof relies on the special properties of the sets $\mathcal{C}$ and $\mathcal{K}$.

Theorem 11.3.4. *Suppose that the locally bounded martingale S satisfies the (NFLVR) property with respect to general admissible integrands. If $f \geq 0$, or more generally if f is bounded below by a constant, then*

$$\sup_{Q \in \mathcal{M}^e(\mathbf{P})} \mathbf{E}_Q[f] = \sup_{Q \in \mathcal{M}(\mathbf{P})} \mathbf{E}_Q[f]$$

$$= \inf\{x \mid \exists h \in \mathcal{K} \ \ x + h \geq f\}$$

and when the expression is finite

$$= \min\{x \mid \exists h \in \mathcal{K} \ \ x + h \geq f\}.$$

Proof. We suppose that $f \geq 0$. The first equality follows again from the density of $\mathcal{M}^e(\mathbf{P})$ in the set $\mathcal{M}(\mathbf{P})$ and Fatou's lemma. The left hand side is smaller than the right hand side exactly as in the proof of the previous theorem. We remark that this already implies that we have equality as soon as $\sup_{Q \in \mathcal{M}^e(\mathbf{P})} \mathbf{E}_Q[f] = \infty$. Let now z be a real number such that $z > \sup_{Q \in \mathcal{M}^e(\mathbf{P})} \mathbf{E}_Q[f]$. For all natural numbers we therefore have that $z > \sup_{Q \in \mathcal{M}^e(\mathbf{P})} \mathbf{E}_Q[f \wedge n]$. The theorem for bounded functions now implies the existence of $h_n \in \mathcal{K}$ and $0 \leq x_n < z$ such that $f \wedge n \leq x_n + h_n$. We may extract subsequences and suppose that the bounded sequence x_n converges to a real number $x \leq z$. The functions h_n are bigger than $-x_n$ and therefore the result of an x_n and hence a z-admissible strategy H^n. The sequence of functions h_n is in $\mathcal{K}_z$, a bounded convex set of $L^0(\mathbf{P})$. Using Lemma 9.8.1 we may take convex combinations of h_n that converge almost everywhere to a function h. We still have that $h + x \geq f$. The properties of $\mathcal{K}_z$ listed above (see Theorem 11.2.6 (2)), imply that there is an element $g \in \mathcal{K}_z$ such that $g \geq h$. This element clearly satisfies $x + g \geq f$ and hence we obtain

$$z \geq \inf\{x \mid \exists h \in \mathcal{K} \ \ x + h \geq f\}.$$

We therefore see that

$$\sup_{Q \in \mathcal{M}^e(\mathbf{P})} \mathbf{E}_Q[f] = \sup_{Q \in \mathcal{M}(\mathbf{P})} \mathbf{E}_Q[f] = \inf\{x \mid \exists h \in \mathcal{K} \ \ x + h \geq f\}.$$

To see that the infimum is a minimum we take a sequence x_n tending to the infimum and a corresponding sequence of outcomes h_n. We can apply the same reasoning to see that the infimum is attained. $\qquad \square$

Corollary 11.3.5. *Suppose that the locally bounded semi-martingale S satisfies the (NFLVR) property with respect to general admissible integrands. If $f \geq 0$ and if $x = \sup_{Q \in \mathcal{M}^e(\mathbf{P})} \mathbf{E}_Q[f] < \infty$, then there is a maximal element $g \in \mathcal{K}$ such that $f \leq x + g$.*

Proof. This follows from the proof of the theorem. $\qquad \square$

11.4 Hedging and Change of Numéraire

Before we give a martingale characterisation of maximal elements of $\mathcal{K}$, we first study the *(NA)* property under the change of numéraire. Since we want to apply it in a fairly general setting, we will work with an abstract $\mathbb{R}^d$-valued semi-martingale W. In this section we do not even require the semi-martingale to be locally bounded. When we change the numéraire from the constant 1 into the process V we will have to rescale the process W. The best way to do this is to introduce the $(d+2)$-dimensional process $(W, 1, V)$. The constant 1, which corresponds to the original numéraire was added, because under the new numéraire V, this will not be constant anymore but will be replaced by $\frac{1}{V}$. On the other hand, the process V will be replaced by 1. By adding this constant process, we obtain more symmetry. Under the new numéraire the system is described by the process $(\frac{W}{V}, \frac{1}{V}, 1)$. Before proving the change of numéraire theorem, a theorem that relates the *(NA)* property of both systems, let us give an example of what happens in a discrete time setting and when $d = 0$, the simplest possible case.

Example 11.4.1. The semi-martingale V which describes the price of the new numéraire (in terms of the old one) is supposed to satisfy $V_0 = 1$, a pure normalisation assumption, $V_t > 0$, a.s. and $\lim_{t \to \infty} V_t = V_\infty$ exists a.s. and is strictly positive a.s.. Note that the symmetry in these assumptions if we pass from V to $\frac{1}{V}$, i.e. they are invariant whether we consider the new numéraire in terms of the old one or vice versa. The process is driven by a sequence of independent identically distributed Bernoulli variables $(\varepsilon_n)_{n \geq 1}$. They are such that $\mathbf{P}[\varepsilon_n = 1] = \mathbf{P}[\varepsilon_n = -1] = \frac{1}{2}$. To facilitate the writing, we call the two currencies € and \$. The process V describes the value of the \$ in terms of the €. Let us now fix α such that $0 < \alpha < 1$. At time $n = 0$, we require that $V_0 = 1$. Let us suppose that V_{n-1} is already defined. If the Bernoulli variable $\varepsilon_n = 1$ then we put $V_n = \alpha$. If $\varepsilon_n = -1$, then we put $V_n = 2V_{n-1} - \alpha$. In such a way the process V remains strictly positive, in fact greater than α, it becomes eventually equal to α and the limit $V_\infty = \alpha$ therefore exists. The process V is also a non-uniformly integrable martingale with respect to the measure $\mathbf{P}$. Remark that once the process hits the level α it remains at that level forever. In economic terms we may say that an investment in \$ seems to be a fair game, since V is a martingale, but that at the end it was not a good choice. Indeed, since $\alpha < 1$, the investment is, in the long run, a losing one. An economic agent might try to get a profit out of it by selling short the \$. But here is an obstruction. Indeed by going short on \$, the € investor will realise that he is using a non-admissible strategy. Therefore she will not be able to take advantage of this special situation. A \$ investor on the contrary is able to buy € at an initial price of 1\$ and then in the long run sell this € for $\frac{1}{\alpha}$, making arbitrage profits! As a last point let us observe that the 0-variable dominates the outcome $V_\infty - 1 = \alpha - 1$ and hence the variable $V_\infty - 1$ is not maximal. The example is simple but it has all the features that appear in greater generality in the theorem.

Theorem 11.4.2. *Let W be a semi-martingale, taking values in $\mathbb{R}^d$. Let V be a strictly positive semi-martingale such that $V_\infty = \lim_{t \to \infty} V_t$ exists and is strictly positive a.s.. The semi-martingale X is the $(d + 2)$-dimensional process $X = (W, 1, V)$. The process Z defined as $Z = (\frac{W}{V}, \frac{1}{V}, 1)$ is a $(d + 2)$-dimensional semi-martingale. It satisfies the (NA) property with respect to general admissible integrands if and only if $V_\infty - 1$ is maximal in the set of outcomes of 1-admissible integrands for X.*

Proof. Using the symmetry between the processes X and Z we first reformulate the statement of the theorem. We can regard the process X as obtained from Z by dividing it by the process $\frac{1}{V}$. The process $\frac{1}{V}$ is also strictly positive and at infinity its limit exists a.s. and is still strictly positive. If we change the role of X and Z, resp. V and $\frac{1}{V}$, we see that the proof of the theorem is equivalent to the proof of the following two statements

(1) If X satisfies the *(NA)* property with respect to general admissible integrands then $\frac{1}{V_\infty} - 1$ is maximal in the set of outcomes of 1-admissible integrands for Z.

(2) If Z permits arbitrage with respect to general admissible integrands then $V_\infty - 1$ is not maximal in the set of outcomes of 1-admissible integrands for X.

The proof depends on the following calculation from vector stochastic calculus. From $X = VZ$ we deduce that

$$dX_t = dV_t\, Z_{t-} + V_{t-}\, dZ_t + d[V, Z]_t\,.$$

If K is a $(d + 2)$-dimensional predictable process that is a 1-admissible integrand for the system $Z = (\frac{W}{V}, \frac{1}{V}, 1)$ then we let $Y = (1 + K \cdot Z)V$. Remark that Y is a process that describes a portfolio obtained by using an investment described by the system Z that afterwards is converted, through the change of numéraire V, into values that fit in the system X. We have that

$$dY_t = dV_t\,(1 + (K \cdot Z)_{t-}) + V_{t-}\, K_t\, dZ_t + K_t\, d[V, Z]_t\,.$$

Using the expression for dX we may convert this into

$$dY_t = dV_t\,(1 + (K \cdot Z)_{t-}) + K_t\, dX_t - dV_t\, K_t Z_{t-}$$

which is of the form

$$dY_t = L_t\, dX_t$$

for some $(d + 2)$-dimensional predictable and X-integrable process L. Since K was 1-admissible for Z, we have that Y is positive and therefore L is 1-admissible for X. We now apply the above equality in two different cases. To prove (1) we suppose that $\frac{1}{V_\infty} - 1$ is not maximal. Take K a 1-admissible integrand for Z such that the limit at infinity exists and such that $1 + (K \cdot Z)_\infty \geq \frac{1}{V_\infty}$, with strict inequality on a non-negligible set. In that case we have that $Y_\infty - 1 = (L \cdot X)_\infty$ is non-negative and strictly positive on a non-negligible set. This should produce arbitrage for X.

The second part is proved in a similar way. Suppose that Z allows arbitrage and that K is the 1-admissible integrand responsible for it. The outcome $Y_\infty - 1$ is now greater than $V_\infty - 1$, with strict inequality on a non-negligible set. A contradiction to its maximality.   □

Corollary 11.4.3. *Using the same notation as in the theorem we see that X satisfies the (NA) property with respect to general admissible integrands and $V_\infty - 1$ is maximal "for X" if and only if Z satisfies the (NA) property with respect to general admissible integrands and $\frac{1}{V_\infty} - 1$ is maximal "for Z".*

Proof. This is a straightforward application of the previous theorem. The only difference lies in the statement that $V_\infty - 1$ is maximal in the set of all outcomes of admissible integrand and not just in the set of outcomes of 1-admissible integrands. If X satisfies *(NA)* and $V_\infty - 1$ is maximal then we can apply both parts of the theorem. In this case we know, from Sect. 11.2, that $f = \frac{1}{V_\infty} - 1$ is maximal in the set of outcomes of all admissible integrands. This proves the *if* statement. The *only if* part is the same statement as the *if* part because X is obtained from Z by multiplying with V^{-1}.   □

We can now apply the above reasoning to the original setting of this paper. Given a locally bounded semi-martingale S that satisfies the *(NFLVR)* property with respect to general admissible integrands, we use a process of the form $V = 1 + H \cdot S$ for the new numéraire. If H is admissible and $V_\infty > 0$ a.s., then we can apply the previous theorem. In this case we certainly have that the system $(S, 1, V)$ has the *(NA)* property with respect to general admissible integrands. With the assumption that $\mathbf{P}$ was a local martingale measure for S, the system $(S, 1, V)$ becomes in fact a local martingale for $\mathbf{P}$. The previous theorem then yields

Theorem 11.4.4. *Suppose that S is a locally bounded semi-martingale that satisfies the (NFLVR) property with respect to general admissible integrands. Suppose that H is admissible and that the process $V = 1 + (H \cdot S)$ satisfies $f = V_\infty = 1 + (H \cdot S)_\infty > 0$ a.s.. Then the following are equivalent:*

(1) $(H \cdot S)_\infty$ *is maximal in the set $\mathcal{K}$.*
(2) *The process $\widetilde{S} = (\frac{S}{V}, \frac{1}{V})$ satisfies (NA) with respect to general admissible integrands.*
(3) *There is $\mathbf{Q} \in \mathcal{M}^e(\mathbf{P})$ such that $H \cdot S$ is a $\mathbf{Q}$-uniformly integrable martingale.*

If V^{-1} is locally bounded then these statements are equivalent to:

(4) *The process $\widetilde{S}$ has an equivalent local martingale measure.*

Remark 11.4.5. We conjecture that the assumption that V^{-1} is locally bounded can be removed.[†]

Proof. (1) *and* (2) *are equivalent*: Since S satisfies the *(NFLVR)* property with respect to general admissible integrands, there is an equivalent local martingale measure, $\mathbf{Q}$ for S. Because the stochastic integral $H \cdot S$ is bounded below, the theorem of Ansel-Stricker, see [AS 94], implies that it, and hence also V, is a local martingale. Since the final value V_∞ of V is strictly positive, the result in Dellacherie-Meyer [DM 80, Theorem 17, p. 85] implies that the process V is bounded away from zero a.s.. We can now apply Theorem 11.4.2 to see that (1) and (2) are already equivalent.

(1) *implies* (4): In case V^{-1} is locally bounded we have that $\widetilde{S}$ is also locally bounded. It has the *(NA)* property and the product $V\,\widetilde{S}$ is a local martingale. Therefore the process has the *(NFLVR)* property and by Theorem 11.2.3 and Theorem 11.2.9 it has an equivalent local martingale measure.

(1) *and/or* (2) *imply* (3): Now we apply the statement that (1) implies (4) on the process $V' = \frac{1}{2}(1+V)$. This process is defined using $\frac{H}{2}$ instead of H. It has the advantage that $\frac{1}{V'}$ is bounded. Let $\widetilde{\mathbf{Q}}$ be an equivalent local martingale measure for $(\frac{S}{V'}, \frac{1}{V'})$. Since the last coordinate $X = \frac{1}{V'}$ is bounded and is a $\widetilde{\mathbf{Q}}$-local martingale it is a strictly positive bounded martingale, starting at 1. When we define the probability measure $\mathbf{Q}$ by $d\mathbf{Q} = X_\infty\, d\widetilde{\mathbf{Q}}$, we obtain that $S = \frac{S}{V'}\, V'$ is a $\mathbf{Q}$-local martingale and V' is a $\mathbf{Q}$-uniformly integrable martingale. This implies that $H \cdot S$ is a $\mathbf{Q}$-uniformly integrable martingale. The proof that (1) and/or (2) implies (3) is complete.

(3) *implies* (1): If $H \cdot S$ is a $\mathbf{Q}$-uniformly integrable martingale for some $\mathbf{Q} \in \mathcal{M}^e(\mathbf{P})$ then $(H \cdot S)_\infty$ is necessarily maximal. Indeed if say $(K \cdot S)_\infty \geq (H \cdot S)_\infty$ for some admissible K, then by taking expectations with respect to $\mathbf{Q}$, applying the super-martingale property of $K \cdot S$ and the martingale property of $(H \cdot S)$ we see

$$0 = \mathbf{E}_{\mathbf{Q}}[(H \cdot S)_\infty] \leq \mathbf{E}_{\mathbf{Q}}[(K \cdot S)_\infty] \leq 0\,.$$

It follows that $\mathbf{E}_{\mathbf{Q}}[(K \cdot S)_\infty] = 0$ and $(H \cdot S)_\infty = (K \cdot S)_\infty$. This completes the proof that (3) implies (1).

(4) *implies* (2): Since the existence of an equivalent local martingale measure implies the *(NA)* property with respect to general admissible integrands, this is trivial.   □

Corollary 11.4.6. *If the locally bounded semi-martingale S satisfies (NFLVR) with respect to general admissible integrands then for an admissible integrand H the following are equivalent:*

[†] Note added in this reprint: The hypothesis of local boundedness is not needed since the process V can be used as a martingale measure density. If (4) is satisfied then even the existence of an equivalent sigma-martingale measure implies the *(NA)* propertry for $(\widetilde{S})$.

(1) $(H \cdot S)_\infty$ is maximal in $\mathcal{K}$

(2) there is $\mathbf{Q} \in \mathcal{M}^e(\mathbf{P})$ such that $\mathbf{E_Q}[(H \cdot S)_\infty] = 0$

(3) there is $\mathbf{Q} \in \mathcal{M}^e(\mathbf{P})$ such that $H \cdot S$ is a $\mathbf{Q}$-uniformly integrable martingale.

The theorem also allows us to give a characterisation of strict local martingales as studied by Elworthy, Li and Yor, [ELY 99]. They define a strict local martingale as a local martingale that is not a uniformly integrable martingale.

Corollary 11.4.7. Let $S = L$ be a strictly positive locally bounded local martingale such that $L_\infty > 0$ a.s.. Let

$$\mathcal{M}^e(\mathbf{P}) = \left\{ \mathbf{Q} \;\middle|\; \begin{array}{l} \mathbf{Q} \text{ is equivalent to } \mathbf{P} \\ \text{and the process } L \text{ is a } \mathbf{Q}\text{-local martingale} \end{array} \right\}.$$

The process $\frac{1}{L}$ satisfies the (NA) property with respect to general admissible integrands if and only if L is a uniformly integrable martingale for some $\mathbf{Q}$ in $\mathcal{M}^e(\mathbf{P})$.

Remark 11.4.8. From Schachermayer [S 93] (see Chap. 10 for an easier example) it follows that under the assumptions of the corollary, the process L need not be a uniformly integrable martingale under all elements of $\mathcal{M}^e(\mathbf{P})$.

Remark 11.4.9. In the case that $R = \frac{1}{L^1}$ equals the Bessel(3) process with its natural filtration, stopped at time 1, we have that L^1 is a local martingale for $\mathbf{P}$. This is the only candidate for a martingale measure and hence we deduce that R has arbitrage with respect to general admissible integrands. The preceding corollary is a generalisation of this phenomenon to the case that $\mathcal{M}^e(\mathbf{P})$ is not a singleton, see also Sect. 11.2.

Definition 11.4.10. If S is a locally bounded semi-martingale that satisfies the (NFLVR) property with respect to general admissible integrands, then we say that a positive random variable (or contingent claim) f can be hedged if there is $x \in \mathbb{R}$ and a maximal element $h \in \mathcal{K}$ such that $f = x + h$.

There is a good reason to require the use of maximal elements. If h is not maximal then there is a maximal element $g \in \mathcal{K}$, $g \neq h$ such that $g \geq h$. An investor who would try to hedge f by using an admissible strategy, would be better off to use a strategy that gives her the outcome g instead of h. Starting with the same initial investment x, she will obtain something better than f and since $g > h$ on a set of positive measure, she will be strictly better off in some cases. In such a case the contingent claim f is not the result of a good optimal hedging policy.

The following theorem is due to Ansel-Stricker [AS 94] and, independently, to Jacka [J 92]. They proved it using H^1-BMO duality. We shall see that it is also a consequence of the characterisation of maximal elements.

Theorem 11.4.11. If S is a locally bounded semi-martingale that satisfies the (NFLVR) property with respect to general admissible integrands then for a random variable $f \geq 0$, the following are equivalent:

(1) *f can be hedged,*
(2) *there is $\mathbf{Q}$ in $\mathcal{M}^e(\mathbf{P})$ such that*

$$\mathbf{E_Q}[f] = \sup\{\mathbf{E_R}[f] \mid \mathbf{R} \in \mathcal{M}^e(\mathbf{P})\} < \infty.$$

Proof. (1) *implies* (2): If f can be hedged, then there is an admissible strategy H and a real number x, such that $f = x + (H \cdot S)_\infty$ and $H \cdot S$ is a uniformly integrable martingale for some $\mathbf{Q} \in \mathcal{M}^e(\mathbf{P})$. For all $\mathbf{R} \in \mathcal{M}^e(\mathbf{P})$ we have that $H \cdot S$ is a super-martingale and hence $\mathbf{E_R}[f] \leq x = \mathbf{E_Q}[f]$.

(2) *implies* (1): If we have $\mathbf{E_Q}[f] = \sup\{\mathbf{E_R}[f] \mid \mathbf{R} \in \mathcal{M}^e(\mathbf{P})\} < \infty$, then clearly we have that

$$x = \mathbf{E_Q}[f] = \min\{z \mid \exists\, h \in \mathcal{K} \text{ such that } z + h \geq f\} < \infty.$$

The duality relation of Sect. 3 now implies that there is an admissible integrand H such that $f \leq x + (H \cdot S)_\infty$. Since $H \cdot S$ is a super-martingale for $\mathbf{Q}$ we have that

$$x = \mathbf{E_Q}[f] \leq x + \mathbf{E_Q}[(H \cdot S)_\infty] \leq x$$

and hence $\mathbf{E_Q}[(H \cdot S)_\infty] = 0$. This implies that $f = x + (H \cdot S)_\infty$ and that $H \cdot S$ is uniformly integrable for $\mathbf{Q}$. Therefore $(H \cdot S)_\infty$ is maximal in $\mathcal{K}$. $\square$

The Existence of Absolutely Continuous Local Martingale Measures (1995)

Abstract. We investigate the existence of an absolutely continuous martingale measure. For continuous processes we show that the absence of arbitrage for general admissible integrands implies the existence of an absolutely continuous (not necessarily equivalent) local martingale measure. We also rephrase Radon-Nikodým theorems for predictable processes.

12.1 Introduction

In Chap. 9 we showed that for locally bounded finite dimensional stochastic price processes S, the existence of an equivalent (local) martingale measure, sometimes called risk neutral measure, is equivalent to a property called no free lunch with vanishing risk *(NFLVR)*. We also proved that if the set of (local) martingale measures contains more than one element, then necessarily, there are non-equivalent absolutely continuous local martingale measures for the process S. We also gave an example, see Example 9.7.7, of a process that does not admit an equivalent (local) martingale measure but for which there is a martingale measure that is absolutely continuous. The example moreover satisfies the weaker property of no-arbitrage with respect to general admissible integrands. We were therefore led to investigate the relationship between the two properties, the existence of an absolutely continuous martingale measure *(ACMM)* and the absence of arbitrage for general admissible integrands *(NA)*.

From an economic viewpoint a local martingale measure $\mathbf{Q}$, that gives zero measure to a non-negligible event, say F, poses some problems. The price of the contingent claim that pays one unit of currency subject to the occurrence of the event F, is given by the probability $\mathbf{Q}[F]$. Since F is negligible for this probability, the price of the commodity becomes zero. In most economic

[DS 95a] The Existence of Absolutely Continuous Local Martingale Measures. *Annals of Applied Probability*, vol. 5, no. 4, pp. 926–945, (1995).

* Part of this research was supported by the European Community Stimulation Plan for Economic Science contract Number SPES-CT91-0089.

models preference relations are supposed to be strictly monotone and hence there would be an infinite demand for this commodity. At first sight the property *(ACMM)* therefore seems meaningless in the study of general equilibrium models. However, as the present paper shows, for continuous processes it is a consequence of the absence of arbitrage *(NA)*. We therefore think that the *(ACMM)* property has some interest also from the economic viewpoint.

Throughout this paper all variables and processes are defined on a probability space $(\Omega, \mathcal{F}, \mathbf{P})$. The space of all measurable functions, equipped with the topology of convergence in probability is denoted by $L^0(\Omega, \mathcal{F}, \mathbf{P})$ or simply $L^0(\Omega)$ or L^0. If $F \in \mathcal{F}$ has non-zero measure, then the closed subspace of functions, vanishing on the complement F^c of F is denoted by $L^0(F)$. The conditional probability with respect to a non-negligible event F is denoted by $\mathbf{P}_F$ and is defined as $\mathbf{P}_F[B] = \frac{\mathbf{P}[F \cap B]}{\mathbf{P}[F]}$. To simplify terminology we say that a probability $\mathbf{Q}$ that is absolutely continuous with respect to $\mathbf{P}$ is supported by the set F if $\mathbf{Q}$ is equivalent to $\mathbf{P}_F$, in particular we then have $\mathbf{Q}[F] = 1$. Indicator functions of sets F and so forth are denoted by $\mathbf{1}_F$ and so on. The probability space Ω is equipped with a filtration $(\mathcal{F}_t)_{0 \le t < \infty}$. We use the time set $[0, \infty[$ as this is the most general case. Discrete time sets and bounded time sets can easily be imbedded in this framework. We will mainly study continuous processes and in this case the discrete time set makes no sense at all. However, Sect. 12.2 contains some results that remain valid for processes with jumps.

We assume that the filtration $(\mathcal{F}_t)_{0 \le t < \infty}$ satisfies the usual conditions, i.e. it is right continuous and saturated for $\mathbf{P}$-null sets. Stopping times are with respect to this filtration. We draw the attention of the reader to the problem that when $\mathbf{P}$ is replaced by an absolutely continuous measure $\mathbf{Q}$, these usual hypotheses will no longer hold. In particular we will have to saturate the filtration with the $\mathbf{Q}$-null sets.

The process S, sometimes denoted as $(S_t)_{0 \le t < \infty}$, is a fixed càdlàg, locally bounded process that is a semi-martingale with respect to $(\Omega, (\mathcal{F}_t)_{0 \le t < \infty}, \mathbf{P})$. The process S is supposed to take values in the d-dimensional space $\mathbb{R}^d$ and may be interpreted as the (discounted) price process of d stocks. If T_1 and T_2 are two stopping times such that $T_1 \le T_2$ then $]\!]T_1, T_2]\!]$ is the stochastic interval $\{(t, \omega) \mid t < \infty, \, T_1(\omega) < t \le T_2(\omega)\} \subset [0, \infty[\times \Omega$. Other intervals are denoted in a similar way.

If H is a predictable process we say that H is simple if it is a linear combination of elements of the form $f \mathbf{1}_{]\!]T_1, T_2]\!]}$ where $T_1 \le T_2$ are stopping times and f is $\mathcal{F}_{T_1}$-measurable. For the theory of stochastic integration we refer to [P 90] and for vector stochastic integration we refer to [J 79]. The reader who is not familiar with vector stochastic integration can think of S as being one-dimensional, i.e. $d = 1$. If H is a d-dimensional predictable process that is S-integrable, then the process obtained by stochastic integration is denoted $H \cdot S$, its value at time t is $(H \cdot S)_t$.

A strategy is a predictable process that is integrable with respect to the semi-martingale S and that satisfies $H_0 = 0$. As in Chap. 9, we will need the concept of admissible strategy.

Definition 12.1.1. *An S-integrable predictable strategy H is k-admissible, for $k \in \mathbb{R}_+$, if the process $H \cdot S$ is always bigger than $-k$ and if the limit $\lim_{t \to \infty} (H \cdot S)_t$ exists almost surely. In particular, if H is 1-admissible then $H \cdot S \geq -1$.*

For a discussion of this topic and its origin in mathematical finance we refer to [HP 81].

We also refer to [HP 81] for a discussion of the fact that, by considering the discounted values of the stock price S, there is no loss of generality in assuming that the "riskless interest rate r" is assumed to be zero, as we shall assume throughout the paper to alleviate notation. The outcome $(H \cdot S)_\infty$ can be seen as the net profit (or loss) by following the strategy H. If the time set is bounded, then of course the condition on the existence of the limit at infinity becomes vacuous. As shown in Theorem 9.3.3, the existence of the limit at infinity follows from arbitrage properties.

Fundamental in the proof of the existence of an equivalent local martingale measure are the sets

$$\mathcal{K}_1 = \{ (H \cdot S)_\infty \mid H \text{ is a 1-admissible strategy } \} \text{ and}$$
$$\mathcal{K} = \{ (H \cdot S)_\infty \mid H \text{ is admissible } \} .$$

From Corollary 9.3.8 we recall the following definition.

Definition 12.1.2. *We say that the semi-martingale S satisfies the condition no-arbitrage (NA) with respect to general admissible integrands if*

$$\mathcal{K} \cap L_+^0(\Omega) = \{0\} .$$

We say that the semi-martingale S satisfies the no free lunch with vanishing risk property (NFLVR) with respect to general admissible integrands if, for a sequence of S-integrable strategies $(H_n)_{n \geq 1}$ such that each H_n is a δ_n-admissible strategy and where δ_n tends to zero, we have that $(H \cdot S)_\infty$ tends to zero in probability $\mathbf{P}$.

The following theorem describes the relation between the *(NFLVR)* property and the existence of a local martingale measure. The equivalence of these two properties ((a) resp. (d) below) is the subject of Corollary 9.3.9 and Theorem 9.1.1. The equivalence with properties (b) and (c) below was proved in Theorem 11.2.9, see also [DS 95c].

Theorem 12.1.3. *For a locally bounded semi-martingale S the following properties are equivalent:*

(a) *S satisfies (NFLVR).*

(b) (i) S satisfies the property (NA) and (ii) $\mathcal{K}_1$ is bounded in L^0.
(c) (i) S satisfies the property (NA) and (ii) there is a strictly positive local
    martingale L such that at infinity $L_\infty > 0$, **P**-a.s. and such that LS is
    a local martingale.
(d) S admits an equivalent local martingale measure **Q**.

In the present paper we will enlarge the scope of the preceding theorem by
giving conditions for the existence of an absolutely continuous local martingale
measure. In particular we shall prove in Sect. 12.4 the following central result
of the paper.

Main Theorem 12.1.4. *If the continuous semi-martingale S satisfies the
no-arbitrage property with respect to general admissible integrands, then there
is an absolutely continuous local martingale measure for the process S.*

The paper is organised as follows. Sect. 12.2 contains some well-known
material on the existence of predictable Radon-Nikodým derivatives. The results are mainly due to C. Doléans and are scattered in the "Séminaires".
We need a more detailed version for finite dimensional processes. More precisely we treat the case of a predictable measure taking values in the set of
positive operators on the space $\mathbb{R}^d$, and we investigate under what conditions a vector measure has a Radon-Nikodým derivative with respect to this
operator-valued measure. In this context we say that an operator is positive
when it is positive definite. (If we were aiming for a coordinate-free approach,
we would rather interpret such an operator-valued measure as taking values
in the set of semi-positive bilinear forms on $\mathbb{R}^d$). This Radon-Nikodým problem, even for deterministic processes, is not treated in the literature (to the
best of our knowledge). The proofs are straightforward generalisations of the
one-dimensional case. For completeness we give full details.

We need these techniques to prove in Sect. 12.3 the fact that if the continuous semi-martingale $S = M + A$ does not allow arbitrage (with respect
to general admissible integrands) then dA may be written as $dA = d\langle M, M \rangle h$
for some predictable $\mathbb{R}^d$-valued process h. This result seems well-known to
people working in Mathematical Finance, but to the best of our knowledge at
least the d-dimensional version of this theorem has not been presented in the
literature. In Sect. 12.3 we then investigate the no-arbitrage properties and
we introduce the concept of immediate arbitrage. We also give an example
that illustrates this phenomenon.

In Sect. 12.4 we prove the main theorem stated above.

After finishing this paper we were informed of the paper of Levental and
Skorohod [LS 94], which has a very significant overlap with our results here.
In particular, although our framework is more general, the content and the
probabilistic approach we give here to proving Theorem 12.3.7 are essentially
identical to that of [LS 94, Lemma 2]. Their proof appears to have been constructed earlier than ours, although this theorem based on a rather more

complicated analytic proof had already been presented by the present authors during the SPA conference in Amsterdam in June 1993 [DS 93] and in the seminar of Tokyo University in September 93. Also [LS 94, Theorem 1] corresponds to our Main Theorem 12.1.4 under the additional assumption that the local martingale part M of the continuous semi-martingale S is of the form $M = \Sigma \cdot W$, where W is a d-dimensional Brownian motion defined on its (saturated) natural filtration and $\Sigma = (\Sigma_t)_{0 \le t \le 1}$ an adapted matrix valued process such that each Σ_t is invertible.

12.2 The Predictable Radon-Nikodým Derivative

In this section we will prove Radon-Nikodým theorems for stochastic measures. We first deal with the case of one-dimensional processes. A stochastic measure on $\mathbb{R}_+$ is described by a stochastic process of finite variation. In our setting it is convenient to require that the measure has no mass at zero, i.e. the initial value of the process is 0. If we have two predictable stochastic measures defined by the finite variation processes A and B, respectively, we can for almost every ω in Ω decompose the A-measure in a part absolutely continuous with respect to the B-measure and a component that is singular to it. We are interested in the problem whether such a decomposition can be done in a measurable or even predictable way. Similar problems can be stated for the optional and for the measurable case. For applications in Sect. 12.3, we only need the case of continuous processes. However, the more general case is almost the same and therefore we treat, at little extra cost, processes with jumps.

Theorem 12.2.1. (i)   *If $A \colon \mathbb{R}_+ \times \Omega \to \mathbb{R}$ is a predictable, càdlàg process of finite variation on finite intervals, then the process V, defined by setting V_t equal to the variation of A on the interval $[0, t]$, is càdlàg and predictable.*

(ii)   *If $A \colon \mathbb{R}_+ \times \Omega \to \mathbb{R}$ is a predictable, càdlàg process of finite variation on finite intervals, if V is defined as in (i), there is a decomposition of $\mathbb{R}_+ \times \Omega$ into two disjoint, predictable subsets, D^+ and D^-, such that*

$$A_t = \int_0^t (\mathbf{1}_{D^+} - \mathbf{1}_{D^-})\, dV .$$

(iii) *If $A \colon \mathbb{R}_+ \times \Omega \to \mathbb{R}$ is a predictable, càdlàg process of finite variation on finite intervals, if V is càdlàg, predictable and increasing, then there are predictable $\varphi \colon \mathbb{R}_+ \times \Omega \to \mathbb{R}$ and a predictable subset N of $\mathbb{R}_+ \times \Omega$ such that*

$$A_t = \int_{[0,t]} \varphi_u\, dV_u + \int_{[0,t]} \mathbf{1}_N(u)\, dA_u \quad and \quad \int_{\mathbb{R}_+} \mathbf{1}_N\, dV_u = 0 .$$

Proof. (i) We give the proofs only in the case $A_0 = V_0 = 0$. For the proof we need some results from the general theory of stochastic processes (see

[DM 80]). One of these results says that there is a sequence of predictable stopping times $(T_n)_{n \geq 1}$ that exhausts all the jumps of A. Fix n and let $(\tau_k)_{0 \leq k \leq N_n}$ be the finite ordered sequence of stopping times obtained from the set $\{0, \frac{1}{2^n}, \ldots, \frac{n}{2^n}, T_1, \ldots, T_n\}$.

Put $V^n = \sum_{k=0}^{N_n-1} |A_{\tau_{k+1}} - A_{\tau_k}| \, 1_{[\tau_{k+1}, \infty[}$.

Because A is predictable, the variables A_{τ_k} are $\mathcal{F}_{\tau_k-}$-measurable and hence the processes V^n are predictable. Because V^n tends pointwise to V, this process is also predictable.

(ii) The second part is proved using a constructive proof of the Hahn-Jordan decomposition theorem. It could be left as an exercise but we promised to give details. Let $V = \text{var}(A)$ as obtained in the first part. Being predictable and càdlàg, the process is locally bounded ([DM 80]) and hence there is an increasing sequence $(T_n)_{n \geq 1}$ of stopping times such that $T_n \nearrow \infty$ and $V_{T_n} \leq n$. Define now

$$\mathcal{H} = \left\{ \varphi \,\middle|\, \varphi \text{ predictable and } \mathbf{E}\left[\int_{\mathbb{R}_+} \varphi^2 \, dV_u \right] < \infty \right\}.$$

With the obvious inner product $\langle \varphi, \psi \rangle = \mathbf{E}[\int \varphi_u \psi_u \, dV_u]$, the space $\mathcal{H}$ divided by the obvious subspace $\{\varphi \mid \mathbf{E}[\int \varphi^2 \, dV_u] = 0\}$, is a Hilbert space. For each n we define the linear functional L^n on $\mathcal{H}$ as

$$L^n(\varphi) = \mathbf{E}\left[\int_{[0,T_n]} \varphi_u \, dA_u \right].$$

Since

$$\left| \int_{[0,T_n]} \varphi_u \, dA_u \right| \leq \int_{[0,T_n]} |\varphi_u| \, dV_u \leq \sqrt{n} \left(\int_{[0,T_n]} \varphi_u^2 \, dV_u \right)^{\frac{1}{2}},$$

the functional L^n is well-defined. Therefore there is ψ^n such that

$$L^n(\varphi) = \mathbf{E}\left[\int_{[0,T_n]} \varphi_u \psi_u^n \, dV_u \right].$$

Clearly the elements ψ^n and ψ^{n+1} agree for functions φ supported on $[0, T_n]$. Hence (with the convention that $T_0 = 0$) we have that $\psi = \sum_{n \geq 1} \psi^n 1_{]T_{n-1}, T_n]}$ is predictable and satisfies for all n:

$$L^n(\varphi) = \mathbf{E}\left[\int_{[0,T_n]} \varphi \psi \, dV \right].$$

Let now $C_t = A_t - \int_0^t \psi_u \, dV_u$. We will show that $C = 0$. First we show that C is continuous. Let τ be a predictable stopping time. Define $\varphi = \Delta C_\tau 1_{[\tau]}$. By definition of C and by the property of ψ we have for all n that $\mathbf{E}[(\Delta C)^2_{\tau \wedge T_n}] = 0$.

This shows that C is continuous. Next we put $\varphi = C\mathbf{1}_{[\![0, T_n \wedge t]\!]}$ and we find that $\mathbf{E}[C^2_{T_n \wedge t}] = 0$. From this it follows that for all t we have that $C_t = 0$. Because C is càdlàg, this implies that the process C vanishes identically.

So far we proved that in a predictable way $dA = \psi \, dV$. Let now $D^+ = \{\psi = 1\}$ and let $D^- = \mathbb{R}_+ \times \Omega \setminus D^+$. Both sets are predictable and from ordinary measure theory we deduce that $A_t = \int_0^t (\mathbf{1}_{D^+} - \mathbf{1}_{D^-}) \, dV$. This gives us the desired Hahn-Jordan decomposition.

(iii) The third part is again standard, a constructive proof of Lebesgue's decomposition theorem. Let A and V be given. As in ordinary measure theory we decompose A into its positive and its negative part. Part (ii) shows that this can be done in a predictable way. It is therefore sufficient to prove the claim for A increasing. We define $B = A + V$. We now repeat the proof of the second part and we find a predictable ψ, $0 \leq \psi \leq 1$ and $dA = \psi dB$. Let $N = \{\psi = 1\}$, a predictable set. We find $dA = \psi \, dA + \psi \, dV$. As in the classical proof we deduce from this equality that $dA = \mathbf{1}_N \, dA + \varphi \, dV$ where $\int \mathbf{1}_N \, dV = 0$ and where φ is predictable. $\qquad\square$

Corollary 12.2.2. *If A and V are as in part (iii) of Theorem 12.2.1, if $dA \ll dV$ with respect to the predictable σ-algebra, i.e. for each predictable set N the property $\int \mathbf{1}_N \, dV = 0$ implies that also $\int \mathbf{1}_N \, dA = 0$, then for almost all ω the measure $dA(\omega)$ is absolutely continuous with respect to $dV(\omega)$ on $\mathbb{R}_+$.*

For applications in finance we need a vector measure generalisation of the preceding results. The theory was developed by [J 79]. We need two kinds of vector measures. The first kind is an ordinary vector measure taking values in $\mathbb{R}^d$. The second kind is an operator-valued measure that takes values in the set of all operators on $\mathbb{R}^d$; in daily language, in the space of all $d \times d$ matrices. Positive measures on $\mathbb{R}_+$ are generalised as measures that take values in the cone $\mathrm{Pos}(\mathbb{R}^d)$ of all positive semi-definite operators on $\mathbb{R}^d$. In this setting the variation process V becomes a predictable, càdlàg, increasing process $V \colon \mathbb{R}_+ \times \Omega \to \mathrm{Pos}(\mathbb{R}^d)$. On the set of all operators we put the nuclear norm; for positive operators this simply means the trace of the operator. Let now $\lambda_t = \mathrm{trace}(V_t)$. The process λ is predictable, càdlàg and increasing. Again we assume $V_0 = 0$, which results in $\lambda_0 = 0$. We have that $dV \ll d\lambda$ in the sense that all elements of the matrix function define measures that are absolutely continuous with respect to λ. If we calculate the Radon-Nikodým derivative using dyadic approximations we see that $dV = \sigma \, d\lambda$, where σ is a predictable process taking values in $\mathrm{Pos}(\mathbb{R}^d)$.

For a positive operator a we have that the range $R(a)$ is invariant under a and that on $R(a)$ the operator a is invertible. If we define P_a as the orthogonal projection on $R(a)$ we see that $a^{-1} = a^{-1} \circ P_a$ is a generalised inverse of a. More precisely we have $a \circ a^{-1} := a^{-1} \circ a = P_a$. The correspondence between a, a^{-1} and P_a can be described in a Borel-measurable way. This is an easy exercise but we promised to give details.

First note that, for each strictly positive operator b_0, the map $b \to b^{-1}$ is continuous at b_0.

To calculate P_a we simply take the limit

$$\lim_{\varepsilon \searrow 0} a \circ (a + \varepsilon \,\mathrm{id})^{-1}.$$

This constructive definition shows that the mapping $a \to P_a$ is a Borel-measurable mapping. The same trick is used to obtain the generalised inverse

$$a^{-1} = \lim_{\varepsilon \searrow 0} a \circ (a + \varepsilon \,\mathrm{id})^{-2}.$$

The processes σ^{-1} and P_σ are therefore still predictable since they are the composition of a predictable and a Borel-measurable mapping.

We will now describe a kind of absolute continuity of a vector measure with respect to an operator-valued measure. Let ν be a measure defined on the σ-ring of relatively compact Borel sets of $\mathbb{R}_+$ and taking values in $\mathbb{R}^d$. Let μ be a measure defined on the same σ-ring and taking values in $\mathrm{Pos}(\mathbb{R}^d)$. We say that $\nu \ll \mu$, if whenever $f \colon \mathbb{R}_+ \to \mathbb{R}^d$ is a Borel function such that either $f(t) = 0$ or $\|f(t)\| = 1$, the expression $d\mu\,f = 0$ (as a vector measure) implies $f'd\nu = 0$ (as a scalar measure). (Here f' is the transpose of f). One can show that in this case the measure ν has a Radon-Nikodým derivative with respect to μ. Again we will need a predictable version of this theorem, so we give details.

Suppose that $A \colon \mathbb{R}_+ \times \Omega \to \mathbb{R}^d$ is predictable, càdlàg and of finite variation on finite intervals. Suppose that $A_0 = 0$. Let V be as above, predictable, càdlàg taking values in $\mathrm{Pos}(\mathbb{R}^d)$ and increasing. Suppose that for every predictable process $f \colon \mathbb{R}_+ \times \Omega \to \mathbb{R}^d$, such that $\|f(t, \omega)\|$ is either 0 or 1, the relation $dV\,f = 0$ implies that $f'dA = 0$. This means that $dA \ll dV$ in a predictable way. Let $\lambda = \mathrm{trace}(V)$ and let N be a predictable null set for λ, i.e. $\mathbf{1}_N\,d\lambda = 0$. For such a predictable set N and for each predictable k we have $\mathbf{1}_N\,dV\,k = 0$. The hypothesis on A then implies that $\mathbf{1}_N\,k'dA = 0$. This shows that $dA \ll d\lambda$ and the predictable Radon-Nikodým theorem (applied for each coordinate) shows the existence of a predictable $\mathbb{R}^d$-valued process g such that $dA = g\,d\lambda$. Now $(\mathrm{id} - \sigma \circ \sigma^{-1})\,dV = dV\,(\mathrm{id} - \sigma \circ \sigma^{-1}) = (\mathrm{id} - \sigma \circ \sigma^{-1})\sigma\,d\lambda = 0$ and by the assumption on A we have $(\mathrm{id} - \sigma \circ \sigma^{-1})\,dA = 0$. This implies that $(\mathrm{id} - \sigma \circ \sigma^{-1})\,g\,d\lambda = 0$ and that up to null sets for λ, we have $g \in R(\sigma)$. Now let $h = \sigma^{-1}(g)$. Then obviously $\sigma(h) = g$ (because $g \in R(\sigma)$), $h \in R(\sigma)$ and $dA = \sigma\,h\,d\lambda = dV\,h$. The range $R(\sigma)$ could have been called the infinitesimal range $R(dV)$ of the measure V. It is easy to show that it does not depend on the control measure. We completed the proof of the following theorem.

Theorem 12.2.3. *If V is an increasing predictable càdlàg process, taking values in the cone of the positive semi-definite operators on $\mathbb{R}^d$, then the vector measure defined by the predictable $\mathbb{R}^d$-valued càdlàg process A of finite variation is of the form $dA = dV\,h$, for some predictable $\mathbb{R}^d$-valued process h, if* *and only if for each predictable $\mathbb{R}^d$-process f, such that $\|f(t,\omega)\|$ is either 0 or 1, the relation $dV f = 0$ implies $f' dA = 0$.*

Remark 12.2.4. If S is a semi-martingale with values in $\mathbb{R}^d$, then the bracket $[S, S]$ and (if it exists) also the bracket $\langle S, S \rangle$ define increasing processes with values in $\mathrm{Pos}(\mathbb{R}^d)$. The fact that values are taken in $\mathrm{Pos}(\mathbb{R}^d)$ is a reformulation of the Kunita-Watanabe inequalities:

$$|d[S^i, S^j]| \leq \sqrt{d[S^i, S^i]\, d[S^j, S^j]}\,,$$
$$|d\langle S^i, S^j \rangle| \leq \sqrt{d\langle S^i, S^i \rangle\, d\langle S^j, S^j \rangle}\,.$$

12.3 The No-Arbitrage Property and Immediate Arbitrage

We now turn to the main theme of the paper, a detailed analysis of the notion of *no-arbitrage*. We start with an easy lemma, which turns out to be very useful. It shows that the general case of an arbitrage may be reduced to two special kinds of arbitrage.

Lemma 12.3.1. *If the càdlàg semi-martingale S does not satisfy the no-arbitrage property with respect to general admissible integrands then at least one of the two following statements holds*

(i) *There is an S-integrable strategy H and a stopping time T, $\mathbf{P}[T < \infty] > 0$ such that H is supported by $[\![T, T+1[\![$, $H \cdot S \geq 0$ and $(H \cdot S)_t > 0$, for $t > T$.*

(ii) *There is an S-integrable 1-admissible strategy K, $\varepsilon > 0$ and two stopping times $T_1 \leq T_2$ such that $T_2 < \infty$ on $\{T_1 < \infty\}$, $\mathbf{P}[T_2 < \infty] > 0$, $K = K\mathbf{1}_{]\!]T_1,T_2]\!]}$ and $(K \cdot S)_{T_2} \geq \varepsilon$ on the set $\{T_2 < \infty\}$.*

Proof. Let S allow arbitrage and let H be a 1-admissible strategy that produces arbitrage, i.e., $(H \cdot S)_\infty \geq 0$ with strict inequality on a set of strictly positive probability. We now distinguish two cases. Either the process $H \cdot S$ is never negative or the process $H \cdot S$ becomes negative with positive probability. In the first case let $T = \inf\{t \mid (H \cdot S)_t > 0\}$.

Let $(\vartheta_n)_{n=1}^\infty$ be a dense in $]0, 1[$ and let $\widetilde{H} = \sum_{n=1}^\infty 2^{-n} H \mathbf{1}_{]\!]T,T+\vartheta_n[\![}$. Then $\widetilde{H}$ satisfies (i). We thank an anonymous referee for correcting a slip in a previous version of this paper at this point.

In the second case we first look for $\varepsilon > 0$ such that $\mathbf{P}[\inf_t (H \cdot S)_t < -2\varepsilon] > 0$. We then define T_1 as the first time the process $H \cdot S$ goes under -2ε, i.e.

$$T_1 = \inf\{t \mid (H \cdot S)_t < -2\varepsilon\}\,.$$

On the set $\{T_1 < \infty\}$ we certainly have that the process $H \cdot S$ has to gain at least 2ε. Indeed at the end the process $H \cdot S$ is positive and therefore the time

$$T_2 = \inf\{t \mid t > T_1,\ (H \cdot S)_t \geq -\varepsilon\}$$

is finite on the set $\{T_1 < \infty\}$. We now put $K = H\mathbf{1}_{]\!]T_1, T_2]\!]}$. The process K is 1-admissible since $(K \cdot S)_t \geq -1 + 2\varepsilon$ on the set $\{T_1 < \infty\}$. Also $(K \cdot S)_{T_2} \geq \varepsilon$ on the set $\{T_1 < \infty\}$. $\qquad\square$

Definition 12.3.2. *We say that the semi-martingale S admits immediate arbitrage at the stopping time T, where we suppose that $\mathbf{P}[T < \infty] > 0$, if there is an S-integrable strategy H such that $H = H\mathbf{1}_{]\!]T, \infty]\!]}$, and $(H \cdot S)_t > 0$ for $t > T$.*

Remark 12.3.3. (a): Let us explain why we use the term immediate arbitrage. Suppose S admits immediate arbitrage at T and that H is the strategy that realises this arbitrage opportunity. Clearly $H \cdot S \geq 0$ and $(H \cdot S)_{T+t} > 0$ for all $t > 0$ almost surely on $\{T < \infty\}$. Hence we can make an arbitrage almost surely *immediately* after the stopping time T has occurred.

(b): Lemma 12.3.1 shows that either we have an immediate arbitrage opportunity or we have a more conventional form of arbitrage. In the second alternative the strategy to follow is also quite easy. We wait until time T_1 and then we start our strategy K. If the strategy starts at all (i.e., if $T_1 < \infty$) then we are sure to collect at least the amount ε in a finite time. It is clear that such a form of arbitrage is precisely what one wants to avoid in economic models. The immediate arbitrage seems, at first sight, to be some mathematical pathology that can never occur. However, the concept of immediate arbitrage can occur as the following example shows. In model building one therefore cannot neglect the phenomenon.

Example 12.3.4. Take the one-dimensional Brownian motion $W = (W_t)_{t \in [0,1]}$ with its usual filtration. For the price process S we take $S_t = M_t + A_t = W_t + \sqrt{t}$ which satisfies the differential equation $dS_t = dW_t + \frac{dt}{2\sqrt{t}}$. We will show that such a situation leads to "immediate" arbitrage at time $T = 0$. Take $H_t = \frac{1}{\sqrt{t}(\ln t)^2}$. With this choice the integral on the drift-term $\int_0^t H_u \frac{du}{2\sqrt{u}} = \frac{1}{2}(\ln(t^{-1}))^{-1}$ is convergent.

As for the martingale part, the random variable $\int_0^t H_u\, dW_u$ has variance $\int_0^t \frac{1}{u(\ln u)^4}\, du$ which is of the order $\ln(t^{-1})^{-3}$. The iterated logarithm law implies that, for $t = t(\omega)$ small enough,

$$|(H \cdot W)_t(\omega)| \leq C\sqrt{(\ln(t^{-1}))^{-3} \ln\ln((\ln(t^{-1}))^3)} \leq C'(\ln(t^{-1}))^{-\frac{5}{4}}.$$

It follows that, for t small enough, we necessarily have that $(H \cdot S)_t(\omega) > 0$. We now define the stopping time T as $T = \inf\{t > 0 \mid (H \cdot S)_t = 0\}$ and, for $n > 0$, $T_n = T \wedge n^{-1}$. Clearly $(H \cdot S)^T \geq 0$ and $\mathbf{P}[(H \cdot S)_{T_n} > 0]$ tends to 1 as n tends to infinity. By considering the integrand $L = \sum_{n=1}^{\infty} \alpha_n H\mathbf{1}_{[\![0, T_n]\!]}$ for a sequence $\alpha_n > 0$ tending to zero sufficiently fast, we can even obtain that $(L \cdot S)_t$ is almost surely strictly positive for each $t > 0$.

We now give some more motivation why such a form of arbitrage is called immediate arbitrage. In the preceding example, for each stopping time $T > 0$ the process $S - S^T$ admits an equivalent martingale measure $\mathbf{Q}(T)$ given by the density $f_T = \exp(-\frac{1}{2} \int_T^1 \frac{1}{\sqrt{u}} \, dW_u - \frac{1}{8} \int_T^1 \frac{1}{u} \, du)$. We can check this by means of the Girsanov-Maruyama formula or we can check it even more directly via Itô's rule. This statement shows that if one wants to make an arbitrage profit, one has to be very quick since a profit has to be the result of an action taken before time T.

Let us also note that the process S also satisfies the *(NA)* property for simple integrands. As is well-known it suffices to consider integrands of the form $f\mathbf{1}_{]T_0,T_1]}$ where f is $\mathcal{F}_{T_0}$-measurable (see Chap. 9). Let us show that such an integrand does not allow an arbitrage. Take two stopping times $T_0 \leq T_1$. We distinguish between $\mathbf{P}[T_0 > 0] = 1$ and $T_0 = 0$. (The 0-1-law for $\mathcal{F}_0$ (Blumenthal's theorem) shows that one of the two holds).

If $T_0 > 0$, $\mathbf{P}$-a.s., then the result follows immediately from the existence of the martingale measure $\mathbf{Q}(T_0)$ for the process $S - S^{T_0}$.

If $T_0 = 0$, we have to prove that $S_{T_1} \geq 0$ (or $S_{T_1} \leq 0$) implies that $S_{T_1} = 0$ a.s..

We concentrate on the first case and assume to the contrary that $S_{T_1} \geq 0$ and $\mathbf{P}\{S_{T_1} > 0\} > 0$. Note that it follows from the law of the iterated logarithm that $\inf\{t|S_t < 0\} = 0$ almost surely, hence the stopping times

$$T_\varepsilon = \inf\{t > \varepsilon \mid S_t < -\varepsilon\}$$

tend to zero a.s. as ε tends to zero. Let $\varepsilon > 0$ be small enough such that $\{T_\varepsilon < T_1\}$ has positive measure to arrive at a contradiction:

$$0 > \mathbf{E}_{\mathbf{Q}(T_\varepsilon)}[S_{T_\varepsilon} \mathbf{1}_{\{T_\varepsilon < T_1\}}] = \mathbf{E}_{\mathbf{Q}(T_\varepsilon)}[S_{T_1} \mathbf{1}_{\{T_\varepsilon < T_1\}}] \geq 0.$$

The following theorem, which is based on the material developed in Sect. 12.2, is well-known and has been around for some time. At least in dimension $d = 1$ the result should be known for a long time. For dimension $d > 1$, the presentation below is, we guess, new.

Theorem 12.3.5. *If the d-dimensional, locally bounded semi-martingale S satisfies the (NA) property for general admissible integrands, then the Doob-Meyer decomposition $S = M + A$ satisfies $dA = d\langle M,M \rangle \, h$, where h is a d-dimensional predictable process and where $d\langle M,M \rangle$ denotes the operator-valued measure defined by the $(d \times d)$-matrix process $(\langle M,M \rangle)_{i,j \leq d}$. The process h may be chosen to take its values in the infinitesimal range $R(d\langle M, M \rangle)$.*

Proof. We apply the criterion of Sect. 12.2. Take f a d-dimensional predictable process such that the measure $d\langle M,M \rangle \, f$ is zero and such that either f has norm 1 or norm 0. It is obvious that the stochastic integral $f' \cdot M$ exists and results in the zero process. If the process $f' \cdot A$ is not zero then we replace f by the sign function coming from the Jordan-Hahn decomposition of $f' \cdot A$. This sign function ϕ is a predictable process equal to $+1$ or -1. The predictable integrand $g = \phi\,f$ still satisfies $g \cdot M = 0$ but the component $g' \cdot A$ now results in an arbitrage profit. This contradiction shows that the criterion of Sect. 12.2 is fulfilled and hence the existence of the process h is proved. If we write $d\langle M,M\rangle$ as $\sigma\,d\lambda$ for some control measure λ and an operator-valued predictable process σ, then we may, by the results of Sect. 12.2, suppose that h_t is in the range of the operator σ_t. $\square$

The following theorem is the basic theorem in dealing with the *(NA)* property in the case of continuous price processes.

Theorem 12.3.6. *If the continuous semi-martingale S with Doob-Meyer decomposition $S = M + A$ satisfies the (NA) property for general admissible integrands, then we have $dS = dM + d\langle M,M\rangle\,h$, where the predictable process h satisfies:*

(i) $T = \inf \left\{ t \,\Big|\, \int_0^t h'_u\,d\langle M,M\rangle_u\,h_u = \infty \right\} > 0$ *a.s..*

(ii) *The $[0,\infty]$-valued increasing process $\int_0^t h'_u\,d\langle M,M\rangle_u\,h_u$ is continuous; in particular it does not jump to ∞.*

Proof. The existence of the process h follows from the preceding theorem. The stopping time T is well-defined. The first claim on the stopping time T follows from the second, so we limit the proof to the second statement. We will prove that the set

$$F = \{T < \infty\} \cap \left\{ \int_T^{T+\varepsilon} h'_t\,d\langle M, M\rangle_t\,h_t = \infty, \ \ \forall \varepsilon > 0 \right\}$$

has zero measure. Clearly F is, by right continuity of the filtration, an element of the σ-algebra $\mathcal{F}_T$. As the process $\langle M, M\rangle_t$ is continuous, assertion (ii) will follow from the fact that $\mathbf{P}[F] = 0$. Suppose now to the contrary that F has strictly positive measure. We then look at the process $\mathbf{1}_F(S - S^T)$, adapted to the filtration $(\mathcal{F}_{T+t})_{t\geq0}$ and we replace the probability $\mathbf{P}$ by $\mathbf{P}_F$. With this notation the theorem is reduced to the case $T = 0$. This case is treated in the following theorem. It is clear that this will complete the proof. $\square$

Immediate Arbitrage Theorem 12.3.7. *Suppose the d-dimensional continuous semi-martingale S has a Doob-Meyer decomposition given by*

$$dS_t = dM_t + d\langle M, M\rangle_t\,h_t$$

where h is a d-dimensional predictable process. Suppose that a.s.

$$\int_0^\varepsilon h'_t\,d\langle M, M\rangle_t\,h_t = \infty, \ \ \forall\,\varepsilon > 0. \tag{12.1}$$

Then for all $\varepsilon > 0$, there is an S-integrable strategy H such that $H = H\mathbf{1}_{[\![0,\varepsilon]\!]}$, $H \cdot S \geq 0$ and $\mathbf{P}[(H \cdot S)_t > 0] = 1$, for each $t > 0$. In other words, S admits immediate arbitrage at time $T = 0$.

The proof of the theorem is based on the following lemma:

Lemma 12.3.8. *If (12.1) holds almost surely, then for any $a, \varepsilon, \eta > 0$ we can find $0 < \delta < \frac{\varepsilon}{2}$ and an a-admissible integrand H with*

$$H = H \mathbf{1}_{]\![\delta, \varepsilon]\!]} \, ,$$

$$\int_\delta^\varepsilon |H'_s \, dA|_s + \int_\delta^\varepsilon H'_s \, d\langle M, M \rangle_s \, H_s < 2 + a \, ,$$

$$\mathbf{P}[(H \cdot S)_\varepsilon \geq 1] \geq 1 - \eta \, .$$

Proof of the lemma. Fix $a, \varepsilon, \eta > 0$ and let $R \geq \max\{\frac{8}{\eta} \left(\frac{1+a}{a}\right)^2, (1+a)^2\}$. Since (i) is satisfied almost surely, we have that

$$\lim_{\substack{K \nearrow \infty \\ \delta \searrow 0}} \mathbf{P}\left[\int_\delta^\varepsilon \mathbf{1}_{\{|h| \leq K\}} \, h'_t \, d\langle M, M \rangle_t \, h_t \geq R\right] = 1 \, .$$

Hence we can find a $K > 0$ and a $0 < \delta < \frac{\varepsilon}{2}$ such that

$$\infty > \int_\delta^\varepsilon \mathbf{1}_{\{|h| \leq K\}} \, h'_t \, d\langle M, M \rangle_t \, h_t \geq R$$

on a $\mathcal{F}_\varepsilon$-measurable set Λ with $\mathbf{P}[\Lambda] \geq 1 - \frac{\eta}{2}$. Let

$$T = \inf\left\{t > 0 \, \middle| \, \int_\delta^t \mathbf{1}_{\{|h| \leq K\}} \, h'_t \, d\langle M, M \rangle_t \, h_t \geq R\right\} \wedge \varepsilon$$

and let $H = \frac{1+a}{R} h \mathbf{1}_{]\![\delta, T]\!]} \mathbf{1}_{\{|h| \leq K\}}$. Then

$$\int_0^\varepsilon H'_s \, d\langle M, M \rangle_s \, H_s \leq \frac{(1+a)^2}{R}$$

and

$$\int_0^\varepsilon |H_s \, dA|_s \leq (1+a) \quad \text{a.s..}$$

Therefore H is S-integrable. Moreover, $(H \cdot A)_\varepsilon = 1 + a$ on Λ.

Since $\|H \cdot M\|_2^2 = \mathbf{E}[\int_0^\varepsilon H'_s d\langle M, M \rangle_s H_s] \leq \frac{(1+a)^2}{R}$ we obtain from Doob's inequality together with Tchebycheff's inequality (both in their L^2-version)

$$\mathbf{P}\left[(H \cdot M)^* \geq a\right] \leq 4 \left(\tfrac{1+a}{a}\right)^2 \tfrac{1}{R} \leq \tfrac{\eta}{2} \, . \tag{12.2}$$

We now localise H to be a-admissible. Let

$$T_2 = \inf\{t > 0 \mid (H \cdot M)_t < -a\} \wedge T \, .$$

Then $T_2 = T$ on $\{(H \cdot M)^* \leq a\}$ and from (12.2) we obtain

$$\mathbf{P}\left[(H\,\mathbf{1}_{[\![0,T_2]\!]}\cdot S)_\varepsilon \geq 1\right] \geq \mathbf{P}\left[\{(H\cdot A)_\varepsilon \geq 1+a\} \cap \{(H\cdot M)^* < a\}\right]$$
$$\geq \mathbf{P}\left[\Lambda\right] - \mathbf{P}\left[(H\cdot M)^* \geq a\right] \geq 1-\eta$$

which proves the lemma.   □

Proof of the Immediate Arbitrage Theorem 12.3.7. Assume that (12.1) is valid for almost every $\omega \in \Omega$. We will now construct an integrand which realises immediate arbitrage. Let $\varepsilon_0 > 0$ be such that $\varepsilon_0 \leq \min(\varepsilon, \frac{1}{2})$. By Lemma 12.3.8 we can find a strictly decreasing sequence of positive numbers $(\varepsilon_n)_{n\geq 0}$ with $\lim_{n\to\infty} \varepsilon_n \to 0$ and integrands $H_n = H_n \mathbf{1}_{]\![\varepsilon_{n+1},\varepsilon_n]\!]}$ such that H_n is 4^{-n}-admissible, $\int_{\varepsilon_{n+1}}^{\varepsilon_n} |(H_n)'_s\,dA_s| + \int_{\varepsilon_{n+1}}^{\varepsilon_n} (H_n)'_s\,d\langle M,M\rangle_s\,(H_n)_s < \frac{3}{2^n}$ and $\mathbf{P}\left[(H_n\cdot S)_{\varepsilon_n} \geq 2^{-n}\right] \geq 1-2^{-n}$. Let $\widehat{H} = \sum_{n=1}^{\infty} H_n$. Then $\widehat{H}$ is S-integrable. Define

$$T = \inf\left\{t > 0 \mid (\widehat{H}\cdot S)_t = 0\right\}.$$

We claim that $T(\omega) > 0$ for almost every $\omega \in \Omega$. Since $\mathbf{P}\left[(H_n\cdot S)_{\varepsilon_n} < 2^{-n}\right] \leq 2^{-n}$, we obtain from the Borel-Cantelli lemma that for almost every $\omega \in \Omega$ there is a $N(\omega) \in \mathbb{N}$ with $(H_n\cdot S)_{\varepsilon_n}(\omega) > 2^{-n}$ for all $n > N(\omega)$. If $n > N(\omega)$ and $\varepsilon_{n+1} < t \leq \varepsilon_n$ then

$$(\widehat{H}\cdot S)_t(\omega) = \underbrace{\sum_{k>n}^{\infty}(H_k\cdot S)_{\varepsilon_k}(\omega)}_{\geq 2^{-n}} + \underbrace{(H_n\cdot S)_t(\omega)}_{\geq -2^{-(n+1)}} \geq \frac{1}{2^{n+1}}$$

and we have verified the claim. Hence

$$\lim_{t\to 0}\mathbf{P}\left[\left(\widehat{H}\,\mathbf{1}_{[\![0,T]\!]}\cdot S\right)_t > 0\right] = 1.$$

Finally let

$$H = \sum_{n=1}^{\infty} 2^{-n}\widehat{H}\mathbf{1}_{]\![0,T\wedge\varepsilon_n[\![}$$

to find an S-integrable predictable process supported by $[0,\varepsilon]$ such that $(H\cdot S)_t > 0$ for each $t > 0$.   □

12.4 The Existence of an Absolutely Continuous Local Martingale Measure

We start this section with the investigation of the support of an absolutely continuous risk neutral measure. The theory is based on the analysis of the density given by a Girsanov-Maruyama transformation. If $dS_t = dM_t + d\langle M,M\rangle_t\,h_t$ defines the Doob-Meyer decomposition of a continuous semi-martingale, where h is a d-dimensional predictable process and where M is a d-dimensional continuous local martingale, then the Girsanov-Maruyama transformation is, at least formally, given by the local martingale $L_t = \exp(\int_0^t -h'_u \, dM_u - \frac{1}{2} \int_0^t h'_u \, d\langle M, M \rangle_u \, h_u)$, $L_0 = 1$. Formally one can verify that LS is a local martingale. However, things are not so easy. First of all, there is no guarantee that the process h is M-integrable, so L need not be defined. Second, even if L is defined, it may only be a local martingale and not a uniformly integrable martingale. The examples in [S 93] and in Chap. 10 show that even when an equivalent risk neutral measure exists, the local martingale L need not to be uniformly integrable. In other words a risk neutral measure need not be given by L. Third, in case the two previous points are fulfilled, the density L_∞ need not be different from zero a.s..

What can we save in our setting? In any case, Theorem 12.3.6 shows that in the case when S satisfies the no-arbitrage property for general admissible integrands, the process h satisfies the properties:

(1) $T = \inf\{t \mid \int_0^t h' d\langle M, M\rangle \, h = \infty\} > 0$ a.s..
(2) The $[0, \infty]$-valued process $\int_0^t h'_u \, d\langle M, M\rangle_u \, h_u$ is continuous; in particular, it does not jump to ∞.

In this case the stochastic integrals $h \cdot M$ and $h \cdot S$ can be defined on the interval $[\![0, T[\![$ and at time T we have that L_T can be defined as the left limit. The theory of continuous martingales ([RY 91]) shows that

$$\{L_T = 0\} = \left\{ \int_0^T h'_t \, d\langle M, M \rangle_t \, h_t = \infty \right\}.$$

If after time T, i.e. for $t > T$, we put $L_t = 0$, the process L is well-defined, it is a continuous local martingale, it satisfies $dL_t = -L_t \, h'_t \, dM_t$ and LS is a local martingale. The process $X = \frac{1}{L} - 1$ is also defined on the interval $[\![0, T[\![$ and on the set $\{L_T = 0\}$ its left limit equals infinity. The crucial observation is now that on the interval $[\![0, T[\![$, we have that $dX_t = \frac{1}{L_t} h'_t \, dS_t$.

This follows simply by plugging in Itô's formula (compare Chap. 11).

For each $\varepsilon > 0$ let τ^ε be the stopping time defined by $\tau^\varepsilon = \inf\{t \mid L_t \leq \varepsilon\}$. Because the process X is always larger than -1, the stopped processes X^{τ^ε} are outcomes of admissible integrands. If $\mathbf{Q}$ is an absolutely continuous probability measure such that S becomes a local martingale thae, by Theorem 12.1.3 we have that the set $\mathcal{H} = \left\{ X_\infty^{\tau^\varepsilon} \mid \varepsilon > 0 \right\}$ is bounded in $L^0(\{\frac{d\mathbf{Q}}{d\mathbf{P}} > 0\})$. But it is clear that on the set $\{L_T = 0\}$, the set $\mathcal{H}$ is unbounded.

As a consequence we obtain the following lemma.

Lemma 12.4.1. *If the continuous semi-martingale S satisfies the no-arbitrage condition with respect to general admissible integrands and if $\mathbf{Q}$ is an absolutely continuous local martingale measure for S, then $\{\frac{d\mathbf{Q}}{d\mathbf{P}} > 0\} \subset \{L_T > 0\}$.*

In order to prove the existence of an absolutely continuous local martingale measure $\mathbf{Q}$ we therefore should restrict ourselves to measures supported by

$$F = \{L_T > 0\}.$$

Note that the no-arbitrage condition implies that $\mathbf{P}[F] > 0$. Indeed, suppose that $\mathbf{P}[F] = 0$ and let

$$U = \inf\{t \mid L_t \leq \tfrac{1}{2}\}.$$

We than have that $\mathbf{P}[U < \infty] = 1$, $L_U \equiv \tfrac{1}{2}$ and therefore $X_U \equiv 1$. Hence $H = \tfrac{1}{L}h'\mathbf{1}_{[\![0,U]\!]}$ is a 1-admissible integrand such that $(H \cdot S)_\infty \equiv X_U \equiv 1$, a contradiction to (NA).

So we will look at the process S under the conditional probability measure $\mathbf{P}_F$.

Our strategy will be to verify that S satisfies the property $(NFLVR)$ with respect to $\mathbf{P}_F$ which will imply the existence of a local martingale measure $\mathbf{Q}$ for S which is equivalent to $\mathbf{P}_F$ and therefore absolutely continuous with respect to $\mathbf{P}$. However, there are difficulties: under the measure $\mathbf{P}_F$ the Doob-Meyer decomposition will change, there will be more admissible integrands and the verification of the no free lunch with vanishing risk property for general admissible integrands (under $\mathbf{P}_F$) is by no means trivial.

We are now ready to reformulate the main theorem stated in the Introduction 12.1 in a more precise way and to commence the proof:

Main Theorem 12.4.2. *If the continuous semi-martingale S satisfies the no-arbitrage property with respect to general admissible integrands, then with the notation introduced above, it satisfies the no free lunch with vanishing risk property with respect to $\mathbf{P}_F$.*

As a consequence there is an absolutely continuous local martingale measure that is equivalent to $\mathbf{P}_F$, i.e. it is precisely supported by the set F.

The proof of the theorem still needs some auxiliary steps which will be stated below.

We first deal with the problem of the usual hypotheses under the measure $\mathbf{P}_F$. The σ-algebras $\widetilde{\mathcal{F}}_t$ of the $\mathbf{P}_F$-augmented filtration are obtained from $\mathcal{F}_t$ by adding all $\mathbf{P}_F$ null sets. It is easily seen that the new filtration is still right continuous and satisfies the usual hypotheses for the new measure $\mathbf{P}_F$. The following technical results are proved in [DS 95c].

Proposition 12.4.3. *If $\widetilde{\tau}$ is a stopping time with respect to the filtration $(\widetilde{\mathcal{F}}_t)_{t\geq 0}$ then there is a stopping time τ with respect to the filtration $(\mathcal{F}_t)_{t\geq 0}$ such that $\mathbf{P}_F$-a.s. we have $\widetilde{\tau} = \tau$. If $\widetilde{\tau}$ is finite or bounded, then τ may be chosen to be finite or bounded.*

Proposition 12.4.4. *If $\widetilde{H}$ is a predictable process with respect to the filtration $(\widetilde{\mathcal{F}}_t)_{t\geq 0}$ then there is a predictable process H with respect to the filtration $(\mathcal{F}_t)_{t\geq 0}$, such that $\mathbf{P}_F$-a.s. we have $\widetilde{H} = H$.*

This settles the problem of the usual hypotheses. Each time we need an $\widetilde{\mathcal{F}}$-predictable process, we can without danger replace it by a predictable process for $\mathcal{F}$. Without further notice we will do this.

The process S is a semi-martingale with respect to the system $(\widetilde{\mathcal{F}}, \mathbf{P}_F)$. This is well-known, see [P 90].

Note also that for $\mathbf{P}_F$ we have that $\int_0^\infty h'_u \, d\langle M, M \rangle_u \, h_u < \infty$ a.s.. We will need this later on.

As a first step we will decompose S into a sum of a $\mathbf{P}_F$-local martingale and a predictable process of finite variation. Because $\mathbf{P}_F$ is only absolutely continuous with respect to $\mathbf{P}$ we need an extension of the Girsanov-Maruyama formula for this case. The generalisation was given by [L 77]. We need the càdlàg martingale U defined as

$$U_t = \mathbf{E}\left[\left.\frac{\mathbf{1}_F}{\mathbf{P}[F]}\,\right|\,\mathcal{F}_t\right].$$

Note that U is not necessarily continuous, as we only assumed that S is continuous and not that each $\mathcal{F}_t$-martingale is continuous.

Together with the process U we need the stopping time

$$\nu = \inf\{t \mid U_t = 0\} = \inf\{t > 0 \mid U_{t-} = 0\}$$

(see [DM 80] for this equality).

Lemma 12.4.5. *We have $\nu = T$ $\mathbf{P}$-almost surely.*

Proof of Lemma. We first show that for an arbitrary stopping time σ we have that $L_\sigma > 0$ on the set $\{U_\sigma > 0\}$. Let A be a set in $\mathcal{F}_\sigma$ such that $\mathbf{P}[A \cap \{U_\sigma > 0\}] > 0$. This already implies that $\mathbf{P}[A \cap F] > 0$. Indeed we have that

$$\mathbf{E}[\mathbf{1}_A \mathbf{1}_F \mid \mathcal{F}_\sigma] = \mathbf{P}[F] \mathbf{1}_A U_\sigma$$

and hence we necessarily have that $\mathbf{P}[A \cap F] > 0$. The following chain of equalities is almost trivial

$$\int_{A \cap \{U_\sigma > 0\}} L_\sigma = \int_A L_\sigma \mathbf{1}_{\{U_\sigma > 0\}} \geq \mathbf{P}[F] \int_A L_\sigma U_\sigma = \int_A L_\sigma \mathbf{1}_F = \int_{A \cap F} L_\sigma .$$

The last term is strictly positive since $L_\sigma > 0$ on F. This proves that for each set A such that $\mathbf{P}[A \cap \{U_\sigma > 0\}] > 0$ we must have $\int_{A \cap \{U_\sigma > 0\}} L_\sigma > 0$. This implies that $L_\sigma > 0$ on the set $\{U_\sigma > 0\}$, hence $\nu \leq T$.

The converse inequality is less trivial and requires the use of the *(NA)* property of S. We proceed in the same way. Take $G \in \mathcal{F}_\sigma$ such that $G \subset \{L_\sigma > 0\}$ and $\mathbf{P}[G] > 0$. Suppose that $U_\sigma = 0$ on G. We will show that this leads to a contradiction. If $U_\sigma = 0$ on G then clearly $G \cap F = \emptyset$. However, on F^c we have that L_t tends to zero and hence $\frac{1}{L_t}$ tends to ∞. We know that $\frac{1}{L_t} - 1$ can be obtained as a stochastic integral with respect to S. We take the stopping time $\mu = \infty$ on G^c and equal to $\inf\{t \mid L_t \leq \frac{1}{2}L_\sigma\}$ on the set G. The outcome

$$\mathbf{1}_G = \left(\frac{1}{L_\mu} - \frac{1}{L_\sigma}\right) L_\sigma \mathbf{1}_G$$

is the result of a 1-admissible strategy and clearly produces arbitrage. We may therefore suppose that $\mathbf{P}[G \cap F] > 0$ and hence we also have $\int_G U_\sigma > 0$. Again this suffices to show that $U_\sigma > 0$ on the set $\{L_\sigma > 0\}$ and again implies that $T \leq \nu$. The proof of the lemma is complete now. $\square$

Proof of the Main Theorem 12.4.2. We now calculate the decomposition of the continuous semi-martingale S under $\mathbf{P}_F$. If $S = M + A$ is the Doob-Meyer decomposition of S under $\mathbf{P}$ then, under $\mathbf{P}_F$ we write $S = \widetilde{M} + \widetilde{A}$ where $\widetilde{A}_t = A_t + \int \frac{d\langle M, U\rangle_s}{U_s}$, see [L 77]. This integral exists for the measure $\mathbf{P}_F$ since on F the process U is bounded away from 0. A more explicit formula for $\widetilde{A}$ can be found if we use the structure of $\langle M, U\rangle$. We thereto use the Kunita-Watanabe decomposition of the L^2-martingale U with respect to the martingale M. This is done in the following way (see [J 79]). The space of all L^2-martingales of the form $\alpha \cdot M$ is a stable space and in fact we have $\|(\alpha \cdot M)_\infty\|_2 = \mathbf{E}[\int \alpha' \, d\langle M, M\rangle \alpha]$. The orthogonal projection of U_∞ on this space is given by $(\beta \cdot M)_\infty$ for some predictable process β, where of course

$$\mathbf{E}\left[\int \beta' \, d\langle M, M\rangle \beta\right] < \infty.$$

In this notation we may write:

$$d\langle M, U\rangle = d\langle M, M\rangle \beta.$$

It follows that also $\int \beta' \, d\langle M, M\rangle \beta < \infty$ a.s. for the measure $\mathbf{P}_F$ and the measure $d\widetilde{A}$ can be written as

$$d\widetilde{A} = d\langle M, M\rangle \left(h_t + \frac{\beta_t}{U_t}\right) = d\langle M, M\rangle k_t.$$

Here we have put $k = h + \frac{\beta}{U}$ to simplify notation.

To prove the *(NFLVR)* property for S under $\mathbf{P}_F$ we use the criterion of Theorem 12.1.3 above.

Step 1: the set of 1-admissible integrands for $\mathbf{P}_F$ is bounded in $L^0(F)$. From the properties of β and h we deduce that, for the measure, $\mathbf{P}_F$, the integral

$$\int_0^\infty k_t' d\langle M, M\rangle k_t < \infty \qquad \mathbf{P}_F\text{-a.s.}.$$

The $\mathbf{P}_F$-local martingale $\widetilde{L}$ is now defined as

$$\widetilde{L}_t = \exp\left(-\int_0^t k_u' \, d\widetilde{M}_u - \frac{1}{2}\int_0^t k_u' \, d\langle M, M\rangle_u k_u\right).$$

It follows that
$$\widetilde{L}_\infty > 0 \qquad \mathbf{P}_F\text{-a.s.}.$$

It is chosen in such a way that $\widetilde{L}S$ is a $\mathbf{P}_F$-local martingale and therefore the set $\widetilde{\mathcal{K}}_1$ constructed with the 1-admissible, with respect to $\mathbf{P}_F$, integrands, is bounded in $L^0(\mathbf{P}_F)$.

In particular this also excludes the possibility of immediate arbitrage for S with respect to $\mathbf{P}_F$.

Step 2: S satisfies *(NA)* with respect to $\mathbf{P}_F$ (and with respect to general admissible integrands).

Since by step 1 immediate arbitrage is excluded, the violation of the *(NA)* property would, by Lemma 12.3.1, give us a predictable integrand H such that for $\mathbf{P}_F$ the integrand is of finite support, is S-integrable and 1-admissible. When the support of H is contained in $]\!]\sigma_1, \sigma_2]\!]$ it gives an outcome at least ε on the set $\{\sigma_1 < \infty\}$. All this, of course, with respect to $\mathbf{P}_F$.

The rest of the proof is devoted to the transformation of this phenomenon to a situation valid for $\mathbf{P}$.

Without loss of generality we may suppose that for the measure $\mathbf{P}$ we have $\sigma_1 \leq \sigma_2 \leq T$, we replace, e.g., the stopping time σ_2 by $\max(\sigma_1, \sigma_2)$ and then we replace σ_1 and σ_2 by, respectively, $\min(T, \sigma_1)$ and $\min(T, \sigma_2)$. All these substitutions have no effect when seen under the measure $\mathbf{P}_F$. Since $\mathbf{P}_F[\{\sigma_1 < \sigma_2 < \infty\}] > 0$, we certainly have that $\mathbf{P}[\{\sigma_1 < \sigma_2 < T\}] > 0$.

Roughly speaking we will now use the strategy H to construct arbitrage on the set F and we use the process $\frac{1}{L}$ to construct a sure win on the set F^c, as on the interval $[\![0, T[\![$, the process $\frac{1}{L} - 1$ equals $K \cdot S$ for a well-chosen integrand K. When we add the two integrands, H and K, we should obtain an integrand that gives arbitrage on Ω with respect to $\mathbf{P}$ and this will provide the desired contradiction.

Let the sequence of stopping times τ_n be defined as

$$\tau_n = \inf\left\{ t \,\middle|\, L_t \leq \frac{1}{n} \right\}.$$

We have that $\tau_n \nearrow T$ for $\mathbf{P}$ and $\tau_n \nearrow \infty$ for the measure $\mathbf{P}_F$. Since we have that $L_{\tau_n} > 0$ a.s., we also have that $U_{\tau_n} > 0$ a.s.. It follows that on the σ-algebra $\mathcal{F}_{\tau_n}$ the two measures, $\mathbf{P}$ and $\mathbf{P}_F$ are equivalent. We can therefore conclude that for each n the integrand $H\mathbf{1}_{[\![0,\tau_n]\!]}$ as well as the integrand $K\mathbf{1}_{[\![0,\tau_n]\!]}$ is S-integrable and 1-admissible for $\mathbf{P}$. The last integrand still has to be renormalised.

In fact on the set F itself, the lower bound -1 for the process $K \cdot S$ is too low since it will be compensated at most by ε. We therefore transform K in such a way that it will stay above $\frac{\varepsilon}{2}$ but will nevertheless give outcomes that are very big on the set F^c. Let us define

$$\widetilde{K} = K\mathbf{1}_{\{\sigma_1 < T\}}\frac{\varepsilon}{2}L_{\sigma_1},$$

$$\widetilde{K}^n = \widetilde{K}\,\mathbf{1}_{[\![0,\tau_n]\!]},$$

$$\widetilde{H} = H\mathbf{1}_{\{\sigma_1 < T\}},$$
$$\widetilde{H}^n = \widetilde{H}\,\mathbf{1}_{[\![0,\tau_n]\!]}.$$

From the preceding considerations it follows that the integrands $\widetilde{H}^n$ are all 1-admissible for $\mathbf{P}$ and that the integrands $\widetilde{K}^n$ are $\frac{\varepsilon}{2}$-admissible for $\mathbf{P}$. The outcomes $(\widetilde{K}^n \cdot S)_{\tau_n}$ tend to ∞ on $F^c \cap \{\sigma_1 < T\}$, and the outcomes $(\widetilde{H}^n \cdot S)_{\tau_n}$ become larger than ε on the set $F \cap \{\sigma_1 < T\}$. When we add them we see that on the set $\{\sigma_1 < T\}$ we have

$$\liminf_{n\to\infty}((\widetilde{H} + \widetilde{K}) \cdot S)_{\tau_n} = \liminf_{n\to\infty}((\widetilde{H}^n + \widetilde{K}^n) \cdot S)_{\tau_n} \geq \frac{\varepsilon}{2}.$$

Define now the stopping time μ as

$$\mu = \tau_n \text{ if } n \text{ is the first number such that } ((\widetilde{H}^n + \widetilde{K}^n) \cdot S)_{\tau_n} \geq \frac{\varepsilon}{4}.$$

The stopping time μ is finite on the set $\{\sigma_1 < \tau\}$. The integrand $J = (\widetilde{H} + \widetilde{K})\mathbf{1}_{[\![0,\mu]\!]}$ is now S-integrable and is certainly $1+\frac{\varepsilon}{2}$ admissible. By the definition of the stopping time μ we have that $(J \cdot S)_\mu \geq \frac{\varepsilon}{4}\mathbf{1}_{\{\sigma_1 < T\}}$, producing arbitrage. Since the process S satisfied the *(NA)* property, we arrived at a contradiction.

Step 2 is therefore completed and this ends the proof of the theorem. $\qquad \square$

Acknowledgement

The results of this paper were presented at the seminar at Tokyo University in summer 1993. We thank Professor Kotani and Professor Kusuoka for the invitation and for discussions on this topic. We also thank W. Brannath for discussions on the current proof of Theorem 12.3.7, and an anonymous referee for valuable suggestions.

13

The Banach Space of Workable Contingent Claims in Arbitrage Theory (1997)

Abstract. For a locally bounded local martingale S, we investigate the vector space generated by the convex cone of maximal admissible contingent claims. By a maximal contingent claim we mean a random variable $(H \cdot S)_\infty$, obtained as a final result of applying the admissible trading strategy H to a price process S and which is optimal in the sense that it cannot be dominated by another admissible trading strategy. We show that there is a natural, measure-independent, norm on this space and we give applications in Mathematical Finance.

Résumé. Si S est une martingale locale, localement bornée, on étudie l'espace vectoriel engendré par le cône des actifs contingents maximaux. Une variable aléatoire est un actif contingent maximal si elle peut s'écrire sous la forme $(H \cdot S)_\infty$, où la stratégie H est admissible et optimale dans le sens qu'elle n'est pas dominée par une autre stratégie admissible. Sur cet espace, on introduit une norme naturelle, invariante par changement de mesure, et on donne des applications en finance mathématique.

13.1 Introduction

A basic problem in Mathematical Finance is to see under what conditions the price of an asset, e.g. an option, is given by the expectation with respect to a so-called risk neutral measure. The existence of such a measure follows from no-arbitrage properties on the price process S of given assets, see [HK 79], [HP 81], [K 81] for the first papers on the topic and see [DS 94] (Chap. 9 above) for a general form of this theory and for references to earlier papers.

Investment strategies H are described by S-integrable predictable processes and the outcome of the strategy is described by the value at infinity $(H \cdot S)_\infty$. In order to avoid doubling strategies one has to introduce lower bounds on the losses incurred by the economic agent. Mathematically this is

[DS 97] The Banach Space of Workable Contingent Claims in Arbitrage Theory. *Annales de l'IHP*, vol. 33, no. 1, pp. 114–144, (1997).

translated by the property that $H \cdot S$ is bounded below by some constant. In this case we say that H is admissible, see [HP 81]. It turns out that for some admissible strategies H the contingent claim $(H \cdot S)_\infty$ is not optimal in the sense that it is dominated by the outcome of another admissible strategy K. In this case there is no reason for the economic agent to follow the strategy H since at the end she can do better by following K. Let us say that H is maximal if the contingent claim $(H \cdot S)_\infty$ cannot be dominated by another outcome of an admissible strategy K in the sense that $(H \cdot S)_\infty \leq (K \cdot S)_\infty$ a.s. but $\mathbf{P}[(H \cdot S)_\infty < (K \cdot S)_\infty] > 0$.

In Chaps. 9 and 11 we have used such maximal contingent claims in order to show that under the condition of no free lunch with vanishing risk, a locally bounded semi-martingale S admits an equivalent local martingale measure. In Chap. 11 we encountered a close relation between the existence of a martingale measure (not just a local martingale measure) for the process $H \cdot S$ and the maximality of the contingent claim $(H \cdot S)_\infty$. These results generalised results previously obtained by Ansel-Stricker [AS 94] and Jacka [J 92]. We related this connection to a characterisation of good numéraires and to the hedging problem.

In this paper we show that the set of maximal contingent claims forms a convex cone in the space $L^0(\Omega, \mathcal{F}, \mathbf{P})$ of measurable functions and that the vector space generated by this cone can be characterised as the set of contingent claims of what we might call workable strategies. The vector space of these contingent claims, will be denoted by $\mathcal{G}$. It carries a natural norm for which it becomes a Banach space. These properties solve some arbitrage problems when constructing multi-currency models. We refer to a paper of the first named author with Shirakawa on this subject, [DSh 96].

The paper is organised as follows. The rest of this introduction is devoted to the basic notations and assumptions. Sect. 13.2 deals with the concept of acceptable contingent claims and it is shown that the set of maximal admissible contingent claims forms a convex cone. In Sect. 13.3 we introduce the vector space spanned by the maximal admissible contingent claims and we show that there is a natural norm on it. The norm can also be interpreted as the maximal price that one is willing to pay for the absolute value of the contingent claim. Sect. 13.4 gives some results that are related to the geometry of the Banach space $\mathcal{G}$. In the complete market case it is an L^1-space, but we also give an example showing that it can be isomorphic to an L^∞-space. The precise interpretation of these properties in mathematical finance remains a challenging task. In Sect. 13.5 we show that for a given maximal admissible contingent claim f, the set of equivalent local martingale measures $\mathbf{Q}$ such that $\mathbf{E}_\mathbf{Q}[f] = 0$ forms a dense subset in the set of all absolutely continuous local martingale measures. That not all equivalent local martingale measures $\mathbf{Q}$ satisfy the equality $\mathbf{E}_\mathbf{Q}[f] = 0$, is illustrated by a counter-example. The main theorem in Sect. 13.6 states that in a certain way the space of workable contingent claims is invariant for numéraire changes. In Sect. 13.7 we use finitely additive measures in order to describe the closure of the space of bounded workable contingent claims.

Part of the results were obtained when the first named author was visiting the University of Tsukuba in January 1994 and when the second author was visiting the University of Tokyo in January 1995. Discussions with Professor Kusuoka and Professor Shirakawa are gratefully acknowledged.

The setup in this paper is the usual setup in mathematical finance. A probability space $(\Omega, \mathcal{F}, \mathbf{P})$ with a filtration $(\mathcal{F}_t)_{0 \le t}$ is given. The time set is supposed to be $\mathbb{R}_+$, the other cases, e.g. finite time interval or discrete time set, can easily be imbedded in our more general approach. The filtration is assumed to satisfy the "usual conditions", i.e. it is right continuous and $\mathcal{F}_0$ contains all null sets of $\mathcal{F}$.

A price process S, describing the evolution of the discounted price of d assets, is defined on $\mathbb{R}_+ \times \Omega$ and takes values in $\mathbb{R}^d$. We assume that the process S is locally bounded, e.g. continuous. As shown under a wide range of hypothesis, the assumption that S is a semi-martingale follows from arbitrage considerations, see Chap. 9 and references given there. We will therefore assume that the process S is a locally bounded semi-martingale. In order to avoid cumbersome notation and definitions, we will always suppose that measures are absolutely continuous with respect to $\mathbf{P}$. Stochastic integration is used to describe outcomes of investment strategies. When dealing with more dimensional processes it is understood that vector stochastic integration is used. We refer to Protter [P 90] and Jacod [J 79] for details on these matters.

Definition 13.1.1. *An $\mathbb{R}^d$-valued predictable process H is called a-admissible if it is S-integrable, if $H_0 = 0$, if the stochastic integral satisfies $H \cdot S \ge -a$ and if $(H \cdot S)_\infty = \lim_{t \to \infty}(H \cdot S)_t$ exists a.s.. A predictable process H is called admissible if it is a-admissible for some a.*

Remark 13.1.2. We explicitly required that $H_0 = 0$ in order to avoid the contribution of the integral at zero.

The following notations will be used:

$$\mathcal{K} = \{(H \cdot S)_\infty \mid H \text{ is admissible}\}$$
$$\mathcal{K}_a = \{(H \cdot S)_\infty \mid H \text{ is a-admissible}\}$$
$$\mathcal{C}_0 = \mathcal{K} - L_+^0$$
$$\mathcal{C} = \mathcal{C}_0 \cap L^\infty.$$

The basic Theorem 9.1.1 above uses the concept of no free lunch with vanishing risk, *(NFLVR)* for short. This is a rather weak hypothesis of no-arbitrage type and it is stated in terms of L^∞-convergence. The *(NFLVR)* property is therefore independent of the choice of the underlying probability measure, i.e. it does not change if we replace $\mathbf{P}$ by an equivalent probability measure $\mathbf{Q}$. Only the class of negligible sets comes into play. We also recall the definition of the property of no-arbitrage, *(NA)* for short.

Definition 13.1.3. *The locally bounded semi-martingale S satisfies the no-arbitrage or (NA) property if*

$$\mathcal{C} \cap L_+^\infty = \{0\}.$$

Definition 13.1.4. *We say that the locally bounded semi-martingale S satisfies the no free lunch with vanishing risk or (NFLVR) property if*

$$\overline{\mathcal{C}} \cap L_+^\infty = \{0\},$$

where the bar denotes the closure in the sup-norm topology of L^∞.

The fundamental theorem of asset pricing, as in Chap. 9, can now be formulated as follows:

Theorem 13.1.5. *The locally bounded semi-martingale S satisfies the property (NFLVR) if and only if there is an equivalent probability measure $\mathbf{Q}$ such that S is a $\mathbf{Q}$-local martingale. In this case the set $\mathcal{C}$ is already weak-star (i.e. $\sigma(L^\infty, L^1)$) closed in L^∞.*

Remark 13.1.6. If $\mathbf{Q}$ is an equivalent local martingale measure for S and if the integrand or strategy H satisfies $H \cdot S \geq -a$, i.e. H is a-admissible, then by a result of Émery [E 80] and Ansel-Stricker [AS 94], the process $H \cdot S$ is still a local martingale and hence, being bounded below, is a super-martingale. It follows that the limit $(H \cdot S)_\infty$ exists a.s. and that $\mathbf{E}_\mathbf{Q}[(H \cdot S)_\infty] \leq 0$.

We also need the following equivalent reformulations of the property of no free lunch with vanishing risk, see Chap. 9 for more details.

Theorem 13.1.7. *The locally bounded semi-martingale S satisfies the no free lunch with vanishing risk property or (NFLVR) if for any sequence of S-integrable strategies $(H_n, \delta_n)_{n \geq 1}$ such that each H_n is a δ_n-admissible strategy and where δ_n tends to zero, we have that $(H \cdot S)_\infty$ tends to zero in probability $\mathbf{P}$.*

Theorem 13.1.8. *The locally bounded semi-martingale S satisfies the property (NFLVR) if and only if*

(1) *it satisfies the property (NA)*
(2) *$\mathcal{K}_1$ is bounded in L^0, for the topology of convergence in measure.*

Theorem 13.1.9. *The locally bounded semi-martingale S satisfies the property (NFLVR) if and only if*

(1) *it satisfies the property (NA)*
(2) *There is a strictly positive local martingale L, $L_0 = 1$, such that at infinity $L_\infty > 0$, $\mathbf{P}$-a.s. and such that LS is a local martingale.*

We suppose from now on that the process S is a fixed d-dimensional locally bounded semi-martingale and that it satisfies the property *(NFLVR)*. The set of local martingale measures is therefore, according to the previous theorems not empty. In Sect. 13.7, we will also make use of finitely additive measures. So we let $ba\,(\Omega, \mathcal{F}, \mathbf{P})$ be the Banach space of all finitely additive measures that are absolutely continuous with respect to $\mathbf{P}$, i.e. $ba\,(\Omega, \mathcal{F}, \mathbf{P})$ is the dual of $L^\infty\,(\Omega, \mathcal{F}, \mathbf{P})$. We will use Roman letters $\mathbf{P}, \mathbf{Q}, \mathbf{Q}^0, \ldots$ for σ-additive measures and Greek letters for elements of ba which are not necessarily σ-additive. We say that a finitely additive measure μ is absolutely continuous with respect to the probability measure $\mathbf{P}$ if $\mathbf{P}[A] = 0$ implies $\mu[A] = 0$ for any set $A \in \mathcal{F}$.

Let us put:

$$\mathcal{M}^e = \left\{ \mathbf{Q} \,\middle|\, \begin{array}{l} \mathbf{Q} \text{ is equivalent to } \mathbf{P} \\ \text{and the process } S \text{ is a } \mathbf{Q}\text{-local martingale} \end{array} \right\}$$

$$\mathcal{M} = \left\{ \mathbf{Q} \,\middle|\, \begin{array}{l} \mathbf{Q} \text{ is absolutely continuous with respect to } \mathbf{P} \\ \text{and the process } S \text{ is a } \mathbf{Q}\text{-local martingale} \end{array} \right\}$$

$$\mathcal{M}^{ba} = \left\{ \mu \,\middle|\, \begin{array}{l} \mu \text{ is in } ba\,(\Omega, \mathcal{F}_\infty, \mathbf{P}) \\ \text{and for every element } h \in \mathcal{C}: \ \mathbf{E}_\mu[h] \leq 0 \end{array} \right\}$$

We identify, as usual, absolutely continuous measures with their Radon-Nikodým derivatives. It is clear that, under the hypothesis *(NFLVR)*, the set $\mathcal{M}^e(\mathbf{P})$ is dense in $\mathcal{M}(\mathbf{P})$ for the norm of $L^1(\Omega, \mathcal{F}, \mathbf{P})$. This density together with Fatou's lemma imply that for random variables g that are bounded below we have the equality

$$\sup\{\mathbf{E}_\mathbf{Q}[g] \mid \mathbf{Q} \in \mathcal{M}^e\} = \sup\{\mathbf{E}_\mathbf{Q}[g] \mid \mathbf{Q} \in \mathcal{M}\}.$$

We will use this equality freely.

As shown in Remark 9.5.10 the set $\mathcal{M}^e$ is weak-star-dense, i.e. for the topology $\sigma\,(ba, L^\infty)$, in the set $\mathcal{M}^{ba}$.

The first two sets are sets of σ-additive measures, the third set is a set of finitely additive measures. Clearly $\mathcal{M}^e \subset \mathcal{M} \subset \mathcal{M}^{ba}$ and since S is locally bounded the set $\mathcal{M}$ is closed in $L^1\,(\Omega, \mathcal{F}, \mathbf{P})$. If needed we will add the process S in parenthesis, e.g. $\mathcal{M}^e(S)$, to make clear that we are dealing with a set of local martingale measures for the process S.

13.2 Maximal Admissible Contingent Claims

We now give the definition of a maximal admissible contingent claim and its relation to the existence of an equivalent martingale measure. As mentioned above we always suppose that S is a d-dimensional locally bounded semi-martingale that satisfies the *(NFLVR)* property.

Definition 13.2.1. *If $\mathcal{U}$ is a non-empty subset of L^0, then we say that a contingent claim $f \in \mathcal{U}$ is maximal in $\mathcal{U}$, if the properties $g \geq f$ a.s. and $g \in \mathcal{U}$ imply that $g = f$ a.s..*

The *(NA)* property can be rephrased as the property that 0 is maximal in $\mathcal{K}$. It is clear that if S satisfies the no-arbitrage property, then the fact that f is maximal in $\mathcal{K}_a$ already implies that f is maximal in $\mathcal{K}$. Indeed if $g = (H \cdot S)_\infty \in \mathcal{K}$ and $g \geq f$ a.s., then $g \geq -a$. From Proposition 9.3.6 it then follows that g is a-admissible and hence the maximality of f in $\mathcal{K}_a$ implies that $g = f$ a.s..

Definition 13.2.2. *A maximal admissible contingent claim is a maximal element of $\mathcal{K}$. The set of maximal admissible contingent claims is denoted by $\mathcal{K}^{\max}$. The set of maximal a-admissible contingent claims is denoted by $\mathcal{K}_a^{\max}$.*

The proof of the Theorem 13.1.5 uses the following intermediate results, see Sect. 9.4:

Theorem 13.2.3. *If S is a locally bounded semi-martingale and if $(f_n)_{n \geq 1}$ is a sequence in $\mathcal{K}_1$, then*

(1) *there is a sequence of convex combinations $g_n \in \mathrm{conv}\{f_n, f_{n+1}, \ldots\}$ such that g_n tends in probability to a function g, taking finite values a.s.,*
(2) *there is a maximal contingent claim h in $\mathcal{K}_1$ such that $h \geq g$ a.s..*

Corollary 13.2.4. *Under the hypothesis of Theorem 13.2.3, maximal contingent claims of the closure L^0-closure $\overline{\mathcal{K}}_1$ of $\mathcal{K}_1$, are already in $\mathcal{K}_1$. By L^0-closure we mean the closure with respect to convergence in measure.*

Using a change of numéraire technique, the following result was proved in Chap. 11. We refer also to Ansel-Stricker [AS 94] for an earlier proof of the equivalence of (2) and (3) below.

Theorem 13.2.5. *If S is a locally bounded semi-martingale that satisfies the (NFLVR) property then for a contingent claim $f \in \mathcal{K}$ the following are equivalent*

(1) *f is maximal admissible,*
(2) *there is an equivalent local martingale measure $\mathbf{Q} \in \mathcal{M}^e$ such that $\mathbf{E}_\mathbf{Q}[f] = 0$,*
(3) *if $f = (H \cdot S)_\infty$ for some admissible strategy H, then $H \cdot S$ is a uniformly integrable martingale for some $\mathbf{Q} \in \mathcal{M}^e$.*

Corollary 13.2.6. *Suppose that the hypothesis of theorem 13.2.5 is valid. If f is maximal admissible and $f = (H \cdot S)_\infty$ for some admissible strategy H, then for every stopping time T, the contingent claim $(H \cdot S)_T$ is also maximal.*

Proof. If f is maximal and $f = (H \cdot S)_\infty$ where H is a-admissible, then there is $\mathbf{Q} \in \mathcal{M}^e$ such that $\mathbf{E}_\mathbf{Q}[f] = 0$, i.e. $\mathbf{E}_\mathbf{Q}[(H \cdot S)_\infty] = 0$. Because H is admissible, the process $H \cdot S$ is, see [AS 94], a $\mathbf{Q}$-local martingale and hence a $\mathbf{Q}$-super-martingale. Because $\mathbf{E}_\mathbf{Q}[(H \cdot S)_\infty] = 0$, we necessarily have that $H \cdot S$ is a $\mathbf{Q}$-uniformly integrable martingale. It follows that $\mathbf{E}_\mathbf{Q}[(H \cdot S)_T] = 0$ and consequently $(H \cdot S)_T$ is maximal. $\qquad\square$

Remark 13.2.7. The corollary also shows that if $f = (H \cdot S)_\infty$ is maximal admissible, then the strategy that produces f is uniquely determined in the sense that any other admissible strategy K that produces f necessarily satisfies $H \cdot S = K \cdot S$. The following definitions therefore make sense.

Definition 13.2.8. *If H is an admissible strategy such that $f = (H \cdot S)_\infty$ is a maximal admissible contingent claim, then we say that H is a maximal admissible strategy.*

Definition 13.2.9. *We say that a strategy K is acceptable if there is a positive number a and a maximal admissible strategy L such that $(K \cdot S) \geq -(a + (L \cdot S))$.*

Remark 13.2.10. If we take a big enough, the process $V = a + L \cdot S$ stays bounded away from zero and can be used as a new numéraire. Under this new currency unit, the process $K \cdot S$, where K is acceptable, has to be replaced by the process $\frac{K \cdot S}{V}$. The latter process is a stochastic integral with respect to the process $\left(\frac{S}{V}, \frac{1}{V}\right)$, more precisely, see Chap. 11 for the details of this calculation, $\frac{K \cdot S}{V} = (K, (K \cdot S)_- - KS_-) \cdot \left(\frac{S}{V}, \frac{1}{V}\right) = K' \cdot \left(\frac{S}{V}, \frac{1}{V}\right)$ remains bigger than a constant, i.e. the strategy $K' = (K, (K \cdot S)_- - KS_-)$ is admissible. Another way of saying that K is acceptable, is to say that K' is admissible in a new numéraire. In Chap. 11 we proved that the only numéraires that do not destroy the no-arbitrage properties are the numéraires given by maximal strategies. The definition of acceptable strategies is therefore very natural. The outcomes of acceptable strategies are the numéraire invariant version of the outcomes of admissible strategies.

Lemma 13.2.11. *If S is a locally bounded semi-martingale that satisfies the (NFLVR) property and if K is acceptable then $\lim_{t \to \infty} (K \cdot S)_t$ exists a.s..*

Proof. Suppose that $K \cdot S \geq -(a + L \cdot S)$ where L is admissible and maximal. Clearly we have that $K + L$ is a-admissible and hence by the results of Chap. 9 $\lim_{t \to \infty} ((K + L) \cdot S)_t$ exists a.s.. Because $\lim_{t \to \infty} (L \cdot S)_t$ exists a.s., we necessarily have that $\lim_{t \to \infty} (K \cdot S)_t$ also exists a.s.. □

The set of outcomes of acceptable strategies, which is a convex cone in L^0, is denoted by

$$\mathcal{J} = \{(K \cdot S)_\infty \mid K \text{ acceptable}\}.$$

We now prove some elementary properties of acceptable contingent claims. Most of these properties are generalisations of no-arbitrage concepts for admissible contingent claims.

Proposition 13.2.12. *Suppose that S is a locally bounded semi-martingale that satisfies the (NFLVR) property. If K is acceptable and if $(K \cdot S)_\infty \geq 0$, then $(K \cdot S)_\infty = 0$.*

Proof. Suppose that $K \cdot S \geq -(a + L \cdot S)$ where L is admissible and maximal. Clearly we have that $K + L$ is a-admissible. But at infinity we have that $((K + L) \cdot S)_\infty \geq (L \cdot S)_\infty$ and by maximality of L we obtain the equality $((K + L) \cdot S)_\infty = (L \cdot S)_\infty$, which is equivalent to $(K \cdot S)_\infty = 0$ a.s.. $\qquad \square$

In the same way we prove the subsequent result.

Proposition 13.2.13. *Suppose that S is a locally bounded semi-martingale that satisfies the (NFLVR) property. If K is acceptable and $(K \cdot S)_\infty \geq -c$ for some positive real constant c, then the strategy K is already c-admissible.*

Proof. Take $\varepsilon > 0$ and let

$$T_1 = \inf\{t \mid (K \cdot S)_t < -c - \varepsilon\}.$$

We then define

$$T_2 = \inf\{t > T_1 \mid (K \cdot S)_t \geq -c\}.$$

By assumption we have that on $\{T_1 < \infty\}$ the strategy $K\mathbf{1}_{\rrbracket T_1, T_2 \rrbracket}$ produces an outcome $(K \cdot S)_{T_2} - (K \cdot S)_{T_2} \geq \varepsilon$. This strategy is easily seen to be acceptable. Indeed

$$(K\mathbf{1}_{\rrbracket T_1, T_2 \rrbracket}) \cdot S \geq c + \varepsilon + (-a - H \cdot S)$$

for some real number a and some maximal strategy H. By the previous lemma we necessarily have that the contingent claim is zero a.s. and hence $T_1 = \infty$ a.s.. $\qquad \square$

We now turn again to the analysis of maximal admissible contingent claims.

Theorem 13.2.14. *If S is a locally bounded semi-martingale that satisfies the (NFLVR) property, if f and g are maximal admissible contingent claims, then $f + g$ is also a maximal contingent claim. It follows that the set $\mathcal{K}^{\max}$ of maximal contingent claims is a convex cone.*

Proof. Let $f = (H^1 \cdot S)_\infty$ and $g = (H^2 \cdot S)_\infty$, where H^1 and H^2 are maximal strategies and are respectively a_1- and a_2-admissible. Suppose that K is a k-admissible strategy such that $(K \cdot S)_\infty \geq f + g$. From the inequalities $(K - H^2) \cdot S = K \cdot S - H^2 \cdot S \geq -k - H^2 \cdot S$, it follows that $K - H^2$ is acceptable. Since also $((K - H^2) \cdot S)_\infty \geq f \geq -a_1$, the Proposition 13.2.13 shows that $K - H^2$ is a_1-admissible. Because f was maximal we have that $((K - H^2) \cdot S)_\infty = f$ and hence we have that $(K \cdot S)_\infty = f + g$. This shows that $f + g$ is maximal. Since the set $\mathcal{K}^{\max}$ is clearly closed under multiplication with positive scalars, it follows that it is a convex cone. $\qquad \square$

Corollary 13.2.15. *If S is a locally bounded semi-martingale that satisfies the (NFLVR) property and if $(f_n)_{1 \leq n \leq N}$ is a finite sequence of contingent claims in $\mathcal{K}$ such that for each n there is an equivalent risk neutral measure $\mathbf{Q}^n \in \mathcal{M}^e$ with $\mathbf{E}_{\mathbf{Q}^n}[f_n] = 0$, then there is an equivalent risk neutral measure $\mathbf{Q} \in \mathcal{M}^e$ such that $\mathbf{E}_{\mathbf{Q}}[f_n] = 0$ for each $n \leq N$.*

Proof. This is a rephrasing of the Theorem 13.2.14 since by Theorem 13.2.5, the condition on the existence of an equivalent risk neutral measure is equivalent with the maximality property.   □

The previous Corollary 13.2.15 will be generalised to sequences (see Corollary 13.2.18 below). We first prove the following Proposition.

Proposition 13.2.16. *Suppose that S is a locally bounded semi-martingale that satisfies the (NFLVR) property. If $(f_n)_{n \geq 1}$ is a sequence in $\mathcal{K}_1^{\max}$, such that*

(1) *The sequence $f_n \to f$ in probability*
(2) *for all n we have $f - f_n \geq -\delta_n$ where δ_n is a sequence of strictly positive numbers tending to zero,*

then f is in $\mathcal{K}^{\max}$ too, i.e. it is maximal admissible.

Proof. If g is a maximal contingent claim such that $g \geq f$, then we have $g - f_n \geq -\delta_n$. Since each f_n is maximal we find that $g - f_n$ is acceptable and hence δ_n-admissible by Proposition 13.2.13. Since δ_n tends to zero, we find that the *(NFLVR)* property implies that $g - f_n$ tends to zero in probability. This means that $g = f$ and hence f is maximal.   □

Corollary 13.2.17. *If S is a locally bounded semi-martingale that satisfies the (NFLVR) property, if $(a_n)_{n \geq 1}$ is a sequence of strictly positive real numbers such that*

$$\sum_{n=1}^{\infty} a_n < \infty \,,$$

if for each n, H^n is an a_n-admissible maximal strategy, then we have that the series

$$f = \sum_{n=1}^{\infty} (H^n \cdot S)_{\infty}$$

converges in probability to a maximal contingent claim.

Proof. Let $h_n = (H^n \cdot S)_{\infty}$, the partial sums $f_N = \sum_{n=1}^{N} h_n$ are outcomes of $\sum_{n=1}^{\infty} a_n$-admissible strategies. For an arbitrary element $\mathbf{Q} \in \mathcal{M}^e$ we have that

$$\mathbf{E_Q} \left[(h_n + a_n) \right] \leq a_n \,.$$

It follows that the series of positive functions $\sum_{n=1}^{\infty} (h_n + a_n)$ converges in $L^1(\mathbf{Q})$ and hence the series $\sum_{n=1}^{\infty} h_n$ also converges in $L^1(\mathbf{Q})$. The series $f = \sum_{n=1}^{\infty} h_n = \lim f_N$ therefore also converges to a contingent claim f in $\mathbf{P}$. From the Proposition 13.2.16, we now deduce that f is maximal.   □

Corollary 13.2.18. *If S is a locally bounded semi-martingale that satisfies the (NFLVR) property, if $(f_n)_{n\geq 1}$ is a sequence of contingent claims in $\mathcal{K}$ such that for each n there is an equivalent risk neutral measure $\mathbf{Q}^n \in \mathcal{M}^e$ with $\mathbf{E}_{\mathbf{Q}^n}[f_n] = 0$, then there is an equivalent risk neutral measure $\mathbf{Q} \in \mathcal{M}^e$ such that $\mathbf{E}_{\mathbf{Q}}[f_n] = 0$ for each $n \geq 1$.*

Proof. We may without loss of generality suppose that f_n is the result of an a_n-admissible and maximal strategy where the series $\sum_{n=1}^{\infty} a_n$ converges. If not we replace f_n by a suitable multiple $\lambda_n f_n$, with λ_n strictly positive and small enough. The Corollary 13.2.17 then shows that the sum $f = \sum_{n=1}^{\infty} f_n$ is still maximal and hence there is an element $\mathbf{Q} \in \mathcal{M}^e$ such that $\mathbf{E}_{\mathbf{Q}}[f] = 0$. As observed in the proof of the theorem, we have that the series $\sum_{n=1}^{\infty} f_n$ converges to f in $L^1(\mathbf{Q})$. For each n we already have that $\mathbf{E}_{\mathbf{Q}}[f_n] \leq 0$. From this it follows that for each n we need to have $\mathbf{E}_{\mathbf{Q}}[f_n] = 0$. $\square$

Corollary 13.2.19. *If S is a locally bounded semi-martingale that satisfies the (NFLVR) property, if $(f_n)_{n\geq 1}$ is a sequence of 1-admissible maximal contingent claims, if f is a random variable such that for each element $\mathbf{Q} \in \mathcal{M}^e$ we have $f_n \to f$ in $L^1(\mathbf{Q})$, then f is a 1-admissible maximal contingent claim.*

Proof. From Theorem 13.2.3 we deduce the existence of a maximal contingent claim g such that $g \geq f$. From the previous corollary we deduce the existence of an element $\mathbf{Q} \in \mathcal{M}^e$ such that for all n we have $\mathbf{E}_{\mathbf{Q}}[f_n] = 0$. It is straightforward to see that $\mathbf{E}_{\mathbf{Q}}[f] = 0$ and that $\mathbf{E}_{\mathbf{Q}}[g] \leq 0$. This can only be true if $f = g$, i.e. if f is 1-admissible and maximal. $\square$

We now extend the *no free lunch with vanishing risk*-property which was phrased in terms of admissible strategies, to the framework of acceptable strategies. As always it is assumed that S is locally bounded and satisfies *(NFLVR)*.

Theorem 13.2.20. *Suppose that S is a locally bounded semi-martingale that satisfies the (NFLVR) property. Let $f_n = (L^n \cdot S)_{\infty}$ be a sequence of outcomes of acceptable strategies such that $L^n \cdot S \geq -a_n - H^n \cdot S$, with H^n maximal and a_n-admissible. If $\lim_{n\to\infty} a_n = 0$, then $\lim_{n\to\infty} f_n = 0$ in probability $\mathbf{P}$.*

Proof. The strategies $H^n + L^n$ are a_n-admissible and by the *(NFLVR)* property of S we therefore have that $((H^n + L^n) \cdot S)_{\infty}$ tends to zero in probability $\mathbf{P}$. Because each H^n-admissible and $\lim_{n\to\infty} a_n = 0$ the *(NFLVR)* property of S implies that $(H^n \cdot S)_{\infty}$ tends to zero in probability $\mathbf{P}$. It follows that also $(L^n \cdot S)_{\infty}$ tends to zero in probability $\mathbf{P}$. $\square$

13.3 The Banach Space Generated by Maximal Contingent Claims

In this section we show that the subspace $\mathcal{G}$ of L^0, generated by the convex cone $\mathcal{K}^{\max}$ of maximal admissible contingent claims can be endowed with a natural norm. We start with a definition.

Definition 13.3.1. *A predictable process H is called workable if both H and $-H$ are acceptable.*

Proposition 13.3.2. *Suppose that S is a locally bounded semi-martingale that satisfies the (NFLVR) property. The vector space $\mathcal{G}$ or, if there is danger of confusion and the price process S is important, $\mathcal{G}(S)$, generated by the cone of maximal admissible contingent claims, satisfies*

$$\begin{aligned}
\mathcal{G} &= \mathcal{K}^{\max} - \mathcal{K}^{\max} \\
&= \{(H \cdot S)_\infty \mid H \text{ is workable}\} \\
&= \mathcal{J} \cap (-\mathcal{J}) .
\end{aligned}$$

Proof. The first statement is a trivial exercise in linear algebra. If H is workable then there are a real number a and maximal strategies L^1 and L^2 such that $-a - L^1 \cdot S \le H \cdot S \le a + L^2 \cdot S$. Take now $\mathbf{Q} \in \mathcal{M}^e$ such that both $L^1 \cdot S$ and $L^2 \cdot S$ are $\mathbf{Q}$-uniformly integrable martingales. The strategy $H + L^1$ is a-admissible and satisfies $(H + L^1) \cdot S \le a + (L^1 + L^2) \cdot S$. It follows that $(H + L^1) \cdot S$ is a $\mathbf{Q}$-uniformly integrable martingale, i.e. $(H + L^1)$ is a maximal strategy. Since $H = (H + L^1) - L^1$ we obtain that $(H \cdot S)_\infty \in (\mathcal{K}^{\max} - \mathcal{K}^{\max})$. If conversely $H = H^1 - H^2$, where both terms are maximal, then we have to show that H is workable. This is quite obvious, indeed if H^1 is a-admissible we have that $H \cdot S \ge -a - H^2 \cdot S$. A similar reasoning applies to $-H$. $\qquad \square$

Proposition 13.3.3. *Suppose that S is a locally bounded semi-martingale that satisfies the (NFLVR) property. If H is workable then there is an element $\mathbf{Q} \in \mathcal{M}^e$ such that the process $H \cdot S$ is a $\mathbf{Q}$-uniformly integrable martingale. Hence for every stopping time T, the random variable $(H \cdot S)_T$ is in $\mathcal{G}$. The process $H \cdot S$ is uniquely determined by $(H \cdot S)_\infty$.*

Proof. If H is workable then there are maximal admissible strategies K and K' such that $H = K - K'$. From Theorem 13.2.5 and Corollary 13.2.15 it follows that there is an equivalent local martingale measure $\mathbf{Q} \in \mathcal{M}^e$ such that both $K \cdot S$ and $K' \cdot S$ are $\mathbf{Q}$-uniformly integrable martingales. The rest is obvious. $\qquad \square$

Proposition 13.3.4. *Suppose that S is a locally bounded semi-martingale that satisfies the (NFLVR) property. If $g \in \mathcal{G}$ satisfies $\|g^-\|_\infty < \infty$, then $g \in \mathcal{K}^{\max}$.*

Proof. Put $L = (H^1 - H^2)$, where H^1 and H^2 are both maximal, and so that $g = (L \cdot S)_\infty$. Since L is acceptable and $(L \cdot S)_\infty \geq -\|g^-\|_\infty$ we find by proposition 13.2.13 that L is admissible. For a well-chosen element $\mathbf{Q} \in \mathcal{M}^e$, the process $L \cdot S$ is a uniformly integrable martingale and hence L is maximal. $\square$

Corollary 13.3.5. *Suppose that S is a locally bounded semi-martingale that satisfies the (NFLVR) property. If V and W are maximal admissible strategies, if $((V - W) \cdot S)_\infty$ is uniformly bounded from below, then $V - W$ is admissible and maximal.*

Corollary 13.3.6. *Suppose that S is a locally bounded semi-martingale that satisfies the (NFLVR) property. Bounded contingent claims in $\mathcal{G}$ are characterised as*

$$\mathcal{G}^\infty = \mathcal{G} \cap L^\infty = \mathcal{K}^{\max} \cap L^\infty$$
$$= \{(H \cdot S)_\infty \mid H \cdot S \text{ is bounded}\}.$$

Remark 13.3.7. The vector space $\mathcal{G}^\infty$ should not be mixed up with the cone $\mathcal{K} \cap L^\infty$. As shown in Chap. 9 and [DS 94a], the contingent claim -1 can be in $\mathcal{K}$ but by the no-arbitrage property, the contingent claim $+1$ cannot be in $\mathcal{K}$. The vector space $\mathcal{G}^\infty$ was used in the study of the convex set $\mathcal{M}(S)$, see Chap. 9, [AS 94] and [J 92].

Definition 13.3.8 (Notation). *We define the following norm on the space $\mathcal{G}$:*

$$\|g\| = \inf\{ a \mid g = (H^1 \cdot S)_\infty - (H^2 \cdot S)_\infty ,$$
$$H^1, H^2 \ a\text{-admissible and maximal}\} .$$

The norm on the space $\mathcal{G}$ is quite natural and is suggested by its definition. It is easy to verify that $\| . \|$ is indeed a norm. We will investigate the relation of this norm to other norms, e.g. L^∞ and L^1-norms.

Proposition 13.3.9. *Suppose that S is a locally bounded semi-martingale that satisfies the (NFLVR) property. If $g = (H \cdot S)_\infty$ where H is workable then for every stopping time T, $g_T = (H \cdot S)_T \in \mathcal{G}$ and $\|g_T\| \leq \|g\|$.*

Proof. Follows immediately from the definition and the proof of Corollary 13.2.6 above. $\square$

Proposition 13.3.10. *Suppose that S is a locally bounded semi-martingale that satisfies the (NFLVR) property. If $g \in \mathcal{G}^\infty$ then, as shown above, $g \in \mathcal{K}_{\|g^-\|_\infty}$ and $-g \in \mathcal{K}_{\|g^+\|_\infty}$. Hence*

$$\|g\| \leq \min(\|g^+\|_\infty, \|g^-\|_\infty) \leq \max(\|g^+\|_\infty, \|g^-\|_\infty) = \|g\|_\infty .$$

The following lemma is an easy exercise in integration theory and immediately gives the relation with the L^1-norm.

Proposition 13.3.11. *If $f \in L^1(\Omega, \mathcal{F}, \mathbf{Q})$ for some probability measure $\mathbf{Q}$, if $\mathbf{E}_\mathbf{Q}[f] = 0$, if $f = g - h$, where both $\mathbf{E}_\mathbf{Q}[g] \leq 0$ and $\mathbf{E}_\mathbf{Q}[h] \leq 0$, then*

$$\|f\|_{L^1(\mathbf{Q})} = 2\,\mathbf{E}_\mathbf{Q}[f^+] = 2\max(\mathbf{E}_\mathbf{Q}[f^+], \mathbf{E}_\mathbf{Q}[f^-])$$
$$\leq 2\max(\|g^-\|_\infty, \|h^-\|_\infty)\ .$$

Proof. The first line is obvious and shows that the obvious decomposition $f = (f^+ - \mathbf{E}[f^+]) - (f^- - \mathbf{E}[f^-])$ is best possible. So let us concentrate on the last line. If $f = g - h$ then we have the following inequalities:

$$f + \|g^-\|_\infty - \|h^-\|_\infty = g + \|g^-\|_\infty - (h + \|h^-\|_\infty)$$
$$\left(f + \|g^-\|_\infty - \|h^-\|_\infty\right)^+ \leq g + \|g^-\|_\infty$$
$$\left(f + \|g^-\|_\infty - \|h^-\|_\infty\right)^- \leq h + \|h^-\|_\infty\ .$$

These inequalities together with $\mathbf{E}_\mathbf{Q}[g] \leq 0$ and $\mathbf{E}_\mathbf{Q}[h] \leq 0$, imply that

$$\|f + \|g^-\|_\infty - \|h^-\|_\infty\|_{L^1(\mathbf{Q})} \leq \|g^-\|_\infty + \|h^-\|_\infty\ .$$

It is now easy to see that

$$\|f\|_{L^1(\mathbf{Q})} \leq \|g^-\|_\infty + \|h^-\|_\infty + \left|\|g^-\|_\infty - \|h^-\|_\infty\right|$$
$$\leq 2\max\left(\|g^-\|_\infty, \|h^-\|_\infty\right)\ . \qquad \square$$

Corollary 13.3.12. *Suppose that S is a locally bounded semi-martingale that satisfies the (NFLVR) property. If $g \in \mathcal{G}$ then*

$$2\|g\| \geq \sup\{\|g\|_{L^1(\mathbf{Q})} \mid \mathbf{Q} \in \mathcal{M}\}$$

Proof. Take $g = (H^1 \cdot S)_\infty - (H^2 \cdot S)_\infty \in \mathcal{G}$ where H^1 and H^2 are both maximal and a-admissible. For every $\mathbf{Q} \in \mathcal{M}$ we have that $\mathbf{E}_\mathbf{Q}[(H^1 \cdot S)_\infty] \leq 0$ and $\mathbf{E}_\mathbf{Q}[(H^2 \cdot S)_\infty] \leq 0$. The lemma shows that

$$\|g\|_{L^1(\mathbf{Q})} \leq 2\max(\|(H^1 \cdot S)_\infty\|_{L^1(\mathbf{Q})}, \|(H^2 \cdot S)_\infty\|_{L^1(\mathbf{Q})}) \leq 2a\ .$$

By taking the infimum over all decompositions and by taking the supremum over all elements in $\mathcal{M}$ we find the desired inequality. $\qquad \square$

The next theorem shows that in some sense there is an optimal decomposition. The proof relies on Theorem 13.2.3 above and on the technical Lemma 9.8.1.

Theorem 13.3.13. *Suppose that S is a locally bounded semi-martingale that satisfies the (NFLVR) property. If $g \in \mathcal{G}$ then there exist two $\|g\|$-admissible maximal strategies R and U such that $g = (R \cdot S)_\infty - (U \cdot S)_\infty$.*

Proof. Take a sequence of real numbers such that $a_n \searrow \|g\|$. For each n we take H^n and K^n maximal and a_n-admissible such that $g = (H^n \cdot S)_\infty - (K^n \cdot S)_\infty$. From the Theorem 13.2.3 cited above we deduce that there are convex combinations $V_n \in \text{conv}\{H^n, H^{n+1}, \ldots\}$ and $W_n \in \text{conv}\{K^n, K^{n+1}, \ldots\}$ such that $(V^n \cdot S)_\infty \to h$ and $(W^n \cdot S)_\infty \to k$. Clearly $g = h - k$, $h \geq -\|g\|$ and $k \geq -\|g\|$. However, at this stage we cannot assert that h and/or k are maximal. Theorem 13.2.3 above, however, allows us to find a maximal strategy R such that $(R \cdot S)_\infty \geq h \geq -\|g\|$. The strategy $R - H^1 + K^1$ is acceptable and satisfies

$$((R - H^1 + K^1) \cdot S)_\infty = (R \cdot S)_\infty - g \geq h - g = k \geq -\|g\| \,.$$

From the Proposition 13.2.13 above it follows that $U = R - H^1 + K^1$ is $\|g\|$-admissible and maximal. By definition of U and R we have that $g = (R \cdot S)_\infty - (U \cdot S)_\infty$. □

Corollary 13.3.14. *With the notation of the above Theorem 13.3.13:* $(R \cdot S)_\infty + \|g\| \geq g^+$ *and* $(U \cdot S)_\infty + \|g\| \geq g^-$. *Hence we find*

$$\sup\{\mathbf{E_Q}\left[g^+\right] \mid \mathbf{Q} \in \mathcal{M}\} \leq \|g\|$$
$$\sup\{\mathbf{E_Q}\left[g^-\right] \mid \mathbf{Q} \in \mathcal{M}\} \leq \|g\| \,.$$

Theorem 13.3.15. *Suppose that S is a locally bounded semi-martingale that satisfies the (NFLVR) property. If $g \in \mathcal{G}$ then*

$$\|g\| = \sup\{\mathbf{E_Q}\left[g^+\right] \mid \mathbf{Q} \in \mathcal{M}\} = \sup\{\mathbf{E_Q}\left[g^+\right] \mid \mathbf{Q} \in \mathcal{M}^e\}$$
$$= \sup\{\mathbf{E_Q}\left[g^-\right] \mid \mathbf{Q} \in \mathcal{M}\} = \sup\{\mathbf{E_Q}\left[g^-\right] \mid \mathbf{Q} \in \mathcal{M}^e\} \,.$$

Proof. Put $\beta = \sup\{\mathbf{E_Q}\left[g^+\right] \mid \mathbf{Q} \in \mathcal{M}\}$, where the random variable g is decomposed as $g = (H^1 \cdot S)_\infty - (H^2 \cdot S)_\infty$ with H^1 and H^2 maximal. From Corollary 11.3.5 to Theorem 11.3.4, we recall that there is a maximal strategy K^1 such that $g^+ \leq \beta + (K^1 \cdot S)_\infty$, implying that K^1 is β-admissible. The strategy $K^2 = K^1 - H^1 + H^2$ is also β-admissible and by Proposition 13.2.13 therefore maximal. Since $K^1 - K^2 = H^1 - H^2$ we obtain that $\|g\| \leq \beta$. Since the opposite inequality is already shown in Corollary 13.3.14, we therefore proved the theorem. □

Remark 13.3.16. If $\mathbf{Q}$ is a martingale measure for the process $(H^1 - H^2) \cdot S$, then of course $\mathbf{E_Q}[g^+] = \mathbf{E_Q}[g^-]$. But not all elements in the set $\mathcal{M}$ are martingale measures for this process and hence the equality of the suprema does not immediately follow from martingale considerations.

Theorem 13.3.17. *Suppose that S is a locally bounded semi-martingale that satisfies the (NFLVR) property. The norm of the space $\mathcal{G}$ is also given by the formula*

$$2\|g\| = \sup\{\mathbf{E_Q}\left[|g|\right] \mid \mathbf{Q} \in \mathcal{M}\} = \sup\{\mathbf{E_Q}\left[|g|\right] \mid \mathbf{Q} \in \mathcal{M}^e\} \,.$$

Proof. As in the previous result, for a contingent claim $g = (H^1 \cdot S)_\infty - (H^2 \cdot S)_\infty$ where H^1 and H^2 are maximal admissible, let us put:

$$\beta = \sup\{\mathbf{E_Q}\left[|g|\right] \mid \mathbf{Q} \in \mathcal{M}\}$$
$$\leq \sup\{\mathbf{E_Q}\left[g^+\right] \mid \mathbf{Q} \in \mathcal{M}\} + \sup\{\mathbf{E_Q}\left[g^-\right] \mid \mathbf{Q} \in \mathcal{M}\}$$
$$= 2\,\|g\|\,.$$

From Chap. 11 it follows that there is a maximal strategy K, such that $|g| \leq \beta + (K \cdot S)_\infty$. This inequality shows that

$$\beta + ((K \cdot S)_\infty) \geq (H^1 \cdot S)_\infty - (H^2 \cdot S)_\infty$$
$$\beta + ((K \cdot S)_\infty) \geq (H^2 \cdot S)_\infty - (H^1 \cdot S)_\infty\,.$$

As in previous result we obtain that $K - H^1 + H^2$ and $K - H^2 + H^1$ are β-admissible and maximal. Since $2(H^1 - H^2) = (K - H^2 + H^1) - (K - H^1 + H^2)$, we obtain the inequality $\|2\,g\| \leq \beta$. $\qquad\square$

Corollary 13.3.18. *Suppose that S is a locally bounded semi-martingale that satisfies the (NFLVR) property. If $g \in \mathcal{G}$, then there is a sequence of elements $\mathbf{Q}_n \in \mathcal{M}^e$ such that*

(1) $\mathbf{E_{Q_n}}\left[g^+\right] \to \sup\{\mathbf{E_Q}\left[g^+\right] \mid \mathbf{Q} \in \mathcal{M}\}$,
(2) $\mathbf{E_{Q_n}}\left[g^-\right] \to \sup\{\mathbf{E_Q}\left[g^-\right] \mid \mathbf{Q} \in \mathcal{M}\}$,
(3) $\mathbf{E_{Q_n}}\left[|g|\right] \to \sup\{\mathbf{E_Q}\left[|g|\right] \mid \mathbf{Q} \in \mathcal{M}\}\,.$

Proof. It suffices to take a sequence that satisfies the third line. $\qquad\square$

Remark and Example 13.3.19. For a contingent claim $f \in \mathcal{K}^{\max}$ we do not necessarily have that

$$\|f\| = \inf\{a \mid f \in \mathcal{K}_a\}\,.$$

Indeed take a process S such that there is only one risk neutral measure $\mathbf{Q}$. In this case the norm on the space $\mathcal{G}$ is (half) the $L^1(\mathbf{Q})$-norm. As is well-known the market is complete (see e.g. Chap. 9) and $\mathcal{G} = \{f \mid f \in L^1(\mathbf{Q}),\ \mathbf{E_Q}[f] = 0\}$. It follows that $\mathcal{K}_a^{\max} = \{f \mid f \in L^1(\mathbf{Q}),\ \mathbf{E_Q}[f] = 0,\ f \geq -a\}$. This cone may contain contingent claims with $\|f^-\|_\infty = a$ and with arbitrary small $L^1(\mathbf{Q})$-norm.

This example also shows that the space $\mathcal{G}$, which in this example is a hyperplane in L^1, can be isomorphic to an L^1-space. It also shows that the cone $\mathcal{K}^{\max}$ is not necessarily closed. Indeed the cone $\mathcal{K}^{\max}$ contains all contingent claims $f \in L^\infty$ with the property $\mathbf{E_Q}[f] = 0$. This set is dense in $\mathcal{G} = \{f \mid f \in L^1(\mathbf{Q}),\ \mathbf{E_Q}[f] = 0\}$. However, we have the following result.

Proposition 13.3.20. *Suppose that S is a locally bounded semi-martingale that satisfies the (NFLVR) property. The cones $\mathcal{K}_a^{\max}$ are closed in the space $\mathcal{G}$.*

Proof. Take a sequence f_n in $\mathcal{K}_a^{\max}$ and tending to f for the norm of $\mathcal{G}$. Since clearly $f \geq -a$, the contingent claim f is the outcome of an admissible and by Corollary 13.2.19, also of a maximal strategy. $\qquad\square$

13.4 Some Results on the Topology of $\mathcal{G}$

We now show that the space $\mathcal{G}$ is complete. This is of course very important if one wants to apply the powerful tools of functional analysis. The proof uses Theorem 13.3.13 and Corollary 13.2.17 above and in fact especially Corollary 13.2.17 suggests that the space is complete. After the proof of the theorem we will give some examples in order to show what kind of space $\mathcal{G}$ can be.

Theorem 13.4.1. *Suppose that S is a locally bounded semi-martingale that satisfies the (NFLVR) property. The space $\mathcal{G}$, $\| . \|$ is complete, i.e. it is a Banach space.*

Proof. We have to show that each Cauchy sequence converges. This is equivalent to the statement that every series of contingent claims whose norms form a convergent series, actually converges. So we start with a sequence $(g_n)_{n\geq 1}$ in $\mathcal{G}$ such that $\sum_{n\geq 1} \|g_n\| < \infty$. For each n we take according to Theorem 13.3.13 above, two $\|g_n\|$-admissible maximal strategies H^n and L^n such that $g_n = (H^n \cdot S)_\infty - (L^n \cdot S)_\infty$. Since $\sum_{n\geq 1} \|g_n\|$ converges, Proposition 13.2.16 above shows that $h = \sum_{n\geq 1}(H^n \cdot S)_\infty$ and $l = \sum_{n\geq 1}(L^n \cdot S)_\infty$ converge and define the maximal contingent claims h and l. Put now $g = h - l$, clearly an element of the space $\mathcal{G}$. We still have to show that the series actually converge to g for the norm defined on $\mathcal{G}$. But this is obvious since

$$g - \sum_{n=1}^{n=N} g_n = \left(\sum_{n>N} (H^n \cdot S)_\infty - \sum_{n>N} (L^n \cdot S)_\infty \right)$$

and each term on the right hand side defines, according to Corollary 13.2.17, a maximal contingent claim that is generated by a $\sum_{n>N} \|g_n\|$-admissible strategy. This remainder series tends to zero which completes the proof of the theorem. $\square$

Theorem 13.4.2. *Suppose that S is a locally bounded semi-martingale that satisfies the (NFLVR) property. If $(f_n)_{n\geq 1}$ is a sequence that converges in $\mathcal{G}$ to a contingent claim f and if for each n, $f_n = (H^n \cdot S)_\infty$ with H^n workable, then there is an element $\mathbf{Q} \in \mathcal{M}^e$ such that all $H^n \cdot S$ are uniformly integrable $\mathbf{Q}$-martingales as well as a workable strategy H such that the martingales $H^n \cdot S$ converge in $L^1(\mathbf{Q})$ to the martingale $H \cdot S$.*

Proof. Take $\mathbf{Q} \in \mathcal{M}^e$ such that all $(H^n \cdot S)_{n\geq 1}$ are $\mathbf{Q}$-uniformly integrable martingales. Such a probability exists by Corollary 13.2.18. The rest is obvious and follows from the inequality $\|g\| \geq \|g\|_{L^1(\mathbf{Q})}$. $\square$

Theorem 13.4.3. *Suppose that S is a locally bounded semi-martingale that satisfies the (NFLVR) property. If $(f_n)_{n\geq 1}$ is a sequence tending to f in the space $\mathcal{G}$, then there are maximal admissible contingent claims $(g_n, h_n)_{n\geq 1}$ in $\mathcal{K}^{\max}$ such that $f_n = g_n - h_n$ and such that $g_n \to g \in \mathcal{K}^{\max}$, $h_n \to h \in \bar{\mathcal{K}}^{\max}$, both convergences hold for the norm of $\mathcal{G}$.*

Proof. We first show that the statement of the theorem holds for a well-chosen subsequence $(n_k)_{k\geq 1}$. Afterwards we will fill in the remaining gaps.

The subsequence n_k is chosen so that for all $N \geq n_k$ we have $\|f - f_N\| \leq 2^{-k-1}$. It follows that $\|f_{n_{k+1}} - f_{n_k}\| \leq 2^{-k}$, for all k. We take, according to Theorem 13.3.13, contingent claims in $\mathcal{K}^{\max}$, denoted by $(\psi_k, \varphi_k)_{k\geq 1}$ such that

$$f_{n_1} = \psi_1 - \varphi_1$$
$$f_{n_{k+1}} - f_{n_k} = \psi_k - \varphi_k,$$

and such that ψ_k and φ_k are 2^{-k}-admissible for $k \geq 2$. Let $g_{n_k} = \sum_{l=1}^{k} \psi_l$ and $h_{n_k} = \sum_{l=1}^{k} \varphi_l$. By Corollary 13.2.17 and the reasoning in the proof of Theorem 13.4.1, these sequences converge in the norm of $\mathcal{G}$ to respectively g and h. Furthermore $f_{n_k} = g_{n_k} - h_{n_k}$ and hence $f = g - h$.

We now fill in the gaps $]n_k, n_{k+1}[$. For $n_k < n < n_{k+1}$ we choose maximal 2^{-k}-admissible contingent claims ρ_n and σ_n such that $f_n - f_{n_k} = \rho_n - \sigma_n$. To complete the proof we just have to check the obvious fact that $g_n = g_{n_k} + \rho_n$ and $h_n = h_{n_k} + \sigma_n$ satisfy the requirements of the theorem. $\qquad\square$

We will now discuss an example that serves as an illustration of what can go wrong in an incomplete market.

Example 13.4.4. The example is a slight modification of the example presented in Chap. 10, see also [S 93]. We start with a two-dimensional standard Brownian motion (B, W), with its natural filtration $(\mathcal{F}_t)_{t\geq 0}$. For the price process S we take a stochastic volatility process defined as $dS_t = (2 + \arctan(W_t))dB_t$. It is clear that the natural filtration of S is precisely $(\mathcal{F}_t)_{t\geq 0}$. Furthermore it is easy to see that the set of stochastic integrals with respect to S is the same as the set of stochastic integrals with respect to B. We will use this fact without further notice. We define $L = \mathcal{E}(B)$ and $Z = \mathcal{E}(W)$, where $\mathcal{E}$ denotes the stochastic exponential. The stopping times τ and σ are defined as $\tau = \inf\{t \mid L_t \leq \frac{1}{2}\}$ and $\sigma = \inf\{t \mid Z_t \geq 2\}$. The process X is defined as $X = L^{\tau\wedge\sigma}$. The measure $\mathbf{Q}$ is nothing else but $d\mathbf{Q} = Z_{\tau\wedge\sigma}d\mathbf{P}$. For the process X and the measure $\mathbf{Q}$, the following hold:

(1) The process X is continuous, strictly positive, also $X_\infty > 0$ a.s. and $X_0 = 1$, it is a local martingale for $\mathbf{P}$, i.e. $\mathbf{P} \in \mathcal{M}^e$,
(2) under $\mathbf{P}$, the process X is a strict local martingale, i.e. $\mathbf{E}_\mathbf{P}[X_\infty] < 1$,
(3) for each $t < \infty$ the stopped process X^t is a $\mathbf{P}$-uniformly integrable martingale,
(4) there is an equivalent probability measure $\mathbf{Q} \in \mathcal{M}^e$ for which X becomes a $\mathbf{Q}$-uniformly integrable martingale.

Let us now verify some additional features.

Proposition 13.4.5. *In the setting of the above example, the space $\mathcal{G}^\infty$ is not dense in $\mathcal{G}$. In fact even the closure of L^∞ for the norm $\|g\| = \frac{1}{2}\sup\{\|g\|_{L^1(\mathbf{Q})} \mid \mathbf{Q} \in \mathcal{M}^e\}$, does not contain $\mathcal{G}$ as a subset.*

Proof. For each $t \leq \infty$ we clearly have that $f_t = X_t - 1 \in \mathcal{G}$. Suppose now that the contingent claim f_∞ is in the closure of the space L^∞ for the norm $\|g\| = \frac{1}{2} \sup\{\|g\|_{L^1(\mathbf{Q})} \mid \mathbf{Q} \in \mathcal{M}^e\}$. For $\varepsilon = -\frac{\mathbf{E}[f_\infty]}{4} > 0$ we can find g bounded such that for all $\mathbf{Q} \in \mathcal{M}^e$ we have $\|g - f_\infty\| \leq \varepsilon$. For the measure $\mathbf{P}$ we find $\mathbf{E}_\mathbf{P}[|f_\infty - g|] \leq \varepsilon$ and hence for each $t \leq \infty$ we have, by taking conditional expectations,

$$\mathbf{E}_\mathbf{P}\Big[\big|f_t - \mathbf{E}_\mathbf{P}[g \mid \mathcal{F}_t]\big|\Big] \leq \varepsilon.$$

In particular, since $\mathbf{E}[f_t] = 0$ for each $t < \infty$, we have $\mathbf{E}_\mathbf{P}[g] = \mathbf{E}\big[\mathbf{E}_\mathbf{P}[g \mid \mathcal{F}_t]\big] \geq -\varepsilon$. This in turn implies that $\mathbf{E}_\mathbf{P}[f_\infty] \geq -2\varepsilon$ a contradiction to the choice of ε. □

Theorem 13.4.6. *In the setting of the above example, the Banach space $\mathcal{G}$ contains a subspace isometric to ℓ^∞. In other words there is an isometry $u\colon \ell^\infty \to \mathcal{G}$. Moreover u can be chosen such that $u(\ell^\infty) \subset \mathcal{G}^\infty$.*

Proof. We start with a partition of Ω into a sequence of pairwise disjoint sets, defined by the process W. More precisely we put $A_1 = \{W_1 \in]-\infty, 1]\}$ and for $n \geq 2$ we put $A_n = \{W_1 \in]n-1, n]\}$. Let M be the stochastic exponential $M = \mathcal{E}(B - B^1)$ and let the stopping time T be defined as

$$T = \inf\{t \mid M_t \geq 2\}.$$

The sequence that we will use to construct the subspace isometric to ℓ^∞ is defined as

$$f_n = 2(M_T - 1)\mathbf{1}_{A_n}.$$

For each n and each $\varepsilon > 0$ there is a real number $\alpha(n, \varepsilon)$ depending only on ε and n such that the random variable $\phi(n, \varepsilon) = \alpha(n, \varepsilon)\mathbf{1}_{A_k} + \varepsilon\mathbf{1}_{\bigcup_{m \neq k} A_m}$ is strictly positive and defines a density for a measure $d\mathbf{Q}_{n,\varepsilon} = \phi(n, \varepsilon)d\mathbf{P}$ which is necessarily in $\mathcal{M}^e$, since the random variable $\phi(n, \varepsilon)$ can be written as a stochastic integral with respect to W. It is clear that $\mathbf{Q}_{n,\varepsilon}[A_n] \geq 1 - \varepsilon$. This shows that for each n, $\sup\{\mathbf{Q}[A_n] \mid \mathbf{Q} \in \mathcal{M}^e\} = 1$.

Clearly each f_n is a 2-admissible maximal contingent claim. Since for each measure $\mathbf{Q} \in \mathcal{M}^e$ we have $\mathbf{Q}[f_n = 2 \mid \mathcal{F}_1] = \mathbf{Q}[f_n = -2 \mid \mathcal{F}_1] = \frac{1}{2}\mathbf{1}_{A_n}$ we obtain that $\|f_n\|_{L^1(\mathbf{Q})} = 2\mathbf{Q}(A_n)$, hence for the $\mathcal{G}$-norm we find $\|f_n\| = 1$.

We now show that for each $x \in \ell^\infty$ we can define a contingent claim

$$u(x) = \sum_{k \geq 1} x_k f_k \in \mathcal{G}.$$

If $x = (x_k)_{k \geq 1}$ is an element of ℓ^∞ and if m is a natural number, we denote by x^m the element defined as $x_k^m = x_k$ if $k \leq m$ and $x_k^m = 0$ otherwise. Let us already put $u(x^m) = \sum_{k=1}^m x_k^m f_k$. Now if x is a positive element in ℓ^∞ then the sequence $(u(x^m))_{m \geq 1}$ is a sequence converging in $L^1(\mathbf{Q})$ to a contingent claim $u(x) = \sum_{k=1}^\infty x_k f_k$ and this for each $\mathbf{Q} \in \mathcal{M}^e$. By Corollary 13.2.19 and

Theorem 13.2.5, the random variable $u(x)$ is in $\mathcal{G}$. For arbitrary x we split into the positive and the negative part. This defines a linear mapping from ℓ^∞ into $\mathcal{G}$. For each $\mathbf{Q} \in \mathcal{M}^e$ we have that $u(x) = \sum_{k=1}^\infty x_k f_k$, where the sum actually converges in $L^1(\mathbf{Q})$. Let us now calculate the norm of $u(x)$. For an arbitrary measure $\mathbf{Q} \in \mathcal{M}^e$ we find

$$\|u(x)\|_{L^1(\mathbf{Q})} \le \int \sum_{k=1}^\infty |x_k|\, \|f_k\|_{L^1(\mathbf{Q})}$$

and hence we have

$$\|u(x)\| \le \sup_k |x_k|\,.$$

Take now for $\varepsilon > 0$ given, an index k such that $\sup_k |x_k| > \|x\|_\infty - \varepsilon$. Take the measure $\mathbf{Q}_{k,\varepsilon}$ as above.

We find that

$$\|u(x)\|_{L^1(\mathbf{Q}_{k,\varepsilon})} \ge \int_{A_k} |u(x)|\phi(k,\varepsilon)\, d\mathbf{P}$$
$$\ge \alpha(k,\varepsilon)|x_k|2\mathbf{P}[A_k]\,.$$

Since clearly $\alpha(k,\varepsilon)\mathbf{P}[A_k] \ge 1 - \varepsilon$ we find that

$$\|u(x)\|_{L^1(\mathbf{Q}_{k,\varepsilon})} \ge (\|x\|_\infty - \varepsilon)\, 2(1-\varepsilon)\,.$$

Because $\varepsilon > 0$ was arbitrary we find that

$$\|u(x)\| = \|x\|_\infty\,.$$

The linear mapping is therefore an isometry. Furthermore it is easily seen that for each $x \in \ell^\infty$ we have $u(x) \in \mathcal{G}^\infty$. $\qquad\square$

Theorem 13.4.7. *In the setting of the above example, there is a contingent claim f in $\mathcal{G}$ such that for each $\mathbf{Q} \in \mathcal{M}^e$ we have $\mathbf{E}_\mathbf{Q}[f] = 0$, but such that f is not in the closure of $\mathcal{G}^\infty$.*

Proof. We will make use of the notation and proof of the preceding theorem. So we take the same sequence $(A_n)_{n \ge 1}$ as above. This time we introduce stopping times

$$T_n = \inf\{t \mid M_t \ge n+1\}$$

and functions

$$f_n = (M_{T_n} - 1)\mathbf{1}_{A_n}\,.$$

Exactly as in the previous proof one shows that the contingent claim $f = \sum_{n=1}^\infty f_n$ is in $\mathcal{G}$ and has norm 1. Suppose now that h is a bounded variable in $\mathcal{G}$. We will show that $\|f - h\| \ge 1$. For each n we take an element $\mathbf{Q}_n \in \mathcal{M}^e$ such that $\mathbf{Q}[A_n] \ge 1 - \frac{1}{n}$; such an element surely exists. Because $\mathbf{Q}[f_n = n \mid \mathcal{F}_n] = \frac{1}{n}\mathbf{1}_{A_n}$ we find for $n > \|h\|_\infty$, that

$$\mathbf{E}_{\mathbf{Q}_n}[(f - h)^+] \geq \left(1 - \frac{1}{n}\right)\frac{1}{n}(n - \|h\|_\infty)$$

$$\geq \left(1 - \frac{1}{n}\right)\left(1 - \frac{\|h\|_\infty}{n}\right).$$

From Theorem 13.3.17 we can now deduce that the distance of f to $\mathcal{G}^\infty$ is precisely equal to 1.   □

This completes the discussion of the Example 13.4.4.

Example 13.4.8. This is an example showing that the space $\mathcal{G}$ can be one-dimensional, whereas the set $\mathcal{M}^e$ remains very big. For this we take a finite time set $[0, 1]$, and we take $\Omega = [0, 1]$ with the Lebesgue measure. For $t < 1$, we put $\mathcal{F}_t$ equal to the σ-algebra generated by the zero sets with respect to Lebesgue-measure. For $t = 1$ we put $\mathcal{F}_t$ equal to the σ-algebra of all Lebesgue-measurable sets. The price process is defined as $S_t = 0$ for $t < 1$ and $S_1(\omega) = \omega - \frac{1}{2}$. Of course $\mathcal{G} = \text{span}(S_1)$. The set $\mathcal{M}^e$ is the set $\{f \mid f > 0 \,,\, \int_0^1 (t - \frac{1}{2})f(t)\,dt = 0\}$. This set is big in the sense that it is not relatively weakly compact in $L^1[0, 1]$.

Example 13.4.9. This example shows that the space $\mathcal{G}$ can actually be isomorphic to an L^∞-space. The example is constructed is the same spirit as the previous one. We take $[0, 2]$ as the time set and $\Omega = [-1, 1] \times [-1, 1]$ with the two-dimensional Lebesgue measure. Let g_1, respectively g_2, be the first and second coordinate projection defined on Ω. For $t < 1$ the σ-algebra $\mathcal{F}_t$ is the σ-algebra generated by the zero sets, for $1 \leq t < 2$ we have $\mathcal{F}_t = \sigma(\mathcal{F}_0, g_1)$ and $\mathcal{F}_2 = \sigma(\mathcal{F}_1, g_2)$, which is also the σ-algebra of Lebesgue-measurable subsets of Ω. The process S is defined as $S_t = 0$ for $t < 1$, $S_t = g_1$ for $1 \leq t < 2$ and $S_2 = g_1 + g_2$. We remark that the filtration is generated by the process S.

Clearly $(H \cdot S)_2 \in \mathcal{G}$ if and only if it is of the form $(H \cdot S)_2 = \alpha g_1 + h\,g_2$, where h is $\mathcal{F}_1$-measurable and bounded. This implies that $\mathcal{G}$ can be identified with $\mathbb{R} \times L^\infty(\Omega, \mathcal{F}_1, \mathbf{P})$. We will not calculate the norm of the space $\mathcal{G}$, but instead we will use the closed graph theorem to see that this norm is equivalent to the norm defined as $\|(\alpha, h)\| = |\alpha| + \|h\|_\infty$. It follows that $\mathcal{G}$ is isomorphic to an L^∞-space.

Example 13.4.10. The following example is in the same style as the process S has exactly one jump. But this time the behaviour of the process S before the jump is such that the space $\mathcal{G}$ is not of L^∞-type.

We start with the one-dimensional Brownian motion W, starting at zero and with its natural filtration $(\mathcal{H}_t)_{0 \leq t \leq 1}$. At time $t = 1$ we add a jump g uniformly distributed over the interval $[-1, 1]$ and independent of the Brownian motion W. So the price process becomes $S_t = W_t$ for $t < 1$ and $S_1 = W_1 + g$. The filtration becomes, up to null sets, $\mathcal{F}_t = \mathcal{H}_t$ for $t < 1$ and $\mathcal{F}_1 = \sigma(\mathcal{H}_1, g)$. For simplicity we assume that this process is defined on the probability space $\Omega \times [-1, +1]$ where Ω is the trajectory space of Brownian motion, equipped with the usual Wiener measure $\mathbf{P}$ and where we take the uniform distribution m on $[-1, +1]$ as the second factor. The measure is therefore $\mathbf{P} \times m$.

The set of equivalent local martingales measures can also be characterised. Since Brownian motion has only one local martingale measure we see that for each $\mathbf{Q} \in \mathcal{M}^e$ and for each $t < 1$ we have that $\mathbf{Q} = \mathbf{P}$ on the σ-algebra $\mathcal{F}_t = \mathcal{H}_t$. Therefore also $\mathbf{Q} = \mathbf{P}$ on $\mathcal{H}_1$. From the existence theorem of conditional distributions, or the desintegration theorem of measures, we then learn that $\mathbf{Q}$ is necessarily of the form $\mathbf{Q}[d\omega \times dx] = \mathbf{P}[d\omega]\mu_\omega[dx]$, where μ is a probability kernel $\mu \colon \Omega \times \mathcal{B}[-1, +1] \to [0, 1]$, measurable for $\mathcal{H}_1$. In order for $\mathbf{Q}$ to be a local martingale measure μ should satisfy $\int_{[-1,+1]} x\,\mu_\omega(dx) = 0$ for almost all ω. In order to be equivalent to $\mathbf{P} \times m$, a.s. the measure μ_ω should be equivalent to m. This can easily be seen by using the density of $\mathbf{Q}$ with respect to $\mathbf{P} \times m$.

If H is a predictable strategy then it is clear that it is predictable with respect to the filtration of the Brownian motion. A strategy H is therefore S-integrable if and only if $\int_0^1 H_t^2\,dt < \infty$ a.s.. It follows that a necessary condition for a predictable process H to be 1-admissible is $H \cdot W \geq -1$. We can change the value of H at time 1 without perturbing the integral $H \cdot W$. In order to obtain a characterisation of 1-admissible integrands for S, we only need a condition on H_1 in order to have, in addition, that $(H \cdot S)_1 \geq -1$. The outcome at time 1 is $(H \cdot S)_1 = (H \cdot W)_1 + H_1\,g$ and this is almost surely bigger than -1 if and only if $|H_1| \leq 1 + (H \cdot W)_1$ almost surely. If we are looking for 1-admissible maximal contingent claims the condition on H becomes

(1) $H \cdot W$ is a uniformly integrable martingale for $\mathbf{P}$ and $f = (H \cdot W)_1 \geq -1$
(2) $|H_1| \leq 1 + f$.

From this it follows that a random variable k is in $\mathcal{G}$ if and only if it is of the form

$$k = f_1 - f_2 + g\,(h_1 - h_2)$$

where

(1) f_1, f_2, h_1, h_2 are $\mathcal{H}_1$-measurable;
(2) $f_1, f_2 \geq -a$ for some positive real number a;
(3) $\mathbf{E_P}[f_1] = \mathbf{E_P}[f_2] = 0$;
(4) $|h_1| \leq a + f_1$ and $|h_2| \leq a + f_2$.

If we want to find a better description we observe that if f is $\mathcal{H}_1$-measurable, integrable and positive then we can take $f_1 = f_2 = f - \mathbf{E_P}[f]$ and hence the condition on h_1 and h_2 becomes $|h_1|, |h_2| \leq f$. It follows that the space $\mathcal{G}$ is the space of all functions k of the form

$$f + g\,h$$

where

$\mathbf{E_P}[f] = 0$ and where h, f are both $\mathcal{H}_1$-measurable and integrable.

The norm on the space $\mathcal{G}$ can be calculated using Theorem 13.3.17 above and using the characterisation of the measures in $\mathcal{M}^e$. We find

$$2\|f + g\,h\| = \sup_{\mu} \mathbf{E_P} \left[\int_{[-1,+1]} |f + x\,h|\, \mu_\omega(dx) \right].$$

For given ω the measure $\mu_\omega(dx)$ on $[-1, +1]$ that maximises $\int_{[-1,+1]} |f + x\,h|$ $\times \mu_\omega(dx)$ and that satisfies $\int_{[-1,+1]} x\, \mu_\omega(dx) = 0$ is according to balayage arguments (repeated application of Jensen's inequality) the measure that gives mass $\frac{1}{2}$ to both -1 and $+1$. This measure does not satisfy the requirements since it is not equivalent to the measure m on $[-1, 1]$. But an easy approximation argument shows nevertheless that

$$2\|f + g\,h\| = \mathbf{E_P} \left[\frac{|f + h| + |f - h|}{2} \right].$$

This can be rewritten as

$$2\|f + g\,h\| = \mathbf{E_P} \left[\max(|f|, |h|) \right].$$

This equality shows that $\mathcal{G}$ is isomorphic to an L^1-space.

13.5 The Value of Maximal Admissible Contingent Claims on the Set $\mathcal{M}^e$

As shown in Example 13.4.4, maximal contingent claims f may have different expected values for different measures in $\mathcal{M}^e$. In Chap. 10 we showed that under rather general conditions such a phenomenon is generic for incomplete markets. More precisely we have:

Theorem 13.5.1. (Theorem 10.3.1) *Suppose that S is a continuous d-dimensional semi-martingale with the (NFLVR) property. If there is a continuous local martingale W such that $\langle W, S \rangle = 0$ but $d\langle W, W \rangle$ is not singular to $d\langle S, S \rangle$, then for each $\mathbf{R}$ in $\mathcal{M}^e$, there is a maximal contingent claim $f \in \mathcal{K}_1$ such that $\mathbf{E_R}[f] < 0$.*

The preceding theorem brings up the question whether for given $f \in \mathcal{K}^{\max}$, the set of measures $\mathbf{Q} \in \mathcal{M}^e$ such that $\mathbf{E_Q}[f] = 0$ is big.

Theorem 13.5.2. *Suppose that S is a locally bounded semi-martingale that satisfies the (NFLVR) property. If f is a maximal contingent claim i.e. $f \in \mathcal{K}^{\max}$, then the mapping*

$$\phi : \begin{array}{ccc} \mathcal{M}(S) & \longrightarrow & \mathbb{R} \\ \mathbf{Q} & \longmapsto & \mathbf{E_Q}[f] \end{array}$$

is lower semi-continuous for the weak topology $\sigma\big(L^1(\mathbf{P}), L^\infty(\mathbf{P})\big)$. *In particular the set* $\{\mathbf{Q} \mid \mathbf{Q} \in \mathcal{M}; \mathbf{E}_\mathbf{Q}[f] = 0\}$ *is a* G_δ-*set (with respect to the weak and therefore also for the strong topology) in* $\mathcal{M}$. *Furthermore this set is convex and* $\{\mathbf{Q} \mid \mathbf{Q} \in \mathcal{M}^e; \mathbf{E}_\mathbf{Q}[f] = 0\}$ *is strongly dense in* $\mathcal{M}$. *In particular as* $\mathcal{M}$ *is a complete metric space with respect to the strong topology of* $L^1(\mathbf{P})$, *the set* $\{\mathbf{Q} \mid \mathbf{Q} \in \mathcal{M}; \mathbf{E}_\mathbf{Q}[f] = 0\}$ *is of second category.*

Proof. The lower semi-continuity is a consequence of Fatou's lemma and the fact that for convex sets weak and strong closedness are equivalent.

The convexity follows from $\mathbf{E}_\mathbf{Q}[f] \leq 0$ for every $\mathbf{Q} \in \mathcal{M}$.

By the convexity of the set $\{\mathbf{Q} \mid \mathbf{Q} \in \mathcal{M}^e; \mathbf{E}_\mathbf{Q}[f] = 0\}$, it only remains to be shown that the set $\{\mathbf{Q} \mid \mathbf{Q} \in \mathcal{M}^e; \mathbf{E}_\mathbf{Q}[f] = 0\}$ is norm dense in $\mathcal{M}^e$, the latter being norm dense in $\mathcal{M}$.

Take $\mathbf{Q}^0 \in \mathcal{M}^e$ such that $\mathbf{E}_{\mathbf{Q}^0}[f] = 0$. Since f is maximal such a measure exists. Since f is maximal there is a strategy H such that $H \cdot S$ is a $\mathbf{Q}^0$-uniformly integrable martingale and such that $f = (H \cdot S)_\infty$. We may suppose that the process $V = 1 + H \cdot S$ remains bounded away from zero.

Take now $\mathbf{Q} \in \mathcal{M}^e$ and let Z be the càdlàg martingale defined by

$$Z_t = \mathbf{E}\left[\frac{d\mathbf{Q}}{d\mathbf{Q}^0} \,\middle|\, \mathcal{F}_t\right].$$

For each n, a natural number, we define the stopping time

$$T_n = \inf\{t \mid Z_t > n\}.$$

Clearly the process VZ is a $\mathbf{Q}^0$-local martingale and being positive it is a super-martingale. Therefore we have that $V_{T_n} Z_{T_n}$ is in $L^1(\mathbf{Q}^0)$. It follows that $(VZ)^{T_n} \leq nV + V_{T_n} Z_{T_n}$ and hence the process $(VZ)^{T_n}$ is a uniformly integrable martingale. Therefore $\mathbf{E}_{\mathbf{Q}^0}[V_{T_n} Z_{T_n}] = 1$ and the measure $\mathbf{Q}^n$ defined as $d\mathbf{Q}^n = Z_{T_n} d\mathbf{Q}^0$ satisfies $\mathbf{E}_{\mathbf{Q}^n}[V_{T_n}] = 1$. Since $\mathbf{E}_{\mathbf{Q}^n}[V_\infty] = \mathbf{E}_{\mathbf{Q}^n}[V_{T_n}] = 1$ we clearly have $\mathbf{Q}^n \in \{\mathbf{R} \mid \mathbf{R} \in \mathcal{M}^e; \mathbf{E}_\mathbf{R}[f] = 0\}$. Since $\mathbf{Q}^n$ tends to $\mathbf{Q}$ in the L^1-norm, the proof of the theorem is completed. $\qquad\square$

Corollary 13.5.3. *Suppose that S is a locally bounded semi-martingale that satisfies the (NFLVR) property. If $\mathcal{V}$ is a separable subspace of $\mathcal{G}$, then the convex set*

$$\{\mathbf{Q} \mid \mathbf{Q} \in \mathcal{M}^e; \mathbf{E}_\mathbf{Q}[f] = 0 \text{ for all } f \in \mathcal{V}\}$$

is dense in $\mathcal{M}$ *with respect to the norm topology of* $L^1(\Omega, \mathcal{F}, \mathbf{P})$.

Proof. We may and do suppose that there is sequence of maximal contingent claims in $\mathcal{V}$, $(f_n)_{n \geq 1}$ such that the sequence $\{f_n - f_m \mid n \geq 1; m \geq 1\}$ is dense in $\mathcal{V}$, occasionally we enlarge the space $\mathcal{V}$. Obviously

$$\{\mathbf{Q} \mid \mathbf{Q} \in \mathcal{M}; \mathbf{E}_\mathbf{Q}[f] = 0 \text{ for all } f \in \mathcal{V}\}$$
$$= \{\mathbf{Q} \mid \mathbf{Q} \in \mathcal{M}; \mathbf{E}_\mathbf{Q}[f_n] = 0 \text{ for all } n \geq 1\}.$$

For each n the set $\{\mathbf{Q} \mid \mathbf{Q} \in \mathcal{M}; \mathbf{E_Q}[f_n] = 0\}$ is a norm dense and (for the norm topology) a G_δ-set in $\mathcal{M}$. Since $\mathcal{M}$ is a complete space for the L^1-norm, we may apply Baire's category theorem. Therefore the intersection over all n, $\{\mathbf{Q} \mid \mathbf{Q} \in \mathcal{M};$ for all n: $\mathbf{E_Q}[f_n] = 0\}$ is still a dense G_δ-set of $\mathcal{M}$. Because, by corollary 13.2.18, the set $\{\mathbf{Q} \mid \mathbf{Q} \in \mathcal{M}^e;$ for all n: $\mathbf{E_Q}[f_n] = 0\}$ is non-empty, an easy argument using convex combinations yields that $\{\mathbf{Q} \mid \mathbf{Q} \in \mathcal{M}^e; \mathbf{E_Q}[f] = 0$ for all $f \in \mathcal{V}\}$ is dense in $\mathcal{M}$. $\qquad\square$

Corollary 13.5.4. *If S is a continuous d-dimensional semi-martingale with the (NFLVR) property, if there is a continuous local martingale W such that $\langle W, S \rangle = 0$ but $d\langle W, W \rangle$ is not singular to $d\langle S, S \rangle$, then $\mathcal{G}$ is not a separable space.*

Proof. This follows from the previous corollary and from Theorem 13.5.1 above. $\qquad\square$

13.6 The Space $\mathcal{G}$ under a Numéraire Change

If we change the numéraire, e.g. we change from one reference currency to another, what will happen with the space $\mathcal{G}$? Referring to Chap. 11 and especially the proofs of Theorem 11.4.2 and 11.4.4 therein, we expect that there is an obvious transformation which should be the mathematical translation of the change of currency. More precisely we want the contingent claims of $\mathcal{G}$ to be multiplied with the exchange ratio between the two currencies. This section will give some precise information on this problem.

We start with the investigation of how the set of equivalent martingale measures is changed.

Suppose that V is a strictly positive process of the form $V = H \cdot S + 1$ where $1 + (H \cdot S)_\infty$ is strictly positive and where $(H \cdot S)_\infty$ is maximal admissible. Suppose also that the process $\frac{1}{V}$ is locally bounded. This hypothesis allows us to use, without restriction, the theory developed so far. With each element $\mathbf{R}$ of $\mathcal{M}(S)$ we asssociate the measure $\widetilde{\mathbf{R}}$ defined by $d\widetilde{\mathbf{R}} = V_\infty d\mathbf{R}$. Of course this measure is not a probability measure since we do not necessarily have that $\mathbf{E_R}[V_\infty] = 1$. But from Theorem 13.5.2 above it follows, however, that the set $G = \{\mathbf{Q} \in \mathcal{M}(S) \mid \mathbf{E_Q}[V_\infty] = 1\}$ is a dense G_δ-set of $\mathcal{M}(S)$. Likewise the set $\widetilde{G} = \left\{ \widetilde{\mathbf{Q}} \in \mathcal{M}\left(\frac{S}{V}, \frac{1}{V}\right) \mid \mathbf{E}_{\widetilde{\mathbf{Q}}}\left[\frac{1}{V_\infty}\right] = 1 \right\}$ is a dense G_δ-set of $\mathcal{M}\left(\frac{S}{V}, \frac{1}{V}\right)$. The following theorem is obvious.

Theorem 13.6.1. *Suppose that S is a locally bounded semi-martingale that satisfies the (NFLVR) property. With the above notations, the relation $d\widetilde{\mathbf{R}} = V_\infty d\mathbf{R}$, defines a bijection between the sets G and $\widetilde{G}$.*

In the following theorem we make use of the notation introduced in Theorem 13.3.2. The space $\mathcal{G}(S)$ is the space of workable contingent claims that is constructed with the d-dimensional process S, the space $\mathcal{G}\left(\frac{S}{V}, \frac{1}{V}\right)$ is the space of workable contingent claims constructed with the $(d+1)$-dimensional process $\left(\frac{S}{V}, \frac{1}{V}\right)$.

Theorem 13.6.2. *Suppose that S is a locally bounded semi-martingale that satisfies the (NFLVR) property. Suppose that V is a strictly positive process of the form $V = H \cdot S + 1$ where $1 + (H \cdot S)_\infty$ is strictly positive and where $(H \cdot S)_\infty$ is maximal admissible. Suppose that the process $\frac{1}{V}$ is locally bounded. The mapping*

$$\varphi : \quad \mathcal{G}(S) \longrightarrow \mathcal{G}(\tfrac{S}{V}, \tfrac{1}{V})$$
$$g \longmapsto \frac{g}{V_\infty}$$

defines an isometry between $\mathcal{G}(S) = \mathcal{G}(S, 1)$ and $\mathcal{G}(\frac{S}{V}, \frac{1}{V})$.

Proof. Suppose $V = H \cdot S + 1$ where $1 + (H \cdot S)_\infty$ is strictly positive and where $(H \cdot S)_\infty$ is maximal admissible. Take an admissible, with respect to the process S, strategy K. The process $\frac{K \cdot S}{V}$ is the outcome of the strategy $K' = (K, (K \cdot S)_- - KS_-)$, see also the Remark 13.2.10 above and Chap. 11. From Theorem 13.2.5 above it follows that there is an element $\mathbf{Q} \in \mathcal{M}^e$ such that $\mathbf{E}_\mathbf{Q}[(K \cdot S)_\infty] = 0$ and such that $\mathbf{E}_\mathbf{Q}[V_\infty] = 1$. The measure $\widetilde{\mathbf{Q}}$ defined as $d\widetilde{\mathbf{Q}} = V_\infty \, d\mathbf{Q}$ is therefore an element of $\mathcal{M}\left(\frac{S}{V}, \frac{1}{V}\right)$ such that $\mathbf{E}_{\widetilde{\mathbf{Q}}}[\frac{K \cdot S}{V_\infty}] = 0$. It follows that the contingent claim $\frac{1}{V_\infty} - 1$ is maximal and admissible for the process $\left(\frac{S}{V}, \frac{1}{V}\right)$ and hence the contingent claim $\frac{K \cdot S}{V_\infty}$ is workable. It follows that the mapping φ maps $\mathcal{K}^{\max}$, and hence also $\mathcal{G}(S)$, into $\mathcal{G}\left(\frac{S}{V}, \frac{1}{V}\right)$.

If we apply the numéraire $\frac{1}{V}$ to the system $\left(\frac{S}{V}, \frac{1}{V}\right)$ we find the $(d+1)$-dimensional process (S, V). However, because V is given by a stochastic integral with respect to S, we have that $\mathcal{G}(S, V) = \mathcal{G}(S)$. It follows that the mapping that associates with each element $k \in \mathcal{G}\left(\frac{S}{V}, \frac{1}{V}\right)$, the element kV_∞ maps $\mathcal{G}\left(\frac{S}{V}, \frac{1}{V}\right)$ into $\mathcal{G}(S)$. The mapping φ is clearly bijective.

Let $G = \{\mathbf{Q} \in \mathcal{M}(S) \,|\, \mathbf{E}_\mathbf{Q}[V_\infty] = 1\}$ and $\widetilde{G} = \{\widetilde{\mathbf{Q}} \in \mathcal{M}(\frac{S}{V}, \frac{1}{V}) \,|\, \mathbf{E}_{\widetilde{\mathbf{Q}}}[\frac{1}{V_\infty}] = 1\}$. Since both sets are dense in, respectively, $\mathcal{M}(S)$ and $\mathcal{M}\left(\frac{S}{V}, \frac{1}{V}\right)$, it is clear that for every element $g \in \mathcal{G}(S)$,

$$2\|g\| = \sup\{\mathbf{E}_\mathbf{Q}[|g|] \mid \mathbf{Q} \in \mathcal{M}(S)\}$$
$$= \sup\{\mathbf{E}_\mathbf{Q}[|g|] \mid \mathbf{Q} \in G\}$$
$$= \sup\left\{\mathbf{E}_{\widetilde{\mathbf{Q}}}\left[\frac{|g|}{V_\infty}\right] \,\middle|\, \widetilde{\mathbf{Q}} \in \widetilde{G}\right\}$$
$$= \sup\left\{\mathbf{E}_{\widetilde{\mathbf{Q}}}\left[\frac{|g|}{V_\infty}\right] \,\middle|\, \widetilde{\mathbf{Q}} \in \mathcal{M}\left(\frac{S}{V}, \frac{1}{V}\right)\right\} = 2\left\|\frac{g}{V_\infty}\right\|.$$

This shows that φ is also an isometry. $\qquad\qquad\square$

Remark 13.6.3. The previous theorem shows that $\mathcal{G}$ is a numéraire invariant space provided we only accept numéraire changes induced by maximal admissible contingent claims.

13.7 The Closure of $\mathcal{G}^\infty$ and Related Problems

In this section we will study the contingent claims of $\mathcal{K}^{\max}$ that are in the closure of $\mathcal{G}^\infty$. The characterisation is done using either uniform convergence over the set $\mathcal{M}^e$ or using the set $\mathcal{M}^{ba}$. Before we start the program, we first recall some notions from integration theory with respect to finitely additive measures; we refer to Dunford-Schwartz [DS 58] for details.

Let μ be a finitely additive measure that is in $ba(\Omega, \mathcal{F}, \mathbf{P})$. A measurable function f (we continue to identify functions that are equal $\mathbf{P}$-a.s.), defined on Ω is called μ-measurable if for each $\varepsilon > 0$ there is a bounded measurable function g such that $\mu\{\omega \mid |f(\omega) - g(\omega)| > \varepsilon\} < \varepsilon$. The reader can check that since $\mathcal{F}$ is a σ-algebra, this definition coincides with [DS 58, Definition 10]. We say that a μ-measurable function f is μ-integrable if and only if there is sequence $(g_n)_{n\geq 1}$ of bounded measurable functions such that g_n converges in μ-measure to f and such that $\mathbf{E}_\mu[|g_n - g_m|]$ tends to zero if n, m tend to ∞. In this case one defines $\mathbf{E}_\mu[f] = \lim_{n\to\infty} \mathbf{E}_\mu[g_n]$ as the μ-integral $\mathbf{E}_\mu[f]$ of f. In case f is bounded from below the μ-integrability of f implies via the dominated convergence theorem, valid also for finitely additive measures, that $\mathbf{E}[f - f \wedge n]$ tends to zero as n tends to ∞. Contingent claims g of $\mathcal{G}^\infty$ are μ-integrable for all $\mu \in \mathcal{M}^{ba}$ and moreover we trivially have $\mathbf{E}_\mu[g] = 0$ since $\mathbf{E}_\mathbf{Q}[g] = 0$ for all $\mathbf{Q} \in \mathcal{M}^e$.

Proposition 13.7.1. *Suppose that S is a locally bounded semi-martingale that satisfies the (NFLVR) property. If $f \in \mathcal{K}^{\max}$ and $\mu \in \mathcal{M}^{ba}$, then f is μ-integrable and $\mathbf{E}_\mu[f] \leq 0$. Also $\mu[f \geq n] \leq \frac{4\|f\|}{n}$, a uniform bound over $\mu \in \mathcal{M}^{ba}$. In particular for each $\mu \in \mathcal{M}^{ba}$ and each $f \in \mathcal{K}^{\max}$ we find that $f \wedge n$ tends to f in μ-measure and $\mathbf{E}_\mu[f \wedge n]$ tends to $\mathbf{E}_\mu[f]$ as n tends to infinity.*

Proof. We only have to prove the statement for contingent claims f that are 1-admissible and maximal. So suppose that f is such a contingent claim. By the optional stopping theorem, or by the maximal inequality for supermartingales, we find that for all $\mathbf{Q} \in \mathcal{M}^e$, we have that $\mathbf{Q}[f \geq n] \leq \frac{1}{n}$. The set $\mathcal{M}^e$ is $\sigma(ba, L^\infty)$-dense in $\mathcal{M}^{ba}$ (see Remark 9.5.10), hence we obtain that $\mu[f \geq n] \leq \frac{1}{n}$ for all n. Since $\mu[f - f \wedge n > 0] \leq \mu[f \geq n] \leq \frac{1}{n}$, the measurability follows for functions f that are 1-admissible and maximal. The general case follows by splitting f as $f = g - h$ where each g and h are $\|f\|$-admissible and by the fact that $\{|f| > n\} \subset \{|g| > \frac{n}{2}\} \cup \{|h| > \frac{n}{2}\}$.

To see that for $\mu \in \mathcal{M}^{ba}$, the integral $\mathbf{E}_\mu[f]$ exists and is negative, let us first observe that for all n and all $\mathbf{Q} \in \mathcal{M}^e$ we have that $\mathbf{E}_\mathbf{Q}[f \wedge n] \leq 0$. This implies that for all n, necessarily, $\mathbf{E}_\mu[f \wedge n] \leq 0$. The sequence $\mathbf{E}_\mu[f \wedge n]$ is increasing and bounded above, so it converges and since $f \wedge n$ tends to f in μ-measure the $\lim_{n\to\infty} \mathbf{E}_\mu[f \wedge n]$ is necessarily the integral of f with respect to μ. It follows that also $\mathbf{E}_\mu[f] \leq 0$.   □

In the same style we can prove that $f \in \mathcal{K}^{\max}$ is the limit of a sequence obtained by stopping. If f is of the form $f = (H \cdot S)_\infty$ for some S-integrable admissible process H, let for $n \geq 1$:

$$T_n = \inf\{t \mid (H \cdot S)_t > n\}.$$

Proposition 13.7.2. *Suppose that S is a locally bounded semi-martingale that satisfies the (NFLVR) property. If f is 1-admissible and maximal and if $\mu \in \mathcal{M}^{ba}$, then f_{T_n} tends to f in μ-measure.*

Proof. Simply remark that for each $\mathbf{Q} \in \mathcal{M}^e$, we have $\mathbf{Q}[T_n < \infty] \leq \frac{1}{n}$.   □

Theorem 13.7.3. *Suppose that S is a locally bounded semi-martingale that satisfies the (NFLVR) property. If f is in the closure $\overline{\mathcal{G}^\infty}$ of $\mathcal{G}$, then $\mathbf{E}_\mu[f] = 0$ for each $\mu \in \mathcal{M}^{ba}$.*

Proof. Take $(f^n)_{n\geq 1}$ a sequence of bounded contingent claims in $\mathcal{G}$ that tends to f for the topology of $\mathcal{G}$. This means that $\sup\{\|f - f^n\|_{L^1(\mathbf{Q})} \mid \mathbf{Q} \in \mathcal{M}^e\}$ tends to zero. In particular the sequence $(f^n)_{n\geq 1}$ is a Cauchy sequence in $\mathcal{G}$ and hence for all $\mu \in \mathcal{M}^{ba}$ we have that $\mathbf{E}_\mu[|f^n - f^m|]$ tends to zero as n, m tend to infinity. Since, as easily seen, the sequence $(f^n)_{n\geq 1}$ tends to f in μ-measure, we obtain that f is μ-integrable and $\mathbf{E}_\mu[f] = \lim_{n\to\infty} \mathbf{E}_\mu[f^n] = 0$.   □

Proposition 13.7.4. *Suppose that S is a locally bounded semi-martingale that satisfies the (NFLVR) property. Suppose $f \in \mathcal{K}^{\max}$ and $f = (H \cdot S)_\infty$ for a maximal strategy H. If for each $\mu \in \mathcal{M}^{ba}$ the function f satisfies $\mathbf{E}_\mu[f] = 0$, then for each stopping time T and each $\mu \in \mathcal{M}^{ba}$, the function f_T is μ-integrable and satisfies $\mathbf{E}_\mu[f_T] = 0$.*

Proof. We already showed that f_T is in $\mathcal{G}$ and hence is μ-integrable for all $\mu \in \mathcal{M}^{ba}$ and that $\mathbf{E}_\mu[f_T] \leq 0$ for all μ in $\mathcal{M}^{ba}$.

Let us prove the opposite inequality. The sequence $\mathbf{E}_\mu[f \wedge n]$ of continuous functions on $\mathcal{M}^{ba}$ tends increasingly to 0. As follows from Dini's theorem, we have that for each $\delta > 0$ there is a number n such that $\mathbf{E}_\mathbf{Q}[f \wedge n] > -\delta$ for all $\mathbf{Q} \in \mathcal{M}^e$. But for each $\mathbf{Q} \in \mathcal{M}^e$ we have that $\mathbf{E}_\mathbf{Q}[f \mid \mathcal{F}_T] = f_T$ and hence that $\mathbf{E}_\mathbf{Q}[f \wedge n \mid \mathcal{F}_T] \leq f_T \wedge n$. This implies that for all $\mathbf{Q} \in \mathcal{M}^e$ and for all n large enough, we have $\mathbf{E}_\mathbf{Q}[f_T \wedge n] > -\delta$. We therefore obtain that $\mathbf{E}_\mu[f_T] \geq 0$. Since the converse inequality was already shown we obtain $\mathbf{E}_\mu[f_T] = 0$.   □

The converse of Theorem 13.7.3 is less trivial and we need the extra assumption that S is continuous.

Theorem 13.7.5. *Suppose that S is continuous and satisfies the (NFLVR) property. Suppose that $f \in \mathcal{K}^{\max}$ and suppose also that for each $\mu \in \mathcal{M}^{ba}$ we have $\mathbf{E}_\mu[f] = 0$, then $f \in \overline{\mathcal{G}^\infty}$.*

Proof. Let H be a maximal acceptable strategy such that $(H \cdot S)_\infty = f$. For each $n \geq 1$ put $T_n = \inf\{t \mid |(H \cdot S)_t| > n\}$ which is the first time the process $H \cdot S$ exits the interval $[-n, +n]$. Clearly $f^n = (H \cdot S)_{T_n}$ defines a sequence in $\mathcal{G}^\infty$ and we will show that f^n tends to f in the topology of $\mathcal{G}$. Because $-\mathbf{E}_\mu[f \wedge n]$ tends decreasingly to 0 for n tending to infinity we infer from Dini's theorem and Theorem 13.5.2 that $\inf\{\mathbf{E}_\mu[f \wedge n] \mid \mu \in \mathcal{M}^{ba}\}$ tends to zero. It follows that $\sup\{\mathbf{E}_\mathbf{Q}[f - f \wedge n] \mid \mathbf{Q} \in \mathcal{M}^e\}$ tends to zero as n tends to infinity. Because $(f - f^n)^+ = (f - n)^+ = (f - f \wedge n)$ we see that also $\sup\{\mathbf{E}_\mathbf{Q}[(f - f^n)^+] \mid \mathbf{Q} \in \mathcal{M}^e\}$ tends to zero as n tends to infinity. By Theorem 13.3.15 this means that f^n tends to f for the norm on $\mathcal{G}$.   □

Remark 13.7.6. The continuity assumption was only needed to obtain bounded contingent claims and could be replaced by the assumption that the jumps of $H \cdot S$ were bounded.

Example 13.7.7 (Addendum). In the following corollary we use the same notation as in Sect. 13.4, Example 13.4.4 and Theorem 13.4.7. Recall that the contingent claim $f = \sum_{n=1}^\infty f_n$ satisfies $\mathbf{E}_\mathbf{Q}[f] = 0$ for all $\mathbf{Q} \in \mathcal{M}$.

Corollary 13.7.8. *The function $f = \sum_{n=1}^\infty f_n$ is in $\mathcal{K}^{\max}$ but its integral with respect to $\mu \in \mathcal{M}^{ba}$ is not always zero.*

Proof. Indeed if it were, then f would be in $\overline{\mathcal{G}^\infty}$.   □

The Fundamental Theorem of Asset Pricing
for Unbounded Stochastic Processes (1998)

14.1 Introduction

The topic of the present paper is the statement and proof of the subsequent *Fundamental Theorem of Asset Pricing* in a *general version for not necessarily locally bounded semi-martingales*:

Main Theorem 14.1.1. *Let* $S = (S_t)_{t \in \mathbb{R}_+}$ *be an* $\mathbb{R}^d$*-valued semi-martingale defined on the stochastic base* $(\Omega, \mathcal{F}, (\mathcal{F}_t)_{t \in \mathbb{R}_+}, \mathbf{P})$.

Then S *satisfies the condition of* no free lunch with vanishing risk *if and only if there exists a probability measure* $\mathbf{Q} \sim \mathbf{P}$ *such that* S *is a sigma-martingale with respect to* $\mathbf{Q}$.

This theorem has been proved under the additional assumption that the process S is locally bounded in Chap. 9. Under this additional assumption one may replace the term "sigma-martingale" above by the term "local martingale".

We refer to Chap. 9 for the history of this theorem, which goes back to the seminal work of Harrison, Kreps and Pliska ([HK 79], [HP 81], [K 81]) and which is of central importance in the applications of stochastic calculus to Mathematical Finance. We also refer to Chap. 9 for the definition of the concept of *no free lunch with vanishing risk* which is a mild strengthening of the concept of *no-arbitrage*.

On the other hand, to the best of our knowledge, the second central concept in the above theorem, the notion of a *sigma-martingale* (see Definition 14.2.1 below) has not been considered previously in the context of Mathematical Finance. In a way, this is surprising, as we shall see in Remark 14.2.4 that this concept is very well-suited for the applications in Mathematical Finance, where one is interested not so much in the process S itself but rather in the family $(H \cdot S)$ of *stochastic integrals* on the process S, where H runs

[DS 98] The Fundamental Theorem of Asset Pricing for Unbounded Stochastic Processes. *Mathematische Annalen*, vol 312, pp. 215–250, (1998).

through the S-integrable predictable processes satisfying a suitable admissibility condition (see [HP 81], Chap. 9 and Sections 14.4 and 14.5 below). The concept of sigma-martingales, which relates to martingales similarly as sigma-finite measures relate to finite measures, has been introduced by C.S. Chou and M. Émery ([C 77], [E 80]) under the name "semi-martingales de la classe (Σ_m)". We shall show in Sect. 14.2 below (in particular in Example 14.2.3) that this concept is indeed natural and unavoidable in our context if we consider processes S with unbounded jumps.

The paper is organised as follows: In Sect. 14.2 we recall the definition and basic properties of sigma-martingales. In Sect. 14.3 we present the idea of the proof of the main theorem by considering the (very) special case of a two-step process $S = (S_0, S_1) = (S_t)_{t=0}^1$. This presentation is mainly for expository reasons in order to present the basic idea without burying it under the technicalities needed for the proof in the general case. But, of course, the consideration of the two-step case only yields the $(n + 1)$'th proof of the Dalang-Morton-Willinger theorem [DMW 90], i.e., the fundamental theorem of asset pricing in finite discrete time (for alternative proofs see [S 92], [KK 94], [R 94]). We end Sect. 14.3 by isolating in Lemma 14.3.5 the basic idea of our approach in an abstract setting.

Sect. 14.4 is devoted to the proof of the main theorem in full generality. We shall use the notion of the *jump measure* associated to a stochastic process and its *compensator* as presented, e.g., in [JS 87].

Sect. 14.5 is devoted to a generalisation of the duality results obtained in Chap. 11. These results are then used to identify the hedgeable elements as maximal elements in the cone of w-admissible outcomes. The concept of w-admissible integrand is a natural generalisation to the non- locally bounded case of the previously used concept of admissible integrand.

In [K 97] Y.M. Kabanov also presents a proof of our main theorem. This proof is based on Chap. 9 and the ideas of the present paper, but the technical aspects are worked out in a different way.

For unexplained notation and for further background on the main theorem we refer to Chap. 9.

14.2 Sigma-martingales

In this section we recall a concept which has been introduced by C.S. Chou [C 77] and M. Émery [E 80] under the name "semi-martingales de la classe (Σ_m)". This notion will play a central role in the present context. We take the liberty to baptize this notion as *"sigma-martingales"*. We choose this name as the relation between martingales and sigma-martingales is somewhat analogous to the relation between finite and sigma-finite measures (compare [E 80, Proposition 2]). Other researchers prefer the name martingale transform.

Definition 14.2.1. *An $\mathbb{R}^d$-valued semi-martingale $X = (X_t)_{t \geq 0}$ is called a* sigma-martingale *if there exists an $\mathbb{R}^d$-valued martingale M and an M-integrable predictable $\mathbb{R}_+$-valued process φ such that $X = \varphi \cdot M$.*

We refer to [E 80, Proposition 2] for several equivalent reformulations of this definition and we now essentially reproduce the basic example given by M. Émery [E 80, p. 152] which highlights the difference between the notion of a martingale (or, more generally, a local martingale) and a sigma-martingale.

Example 14.2.2 ([E 80]). A sigma-martingale which is not a local martingale.
Let the stochastic base $(\Omega, \mathcal{F}, \mathbf{P})$ be such that there are two independent stopping times T and U defined on it, both having an exponential distribution with parameter 1.
Define M by

$$M_t = \begin{cases} 0 & \text{for } t < T \wedge U \\ 1 & \text{for } t \geq T \wedge U \text{ and } T = T \wedge U \\ -1 & \text{for } t \geq T \wedge U \text{ and } U = T \wedge U. \end{cases}$$

It is easy to verify that M is almost surely well-defined and is indeed a martingale with respect to the filtration $(\mathcal{F}_t)_{t \in \mathbb{R}_+}$ generated by M. The deterministic (and therefore predictable) process $\varphi_t = \frac{1}{t}$ is M-integrable (in the sense of Stieltjes) and $X = \varphi \cdot M$ is well-defined:

$$X_t = \begin{cases} 0 & \text{for } t < T \wedge U \\ \frac{1}{T \wedge U} & \text{for } t \geq T \wedge U \text{ and } T = T \wedge U \\ -\frac{1}{T \wedge U} & \text{for } t \geq T \wedge U \text{ and } U = T \wedge U. \end{cases}$$

But X fails to be a martingale as $\mathbf{E}\left[\|X_t\|\right] = \infty$, for all $t > 0$, and it is not hard to see that X also fails to be a local martingale (see [E 80]), as $\mathbf{E}\left[\|X_T\|\right] = \infty$ for each stopping time T that is not identically zero. But, of course, X is a sigma-martingale. $\qquad\square$

We shall be interested in the class of semi-martingales S which *admit an equivalent measure under which they are a sigma-martingale*. We shall present an example of an $\mathbb{R}^2$-valued process S which admits an equivalent sigma-martingale measure (which in fact is unique) but which does not admit an equivalent local martingale measure. This example will be a slight extension of Émery's example.

The reader should note that in Émery's Example 14.2.2 above one may replace the measure $\mathbf{P}$ by an equivalent measure $\mathbf{Q}$ such that X is a true martingale under $\mathbf{Q}$. For example, choose $\mathbf{Q}$ such that under this new measure T and U are independent and distributed according to a law μ on $\mathbb{R}_+$ such that μ is equivalent to the exponential law (i.e., equivalent to Lebesgue-measure on $\mathbb{R}_+$) and such that $\mathbf{E}_\mu\left[\frac{1}{t}\right] < \infty$.

Example 14.2.3. A sigma-martingale S which does not admit an equivalent local martingale measure.

With the notation of the above example define the $\mathbb{R}^2$-valued process $S = (S^1, S^2)$ by letting $S^1 = X$ and S^2 the compensated jump at time $T \wedge U$ i.e.,

$$S_t^2 = \begin{cases} -2t & \text{for } t < T \wedge U \\ 1 - 2(T \wedge U) & \text{for } t \geq T \wedge U. \end{cases}$$

(Observe that $T \wedge U$ is exponentially distributed with parameter 2).

Clearly S^2 is a martingale with respect to the filtration $(\mathcal{F}_t)_{t \in \mathbb{R}_+}$ generated by S.

Denoting by $(\mathcal{G}_t)_{t \in \mathbb{R}_+}$ the filtration generated by S^2, it is a well-known property of the Poisson-process (see, [J 79, p. 347]) that on $\mathcal{G}$ the restriction of $\mathbf{P}$ to $\mathcal{G} = \bigvee_{t \in \mathbb{R}_+} \mathcal{G}_t$ is the unique probability measure equivalent to $\mathbf{P}$ under which S^2 is a martingale. It follows that $\mathbf{P}$ is the only probability measure on $\mathcal{F} = \bigvee_{t \in \mathbb{R}_+} \mathcal{F}_t$ equivalent to $\mathbf{P}$ under which $S = (S^1, S^2)$ is a sigma-martingale.

As S fails to be a local martingale under $\mathbf{P}$ (its first coordinate fails to be so) we have exhibited a sigma-martingale for which there does not exist an equivalent martingale measure.   □

Remark 14.2.4. In the applications to Mathematical Finance and in particular in the context of *pricing and hedging derivative securities by no-arbitrage arguments* the object of central interest is the set of *stochastic integrals* $H \cdot S$ on a given stock price process S, where H runs through the S-integrable predictable processes such that the process $H \cdot S$ satisfies appropriate regularity condition. In the present context this regularity condition is the admissibility condition $H \cdot S \geq -M$ for some $M \in \mathbb{R}_+$ (see [HP 81], Chap. 9 and Sect. 14.4 below). In different contexts one might impose an $L^p(\mathbf{P})$-boundedness condition on the stochastic integral $H \cdot S$ (see, e.g., [K 81], [DH 86], [Str 90], [DMSSS 97]). In Sect. 14.5, we shall deal with a different notion of admissibility, which is adjusted to the case of big jumps.

Now make the trivial (but nevertheless crucial) observation: passing from S to $\varphi \cdot S$, where φ is a strictly positive S-integrable predictable process, *does not change the set of stochastic integrals*. Indeed, we may write

$$H \cdot S = (H\varphi^{-1}) \cdot (\varphi \cdot S)$$

and, of course, the predictable $\mathbb{R}^d$-valued process H is S-integrable iff $H\varphi^{-1}$ is $\varphi \cdot S$-integrable.

The moral of this observation: when we are interested only in the set of *stochastic integrals* $H \cdot S$ the requirement that S is a sigma-martingale is just as good as the requirement that S is a true martingale.

We end this section with two observations which are similar to the results in [E 80]. The first one stresses the distinction between the notions of a local martingale and a sigma-martingale.

Proposition 14.2.5. *For a semi-martingale X the following assertions are equivalent.*

(i)   *X is a local martingale.*

(ii)   *$X = \varphi \cdot M$ where the M-integrable, predictable $\mathbb{R}_+$-valued process φ is increasing and M is a local martingale.*

(ii') *$X = \varphi \cdot M$ where the M-integrable, predictable $\mathbb{R}_+$-valued process φ is locally bounded and M is a local martingale.*

(iii) *$X = \varphi \cdot M$ where the M-integrable, predictable $\mathbb{R}_+$-valued process φ is increasing and M is a martingale.*

(iii') *$X = \varphi \cdot M$ where the M-integrable, predictable $\mathbb{R}_+$-valued process φ is locally bounded and M is a martingale.*

(iv)   *$X = \varphi \cdot M$ where the M-integrable, predictable $\mathbb{R}_+$-valued process φ is increasing and M is a martingale in $\mathcal{H}^1$.*

(iv') *$X = \varphi \cdot M$ where the M-integrable, predictable $\mathbb{R}_+$-valued process φ is locally bounded and M is a martingale in $\mathcal{H}^1$.*

We will not prove this proposition as its proof is similar to the proof of the next proposition.

Proposition 14.2.6. *For a semi-martingale X the following are equivalent*

(i)   *X is a sigma-martingale.*

(ii)   *$X = \varphi \cdot M$ where the M-integrable, predictable $\mathbb{R}_+$-valued process φ is strictly positive and M is a local martingale.*

(iii) *$X = \varphi \cdot M$ where the M-integrable, predictable $\mathbb{R}_+$-valued process φ is strictly positive and M is a martingale.*

(iv)   *$X = \varphi \cdot M$ where the M-integrable, predictable $\mathbb{R}_+$-valued process φ is strictly positive and M is a martingale in $\mathcal{H}^1$.*

Proof. Since (iv) implies (iii) implies (ii), and since obviously (i) is equivalent to (iii), we only have to prove that (ii) implies (iv). So suppose that there is a local martingale M as well as a non-negative M-integrable predictable process φ such that $X = \varphi \cdot M$. Let $(T_n)_{n \geq 1}$ be a sequence that localises M in the sense that T_n is increasing, tends to ∞ and for each n, M^{T_n} is in $\mathcal{H}^1$. Put $T_0 = 0$ and for $n \geq 1$, define N^n as the $\mathcal{H}^1$-martingale $N^n = (\varphi \mathbf{1}_{\rrbracket T_{n-1}, T_n \rrbracket}) \cdot M^{T_n}$. Let now $N = \sum_{n \geq 1} a_n N^n$, where the strictly positive sequence a_n is chosen such that $\sum a_n \|N^n\|_{\mathcal{H}^1} < \infty$. The process N is an $\mathcal{H}^1$-martingale. We now put $\psi = \mathbf{1}_{\{\varphi = 0\}} + \varphi \sum_n a_n^{-1} \mathbf{1}_{\rrbracket T_{n-1}, T_n \rrbracket}$. It is easy to check that $X = \psi \cdot N$ and that ψ is strictly positive. $\square$

Corollary 14.2.7. *A local sigma-martingale is a sigma-martingale. More precisely, if X is a semi-martingale and if $(T_k)_{k \geq 1}$ is an increasing sequence of stopping times, tending to ∞ such that each stopped process X^{T_k} is a sigma-martingale, then X itself is a sigma-martingale.*

Proof. For each k take φ^k, X^{T_k} integrable such that $\varphi^k > 0$ on $[\![0, T_k]\!]$, $\varphi^k \cdot X^{T_k}$ is a uniformly integrable martingale and $\|\varphi^k \cdot X^{T_k}\|_{\mathcal{H}^1} < 2^{-k}$. Put $T_0 = 0$ and $\varphi^0 = \varphi^1 \mathbf{1}_{[\![0]\!]}$. It is now obvious that $\varphi = \varphi^0 + \sum_{k \geq 1} \varphi^k \mathbf{1}_{]\!]T_{k-1}, T_k]\!]}$ is strictly positive, is X-integrable and is such that $\varphi \cdot X$ is an $\mathcal{H}^1$-martingale.   □

14.3 One-period Processes

In this section we shall present the basic idea of the proof of the main theorem in the easy context of a process consisting only of one jump. Let $S_0 \equiv 0$ and $S_1 \in L^0(\Omega, \mathcal{F}, \mathbf{P}; \mathbb{R}^d)$ be given and consider the stochastic process $S = (S_t)_{t=0}^1$; as filtration we choose $(\mathcal{F}_t)_{t=0}^1$ where $\mathcal{F}_1 = \mathcal{F}$ and $\mathcal{F}_0$ is some sub-σ-algebra of $\mathcal{F}$. At a first stage we shall in addition make the simplifying assumption that $\mathcal{F}_0$ is trivial, i.e., consists only of null-sets and their complements. In this setting the definition of the *no-arbitrage* condition *(NA)* (see [DMW 90] or Chap. 9) for the process S boils down to the requirement that, for $x \in \mathbb{R}^d$, the condition $(x, S) \geq 0$ a.s. implies that $(x, S) = 0$ a.s., where $(. , .)$ denotes the inner product in $\mathbb{R}^d$.

From the theorem of Dalang-Morton-Willinger [DMW 90] we deduce that the no-arbitrage condition *(NA)* implies the existence of an equivalent martingale measure for S, i.e., a measure $\mathbf{Q}$ on $(\Omega, \mathcal{F})$, $\mathbf{Q} \sim \mathbf{P}$, such that $\mathbf{E}_{\mathbf{Q}}[S_1] = 0$.

By now there are several alternative proofs of the Dalang-Morton-Willinger theorem known in the literature ([S 92], [KK 94], [R 94]) and we shall present yet another proof of this theorem in the subsequent lines. While some of the known proofs are very elegant (e.g., [R 94]) our subsequent proof is rather clumsy and heavy. But it is this method which will be extensible to the general setting of an $\mathbb{R}^d$-valued (not necessarily locally bounded) semi-martingale and will allow us to prove the main theorem in full generality.

Let us fix some notation: by Adm we denote the convex cone of *admissible elements* of $\mathbb{R}^d$ which consists of those $x \in \mathbb{R}^d$ such that the random variable (x, S_1) is (almost surely) uniformly bounded from below.

By K we denote the convex cone in $L^0(\Omega, \mathcal{F}, \mathbf{P})$ formed by the admissible stochastic integrals on the process S, i.e.,

$$K = \{(x, S_1) \mid x \in \text{Adm}\}$$

and we denote by C the convex cone in $L^\infty(\Omega, \mathcal{F}, \mathbf{P})$ formed by the uniformly bounded random variables dominated by some element of K, i.e.,

$$C = (K - L^0_+(\Omega, \mathcal{F}, \mathbf{P})) \cap L^\infty(\Omega, \mathcal{F}, \mathbf{P})$$
$$= \{f \in L^\infty(\Omega, \mathcal{F}, \mathbf{P}) \mid \text{there is } g \in K, f \leq g\} .$$

Under the assumption that S satisfies *(NA)*, i.e. $K \cap L^0_+ = \{0\}$, we want to find an equivalent martingale measure $\mathbf{Q}$ for the process S. The first argument is well-known in the present context (compare [S 92] and Theorem 14.4.1

below for a general version of this result; we refer to [S 94] for an account on
the history of this result, in particular on the work of J.A. Yan [Y 80] and
D. Kreps [K 81]):

Lemma 14.3.1. *If S satisfies (NA) the convex cone C is weak-star-closed
in $L^\infty(\Omega, \mathcal{F}, \mathbf{P})$, and $C \cap L^\infty_+(\Omega, \mathcal{F}, \mathbf{P}) = \{0\}$. Therefore the Hahn-Banach
theorem implies that there is a probability measure $\mathbf{Q}_1$ on $\mathcal{F}, \mathbf{Q}_1 \sim \mathbf{P}$ such
that*

$$\mathbf{E}_{\mathbf{Q}_1}[f] \leq 0, \quad \text{for } f \in C.$$

In the case, when S_1 is uniformly bounded, the measure $\mathbf{Q}_1$ is already the
desired equivalent martingale measure. Indeed, in this case the cone Adm of
admissible elements is the entire space $\mathbb{R}^d$ and therefore

$$\mathbf{E}_{\mathbf{Q}_1}[(x, S_1)] \leq 0, \quad \text{for } x \in \mathbb{R}^d,$$

which implies that

$$\mathbf{E}_{\mathbf{Q}_1}[(x, S_1)] = 0, \quad \text{for } x \in \mathbb{R}^d,$$

whence $\mathbf{E}_{\mathbf{Q}_1}[S] = 0$.

But if Adm is only a sub-cone of $\mathbb{R}^d$ (possibly reduced to $\{0\}$), we can only
say much less: first of all, S_1 need not be $\mathbf{Q}_1$-integrable. But even assuming
that $\mathbf{E}_{\mathbf{Q}_1}[S_1]$ exists we cannot assert that this value equals zero; we can only
assert that

$$(x, \mathbf{E}_{\mathbf{Q}_1}[S_1]) = \mathbf{E}_{\mathbf{Q}_1}[(x, S_1)] \leq 0, \quad \text{for } x \in \text{Adm},$$

which means that $\mathbf{E}_{\mathbf{Q}_1}[S_1]$ lies in the cone Adm° polar to Adm, i.e.,

$$\mathbf{E}_{\mathbf{Q}_1}[S_1] \in \text{Adm}^\circ = \{y \in \mathbb{R}^d \mid (x, y) \leq 1, \text{ for } x \in \text{Adm}\}$$
$$= \{y \in \mathbb{R}^d \mid (x, y) \leq 0, \text{ for } x \in \text{Adm}\}.$$

The next lemma will imply that, by passing from $\mathbf{Q}_1$ to an equivalent prob-
ability measure $\mathbf{Q}$ *with distance* $\|\mathbf{Q} - \mathbf{Q}_1\|$ *in total variation norm less than*
$\varepsilon > 0$, we may remedy both possible defects of $\mathbf{Q}_1$: under $\mathbf{Q}$ the expectation
of S_1 is well-defined and it equals zero.

The idea for the proof of this lemma goes back in the special case $d = 1$
and $\text{Adm} = \{0\}$ to the work of D. McBeth [MB 91].

Lemma 14.3.2. *Let $\mathbf{Q}_1$ be a probability measure as in Lemma 14.3.1 and
$\varepsilon > 0$. Denote by B the set of barycenters*

$$B = \{\mathbf{E}_{\mathbf{Q}}[S_1] \mid \mathbf{Q} \text{ probability on } \mathcal{F}, \mathbf{Q} \sim \mathbf{P}, \|\mathbf{Q} - \mathbf{Q}_1\| < \varepsilon,$$
$$\text{and } S_1 \text{ is } \mathbf{Q}\text{-integrable}\}.$$

*Then B is a convex subset of $\mathbb{R}^d$ containing 0 in its relative interior. In
particular, there is $\mathbf{Q} \sim \mathbf{Q}_1, \|\mathbf{Q} - \mathbf{Q}_1\| < \varepsilon$, such that $\mathbf{E}_{\mathbf{Q}}[S_1] = 0$.*

Proof. Clearly B is convex. Let us also remark that it is non-empty. To see this let us take $\delta > 0$ and let us define $\frac{d\mathbf{Q}}{d\mathbf{Q}_1} = \frac{\exp(-\delta\|S_1\|)}{\mathbf{E}_{\mathbf{Q}_1}[\exp(-\delta\|S_1\|)]}$. Clearly S_1 is $\mathbf{Q}$-integrable and from Lebesgue's dominated convergence theorem we deduce that for δ small enough $\|\mathbf{Q} - \mathbf{Q}_1\| < \varepsilon$.

If 0 were not in the relative interior of B we could find by the Minkowski separation theorem, an element $x \in \mathbb{R}^d$, such that B is contained in the halfspace $H_x = \{y \in \mathbb{R}^d \mid (x, y) \geq 0\}$ and such that $(x, y) > 0$ for some $y \in B$. In order to obtain a contradiction we distinguish two cases:

Case 1: x fails to be admissible, i.e., (x, S_1) fails to be (essentially) uniformly bounded from below.

First find, as above, a probability measure $\mathbf{Q}_2 \sim \mathbf{P}, \|\mathbf{Q}_1 - \mathbf{Q}_2\| < \frac{\varepsilon}{2}$ and such that $\mathbf{E}_{\mathbf{Q}_2}[S_1]$ is well-defined.

By assumption the random variable (x, S_1) is not (essentially) uniformly bounded from below, i.e., for $M \in \mathbb{R}_+$, the set

$$\Omega_M = \{\omega \mid (x, S_1(\omega)) < -M\}$$

has strictly positive $\mathbf{Q}_2$-measure. For $M \in \mathbb{R}_+$ define the measure $\mathbf{Q}^M$ by

$$\frac{d\mathbf{Q}^M}{d\mathbf{Q}_2} = \begin{cases} 1 - \frac{\varepsilon}{4} & \text{on } \Omega \setminus \Omega_M \\ 1 - \frac{\varepsilon}{4} + \frac{\varepsilon}{4\mathbf{Q}_2(\Omega_M)} & \text{on } \Omega_M. \end{cases}$$

It is straightforward to verify that $\mathbf{Q}^M$ is a probability measure, $\mathbf{Q}^M \sim \mathbf{P}$, $\|\mathbf{Q}^M - \mathbf{Q}_2\| < \frac{\varepsilon}{2}, \frac{d\mathbf{Q}^M}{d\mathbf{Q}_2} \in L^\infty$ and that

$$(x, \mathbf{E}_{\mathbf{Q}^M}[S_1]) = \mathbf{E}_{\mathbf{Q}^M}\left[(x, S^1)\right] \leq \left(1 - \frac{\varepsilon}{4}\right)\mathbf{E}_{\mathbf{Q}_2}\left[(x, S_1)\right] - \frac{\varepsilon M}{4}.$$

For $M > 0$ big enough the right hand side becomes negative which gives the desired contradiction.

Case 2: x is admissible, i.e., (x, S_1) is (essentially) uniformly bounded from below.

In this case we know from the Beppo-Levi theorem that the random variable (x, S_1) is $\mathbf{Q}_1$-integrable and that $\mathbf{E}_{\mathbf{Q}_1}[(x, S_1)] \leq 0$; (note that, for each $M \in \mathbb{R}_+$, we have that $(x, S_1) \wedge M$ is in C and therefore $\mathbf{E}_{\mathbf{Q}_1}[(x, S_1) \wedge M] \leq 0$).

Also note that (x, S_1) cannot be equal to 0 a.s., because as we saw above there is a $y \in B$ such that $(x, y) > 0$ and hence (x, S_1) cannot equal zero a.s. either. The no-arbitrage property then tells us that $\mathbf{Q}_1[(x, S_1) > 0]$ as well as $\mathbf{Q}_1[(x, S) < 0]$ are both strictly positive.

We next observe that for all $\eta > 0$ the variable $\exp\left(\eta(x, S_1)^-\right)$ is bounded. The measure $\mathbf{Q}_2$, given by $\frac{d\mathbf{Q}_2}{d\mathbf{Q}_1} = \frac{\exp\left(\eta(x, S_1)^-\right)}{\mathbf{E}_{\mathbf{Q}_1}[\exp(\eta(x, S_1)^-)]}$ is therefore well-defined. For η small enough we also have that $\|\mathbf{Q}_2 - \mathbf{Q}_1\| < \varepsilon$. But $\mathbf{Q}_2$ also satisfies:

$$\mathbf{E}_{\mathbf{Q}_2}[(x, S_1)] < 0.$$

Indeed:

$$\mathbf{E}_{\mathbf{Q}_1} \left[\exp \left(\eta(x, S_1)^- \right) (x, S_1) \right]$$
$$= -\mathbf{E}_{\mathbf{Q}_1} \left[\exp \left(\eta(x, S_1)^- \right) (x, S_1)^- \right] + \mathbf{E}_{\mathbf{Q}_1} \left[(x, S_1)^+ \right]$$
$$< -\mathbf{E}_{\mathbf{Q}_1} \left[(x, S_1)^- \right] + \mathbf{E}_{\mathbf{Q}_1} \left[(x, S_1)^+ \right] \leq 0 \,.$$

The measure $\mathbf{Q}_2$ does not necessarily satisfy the requirement that $\mathbf{E}_{\mathbf{Q}_2} \left[\|S_1\| \right] < \infty$. We therefore make a last transformation and we define

$$d\mathbf{Q} = \frac{\exp(-\delta \|S_1\|)}{\mathbf{E}_{\mathbf{Q}_2} \left[\exp(-\delta \|S_1\|) \right]} d\mathbf{Q}_2 \,.$$

For $\delta > 0$ tending to zero we obtain that $\|\mathbf{Q} - \mathbf{Q}_2\|$ tends to 0 and $\mathbf{E}_{\mathbf{Q}} \left[(x, S_1) \right]$ tends to $\mathbf{E}_{\mathbf{Q}_2} \left[(x, S_1) \right]$ which is strictly negative. So for δ small enough we find a probability measure $\mathbf{Q}$ such that $\mathbf{Q} \sim \mathbf{P}$, $\|\mathbf{Q} - \mathbf{Q}_1\| < \varepsilon$, $\mathbf{E}_{\mathbf{Q}} \left[\|S_1\| \right] < \infty$ and $\mathbf{E}_{\mathbf{Q}} \left[(x, S_1) \right] < 0$, a contradiction to the choice of x.   □

Lemma 14.3.2 in conjunction with Lemma 14.3.1 implies in particular that, given the stochastic process $S = (S_t)_{t=0}^1$ with $S_0 \equiv 0$ and $\mathcal{F}_0$ trivial, we may find a probability measure $\mathbf{Q} \sim \mathbf{P}$ such that S is a $\mathbf{Q}$-martingale. We obtained the measure $\mathbf{Q}$ in two steps: first (Lemma 14.3.1) we found $\mathbf{Q}_1 \sim \mathbf{P}$ which took care of the *admissible integrands*, which means that

$$\mathbf{E}_{\mathbf{Q}_1} \left[(x, S_1) \right] \leq 0 \,, \quad \text{for } x \in \text{Adm} \,.$$

In a second step (Lemma 14.3.2) we found $\mathbf{Q} \sim \mathbf{P}$ such that $\mathbf{Q}$ took care of *all integrands*, i.e.,

$$(x, \mathbf{E}_{\mathbf{Q}} \left[S_1 \right]) = \mathbf{E}_{\mathbf{Q}} \left[(x, S_1) \right] \leq 0 \,, \quad \text{for } x \in \mathbb{R}^d$$

and therefore

$$\mathbf{E}_{\mathbf{Q}} \left[S_1 \right] = 0 \,,$$

which means that S is a $\mathbf{Q}$-martingale.

In addition, we could assert in Lemma 14.3.2 that $\|\mathbf{Q}_1 - \mathbf{Q}\| < \varepsilon$, a property which will be crucial in the sequel.

The strategy for proving the main theorem will be similar to the above approach. Given a semi-martingale $S = (S_t)_{t \in \mathbb{R}_+}$ defined on $(\Omega, \mathcal{F}, (\mathcal{F}_t)_{t \in \mathbb{R}_+}, \mathbf{P})$ we first replace $\mathbf{P}$ by $\mathbf{Q}_1 \sim \mathbf{P}$ such that $\mathbf{Q}_1$ "takes care of the admissible integrands", i.e.,

$$\mathbf{E}_{\mathbf{Q}_1} \left[(H \cdot S)_\infty \right] \leq 0 \,, \quad \text{for } H\text{-admissible.}$$

For this first step, the necessary technology has been developed in Chap. 9 and may be carried over almost verbatim.

The new ingredient developed in the present paper is the second step which takes care of the "big jumps" of S. By repeated application of an argument as in Lemma 14.3.2 above we would like to change $\mathbf{Q}_1$ into a measure $\mathbf{Q}, \mathbf{Q} \sim \mathbf{P}$, such that S becomes a **Q**-martingale. A glance at Example 14.2.3 above reveals that this hope is, in the general setting, too optimistic and we can only try to turn S into a **Q**-sigma-martingale. This will indeed be possible, i.e., we shall be able to find **Q** and a strictly positive predictable process φ, such that, *for every — not necessarily admissible — predictable $\mathbb{R}^d$-valued process H satisfying $\|H\|_{\mathbb{R}^d} \leq \varphi$*, we have that $H \cdot S$ is a **Q**-martingale. In particular $\varphi \cdot S$ will be a **Q**-martingale.

In order to complete this program we shall isolate in Lemma 14.3.5 below, the argument proving Lemma 14.3.2 in the appropriate abstract setting. In particular we show that the construction in the proof of Lemma 14.3.2 may be parameterised to depend in a measurable way on a parameter η varying in a measure space $(E, \mathcal{E}, \pi)$. The proof of this lemma is standard but long. One has to check a lot of measurability properties in order to apply the measurable selection theorem. Since the proofs are not really used in the sequel and are standard, the reader can, at a first reading, look at the Definition 14.3.3, convince herself that the two parameterisations given in Lemma 14.3.4 are equivalent and look at Lemma 14.3.5.

Definition 14.3.3. *We say that a probability measure μ on $\mathbb{R}^d$ satisfies the (NA) property if for every $x \in \mathbb{R}^d$ we have $\mu(\{a \mid (x, a) < 0\}) > 0$ as soon as $\mu(\{a \mid (x, a) > 0\}) > 0$.*

We start with some notation that we will keep fixed for the rest of this section. We first assume that $(E, \mathcal{E}, \pi)$ is a probability space that is saturated for the null sets, i.e. if $A \subset B \in \mathcal{E}$ and if $\pi(B) = 0$ then $A \in \mathcal{E}$. The probability π can easily be repaced into a σ-finite positive measure, but in order not to overload the statements we skip this straightforward generalisation. We recall that a Polish space X is a topological space that is homeomorphic to a complete separable metrisable space. The Borel σ-algebra of X is denoted by $\mathcal{B}(X)$. We will mainly be working in a space $E \times X$ where X is a Polish space. The canonical projection of $E \times X$ onto E is denoted by pr. If $A \in \mathcal{E} \otimes \mathcal{B}(X)$ then $pr(A) \in \mathcal{E}$, see [A 65] and [D 72]. Furthermore there is a countable family $(f_n)_{n \geq 1}$ of measurable functions $f_n \colon pr(A) \to X$ such that

(1) for each $n \geq 1$ the graph of f_n is a selection of A, i.e. $\{(\eta, f_n(\eta)) \mid \eta \in pr(A)\} \subset A$,

(2) for each $\eta \in pr(A)$ the set $\{f_n(\eta) \mid n \geq 1\}$ is dense in $A_\eta = \{x \mid (\eta, x) \in A\}$.

We call such a sequence a countable dense selection of A.

The set $\mathcal{P}(\mathbb{R}^d)$ of probability measures on $\mathbb{R}^d$ is equipped with the topology of convergence in law, also called weak-star convergence. It is well-known that $\mathcal{P}(\mathbb{R}^d)$ is Polish. If $F \colon E \to \mathcal{P}(\mathbb{R}^d)$ is a mapping, then the measurability of F can be reformulated as follows: for each bounded Borel function g, we have that the mapping $\eta \mapsto \int_{\mathbb{R}^d} g(y) \, dF_\eta(y)$ is $\mathcal{E}$-measurable. This is easily seen using monotone class arguments. Using such a given measurable function F as a transition kernel, we can define a probability measure λ_F on $E \times \mathbb{R}^d$ as follows. For an element $D \in \mathcal{E} \otimes \mathcal{B}(\mathbb{R}^d)$ of the form $D = A \times B$, we define $\lambda_F(D) = \int_A F_\eta(B) \, \pi(d\eta)$.

For each $\eta \in E$ we define the set $\mathrm{Supp}(F_\eta)$ as the support of the measure F_η, i.e. the smallest closed set of full F_η-measure. The set $\mathcal{S}$ is defined as $\{(\eta, x) \mid x \in \mathrm{Supp}(F_\eta)\}$. The set $\mathcal{S}$ is an element of $\mathcal{E} \otimes \mathcal{B}(\mathbb{R}^d)$. Indeed, take a countable base $(U_n)_{n \geq 1}$ of the topology of $\mathbb{R}^d$ and write the complement as:

$$\mathcal{S}^c = \bigcup_{n \geq 1} \left(\{\eta \mid F_\eta(U_n) = 0\} \times U_n \right).$$

If $x \colon E \to \mathbb{R}^d$ is a measurable function then $\varphi \colon E \to \mathbb{R}_+ \cup \{+\infty\}$ defined as $\varphi(\eta) = \|(x_\eta, \,.\,)^-\|_{L^\infty(F_\eta)}$ is $\mathcal{E}$-measurable. Indeed take a countable dense selection $(f_n)_{n \geq 1}$ of $\mathcal{S}$ and observe that $\varphi(\eta) = \inf\{(x_\eta, f_n(\eta))^- \mid n \geq 1\}$.

For each $\eta \in E$ we denote by $\mathrm{Adm}(\eta)$ the cone in $\mathbb{R}^d$ consisting of elements $x \in \mathbb{R}^d$ so that $(x, \,.\,)^- \in L^\infty(F_\eta)$. The set Adm is then defined as $\{(\eta, x) \mid x \in \mathrm{Adm}(\eta)\}$. This set is certainly in $\mathcal{E} \otimes \mathcal{B}(\mathbb{R}^d)$. Indeed $\mathrm{Adm} = \{(\eta, x) \mid \inf_{n \geq 1}(x, f_n) > -\infty\}$ where the sequence $(f_n)_{n \geq 1}$ is a countable dense selection of $\mathcal{S}$.

Lemma 14.3.4. *If F is a measurable mapping from $(E, \mathcal{E}, \pi)$ into the probability measures on $\mathbb{R}^d$, then the following are equivalent:*

(1) *For almost every $\eta \in E$, the probability measure F_η satisfies the no-arbitrage property.*

(2) *For every measurable selection x_η of Adm, we have $\lambda_F\left[(\eta, a) \mid x_\eta(a) < 0\right] > 0$ as soon as $\lambda_F\left[(\eta, a) \mid x_\eta(a) > 0\right] > 0$.*

Proof. The implication $1 \Rightarrow 2$ is almost obvious since for each $\eta \in E$ we have that $F_\eta\left[\{a \mid (x_\eta, a) < 0\}\right] > 0$ as soon as $F_\eta\left[\{a \mid (x_\eta, a) > 0\}\right] > 0$. Therefore if $\lambda_F\left[(\eta, a) \mid x_\eta(a) > 0\right] > 0$, we have that $\pi(B) > 0$ where B is the set

$$B = \{\eta \in E \mid F_\eta\left[\{a \mid (x_\eta, a) > 0\}\right] > 0\} \,.$$

For the elements $\eta \in B$ we then also have that $F_\eta\left[\{a \mid (x_\eta, a) < 0\}\right] > 0$ and integration with respect to π then gives the result:

$$\lambda_F\left[(\eta, a) \mid (x_\eta, a) < 0\right] = \int_E \pi(d\eta) F_\eta\left[\{a \mid (x_\eta, a) < 0\}\right] > 0 \,.$$

Let us now prove the reverse implication $2 \Rightarrow 1$.

We consider the set

$$A = \{(\eta, x) \mid F_\eta\left[a \mid (x, a) \geq 0\right] = 1 \text{ and } F_\eta\left[a \mid (x, a) > 0\right] > 0\} \,.$$

The reader can check that this set is in $\mathcal{E} \otimes \mathcal{B}(\mathbb{R}^d)$ and therefore the set $B = pr(A) \in \mathcal{E}$. Suppose that $\pi(B) > 0$ and take a measurable selection x_η of A. Outside B we define $x_\eta = 0$. Clearly $\lambda_F\left(\{(\eta, a) \mid (x_\eta, a) > 0\}\right) > 0$ and hence we have that $\lambda_F\left(\{(\eta, a) \mid (x_\eta, a) < 0\}\right) > 0$, a contradiction since $(x_\eta, a) \geq 0$, λ_F a.s.. So we see that $B = \emptyset$ a.s. or what is the same for almost every $\eta \in E$ the measure F_η satisfies the no-arbitrage property. $\qquad\square$

The Crucial Lemma 14.3.5. *Let $(E, \mathcal{E}, \pi)$ be a probability measure space and let $(F_\eta)_{\eta \in E}$ be a family of probability measures on $\mathbb{R}^d$ such that the map $\eta \mapsto F_\eta$ is $\mathcal{E}$-measurable.*

Let us assume that F satisfies the property that for each measurable map $x \colon E \to \mathbb{R}^d$, $\eta \mapsto x_\eta$ with the property that for every $\eta \in E$ we have $(x_\eta, y) \geq -1$, for F_η almost every y, we also have that $\int_{\mathbb{R}^d} F_\eta(dy)(x_\eta, y) \leq 0$.

Let $\varepsilon \colon E \to \mathbb{R}_+ \setminus \{0\}$ be $\mathcal{E}$-measurable and strictly positive.

Then, we may find an $\mathcal{E}$-measurable map $\eta \mapsto G_\eta$ from E to the probability measures on $\mathcal{B}(\mathbb{R}^d)$ such that, for π-almost every $\eta \in E$,

(i) $F_\eta \sim G_\eta$ and $\|F_\eta - G_\eta\| < \varepsilon_\eta$,
(ii) $\mathbf{E}_{G_\eta}[\|y\|] < \infty$ and $\mathbf{E}_{G_\eta}[y] = 0$.

Proof. As observed above the set $\mathcal{P}(\mathbb{R}^d)$ of probability measures on $\mathbb{R}^d$, endowed with the weak-star topology is a Polish space. We will show that the set

$$\left\{ (\eta, \mu) \;\middle|\; \int_{\mathbb{R}^d} \|x\| \, d\mu < \infty; \; \int_{\mathbb{R}^d} x \, d\mu = 0; \; F_\eta \sim \mu; \|\mu - F_\eta\| < \varepsilon(\eta) \right\}$$

is in $\mathcal{E} \otimes \mathcal{B}(\mathcal{P}(\mathbb{R}^d))$. Since, by Lemma 14.3.4, for almost all η, the measure F_η satisfies the no-arbitrage assumption of Definition 14.3.3, we obtain that, for almost all η, the vertical section is non-empty. We can therefore find a measurable selection G_η and this will then end the proof.

The proof of the measurability property is easy but requires some arguments.

First we observe that the set $M = \{\mu \mid \int_{\mathbb{R}^d} \|x\| \, d\mu < \infty\}$ is in $\mathcal{B}(\mathcal{P}(\mathbb{R}^d))$. This follows from the fact that $\mu \mapsto \int \|x\| \, d\mu$ is Borel-measurable as it is an increasing limit of the weak-star continuous functionals $\mu \mapsto \int \min(\|x\|, n) \, d\mu$.

Next we observe that $M \to \mathbb{R}^d; \mu \mapsto \int_{\mathbb{R}^d} x \, d\mu$ is Borel-measurable.

The third observation is that $\{(\eta, \mu) \mid \|\mu - F_\eta\| < \varepsilon(\eta)\}$ is in $\mathcal{E} \otimes \mathcal{B}(\mathcal{P}(\mathbb{R}^d))$.

Finally we show that $\{(\eta, \mu) \mid \mu \sim F_\eta\}$ is also in $\mathcal{E} \otimes \mathcal{B}(\mathcal{P}(\mathbb{R}^d))$. This will then end the proof of the measurability property.

We take an increasing sequence of finite σ-algebras $\mathcal{D}_n$ such that $\mathcal{B}(\mathbb{R}^d)$ is generated by $\bigcup_n \mathcal{D}_n$. For each n we see that the mapping $(\eta, \mu, x) \mapsto \frac{d\mu}{dF_\eta}\Big|_{\mathcal{D}_n}(x) = q_n(\eta, \mu, x)$ is $\mathcal{E} \otimes \mathcal{B}(\mathcal{P}(\mathbb{R}^d)) \otimes \mathcal{B}(\mathbb{R}^d)$-measurable. The mapping

$$q(\eta, \mu, x) = \liminf q_n(\eta, \mu, x)$$

is clearly $\mathcal{E} \otimes \mathcal{B}(\mathcal{P}(\mathbb{R}^d)) \otimes \mathcal{B}(\mathbb{R}^d)$-measurable. By the martingale convergence theorem we have that for each μ, the mapping q defines the Radon-Nikodým density of the part of μ that is absolutely continuous with respect to F_η. Now we have that

$$\{(\eta, \mu) \mid \mu \sim F_\eta\}$$
$$= \left\{ (\eta, \mu) \;\middle|\; \int_{\mathbb{R}^d} q(\eta, \mu, x) \, dF_\eta(x) = 1; \lim_n \int_{\mathbb{R}^d} (nq) \wedge 1 \, dF_\eta(x) = 1 \right\}$$

and this shows that $\{(\eta, \mu) \mid \mu \sim F_\eta\}$ is in $\mathcal{E} \otimes \mathcal{B}(\mathcal{P}(\mathbb{R}^d))$.

In the above arguments we did suppose that $(E, \mathcal{E}, \pi)$ is complete. Now we drop this assumption. In that case we first complete the space $(E, \mathcal{E}, \pi)$ by replacing the σ-algebra $\mathcal{E}$ by $\widetilde{\mathcal{E}}$ generated by $\mathcal{E}$ and all the null sets. We then obtain an $\widetilde{\mathcal{E}}$-measurable mapping $\widetilde{F}_\eta$ which can, by modifying it on a set of measure zero, be replaced by an $\mathcal{E}$-measurable mapping F_η such that π almost surely $F_\eta = \widetilde{F}_\eta$. $\qquad\square$

Remark 14.3.6. We have not striven for maximal generality in the formulation of Lemma 14.3.5: for example, we could replace the probability measures F_η by finite non-negative measures on $\mathbb{R}^d$. In this case we may obtain the G_η in such a way that the total mass $G_\eta(\mathbb{R}^d)$ equals $F_\eta(\mathbb{R}^d)$, π-almost surely.

To illustrate the meaning of the Crucial Lemma we note in the spirit of [MB 91] which shows in particular the limitations of the no-arbitrage-theory when applied e.g. to Gaussian models for the stock returns in finite discrete time.

Proposition 14.3.7. *Let $(S_t)_{t=0}^T$ be an adapted $\mathbb{R}^d$-valued process based on $(\Omega, \mathcal{F}, (\mathcal{F}_t)_{t=0}^T, \mathbf{P})$ such that for every predictable process $(h_t)_{t=1}^T$ we have that $(h \cdot S)_T = \sum_{t=1}^T h_t \Delta S_t$ is unbounded from above and from below as soon as $(h \cdot S)_T \not\equiv 0$. For example, this assumption is satisfied if the $\mathcal{F}_{t-1}$-conditional distributions of the jumps ΔS_t are non-degenerate and normally distributed on $\mathbb{R}^d$.*

Then, for $\varepsilon > 0$, there is a measure $\mathbf{Q} \sim \mathbf{P}, \|\mathbf{Q} - \mathbf{P}\| < \varepsilon$, such that S is a $\mathbf{Q}$-martingale.

As a consequence, the set of equivalent martingale measures is dense with respect to the variation norm in the set of $\mathbf{P}$-absolutely continuous measures.

Proof. Suppose first that $T = 1$. Contrary to the setting of the motivating example at the beginning of this section we do not assume that $\mathcal{F}_0$ is trivial.

Let $(E, \mathcal{E}, \pi)$ be $(\Omega, \mathcal{F}_0, \mathbf{P})$ and denote by $(F_\omega)_{\omega \in \Omega}$ the $\mathcal{F}_0$-conditional distribution of $\Delta S_1 = S_1 - S_0$. The assumption of Lemma 14.3.5 is (trivially) satisfied as by hypothesis the $\mathcal{F}_0$-measurable functions $x(\omega)$ such that $\mathbf{P}$-a.s. we have $(x_\omega, y) \geq -1, F_\omega$-a.s., satisfy $(x_\omega, y) = 0, F_\omega$-a.s., for $\mathbf{P}$-a.e. $\omega \in \Omega$.

Choose $\varepsilon(\omega) \equiv \varepsilon > 0$ and find G_η as in the lemma. To translate the change of the conditional distributions of ΔS_1 into a change of the measure $\mathbf{P}$, find $Y : \Omega \times \mathbb{R}^d \to \mathbb{R}_+$,

$$Y(\omega, x) = \frac{dG_\omega}{dF_\omega}(x), \qquad x \in \mathbb{R}^d, \omega \in \Omega$$

such that, for $\mathbf{P}$-a.e. $\omega \in \Omega, Y(\omega, \,.\,)$ is a version of the Radon-Nikodým derivative of G_ω with respect to F_ω, and such that $Y(\,.\,,\,.\,)$ is $\mathcal{F}_0 \otimes \mathcal{B}(\mathbb{R}^d)$-measurable.

Letting

$$\frac{d\widehat{\mathbf{Q}}}{d\mathbf{P}}(\omega) = Y(\omega, \Delta S_1(\omega))$$

we obtain an $\mathcal{F}_1$-measurable density of a probability measure. Assertion (i) of Lemma 14.3.5 implies that $\widehat{\mathbf{Q}} \sim \mathbf{P}$ and $\|\widehat{\mathbf{Q}} - \mathbf{P}\| < \varepsilon$. Assertion (ii) implies that

$$\mathbf{E}_{\widehat{\mathbf{Q}}} \left[\|\Delta S_1\|_{\mathbb{R}^d} \mid \mathcal{F}_0 \right] < \infty, \quad \text{a.s.}$$

and

$$\mathbf{E}_{\widehat{\mathbf{Q}}} \left[\Delta S_1 \mid \mathcal{F}_0 \right] = 0, \quad \text{a.s..}$$

We are not quite finished yet as this only shows that $(S_t)_{t=0}^1$ is a $\widehat{\mathbf{Q}}$-sigma-martingale but not necessarily a $\widehat{\mathbf{Q}}$-martingale as it may happen that $\mathbf{E}_{\widehat{\mathbf{Q}}} \left[\|\Delta S_1\|_{\mathbb{R}^d} \right] = \infty$. But it is easy to overcome this difficulty: find a strictly positive $\mathcal{F}_0$-measurable function $w(\omega)$, normalised so that $\mathbf{E}_{\widehat{\mathbf{Q}}} [w] = 1$ and such that $\mathbf{E}_{\widehat{\mathbf{Q}}} \left[w(\omega) \mathbf{E} \left[\|\Delta S_1\|_{\mathbb{R}^d} \mid \mathcal{F}_0 \right] \right] < \infty$. We can construct w is such a way that the probability measure $\mathbf{Q}$ defined by

$$\frac{d\mathbf{Q}(\omega)}{d\widehat{\mathbf{Q}}(\omega)} = w(\omega),$$

still satisfies $\|\mathbf{Q} - \mathbf{P}\| < \varepsilon$. Then

$$\mathbf{E}_{\mathbf{Q}} \left[\|\Delta S_1\|_{\mathbb{R}^d} \right] < \infty$$

and

$$\mathbf{E}_{\mathbf{Q}} \left[\Delta S_1 \mid \mathcal{F}_0 \right] = 0, \quad \text{a.s.,}$$

i.e., S is a $\mathbf{Q}$-martingale.

To extend the above argument from $T = 1$ to arbitrary $T \in \mathbb{N}$ we need yet another small refinement: an inspection of the proof of Lemma 14.3.5 above reveals that in addition to assertions (i) and (ii) of Lemma 14.3.5, and given $M > 1$, we may choose G_η such that

$$(\text{iii}) \qquad \left\| \frac{dG_\eta}{dF_\eta} \right\|_{L^\infty(\mathbb{R}^d, F_\eta)} \leq M, \quad \pi\text{-a.s..}$$

We have not mentioned this additional assertion in order not to overload Lemma 14.3.5 and as we shall only need (iii) in the present proof.

Using (iii), with $M = 2$ say, and, choosing w above also uniformly bounded by 2, the argument in the first part of the proof yields a probability $\mathbf{Q} \sim \mathbf{P}$, $\|\mathbf{Q} - \mathbf{P}\| < \varepsilon$, such that $\|\frac{d\mathbf{Q}}{d\mathbf{P}}\|_{L^\infty(\mathbf{P})} \leq 4$.

Now let $T \in \mathbb{N}$ and $(S_t)_{t=0}^T$, based on $(\Omega, (\mathcal{F}_t)_{t=0}^T, \mathcal{F}, \mathbf{P})$, be given. By backward induction on $t = T, \ldots, 1$ apply the first part of the proof to find $\mathcal{F}_t$-measurable densities Z_t such that, defining the probability measure $\mathbf{Q}^{(t)}$ by

$$\frac{d\mathbf{Q}^{(t)}}{d\mathbf{P}} = Z_t,$$

we have that the two-step process $\left(S_u \prod_{v=t+1}^T Z_v \right)_{u=t-1}^t$ is a $\mathbf{Q}^{(t)}$-martingale with respect to the filtration $(\mathcal{F}_u)_{u=t-1}^t$, $\mathbf{Q}^{(t)} \sim \mathbf{P}$, $\|\mathbf{Q}^{(t)} - \mathbf{P}\|_1 < \varepsilon 4^{-T} T^{-1}$, and such that $\|Z_t\|_{L^\infty(\mathbf{P})} \leq 4$.

Defining

$$\frac{d\mathbf{Q}}{d\mathbf{P}} = \prod_{t=1}^{T} Z_t \,,$$

we obtain a probability measure $\mathbf{Q}$, $\mathbf{Q} \sim \mathbf{P}$ such that $(S_t)_{t=0}^{T}$ is a martingale under $\mathbf{Q}$. Indeed,

$$
\begin{aligned}
\mathbf{E_Q}\left[\Delta S_t \mid \mathcal{F}_{t-1}\right] &= \mathbf{E_P}\left[\Delta S_t \prod_{u=1}^{T} Z_u \,\bigg|\, \mathcal{F}_{t-1}\right] \\
&= \left(\prod_{u=1}^{t-1} Z_u\right) \mathbf{E_P}\left[Z_t \Delta S_t \prod_{u=t+1}^{T} Z_u \,\bigg|\, \mathcal{F}_{t-1}\right] \\
&= \left(\prod_{u=1}^{t-1} Z_u\right) \mathbf{E}_{\mathbf{Q}^{(t)}}\left[\Delta S_t \prod_{u=t+1}^{T} Z_u \,\bigg|\, \mathcal{F}_{t-1}\right] = 0
\end{aligned}
$$

and

$$\mathbf{E_Q}\left[\|\Delta S_t\|_{\mathbb{R}^d}\right] \leq \left\|\prod_{u=1}^{t-1} Z_u\right\|_{\infty} \cdot \mathbf{E}_{\mathbf{Q}^{(t)}}\left[\left\|\Delta S_t \prod_{u=t+1}^{T} Z_u\right\|_{\mathbb{R}^d}\right] < \infty\,.$$

Finally we may estimate $\|\mathbf{Q} - \mathbf{P}\|_1$ by

$$
\begin{aligned}
\|\mathbf{Q} - \mathbf{P}\|_1 &= \mathbf{E_P}\left[\left|\prod_{t=1}^{T} Z_t - 1\right|\right] \\
&\leq \mathbf{E_P}\left[\sum_{t=1}^{T}\left|\prod_{u=1}^{t} Z_u - \prod_{u=1}^{t-1} Z_u\right|\right] \\
&\leq \sum_{t=1}^{T}\left\|\prod_{u=1}^{t-1} Z_u\right\|_{L^{\infty}(\mathbf{P})} \mathbf{E_P}\left[|Z_t - 1|\right] \\
&\leq T \cdot 4^T \varepsilon 4^{-T} T^{-1} = \varepsilon\,.
\end{aligned}
$$

The proof of the first part of Proposition 14.3.7 is thus finished and we have shown in the course of the proof that we may find $\mathbf{Q}$ such that, in addition to the assertions of the proposition, $\frac{d\mathbf{Q}}{d\mathbf{P}}$ is uniformly bounded.

As regards the final assertion, let $\mathbf{P}'$ be any $\mathbf{P}$-absolutely continuous measure. For given $\varepsilon > 0$, first take $\mathbf{P}'' \sim \mathbf{P}$ such that $\|\mathbf{P}'' - \mathbf{P}'\| < \varepsilon$. Now apply the first assertion with $\mathbf{P}''$ replacing $\mathbf{P}$. As a result we get an equivalent martingale measure $\mathbf{Q}$ such that $\|\mathbf{Q} - \mathbf{P}''\| < \varepsilon$, hence also $\|\mathbf{Q} - \mathbf{P}'\| < 2\varepsilon$.

This finishes the proof of Proposition 14.3.7. $\qquad\square$

14.4 The General $\mathbb{R}^d$-valued Case

In this section $S = (S_t)_{t\in\mathbb{R}_+}$ denotes a general $\mathbb{R}^d$-valued càdlàg semi-martingale based on $(\Omega, \mathcal{F}, (\mathcal{F}_t)_{t\in\mathbb{R}_+}, \mathbf{P})$ where we assume that the filtration $(\mathcal{F}_t)_{t\in\mathbb{R}_+}$ satisfies the usual conditions of completeness and right continuity.

Similarly as in Chap. 9 we define an S-integrable $\mathbb{R}^d$-valued predictable process $H = (H_t)_{t\in\mathbb{R}_+}$ to be an *admissible integrand* if the stochastic process

$$(H \cdot S)_t = \int_0^t (H_u, dS_u), \qquad t \in \mathbb{R}_+$$

is (almost surely) uniformly bounded from below.

It is important to note that, similarly as in Proposition 14.3.7 above, it may happen that the cone of admissible integrands is rather small and possibly even reduced to zero: consider, for example, the case when S is a compound Poisson process with (two-sided) unbounded jumps, i.e., $S_t = \sum_{i=1}^{N_t} X_i$, where $(N_t)_{t\in\mathbb{R}_+}$ is a Poisson process and $(X_i)_{i=1}^{\infty}$ an i.i.d. sequence of real random variables such that $\|X_i^+\|_\infty = \|X_i^-\|_\infty = \infty$. Clearly, a predictable process H, such that $H \cdot S$ remains uniformly bounded from below, must vanish almost surely.

Continuing with the general setup we denote by K the convex cone in $L^0(\Omega, \mathcal{F}, \mathbf{P})$ given by

$$K = \{f = (H \cdot S)_\infty \mid H \text{ admissible}\}$$

where this definition requires in particular that the random variable $(H \cdot S)_\infty := \lim_{t\to\infty}(H \cdot S)_t$ is (almost surely) well-defined (compare Definition 9.2.7).

Again we denote by C the convex cone in $L^\infty(\Omega, \mathcal{F}, \mathbf{P})$ formed by the uniformly bounded random variables dominated by some element of K, i.e.,

$$C = (K - L_+^0(\Omega, \mathcal{F}, \mathbf{P})) \cap L^\infty(\Omega, \mathcal{F}, \mathbf{P})$$
$$= \{f \in L^\infty(\Omega, \mathcal{F}, \mathbf{P}) \mid \text{ there is } g \in K, f \le g\}. \qquad (14.1)$$

We say (see Definition 8.1.2) that the semi-martingale S satisfies the condition of *no free lunch with vanishing risk (NFLVR)* if the closure $\overline{C}$ of C, taken with respect to the norm-topology $\|.\|_\infty$ of $L^\infty(\Omega, \mathcal{F}, \mathbf{P})$ intersects $L_+^\infty(\Omega, \mathcal{F}, \mathbf{P})$ only in 0, i.e.,

$$S \text{ satisfies } (NFLVR) \Longleftrightarrow \overline{C} \cap L_+^\infty = \{0\}.$$

For the economic interpretation of this concept, which is a very mild strengthening of the "no-arbitrage" concept, we refer to Chap. 8.

The subsequent crucial Theorem 14.4.1 was proved in (9.4.2) under the additional assumption that S is bounded. An inspection of the proof given in Chap. 9 reveals that — for the validity of the subsequent Theorem 14.4.1 — the boundedness assumption on S may be dropped.

Theorem 14.4.1. *Under the assumption (NFLVR) the cone C is weak-star-closed in $L^\infty(\Omega, \mathcal{F}, \mathbf{P})$. Hence there is a probability measure $\mathbf{Q}_1 \sim \mathbf{P}$ such that*

$$\mathbf{E}_{\mathbf{Q}_1}[f] \leq 0, \qquad \text{for } f \in C.$$

Remark 14.4.2. In the case, when S is bounded, $\mathbf{Q}_1$ is already a martingale measure for S, and when S is locally bounded, $\mathbf{Q}_1$ is a local martingale measure for S (compare Theorem 9.1.1 and Corollary 9.1.2).

To take care of the non-locally bounded case we have to take care of the "big jumps" of S. We shall distinguish between the jumps of S occurring at accessible stopping times and those occurring at totally inaccessible stopping times.

We start with an easy lemma which will allow us to change the measure $\mathbf{Q}_1$ countably many times without loosing the equivalence to $\mathbf{P}$.

Lemma 14.4.3. *Let $(\mathbf{Q}_n)_{n=1}^\infty$ be a sequence of probability measures on the probability space $(\Omega, \mathcal{F}, \mathbf{P})$ such that each $\mathbf{Q}_n$ is equivalent to $\mathbf{P}$. Suppose further that the sequence of strictly positive numbers $(\varepsilon_n)_{n\geq 1}$ is such that*

(1) $\|\mathbf{Q}_n - \mathbf{Q}_{n+1}\| < \varepsilon_{n+1}$,
(2) *if $\mathbf{Q}_n[A] \leq \varepsilon_{n+1} 2^n$ then $\mathbf{P}[A] \leq 2^{-n}$.*

Then the sequence $(\mathbf{Q}_n)_{n\geq 1}$ converges with respect to the total variation norm to a probability measure $\mathbf{Q}$, which is equivalent to $\mathbf{P}$.

Proof of Lemma 14.4.3. Clearly the second assumption implies that $\varepsilon_{n+1} \leq 2^{-n}$ and hence the sequence $(\mathbf{Q}_n)_{n\geq 1}$ converges in variation norm to a probability measure $\mathbf{Q}$. We have to show that $\mathbf{Q} \sim \mathbf{P}$. For each n we let q_{n+1} be defined as the Radon-Nikodým derivative of $\mathbf{Q}_{n+1}$ with respect to $\mathbf{Q}_n$. Clearly for each $n \geq 1$ we then have $\int |1 - q_{n+1}| \, d\mathbf{Q}_n \leq \varepsilon_{n+1}$ and hence the Markov inequality implies that $\mathbf{Q}_n[|1 - q_{n+1}| \geq 2^{-n}] \leq 2^n \varepsilon_{n+1}$. The hypothesis on the sequence $(\varepsilon_n)_{n\geq 2}$ then implies that $\mathbf{P}[|1 - q_{n+1}| \geq 2^{-n}] \leq 2^{-n}$. From the Borel-Cantelli lemma it also follows that a.s. the series $\sum_{n\geq 2} |1 - q_n|$ converges and hence the product $\prod_{n\geq 2} q_n$ converges to a function q a.s. different from 0. Clearly $q = \frac{d\mathbf{Q}}{d\mathbf{Q}_1}$ which shows that $\mathbf{Q} \sim \mathbf{Q}_1 \sim \mathbf{P}$. $\qquad\square$

We are now ready to take the crucial step in the proof of the main theorem. To make life easier we make the simplifying assumption that S does not jump at predictable times. In the Proof of the Main Theorem 14.1.1 below we finally shall also deal with the case of the predictable jumps.

Proposition 14.4.4. *Let $S = (S_t)_{t\in\mathbb{R}_+}$ be an $\mathbb{R}^d$-valued semi-martingale which is quasi-left-continuous, i.e., such that, for every predictable stopping time T we have $S_T = S_{T-}$ almost surely.*

Suppose, as in Theorem 14.4.1 above, that $\mathbf{Q}_1 \sim \mathbf{P}$ is a probability measure verifying

$$\mathbf{E}_{\mathbf{Q}_1}[f] \leq 0, \qquad \text{for } f \in C.$$

Then there is, for $\varepsilon > 0$, a probability measure $\mathbf{Q} \sim \mathbf{P}$, $\|\mathbf{Q} - \mathbf{Q}_1\| < \varepsilon$, such that S is a sigma-martingale with respect to $\mathbf{Q}$.

In addition, for every predictable stopping time T, the probabilities $\mathbf{Q}$ and $\mathbf{Q}_1$ on $\mathcal{F}_T$, coincide, conditionally on $\mathcal{F}_{T-}$, i.e.,

$$\frac{d\mathbf{Q}|_{\mathcal{F}_T}}{d\mathbf{Q}_1|_{\mathcal{F}_T}} = \frac{d\mathbf{Q}|_{\mathcal{F}_{T-}}}{d\mathbf{Q}_1|_{\mathcal{F}_{T-}}} \qquad a.s..$$

Proof. Step 1: Define the stopping time T by

$$T = \inf\{t \mid \|\Delta S_t\|_{\mathbb{R}^d} \geq 1\}$$

and first suppose that S remains constant after time T, hence S has at most one jump bigger than 1.

Similarly as in [JS 87, II.2.4] we decompose S into

$$S = X + \check{X}$$

where X equals "S stopped at time $T-$", i.e.,

$$X_t = \begin{cases} S_t & \text{for } t < T \\ S_{T-} & \text{for } t \geq T \end{cases}$$

and $\check{X}$ the jump of S at time T, i.e.,

$$\check{X}_t = \Delta S_T \cdot \mathbf{1}_{[\![T,\infty]\!]}.$$

As X is bounded, it is a special semi-martingale, and we can find its Doob-Meyer decomposition with respect to $\mathbf{Q}_1$

$$X = M + B$$

where M is a local $\mathbf{Q}_1$-martingale and B a predictable process of locally finite variation.

We shall now find a probability measure $\mathbf{Q}_2$ on $\mathcal{F}$, $\mathbf{Q}_2 \sim \mathbf{P}$, s.t.

(i) $\|\mathbf{Q}_2 - \mathbf{Q}_1\| < \frac{\varepsilon}{2}$,
(ii) $\mathbf{Q}_2|_{\mathcal{F}_{T-}} = \mathbf{Q}_1|_{\mathcal{F}_{T-}}$ and $\frac{d\mathbf{Q}_2}{d\mathbf{Q}_1}$ is $\mathcal{F}_T$-measurable,
(iii) S is a sigma-martingale under $\mathbf{Q}_2$.

We introduce the jump measure μ associated to $\check{X}$,

$$\mu(\omega, dt, dx) = \delta_{(T(\omega), \Delta S_T(\omega))},$$

where $\delta_{t,x}$ denotes Dirac-measure at $(t, x) \in \mathbb{R}_+ \times \mathbb{R}^d$ and we denote by ν the $\mathbf{Q}_1$-compensator of μ (see [JS 87, Proposition II.1.6]). Similarly as in [JS 87, Proposition II.2.9] we may find a locally $\mathbf{Q}_1$-integrable, predictable and increasing process A such that

$$B = b \cdot A$$
$$\nu(\omega, dt, dx) = F_{\omega,t}(dx) dA_t(\omega)$$

where $b = (b^i)_{i=1}^d$ is a predictable process and $F_{\omega,t}(dx)$ a transition kernel from $(\Omega \times \mathbb{R}_+, \mathcal{P})$ into $(\mathbb{R}^d, \mathcal{B}(\mathbb{R}^d))$, i.e., a $\mathcal{P}$-measurable map $(\omega, t) \mapsto F_{\omega,t}(dx)$ from $\Omega \times \mathbb{R}_+$ into the non-negative Borel measures on $\mathbb{R}^d$. Since the processes X and $\check{X}$ are quasi left continuous, the processes A and B can be chosen to be continuous, but this is not really needed.

The processes ν and μ are such that for each non-negative $\mathcal{P} \otimes \mathcal{B}(\mathbb{R}^d)$-measurable function g we have that

$$\int_{\Omega \times \mathbb{R}_+ \times \mathbb{R}^d} g(\omega, t, y) \mu(\omega, dt, dy) \mathbf{P}(d\omega) = \int_{\Omega \times \mathbb{R}_+ \times \mathbb{R}^d} g(\omega, t, y) \nu(\omega, dt, dy) \mathbf{P}(d\omega).$$

To stay in line with the notation used in [JS 87], $H_{\omega,t} * F_{\omega,t}$, where H is a predictable $\mathbb{R}^d$-valued process and F is the kernel described above, denotes the predictable $\mathbb{R}$-valued process $E_{F_{\omega,t}}[(H_{\omega,t}, \cdot)] = \int_{\mathbb{R}^d} (H_{\omega,t}, y) F_{\omega,t}[dy]$.

We may assume that A is constant after T, $\mathbf{Q}_1$-integrable and its integral is bounded by one, i.e.,

$$\mathbf{E}_{\mathbf{Q}_1}[A_\infty] = dA(\Omega \times \mathbb{R}_+) \leq 1,$$

where dA denotes the measure on $\mathcal{P}$ defined by $dA(]T_1, T_2]) = \mathbf{E}_{\mathbf{Q}_1}[A_{T_2} - A_{T_1}]$, for stopping times $T_1 \leq T_2$.

We now shall find a $\mathcal{P}$-measurable map $(\omega, t) \mapsto G_{\omega,t}$ such that for dA-almost each (ω, t),

(a) $F_{\omega,t}(dx) \sim G_{\omega,t}(dx), F_{\omega,t}(\mathbb{R}^d) = G_{\omega,t}(\mathbb{R}^d)$ and $\|F_{\omega,t} - G_{\omega,t}\| < \frac{\varepsilon}{2}$,
(b) $\mathbf{E}_{G_{\omega,t}}[\|y\|_{\mathbb{R}^d}] < \infty$ and $\mathbf{E}_{G_{\omega,t}}[y] = -b(\omega, t)$.

This is a task of the type of "martingale problem" or rather "semi-martingale problem" as dealt with, e.g., in [JS 87, Definition III.2.4].

We apply Lemma 14.3.5 and the remark following it: as measure space $(E, \mathcal{E}, \omega)$ we take $(\Omega \times \mathbb{R}_+, \mathcal{P}, dA)$ and we shall consider the map

$$\eta = (\omega, t) \mapsto \widetilde{F}_{\omega,t} := F_{\omega,t} \star \delta_{b(\omega,t)}$$

where $\delta_{b(\omega,t)}$ denotes the Dirac measure at $b(\omega, t) \in \mathbb{R}^d$, $\star$ denotes convolution and therefore $\widetilde{F}_{\omega,t}$ is the measure $F_{\omega,t}$ on $\mathbb{R}^d$ translated by the vector $b(\omega, t)$.

We claim that the family $(\widetilde{F}_{\omega,t})_{(\omega,t) \in \Omega \times \mathbb{R}_+}$ satisfies the assumptions of Lemma 14.3.5 above. Indeed, let $H_{\omega,t}$ be any $\mathcal{P}$-measurable function such that dA-almost surely $H_{\omega,t} \in \text{Adm}(\widetilde{F}_{\omega,t}) = \text{Adm}(F_{\omega,t})$.

By multiplying H with a predictable strictly positive process v, we may eventually assume that $\|H_{\omega,t}\|_{\mathbb{R}^d} \leq 1$ and that, at least dA a.e., also the predictable process $\|\langle H_{\omega,t}, \cdot \rangle^-\|_{L^\infty(F_{\omega,t})}$ is bounded by 1. That the latter process is predictable follows from the discussion preceding the Crucial Lemma and essentially follows from the measurable selection theorem.

The boundedness property translates to the fact that $H = (H_t(\omega))_{t \in \mathbb{R}_+}$ is an *admissible integrand* for the process $\check{X}$. This follows from the definition of the compensator ν in the following way. For each natural number n we have, according to the definition of the compensator that

$$\mathbf{E}\left[\left(\langle H_{\omega,t}, \Delta S_T(\omega)\rangle^-\right)^n \mathbf{1}_{T<\infty}\right]$$

$$= \mathbf{E}\left[\int_{\mathbb{R}_+ \times \mathbb{R}^d} \left(\langle H_{\omega,t}, y\rangle^-\right)^n \mu(\omega, dt, dy)\right]$$

$$= \mathbf{E}\left[\int_{\mathbb{R}_+ \times \mathbb{R}^d} \left(\langle H_{\omega,t}, y\rangle^-\right)^n \nu(\omega, dt, dy)\right]$$

$$= \mathbf{E}\left[\int_{\mathbb{R}_+} \int_{\mathbb{R}^d} \left(\langle H_{\omega,t}, y\rangle^-\right)^n F_{\omega,t}(dy)\, dA_t(\omega)\right]$$

$$\leq \mathbf{E}\left[\int_{\mathbb{R}_+} \int_{\mathbb{R}^d} 1\, F_{\omega,t}(dy)\, dA_t(\omega)\right]$$

$$\leq \mathbf{E}\left[\int_{\mathbb{R}_+ \times \mathbb{R}^d} 1\, \nu(\omega, dt, dy)\right]$$

$$\leq \mathbf{E}\left[\int_{\mathbb{R}_+ \times \mathbb{R}^d} 1\, \mu(\omega, dt, dy)\right] \leq \mathbf{Q}_1[T < \infty].$$

Since the inequality holds for each n we necessarily have that

$$(H \cdot \check{X})_t \geq -1 \qquad \text{a.s., for all } t \in \mathbb{R}_+.$$

Noting that M is a (locally bounded) local martingale and B is of locally bounded variation, we may find a sequence of stopping times $(U_j)_{j=1}^\infty$ increasing to infinity, such that, for each $j \in \mathbb{N}$,

(1) M^{U_j} is a martingale, bounded in the Hardy space $\mathcal{H}^1(\mathbf{Q}_1)$, and
(2) B^{U_j} is of bounded variation.

Hence, for each predictable set P contained in $[\![0, U_j]\!]$, for some $j \in \mathbb{N}$, we have that $H\mathbf{1}_P$ is an admissible integrand for S and $H\mathbf{1}_P \cdot M$ is a martingale bounded in $\mathcal{H}^1(\mathbf{Q}_1)$ and therefore

$$\mathbf{E}_{\mathbf{Q}_1}\left[(H\mathbf{1}_P \cdot M)_\infty\right] = 0.$$

As by hypothesis

$$\mathbf{E}_{\mathbf{Q}_1}\left[(H\mathbf{1}_P \cdot S)_\infty\right] \leq 0$$

we obtain

$$\mathbf{E}_{\mathbf{Q}_1}\left[(H\mathbf{1}_P \cdot (\check{X} + B))_\infty\right] \leq 0.$$

Using the identities

$$\mathbf{E}_{\mathbf{Q}_1}\left[(H\mathbf{1}_P \cdot (\check{X} + B))_\infty\right]$$

$$= \int_{\Omega \times \mathbb{R}_+} (H_{\omega,t} * F_{\omega,t} + (H_{\omega,t}, b_{\omega,t}))\mathbf{1}_P dA(\omega,t)$$

$$= \int_{\Omega \times \mathbb{R}_+} (H_{\omega,t} * \widetilde{F}_{\omega,t})\mathbf{1}_P dA(\omega,t) \leq 0$$

which hold true for each $P \in \mathcal{P}$ contained in $[\![0, U_j]\!]$, for some $j \in \mathbb{N}$, we conclude that for dA-almost each (ω,t) we have

$$H_{\omega,t} * \widetilde{F}_{\omega,t} = \mathbf{E}_{\widetilde{F}_{\omega,t}}\left[(H_{\omega,t}\,,\,.)\right] \leq 0\,.$$

This inequality implies that assumption (2) in Lemma 14.3.4 is satisfied and hence $F_{\omega,t}$ satisfies the no-arbitrage property, i.e. the hypothesis of Lemma 14.3.5 is satisfied.

Hence we may find a transition kernel $\widetilde{G}_{\omega,t}$ as described by Lemma 14.3.5 — with ε replaced by $\frac{\varepsilon}{2}$ — and letting $G_{\omega,t} = \widetilde{G}_{\omega,t} \star \delta_{-b(\omega,t)}$ we obtain a transition kernel satisfying (a) and (b) above.

We now have to translate the change of transition kernels from $F_{\omega,t}$ to $G_{\omega,t}$ into a change of measures from $\mathbf{Q}_1$ to $\mathbf{Q}_2$ on the σ-algebra $\mathcal{F}_T$ which will be done by defining the Radon-Nikodým derivative $\frac{d\mathbf{Q}_2}{d\mathbf{Q}_1}$. We refer to [JS 87, III.3] for a treatment of the relevant version of Girsanov's theorem for random measures.

For (ω,t) fixed, denote by $Y(\omega,t\,,\,.)$ the Radon-Nikodým derivative of $G_{\omega,t}$ with respect to $F_{\omega,t}$, i.e.

$$Y(\omega,t,x) = \frac{dG_{\omega,t}}{dF_{\omega,t}}(x)\,, \qquad x \in \mathbb{R}^d\,,$$

which is $F_{\omega,t}$-almost surely well-defined and strictly positive. We may and do choose for dA-almost each (ω,t), a version $Y(\omega,t,x)$ such that $Y(\,.\,,\,.\,,\,.)$ is $\mathcal{P} \otimes \mathcal{B}(\mathbb{R}^d)$-measurable.

We now define

$$\frac{d\mathbf{Q}_2}{d\mathbf{Q}_1}(\omega) = Z_\infty(\omega) = Y(\omega, T(\omega), \Delta S_{T(\omega)}(\omega))\mathbf{1}_{\{T<\infty\}} + \mathbf{1}_{\{T=\infty\}}$$

$$\text{and} \qquad Z_t(\omega) = Y(\omega, T(\omega), \Delta S_{T(\omega)}(\omega))\mathbf{1}_{\{T\leq t\}} + \mathbf{1}_{\{T>t\}}\,.$$

The intuitive interpretation of these formulas goes as follows: for fixed $\omega \in \Omega$ we look at time $T(\omega)$ which is the unique "big" jump of $(S_t(\omega))_{t\in\mathbb{R}_+}$. The density $Y(\omega, T(\omega), x)$ gives the density of the distribution of the compensated jump measure $G_{\omega,t}$ with respect to $F_{\omega,t}$, if the jump equals x and therefore we evaluate $Y(\omega, T(\omega), x)$ at the point $x = \Delta S_{T(\omega)}(\omega)$ to determine the density of $\mathbf{Q}_2$ with respect to $\mathbf{Q}_1$. If $T(\omega) = \infty$ the density $\frac{d\mathbf{Q}_2}{d\mathbf{Q}_1}(\omega)$ is simply equal to 1. The variable $Y(\omega, T(\omega), \Delta S_T(\omega))$ is certainly integrable. Indeed

$$\mathbf{E}\left[Y(\omega, T(\omega), \Delta S_T(\omega))\mathbf{1}_{T<\infty}\right]$$

$$= \mathbf{E}\left[\int_{\mathbb{R}_+\times\mathbb{R}^d} Y(\omega, t, y)\mu(\omega, dt, dy)\right]$$

$$= \mathbf{E}\left[\int_{\mathbb{R}_+\times\mathbb{R}^d} Y(\omega, t, y)\nu(\omega, dt, dy)\right]$$

$$= \mathbf{E}\left[\int_{\mathbb{R}_+}\int_{\mathbb{R}^d} Y(\omega, t, y)F_{\omega,t}(dy)dA_t(\omega)\right]$$

$$= \mathbf{E}\left[\int_{\mathbb{R}_+}\int_{\mathbb{R}^d} F_{\omega,t}(\mathbb{R}^d)dA_t(\omega)\right]$$

$$= \mathbf{E}\left[\int_{\mathbb{R}_+\times\mathbb{R}^d} \nu(\omega, dt, dy)\right]$$

$$= \mathbf{E}\left[\int_{\mathbb{R}_+\times\mathbb{R}^d} \mu(\omega, dt, dy)\right] = \mathbf{Q}_1[T < \infty].$$

The process Z can also be written as

$$Z = Y(\omega, T, \Delta S_T)\mathbf{1}_{[\![T,\infty[\![} + \mathbf{1}_{[\![0,T[\![}},$$

from which it follows that Z is a process of integrable variation. The maximal function Z^* of Z is therefore integrable.

In order to show that $\mathbf{Q}_2$ is indeed a probability measure and that $Z_t = \frac{d\mathbf{Q}_2|_{\mathcal{F}_t}}{d\mathbf{Q}_1|_{\mathcal{F}_t}}$ we shall show that $(Z_t)_{t\in\mathbb{R}_+}$ is a uniformly integrable martingale closed by Z_∞.

We may write $Z = (Z_t)_{t\in\mathbb{R}_+}$ as

$$Z = 1 + (Y(\omega, t, x) - 1) * \mu.$$

From the definition of the compensator ν ([JS 87, II.1.8]) we deduce that we may write the compensator Z^p of Z

$$Z^p = 1 + (Y(\omega, t, x) - 1) * \nu$$
$$= 1 + ((Y(\omega, t, x) - 1) * F_{\omega,t}) \cdot A.$$

Noting that, for dA-almost each (ω, t) we have that $(Y(\omega, t, x) - 1) * F_{\omega,t} = \mathbf{E}_{F_{\omega,t}}[Y(\omega, t, x)] - 1] = 0$ we deduce that the compensator Z^p is constant. Since $Z - Z^p$ is a martingale, by definition of the compensator of processes of integrable variation, it follows that Z is a martingale as well.

To estimate the distance $\|\mathbf{Q}_2 - \mathbf{Q}_1\|$, note that

$$\|\mathbf{Q}_2 - \mathbf{Q}_1\| = \mathbf{E}_{\mathbf{Q}_1}\left[\left|1 - \frac{d\mathbf{Q}_2}{d\mathbf{Q}_1}\right|\right]$$
$$\leq \mathbf{E}_{\mathbf{Q}_1}\left[(|Y(\omega, t, x) - 1| * \nu)_\infty\right]$$
$$\leq \mathbf{E}_{\mathbf{Q}_1}(\|F_{\omega,t} - G_{\omega,t}\| \cdot A)_\infty \leq \frac{\varepsilon}{2}\mathbf{E}_{\mathbf{Q}_1}[A_\infty] \leq \frac{\varepsilon}{2}.$$

Next observe that $\mathbf{Q}_2|_{\mathcal{F}_{T-}} = \mathbf{Q}_1|_{\mathcal{F}_{T-}}$: indeed, we have to show that $\mathbf{Q}_1$ and $\mathbf{Q}_2$ coincide on the sets of the form $A \cap \{T > t\}$, where $A \in \mathcal{F}_t$, as these sets generate $\mathcal{F}_{T-}$. Noting that Z_t is equal to 1 on $\{T > t\}$, this becomes obvious.

Finally we show that S is a sigma-martingale under $\mathbf{Q}_2$. First note that M remains a local martingale under $\mathbf{Q}_2$ as M is continuous at time T, i.e., $M_{T-} = M_T$, and $\mathbf{Q}_1$ and $\mathbf{Q}_2$ coincide on $\mathcal{F}_{T-}$.

As regards the remaining part $\check{X} + B$ of the semi-martingale S we have by (b) above that, for dA-almost each (ω, t), $\mathbf{E}_{G_{\omega,t}} [\|y\|_{\mathbb{R}^d}] < \infty$ and $\mathbf{E}_{G_{\omega,t}} [y] = -b(\omega, t)$. This *does not necessarily* imply that $\check{X} + B$ is already a martingale (or a local martingale) under $\mathbf{Q}_2$ as a glance at Example 14.2.2 reveals. We may only conclude that $\check{X} + B$ is a $\mathbf{Q}_2$-sigma-martingale, as we presently shall see.

Define

$$\varphi_t(\omega) = (\mathbf{E}_{G_{\omega,t}} [\|y\|_{\mathbb{R}^d}])^{-1} \wedge 1 \,,$$

which is a predictable dA-almost surely strictly positive process. The process $\varphi \cdot (\check{X} + B)$ is a process of $\mathbf{Q}_2$-integrable variation as

$$\mathbf{E}_{\mathbf{Q}_2} \left[\mathrm{var}_{\|\cdot\|_{\mathbb{R}^d}}(\varphi \cdot (\check{X} + B)) \right] = \mathbf{E}_{\mathbf{Q}_2} [\varphi \cdot (\|y\|_{\mathbb{R}^d} * G_{\omega,t} \cdot A + \|b_{\omega,t}\|_{\mathbb{R}^d} \cdot A)]$$
$$\leq 2\mathbf{E}_{\mathbf{Q}_2} [A_\infty] = 2\mathbf{E}_{\mathbf{Q}_1} [A_\infty] \leq 2 \,,$$

where the last equality follows from the fact that $\mathbf{Q}_1$ and $\mathbf{Q}_2$ coincide on $\mathcal{F}_{T-}$ and that, A being predictable, A_∞ is $\mathcal{F}_{T-}$-measurable.

Hence $\varphi \cdot (\check{X} + B)$ is a process of integrable variation whose compensator is constant and therefore $\varphi \cdot (\check{X} + B)$ is a $\mathbf{Q}_2$-martingale of integrable variation, whence in particular a $\mathbf{Q}_2$-martingale. Therefore $\check{X} + B$ as well as S are $\mathbf{Q}_2$-sigma-martingales.

Summing up: We have proved Proposition 14.4.4 under the additional hypothesis that S remains constant after the first time T when S jumps by at least 1 with respect to $\|\cdot\|_{\mathbb{R}^d}$.

Step 2: Now we drop this assumption and assume w.l.g. that $S_0 = 0$. Let $T_0 = 0$, $T_1 = T$ and define inductively the stopping times

$$T_k = \inf\{t > T_{k-1} \mid \|\Delta S_t\|_{\mathbb{R}^d} \geq 1\}, \qquad k = 2, 3, \ldots$$

so that $(T_k)_{k=1}^\infty$ increases to infinity. Let

$$S^{(k)} = \mathbf{1}_{]\!]T_{k-1}, T_k]\!]} \cdot S, \qquad k = 1, 2, \ldots.$$

Note that $S^{(1)}$ satisfies the assumptions of the first part of the proof, where we have shown that there is a measure $\mathbf{Q}_2 \sim \mathbf{P}$, satisfying (i), (ii), (iii) above for $T = T_1$. Now repeat the above argument to choose inductively, for $k = 2, 3, \ldots$, measures $\mathbf{Q}_{k+1} \sim \mathbf{P}$ such that

(i) $\|\mathbf{Q}_{k+1} - \mathbf{Q}_k\| < \frac{\varepsilon}{2^{k-1}} \wedge \inf \left\{ \frac{2^{-k}\mathbf{Q}_k[A]}{\mathbf{P}[A]} \mid A \in \mathcal{F}, \mathbf{P}[A] \geq 2^{-k} \right\}.$

(ii) $\mathbf{Q}_{k+1}|_{\mathcal{F}_{T_k-}} = \mathbf{Q}_k|_{\mathcal{F}_{T_k-}}$ and $\frac{d\mathbf{Q}_{k+1}}{d\mathbf{Q}_k}$ is $\mathcal{F}_{T_k}$-measurable.

(iii) $S^{(k)}$ is a sigma-martingale under $\mathbf{Q}_{k+1}$.

The condition in (i) above is chosen such that we may apply Lemma 14.4.3 to conclude that

$$\mathbf{Q} = \lim_{k\to\infty} \mathbf{Q}_k$$

exists and is equivalent to $\mathbf{P}$. From (ii) and (iii) it follows that each $S^{(k)}$ is a sigma-martingale under $\mathbf{Q}^l$, for each $l \geq k$. It follows that each $S^{(k)}$ is a $\mathbf{Q}$-sigma-martingale and hence S, being a local sigma-martingale is then a sigma-martingale (see Corollary 14.2.7 above). This proves the first part of Proposition 14.4.4.

As regards the final assertion of Proposition 14.4.4 note that, for any predictable stopping time U, the random times

$$U_k = \begin{cases} U & \text{if } T_{k-1} < U \leq T_k \\ \infty & \text{otherwise} \end{cases}$$

are predictable stopping times, for $k = 1, 2, \ldots$. Indeed, as easily seen, the set $\{T_k < U \leq T_{k+1}\}$ is in $\mathcal{F}_{U-}$, showing that U_k is predictable.

By our construction and property (ii) above we infer that, for $k = 1, 2, \ldots$,

$$\frac{d\mathbf{Q}|_{\mathcal{F}_{U_k}}}{d\mathbf{Q}_1|_{\mathcal{F}_{U_k}}} = \frac{d\mathbf{Q}|_{\mathcal{F}_{(U_k)-}}}{d\mathbf{Q}_1|_{\mathcal{F}_{(U_k)-}}} \quad \text{a.s.}$$

which implies that

$$\frac{d\mathbf{Q}|_{\mathcal{F}_U}}{d\mathbf{Q}_1|_{\mathcal{F}_U}} = \frac{d\mathbf{Q}|_{\mathcal{F}_{U-}}}{d\mathbf{Q}_1|_{\mathcal{F}_{U-}}} \quad \text{a.s..}$$

The proof of Proposition 14.4.4 is complete now.   □

Proposition 14.4.4 contains the major part of the proof of the main theorem. The missing ingredient is still the argument for the predictable jumps of S. The argument for the predictable jumps given below will be similar to (but technically easier than) the proof of Proposition 14.4.4.

Proof of the Main Theorem 14.1.1. Let S be an $\mathbb{R}^d$-valued semi-martingale satisfying the assumption *(NFLVR)*. By Theorem 14.4.1 we may find a probability measure $\mathbf{Q}_1 \sim \mathbf{P}$ such that,

$$\mathbf{E}_{\mathbf{Q}_1}[f] \leq 0, \qquad \text{for } f \in C.$$

We also may find a sequence $(T_k)_{k=1}^{\infty}$ of predictable stopping times exhausting the accessible jumps of S, i.e., such that for each predictable stopping time T with $\mathbf{P}[T = T_k < \infty] = 0$, for each $k \in \mathbb{N}$, we have that $S_{T-} = S_T$ almost surely. We may and do assume that the stopping times $(T_k)_{k=1}^{\infty}$ are disjoint, i.e., that $\mathbf{P}[T_k = T_j < \infty] = 0$ for $k \neq j$.

Denote by D the predictable set

$$D = \bigcup_{k \geq 1} [\![T_k]\!] \subseteq \Omega \times \mathbb{R}_+$$

and split S into $S = S^a + S^i$, where

$$S^a = \mathbf{1}_D \cdot S \qquad \text{and} \qquad S^i = \mathbf{1}_{(\Omega \times \mathbb{R}_+) \setminus D} \cdot S$$

where the letters "a" and "i" refer to "accessible" and "inaccessible". S^a and S^i are well-defined semi-martingales and in view of the above construction S^i is quasi-left-continuous.

Denote by C^a and C^i the cones in $L^\infty(\Omega, \mathcal{F}, \mathbf{P})$ associated by (14.1) to S^a and S^i, and observe that C^a and C^i are subsets of C (obtained by considering only integrands supported by D or $(\Omega \times \mathbb{R}_+) \setminus D$ respectively) hence

$$\mathbf{E}_{\mathbf{Q}_1}[f] \leq 0, \qquad \text{for } f \in C^a \text{ and for } f \in C^i.$$

Hence S^i satisfies the assumptions of Proposition 14.4.4 with respect to the probability measure $\mathbf{Q}_1$ and we therefore may find a probability measure, now denoted by $\widehat{\mathbf{Q}}$, $\widehat{\mathbf{Q}} \sim \mathbf{P}$, which turns S^i into a sigma-martingale and such that, for each predictable stopping time T, we have

$$\frac{d\widehat{\mathbf{Q}}|_{\mathcal{F}_T}}{d\mathbf{Q}_1|_{\mathcal{F}_T}} = \frac{d\widehat{\mathbf{Q}}|_{\mathcal{F}_{T-}}}{d\mathbf{Q}_1|_{\mathcal{F}_{T-}}}. \tag{14.2}$$

By assumption we have, for each $k = 1, 2, \ldots$, and for each admissible integrand H supported by $[\![T_k]\!]$, that

$$\mathbf{E}_{\mathbf{Q}_1}[(H \cdot S)_\infty] = \mathbf{E}_{\mathbf{Q}_1}\left[H_{T_k}(S_{T_k} - S_{(T_k)-})\right] \leq 0.$$

Noting that the inequality remains true if we replace H by $H\mathbf{1}_A$, for any $\mathcal{F}_{(T_k)-}$-measurable set A, and using (14.2) we obtain

$$\mathbf{E}_{\widehat{\mathbf{Q}}}[(H \cdot S)_\infty] = \mathbf{E}_{\widehat{\mathbf{Q}}}\left[H_{T_k}(S_{T_k} - S_{(T_k)-})\right] \leq 0 \tag{14.3}$$

for each admissible integrand supported by $[\![T_k]\!]$.

We now shall proceed inductively on k: suppose we have chosen, for $k \geq 0$, probability measures $\widetilde{\mathbf{Q}}_0 = \widehat{\mathbf{Q}}, \widetilde{\mathbf{Q}}_1, \ldots, \widetilde{\mathbf{Q}}_k$ such that

$$\mathbf{E}_{\widetilde{\mathbf{Q}}_k}\left[S_{T_j} \mid \mathcal{F}_{(T_j)-}\right] = S_{(T_j)-}, \qquad j = 1, \ldots, k$$

and such that, for

$$\varepsilon_j = \frac{\varepsilon}{2^{j+1}} \wedge \inf\left\{\left.\frac{2^{-j}\widetilde{\mathbf{Q}}_j[A]}{\mathbf{P}[A]}\right| A \in \mathcal{F}, \mathbf{P}[A] \geq 2^{-j}\right\},$$

we have

$$\|\widetilde{\mathbf{Q}}_j - \widetilde{\mathbf{Q}}_{j+1}\| < \varepsilon_j, \qquad j = 0, \ldots, k-1.$$

In addition we assume that $\widetilde{\mathbf{Q}}_j$ and $\widetilde{\mathbf{Q}}_{j-1}$ agree "before $(T_j)-$ and after T_j"; this means that $\widetilde{\mathbf{Q}}_j$ and $\widetilde{\mathbf{Q}}_{j-1}$ coincide on the σ-algebra $\mathcal{F}_{(T_j)-}$ and that the Radon-Nikodým derivative $\frac{d\widetilde{\mathbf{Q}}_j}{d\widetilde{\mathbf{Q}}_{j-1}}$ is $\mathcal{F}_{T_j}$-measurable.

Now consider the stopping time T_{k+1}: denote on the set $\{T_{k+1} < \infty\}$ by F_ω the jump measure of the jump $S^a_{T_{k+1}} - S^a_{(T_{k+1})-}$ conditional on $\mathcal{F}_{(T_{k+1})-}$. By (14.3) this $(\Omega, \mathcal{F}_{(T_{k+1})-}, \mathbf{P})$-measurable family of probability measures on $\mathbb{R}^d$ satisfies the assumptions of Lemma 14.3.5 and we therefore may find an $\mathcal{F}_{(T_{k+1})-}$-measurable family of probability measures G_ω, a.s. defined on $\{T_{k+1} < \infty\}$, such that

(i) $F_\omega \sim G_\omega$ and $\|F_\omega - G_\omega\| < \varepsilon_k$
(ii) $\mathbf{E}_{G_\omega}[\|y\|_{\mathbb{R}^d}] < \infty$ and $\mathrm{bary}(G_\omega) = \mathbf{E}_{G_\omega}[y] = 0$.

Letting, similarly as in the proof of Proposition 14.4.4 above,

$$Y(\omega, x) = \frac{dF_\omega}{dG_\omega}(x)$$

be a $\mathcal{F}_{(T_{k+1})-} \otimes \mathcal{B}(\mathbb{R}^d)$-measurable version of the Radon-Nikodým derivatives $\frac{dF_\omega}{dG_\omega}$ and defining

$$\frac{d\widetilde{\mathbf{Q}}_{k+1}}{d\widetilde{\mathbf{Q}}_k}(\omega) = \mathbf{1}_{\{T_{k+1}<\infty\}} Y(\omega, \Delta S_{T_{k+1}}(\omega)) + \mathbf{1}_{\{T_{k+1}=\infty\}}$$

we obtain a measure $\widetilde{\mathbf{Q}}_{k+1} \sim \mathbf{P}$, so that $\|\widetilde{\mathbf{Q}}_{k+1} - \widetilde{\mathbf{Q}}_k\| < \varepsilon_k$, $\widetilde{\mathbf{Q}}_{k+1}|_{\mathcal{F}_{(T_{k+1})-}} = \widetilde{\mathbf{Q}}_k|_{\mathcal{F}_{(T_{k+1})-}}$ and $\frac{d\widetilde{\mathbf{Q}}_{k+1}}{d\widetilde{\mathbf{Q}}_k}$ being $\mathcal{F}_{T_{k+1}}$-measurable. For each $M \in \mathbb{R}_+$

$$\mathbf{1}_{[\![T_{k+1}]\!] \cap \{T_{k+1}<\infty \text{ and } \mathbf{E}_{G_\omega}[\|y\|] \leq M\}} \cdot S = \mathbf{1}_{[\![T_{k+1}]\!] \cap \{T_{k+1}<\infty \text{ and } \mathbf{E}_{G_\omega}[\|y\|] \leq M\}} \cdot S^a$$

is a martingale under $\widetilde{\mathbf{Q}}_{k+1}$ and therefore

$$S^{(k+1)} := \mathbf{1}_{[\![T_{k+1}]\!] \cap \{T_{k+1}<\infty\}} \cdot S$$

is a sigma-martingale under $\widetilde{\mathbf{Q}}_{k+1}$.

Letting $\mathbf{Q} = \lim_{k\to\infty} \widetilde{\mathbf{Q}}_k$, each of the semi-martingales $S^{(k)} = \mathbf{1}_{[\![T_k]\!]} \cdot S$ is a $\mathbf{Q}$-sigma-martingale. It follows that

$$S^a = \sum_{k=1}^{\infty} S^{(k)}$$

is a $\mathbf{Q}$-sigma-martingale and therefore

$$S = S^a + S^i$$

is a $\mathbf{Q}$-sigma-martingale too.

The proof of the main theorem is complete now.   □

For later use, let us resume in the subsequent proposition what we have shown in the above proof.

Proposition 14.4.5. *Denote by $\mathcal{M}_s^e$ the set of probability measures $\mathbf{Q}$ equivalent to $\mathbf{P}$ such that, for admissible integrands, the process $H \cdot S$ becomes a super-martingale. More precisely*

$$\mathcal{M}_s^e = \{\mathbf{Q} \mid \mathbf{Q} \sim \mathbf{P} \text{ and for each } f \in C \colon \mathbf{E_Q}[f] \leq 0\}.$$

If S satisfies (NFLVR), then

$$\mathcal{M}_\sigma^e = \{\mathbf{Q} \mid S \text{ is a } \mathbf{Q} \text{ sigma-martingale}\},$$

is dense in $\mathcal{M}_s^e$.

Theorem 14.4.6. *The set $\mathcal{M}_\sigma^e$ is a convex set.*

Proof. Let $\mathbf{Q}_1, \mathbf{Q}_2 \in \mathcal{M}_\sigma^e$ and let φ_1, φ_2 be strictly positive real-valued S-integrable predictable processes, such that for $i = 1, 2$, $\varphi_i \cdot S$ is an $\mathcal{H}^1(\mathbf{Q}_i)$-martingale. Take now $\varphi = \min(\varphi_1, \varphi_2)$. Since $0 < \varphi \leq \varphi_1$, $\varphi \cdot S$ is still an $\mathcal{H}^1(\mathbf{Q}_1)$-martingale. Similarly $\varphi \cdot S$ is still an $\mathcal{H}^1(\mathbf{Q}_2)$-martingale. From this it follows that $\varphi \cdot S$ is an $\mathcal{H}^1\left(\frac{\mathbf{Q}_1 + \mathbf{Q}_2}{2}\right)$-martingale.   □

14.5 Duality Results and Maximal Elements

In this section we suppose without further notice that S is an $\mathbb{R}^d$-valued semi-martingale that satisfies the *(NFLVR)* property, so that the set

$$\mathcal{M}_\sigma^e = \{\mathbf{Q} \mid \mathbf{Q} \sim \mathbf{P} \text{ and } S \text{ is a } \mathbf{Q} \text{ sigma-martingale}\}$$

is non-empty. We remark that when the price process S is locally bounded then the set $\mathcal{M}_\sigma^e$ coincides with the set $\mathcal{M}^e(S)$ as introduced in Chap. 9, i.e. the set of all equivalent local martingale measures for the process S.

In the case of locally bounded processes we showed the following duality equality, (see [D 92, Theorem 6.1] for the case of continuous bounded processes, El Karoui-Quenez [EQ 95] for the L^2 case, Theorem 9.5.8 for the case of bounded functions and Theorem 11.3.4 for the case of positive functions). The duality argument was used by El Karoui-Quenez [EQ 95]. For a non-negative random variable g we have:

$$\sup_{\mathbf{Q} \in \mathcal{M}_\sigma^e} \mathbf{E_Q}[g] = \inf\{\alpha \mid \text{ there is } H \text{ admissible and } g \leq \alpha + (H \cdot S)_\infty\}.$$

Using this equality we were able to derive a characterisation of maximal elements, see Corollary 11.4.6.

In the general case, i.e. when the process S is not necessarily locally bounded, the set of admissible integrands might be restricted to the zero integrand, compare Proposition 14.3.7 above. Below we will show that also in this case the above equality remains valid, at least for positive random variables g. This result does not immediately follow from the results in Chap. 11.

Another approach to the problem is to enlarge the concept of admissible integrands in a similar way as was done in [S 94] and Chap. 15. Here the idea is to allow for integrands H that are such that the process $H \cdot S$ is controlled from below by an appropriate function w, the so-called w-admissible integrands. We will generalise the above duality equality to the setting of such integrands and we will see that even in the locally bounded case this generalisation yields some new results.

If we want to control a process $H \cdot S$ from below by a function w then, of course, the problem is that w cannot be too big, as this would allow doubling strategies and therefore arbitrage. Also w cannot be too small because this could imply that the only such integrand H is the zero integrand. This idea is made precise in the following definitions of w-admissible integrands and of feasible weight functions.

Definition 14.5.1. *If $w \geq 1$ is a random variable, if there is $\mathbf{Q}_0 \in \mathcal{M}_\sigma^e$ such that $\mathbf{E}_{\mathbf{Q}_0} [w] < \infty$, if a is a non-negative number, then we say that the integrand H is (a, w)-admissible if for each element $\mathbf{Q} \in \mathcal{M}_\sigma^e$ and each $t \geq 0$, we have $(H \cdot S)_t \geq -a\mathbf{E}_{\mathbf{Q}} [w \mid \mathcal{F}_t]$. We simply say that H is w-admissible if H is (a, w)-admissible for some non-negative a.*

Remark 14.5.2. If we put $w = 1$ we again find the usual concept of admissible integrands. Of course, we could have defined the concept of w-admissible integrands for general non-negative functions w. We, however, required that $w \geq 1$, so that the admissible integrands become automatically w-admissible. The idea in fact is to allow unbounded functions w and therefore there seems to be no gain in introducing functions w that are too small. Requiring that $w \geq 1$ is by no means a restriction compared to the seemingly more general requirement $\operatorname{ess\,inf}(w) > 0$.

Remark 14.5.3. The present notion of admissible integrand is more suitable for our purposes than the one introduced in Chap. 15.

The next lemma, based on a stability property of the set $\mathcal{M}_\sigma^e$, shows that in the inequality $(H \cdot S)_t \geq -\mathbf{E}_{\mathbf{Q}} [w \mid \mathcal{F}_t]$, it does not harm to restrict to elements $\mathbf{Q} \in \mathcal{M}_\sigma^e$ such that $\mathbf{E}_{\mathbf{Q}} [w] < \infty$.

Lemma 14.5.4. *Let $w \geq 0$ be such that $\mathbf{E}_{\mathbf{Q}_0} [w] < \infty$ for some $\mathbf{Q}_0 \in \mathcal{M}_\sigma^e$. Suppose that ,for some $\mathbf{Q} \in \mathcal{M}_\sigma^e$, $t \geq 0$ and some real constant k, the set $A = \{\mathbf{E}_{\mathbf{Q}} [w \mid \mathcal{F}_t] \leq k\}$ has positive probability, then there is $\mathbf{Q}_1 \in \mathcal{M}_\sigma^e$ such that we have $\mathbf{E}_{\mathbf{Q}_1} [w \mid \mathcal{F}_t] = \mathbf{E}_{\mathbf{Q}} [w \mid \mathcal{F}_t]$ a.s. on the set A and $\mathbf{E}_{\mathbf{Q}_1} [w] < \infty$.*

Proof. Let Z_s be a càdlàg version of the density process $Z_s = \mathbf{E}_{\mathbf{Q}_0}\left[\frac{d\mathbf{Q}}{d\mathbf{Q}_0} \mid \mathcal{F}_s\right]$. Now we put

$$
\begin{aligned}
Z_s^1 &= 1 &&\text{for } s < t \\
Z_s^1 &= 1 &&\text{for } s \geq t \text{ and } \omega \notin A \\
Z_s^1 &= \frac{Z_s}{Z_t} &&\text{for } s \geq t \text{ and } \omega \in A.
\end{aligned}
$$

Clearly the probability measure $\mathbf{Q}_1$ defined by $d\mathbf{Q}_1 = Z_\infty^1 d\mathbf{Q}_0$ is in the set $\mathcal{M}_\sigma^e$ and satisfies the required properties. Indeed on the set A we have $\mathbf{E}_{\mathbf{Q}_1}[w \mid \mathcal{F}_t] = \mathbf{E}_{\mathbf{Q}}[w \mid \mathcal{F}_t]$ and $\mathbf{E}_{\mathbf{Q}_1}[w] \leq \mathbf{E}_{\mathbf{Q}_0}[w] + k < \infty$. $\qquad\square$

In Chap. 7 we recalled Émery's example showing that a stochastic integral with respect to a martingale is not always a local martingale. In [AS 94] Ansel and Stricker gave necessary and sufficient conditions under which a stochastic integral with respect to a local martingale remains a local martingale (see Theorem 7.3.7). We rephrase part of their result in our context of sigma-martingales.

Theorem 14.5.5. *Let H be S-integrable and w-admissible (where $w \geq 1$ is any random variable), then $H \cdot S$ is a local martingale (and hence also a super-martingale) for each $\mathbf{Q} \in \mathcal{M}_\sigma^e$ satisfying $\mathbf{E}_{\mathbf{Q}}[w] < \infty$.*

Proof. Simply write $H \cdot S$ as $(H\varphi^{-1}) \cdot (\varphi \cdot S)$, where the strictly positive predictable real-valued process φ is such that $\varphi \cdot S$ is a $\mathcal{H}^1(\mathbf{Q})$-martingale. Then apply the Ansel-Stricker result. $\qquad\square$

Remark 14.5.6. The statement of the preceding theorem becomes false if we replace the condition $\mathbf{Q} \in \mathcal{M}_\sigma^e$ by $\mathbf{Q} \in \mathcal{M}_s^e$, introduced as in Proposition 14.4.5 above, as the set of equivalent measures, under which $H \cdot S$ is a super-martingale for each admissible H. To see this, take the process S defined as $S_t = 0$ for $t \leq 1$ and $S_t = S_1$, a non-degenerate one-dimensional normal variable, for $t \geq 1$. The filtration is simply the filtration generated by S. As there are no admissible integrands, every equivalent probability measure $\mathbf{Q}$ is in $\mathcal{M}_s^e$. But it is clearly false that S becomes a $\mathbf{Q}$-super-martingale (i.e. $\mathbf{E}_{\mathbf{Q}}[S_1] \leq 0$) as soon as $\mathbf{E}_{\mathbf{Q}}[|S_1|] < \infty$.

Definition 14.5.7. *A random variable $w \colon \Omega \to \mathbb{R}_+$ such that $w \geq 1$ is called a feasible weight function for the process S, if*

(1) *there is a strictly positive bounded predictable process φ such that the maximal function of the $\mathbb{R}^d$-valued stochastic integral $\varphi \cdot S$ satisfies $(\varphi \cdot S)^* \leq w$.*
(2) *there is an element $\mathbf{Q} \in \mathcal{M}_\sigma^e$ such that $\mathbf{E}_{\mathbf{Q}}[w] < \infty$.*

Remark 14.5.8. For feasible weight functions w, it might happen that for some elements $\mathbf{Q} \in \mathcal{M}_\sigma^e$ we have that $\mathbf{E}_{\mathbf{Q}}[w] = \infty$, see the Example 14.5.23 below.

If no confusion can arise to which process the feasibility condition refers, then we will simply say that the weight function is feasible. The first item in the definition requires that w is big enough in order to allow sufficiently many integrands H such that both H and $-H$ are w-admissible. The second item requires w to be not too big, and as we will see, this will avoid arbitrage opportunities. It follows from Proposition 14.2.6 and the assumption that $\mathcal{M}^e_\sigma \neq 0$, that the existence of feasible weight functions is guaranteed. We also not that for locally bounded processes S, a function $w \geq 1$ is feasible as soon as there is $\mathbf{Q} \in \mathcal{M}^e_\sigma$ with $\mathbf{E_Q}[w] < \infty$.

We can now state the generalisations of the duality theorem mentioned above.

Theorem 14.5.9. *If w is a feasible weight function and g is a random variable such that $g \geq -w$ then:*

$$\sup_{\substack{\mathbf{Q} \in \mathcal{M}^e_\sigma \\ \mathbf{E_Q}[w] < \infty}} \mathbf{E_Q}[g]$$
$$= \inf \{\alpha \mid \text{ there is } H \text{ } w\text{-admissible and } g \leq \alpha + (H \cdot S)_\infty\} .$$

If the quantities are finite then the infimum is a minimum.

Remark 14.5.10. The reader can see that even in the case of locally bounded processes S the result yields more precise information. Indeed we restrict the supremum to those measures $\mathbf{Q} \in \mathcal{M}^e_\sigma$ such that $\mathbf{E_Q}[w] < \infty$.

For a feasible weight function w, we denote by $\mathcal{K}_w$ the set

$$\mathcal{K}_w = \{(H \cdot S)_\infty \mid H \text{ is } w\text{-admissible}\} .$$

Definition 14.5.11. *An element $g \in \mathcal{K}_w$ is called maximal if $h \in \mathcal{K}_w$ and $h \geq g$ imply that $h = g$.*

The maximal elements in this set are then characterised as follows:

Theorem 14.5.12. *If $w \geq 1$ is a feasible weight function, if H is w-admissible and if $h = (H \cdot S)_\infty$, then the following are equivalent:*

(1) h is maximal
(2) there is $\mathbf{Q} \in \mathcal{M}^e_\sigma$ such that $\mathbf{E_Q}[w] < \infty$ and $\mathbf{E_Q}[h] = 0$
(3) there is $\mathbf{Q} \in \mathcal{M}^e_\sigma$ such that $\mathbf{E_Q}[w] < \infty$ and $H \cdot S$ is a $\mathbf{Q}$-uniformly integrable martingale.

In the proof of these results we will make frequent use of Theorem 15.D and Corollary 15.4.11. These two results were proved for a slightly more restrictive notion of admissibility, but the reader can go through the proofs and check that the results remain valid for the present notion of w-admissible integrands. Indeed the lower bound $H \cdot S \geq -w$ is only used to control the negative parts of the possible jumps in the stochastic integral. This can also be achieved by the inequality $H \cdot S \geq -\mathbf{E_Q}[w \mid F_t]$ where $\mathbf{E_Q}[w] < \infty$. Compare the formulation of Theorems 15.B and 15.C. For the convenience of the reader let us rephrase the results of Chap. 15 in the present setting.

Theorem 14.5.13 (Theorem 15.D). *Let* $\mathbf{Q}$ *be a probability measure, equivalent to* $\mathbf{P}$. *Let* M *be an* $\mathbb{R}^d$-*valued* $\mathbf{Q}$-*local martingale and* $w \geq 1$ *a* $\mathbf{Q}$-*integrable function.*

Given a sequence $(H^n)_{n \geq 1}$ *of* M-*integrable* $\mathbb{R}^d$-*valued predictable processes such that*

$$(H^n \cdot M)_t \geq -\mathbf{E_Q}[w \mid \mathcal{F}_t], \quad \text{for all } n, t,$$

then there are convex combinations

$$K^n \in \mathrm{conv}\{H^n, H^{n+1}, \ldots\},$$

and there is a super-martingale $(V_t)_{t \in \mathbb{R}_+}, V_0 \leq 0$, *such that*

$$\lim_{\substack{s \searrow t \\ s \in \mathbf{Q}_+}} \lim_{n \to \infty} (K^n \cdot M)_s = V_t, \quad \text{for } t \in \mathbb{R}_+, \text{ a.s.,}$$

and an M-*integrable predictable process* H^0 *such that*

$$((H^0 \cdot M)_t - V_t)_{t \in \mathbb{R}_+} \quad \text{is increasing.}$$

In addition, $H^0 \cdot M$ *is a local martingale and a super-martingale.*

Corollary 14.5.14 (Corollary 15.4.11). *Let* S *be a semi-martingale taking values in* $\mathbb{R}^d$ *such that* $\mathcal{M}^e_\sigma(S) \neq \emptyset$ *and* $w \geq 1$ *a weight function such that there is some* $\mathbf{Q} \in \mathcal{M}^e_\sigma(S)$ *with* $\mathbf{E_Q}[w] < \infty$.

Then the convex cone

$$\{g \mid \text{there is a } (1, w)\text{-admissible integrand } H \text{ such that } g \leq (H \cdot S)_\infty\}$$

is closed in $L^0(\Omega)$ *with respect to the topology of convergence in measure.*

Let $w \geq 1$ be such that there is $\mathbf{Q} \in \mathcal{M}^e_\sigma$, with $\mathbf{E_Q}[w] < \infty$. The set

$$\mathcal{K}_w = \{(H \cdot S)_\infty \mid H \text{ is } w\text{-admissible}\}$$

is a cone in the space of measurable functions L^0. As in Chap. 9 we need the cone of all elements that are dominated by outcomes of w-admissible integrands:

$$\mathcal{C}^0_w = \{g \mid g \leq (H \cdot S)_\infty, \text{ where } H \text{ is } w\text{-admissible}\}.$$

If H is w-admissible and $\mathbf{E_Q}[w] < \infty$ for some $\mathbf{Q} \in \mathcal{M}^e_\sigma$, then it follows from the results in [AS 94] that the process $H \cdot S$ is a $\mathbf{Q}$-super-martingale. Therefore the limit $(H \cdot S)_\infty$ exists and $\mathbf{E_Q}[(H \cdot S)_\infty] \leq 0$. It also follows that for elements $g \in \mathcal{C}^0_w$, we have that $-\infty \leq \mathbf{E_Q}[g] \leq 0$. We will use this result frequently. We also remark that if H is w-admissible and if $(H \cdot S)_\infty \geq -w$ then H is already $(1, w)$-admissible. Indeed because of the super-martingale property of $H \cdot S$ we have that (at least for those $\mathbf{Q} \in \mathcal{M}_\sigma^e$ such that $\mathbf{E}_\mathbf{Q}[w] < \infty$):

$$(H \cdot S)_t \geq \mathbf{E}_\mathbf{Q}[(H \cdot S)_\infty \mid \mathcal{F}_t] \geq \mathbf{E}_\mathbf{Q}[-w \mid \mathcal{F}_t] .$$

By Lemma 14.5.4 this means that H is $(1, w)$-admissible.

Theorem 14.5.15. *If $w \geq 1$ and if there is some $\mathbf{Q} \in \mathcal{M}_\sigma^e$ such that $\mathbf{E}_\mathbf{Q}[w] < \infty$, then*

$$\mathcal{C}_w^\infty = \left\{ h \mid h \in L^\infty \text{ and } hw \in \mathcal{C}_w^0 \right\}$$

is weak-star-closed in $L^\infty(\mathbf{Q})$.

Proof. This is just a reformulation of Corollary 14.5.14 cited above.   □

We now prove the duality result stated in Theorem 14.5.9. The proof is broken up into several lemmata. As we will work with functions $w \geq 1$ that are not necessarily feasible weight functions we will make use of a larger class of equivalent measures namely:

$$\mathcal{M}_{s,w}^e = \left\{ \mathbf{Q} \sim \mathbf{P} \mid \mathbf{E}_\mathbf{Q}[w] < \infty \text{ and for each } h \in \mathcal{C}_w : \mathbf{E}_\mathbf{Q}[h] \leq 0 \right\} .$$

The reader can check that $\mathcal{M}_{s,w}^e$ is the set of equivalent probability measures so that w is integrable and with the property that for a w-admissible integrand H, the process $H \cdot S$ is a super-martingale. When we work with admissible integrands, i.e. with w identically equal to 1, then we simply drop, as in Proposition 14.4.5, the subscript w.

Lemma 14.5.16. *If $w \geq 1$ has a finite expectation for at least one element $\mathbf{Q} \in \mathcal{M}_\sigma^e$, if g is a random variable such that $g \geq -w$ then:*

$$\sup_{\substack{\mathbf{Q} \in \mathcal{M}_{s,w}^e \\ \mathbf{E}_\mathbf{Q}[w] < \infty}} \mathbf{E}_\mathbf{Q}[g] \leq \inf \left\{ \alpha \mid \text{ there is } H \ w\text{-admissible and } g \leq \alpha + (H \cdot S)_\infty \right\} .$$

Proof. The proof follows the same lines as the proof of Theorem 11.3.4. If $w \geq 1$ and $\mathbf{Q} \in \mathcal{M}_{s,w}^e$ then as observed above, the process $H \cdot S$ is a $\mathbf{Q}$-super-martingale for each H that is w-admissible. Therefore the inequality $g \leq \alpha + (H \cdot S)_\infty$ implies that $\mathbf{E}_\mathbf{Q}[g] \leq \alpha$.   □

Remark 14.5.17. If, under the same hypothesis of the Theorem 14.5.9 above, $\sup_{\mathbf{Q} \in \mathcal{M}_\sigma^e; \mathbf{E}_\mathbf{Q}[w] < \infty} \mathbf{E}_\mathbf{Q}[g] = \infty$, then also $\inf\{\alpha \mid$ there is H w-admissible and $g \leq \alpha + (H \cdot S)_\infty\} = \infty$. This simply means that no matter how big the constant a is taken, there is no w-admissible integrand H such that $g \leq a + (H \cdot S)_\infty$.

Lemma 14.5.18. *If* $w \geq 1$*, if for some* $\mathbf{Q}_0 \in \mathcal{M}_\sigma^e$ *we have* $\mathbf{E}_{\mathbf{Q}_0}[w] < \infty$*, if* $\frac{g}{w}$ *is bounded and if*

$$\beta < \inf \{\alpha \mid \text{ there is } H \text{ } w\text{-admissible and } g \leq \alpha + (H \cdot S)_\infty\},$$

then there is a probability measure $\mathbf{Q} \in \mathcal{M}_{s,w}^e$ *such that* $\mathbf{E}_{\mathbf{Q}}[g] > \beta$*.*

Proof. The hypothesis on β means that:

$$\left(\frac{g - \beta}{w} + L_+^\infty\right) \cap \mathcal{C}_w^\infty = \{0\}.$$

Because the set $\mathcal{C}_w^\infty$ is weak-star-closed, we can apply Yan's separation theorem [Y 80] (see also Theorem 5.2.2 above) and we obtain a strictly positive measure μ, equivalent to $\mathbf{P}$ such that

(1) $\mathbf{E}_\mu\left[\frac{g-\beta}{w}\right] > 0$
(2) for all $h \in \mathcal{C}_w^\infty$ we have $\mathbf{E}_\mu[h] \leq 0$.

If we normalise μ so that the measure $\mathbf{Q}$ defined as $d\mathbf{Q} = \frac{1}{w}d\mu$ becomes a probability measure, then we find that

(1) $\mathbf{Q} \sim \mathbf{P}$ and $\mathbf{E}_{\mathbf{Q}}[w] < \infty$,
(2) $\mathbf{E}_{\mathbf{Q}}[g] > \beta$,
(3) for all $h \in \mathcal{C}_w^\infty$ we have that $\mathbf{E}_{\mathbf{Q}}[hw] \leq 0$.

The latter inequality together with the Beppo-Levi theorem then implies that for each w-admissible integrand H we have that $\mathbf{E}_{\mathbf{Q}}[(H \cdot S)_\infty] \leq 0$.   □

Lemma 14.5.19. *If* $w \geq 1$*, if some* $\mathbf{Q}_0 \in \mathcal{M}_\sigma^e$ *we have* $\mathbf{E}_{\mathbf{Q}_0}[w] < \infty$*, if* $g \geq -w$ *then*

$$\sup_{\mathbf{Q} \in \mathcal{M}_{s,w}^e} \mathbf{E}_{\mathbf{Q}}[g] \geq \inf \{\alpha \mid \text{ there is } H \text{ } w\text{-admissible and } g \leq \alpha + (H \cdot S)_\infty\}.$$

Moreover if the quantity on the right hand side is finite, then the infimum is a minimum.

Proof. For each $n \geq 1$, we have that $\frac{g \wedge n}{w}$ is bounded and hence we can apply the previous lemma. This tells us that, for each $n \in \mathbb{N}$,

$$\alpha_n = \sup \{\mathbf{E}_{\mathbf{Q}}[g \wedge n] \mid \mathbf{Q} \in \mathcal{M}_{s,w}^e\}$$
$$\geq \inf \{\alpha \mid \text{ there is } H \text{ } w\text{-admissible and } g \wedge n \leq \alpha + (H \cdot S)_\infty\}.$$

Because there is nothing to prove when $\lim_n \alpha_n = \infty$ we may suppose that $\sup_n \alpha_n = \lim_n \alpha_n = \alpha < \infty$. So, for each n, we take a w-admissible integrand H^n such that $g \wedge n \leq \alpha_n + \frac{1}{n} + (H^n \cdot S)_\infty$. Let us now fix $\mathbf{Q}_0 \in \mathcal{M}_\sigma^e$ such that $\mathbf{E}_{\mathbf{Q}_0}[w] < \infty$. From Theorem 14.5.13, cited above, we deduce the existence of $K^n \in \text{conv}\{H^n, H^{n+1}, \ldots\}$ as well as H^0, such that

(1) $V_t = \lim_{s \searrow t; s \in \mathbf{Q}_+} \lim_{n \to \infty} (K^n \cdot S)_s$ exists a.s., for all $t \geq 0$,

(2) $(H^0 \cdot S)_t - V_t$ is increasing,

(3) $V_0 \leq 0$.

From this it follows that $(H^0 \cdot S)_t \geq -V_0 + V_t \geq V_t$. Since H^n is w-admissible (and hence $(1, w)$-admissible) we have that K^n is $(1, w)$-admissible and hence we find that $V_t \geq -\mathbf{E_Q}[w \mid \mathcal{F}_t]$ for all $\mathbf{Q} \in \mathcal{M}_\sigma^e$ such that $\mathbf{E_Q}[w] < \infty$. It is now clear that H^0 is w-admissible. Since the sequence α_n is increasing, we also obtain that for all t and all $\mathbf{Q} \in \mathcal{M}_\sigma^e$ with $\mathbf{E_Q}[w] < \infty$:

$$(K^n \cdot S)_t + \alpha_n + \frac{1}{n} \geq \mathbf{E_Q}\left[(K^n \cdot S)_\infty \mid \mathcal{F}_t\right] + \alpha_n + \frac{1}{n} \geq \mathbf{E_Q}\left[g \wedge n \mid \mathcal{F}_t\right].$$

This yields that, for all t and all n,

$$(H^0 \cdot S)_t + \alpha_n + \frac{1}{n} \geq V_t + \alpha_n + \frac{1}{n} \geq \mathbf{E_Q}\left[g \wedge n \mid \mathcal{F}_t\right].$$

If t tends to infinity this gives $(H^0 \cdot S)_\infty + \alpha_n + \frac{1}{n} \geq g \wedge n$ for all n. By taking the limit over n we finally find that

$$(H^0 \cdot S)_\infty + \alpha \geq g.$$

This shows the desired inequality and at the same time also shows that the infimum is a minimum.   □

We are now ready to prove the duality results. We start with the case of admissible integrands thus extending Theorem 11.3.4 to the case of non-locally bounded processes S. Recall that we assume throughout this section that S is an $\mathbb{R}^d$-valued semi-martingale satisfying *(NFLVR)*.

Theorem 14.5.20. *For a non-negative random variable g we have:*

$$\sup_{\mathbf{Q} \in \mathcal{M}_\sigma^e} \mathbf{E_Q}[g] = \inf \left\{\alpha \mid \text{there is } H \text{ admissible and } g \leq \alpha + (H \cdot S)_\infty\right\}.$$

Proof. From the previous lemmata it follows that we only have to show that

$$\sup_{\mathbf{Q} \in \mathcal{M}_\sigma^e} \mathbf{E_Q}[g] = \sup_{\mathbf{Q} \in \mathcal{M}_s^e} \mathbf{E_Q}[g].$$

This follows from Proposition 14.4.5 and the fact that g is bounded from below.   □

We now complete the proof for the case of feasible weight functions w and w-admissible integrands:

Proof of Theorem 14.5.9. In this case we show that $\mathcal{M}_\sigma^e = \mathcal{M}_{s,w}^e$. We already observed that $\mathcal{M}_\sigma^e \subset \mathcal{M}_{s,w}^e$. Take now $\mathbf{Q} \in \mathcal{M}_{s,w}^e$.

Since w is now supposed to be a feasible weight function, we have the existence of a strictly positive predictable function φ such that $(\varphi \cdot$

$S)^* \leq w$. It follows that random variables of the form $\mathbf{1}_A \left(\varphi \cdot S_t^i - \varphi \cdot S_s^i \right)$ or $-\mathbf{1}_A \left(\varphi \cdot S_t^i - \varphi \cdot S_s^i \right)$, where $s < t$, $A \in \mathcal{F}_s$ and $(S^i)_{i=1...d}$ are the coordinates of S, are results of w-admissible integrands. Therefore $\varphi \cdot S$ is a $\mathbf{Q}$-martingale and $\mathbf{Q} \in \mathcal{M}_\sigma^e$. $\qquad\square$

Corollary 14.5.21. *If $w \geq 1$ is a feasible weight function then the set*

$$\{ \mathbf{Q} \mid \mathbf{Q} \in \mathcal{M}_\sigma^e, \ \mathbf{E}_\mathbf{Q}[w] < \infty \}$$

is dense in $\mathcal{M}_\sigma^e$ for the variation norm.

Proof. If the set would not be dense then by the Hahn-Banach theorem, there exists $\mathbf{Q}_0 \in \mathcal{M}_\sigma^e$ and a bounded function g such that

$$\mathbf{E}_{Q_0}[g] > \sup \{ \mathbf{E}_\mathbf{Q}[g] \mid \mathbf{Q} \in \mathcal{M}_\sigma^e, \ \mathbf{E}_\mathbf{Q}[w] < \infty \} = \alpha .$$

This, together with Theorem 14.5.9, would then imply

$$
\begin{aligned}
\alpha_0 &= \inf \{ \alpha \mid \text{there is } H \text{ admissible and } g \leq \alpha + (H \cdot S)_\infty \} \\
&= \sup_{\mathbf{Q} \in \mathcal{M}_\sigma^e} \mathbf{E}_\mathbf{Q}[g] \\
&> \sup_{\substack{\mathbf{Q} \in \mathcal{M}_\sigma^e \\ \mathbf{E}_\mathbf{Q}[w] < \infty}} \mathbf{E}_\mathbf{Q}[g] \\
&= \inf \{ \alpha \mid \text{there is } H \ w\text{-admissible and } g \leq \alpha + (H \cdot S)_\infty \} .
\end{aligned}
$$

But a w-admissible integrand H such that $(H \cdot S)_\infty + \alpha \geq g$ is already admissible, proving that the strict inequality cannot hold. Indeed the process $H \cdot S$ is a $\mathbf{Q}$-super-martingale for each element $\mathbf{Q} \in \mathcal{M}_\sigma^e$ such that $\mathbf{E}_\mathbf{Q}[w] < \infty$. Therefore the process $H \cdot S$ is bounded below by $-\alpha - \|g\|_\infty$ and this means that H is admissible.

Remark 14.5.22. An interesting question is, whether by taking the supremum in Theorem 14.5.9, we have, for general unbounded functions g, to restrict to those elements $\mathbf{Q} \in \mathcal{M}_\sigma^e$ such that for the feasible weight function w we have $\mathbf{E}_\mathbf{Q}[w] < \infty$. More precisely is there a contingent claim $g \geq -w$ such that

$$\sup_{\mathbf{Q} \in \mathcal{M}_\sigma^e} \mathbf{E}_\mathbf{Q}[g] > \sup_{\substack{\mathbf{Q} \in \mathcal{M}_\sigma^e \\ \mathbf{E}_\mathbf{Q}[w] < \infty}} \mathbf{E}_\mathbf{Q}[g] .$$

An inspection of the proof of the above theorem shows that we used the $\mathbf{Q}$-integrability of the feasible weight function w in order to conclude that the w-admissible integrand H defined a $\mathbf{Q}$-super-martingale $H \cdot S$.

The next example, however, shows that it might happen that, for some sigma-martingale measure, $H \cdot S$ is a super-martingale, while for other sigma-martingale measures, it fails to be so.

Example 14.5.23. [†] There is a continuous process S, $S_0 = 0$ satisfying *(NFLVR)* and so that

(i) $\mathbf{P} \in \mathcal{M}_\sigma^e$,
(ii) S is not uniformly integrable under $\mathbf{P}$ and $\mathbf{E}_\mathbf{P}[S_\infty] > 0$,
(iii) the maximum function $S^* = \sup_{0 \leq t < \infty} |S_t|$ is not $\mathbf{P}$-integrable,
(iv) there is $\mathbf{Q}_0 \in \mathcal{M}_\sigma^e$ such that S is uniformly $\mathbf{Q}_0$-integrable and, more precisely, $\mathbf{E}_{\mathbf{Q}_0}[(S^*)^\gamma] < \infty$ for some $\gamma > 1$,
(v) the weight function $w = 1 + S^*$ is feasible and

$$\sup_{\mathbf{Q} \in \mathcal{M}_\sigma^e} \mathbf{E}_\mathbf{Q}[S_\infty] > \sup_{\substack{\mathbf{Q} \in \mathcal{M}_\sigma^e \\ \mathbf{E}_\mathbf{Q}[w] < \infty}} \mathbf{E}_\mathbf{Q}[S_\infty] = 0.$$

The example is the same as in Chap. 10 but we need additional properties. The space Ω supports two independent Brownian motions: B and W. We first introduce the two stochastic exponentials $L_t = \exp(B_t - \frac{1}{2}t)$ and $Z_t = \exp(W_t - \frac{1}{2}t)$. As in Chap. 10 we define $\tau = \inf\{t \mid L_t \leq \frac{1}{2}\}$ and $\sigma = \inf\{t \mid Z_t \geq 2\}$. The process Z^σ is bounded by 2 and L^τ is a strict local martingale. The process S is defined as $S = 1 - L^{\tau \wedge \sigma}$ and $\mathbf{Q}_0$ is defined by $\frac{d\mathbf{Q}_0}{d\mathbf{P}} = Z_{\tau \wedge \sigma}$. Since $\tau < \infty$ a.s. $\mathbf{Q}_0$ and $\mathbf{P}$ are equivalent. In Chap. 10 it is shown that (i) and (ii) and hence (iii) hold true. Also it is shown that S is a uniformly $\mathbf{Q}_0$-integrable martingale. We will now show that $\mathbf{E}_{\mathbf{Q}_0}[(S^*)^\gamma] < \infty$ for some $\gamma > 1$. This implies that $w = 1 + S^*$ is a feasible weight function and since $\mathbf{E}_\mathbf{P}[S_\infty] > 0$ and $\mathbf{P} \in \mathcal{M}_\sigma^e$ we get (v) as a consequence.

Hence we still have to show (iv). The estimate on $\mathbf{E}_{\mathbf{Q}_0}[(S^*)^\gamma]$ follows from the statement $\mathbf{E}_{\mathbf{Q}_0}[L_{\tau \wedge \sigma}^\gamma] < \infty$ for some $\gamma > 1$. This in turn follows from the following claim (see also [RY 91, Chap. II, Exercise (3.14)]).

Claim 1: Let W be a Brownian motion. Let $\nu = \inf\{t \mid W_t + \frac{1}{2}t \geq \ln 2\}$, then for $\beta \geq 0$ we have

$$\mathbf{E}[\exp(-\beta\nu)] = 2^{\frac{1 - \sqrt{1+8\beta}}{2}}.$$

This is seen as follows. For $\alpha \geq 0$ take the martingale $X_t = \exp(\alpha W_t - \frac{1}{2}\alpha^2 t)$ stopped at time ν. Since X^ν is bounded we get $\mathbf{E}[X_\nu] = 1$ and this implies $1 = \mathbf{E}[\exp(\alpha \ln 2 - \frac{\alpha}{2}\nu - \frac{1}{2}\alpha^2\nu)] = 2^\alpha \mathbf{E}[\exp(-\frac{\alpha(\alpha+1)}{2}\nu)]$. The equation $\frac{\alpha(\alpha+1)}{2} = \beta$ has one positive root for α, namely $\alpha = \frac{-1+\sqrt{1+8\beta}}{2}$. This gives $\mathbf{E}[\exp(-\beta\nu)] = 2^{\frac{1-\sqrt{1+8\beta}}{2}}$.

Claim 2: Let $f \geq 0$ be a random variable and for $\beta \in \mathbf{C}$, $\mathrm{Re}(\beta) \geq 0$ let $\varphi(\beta) = \mathbf{E}[\exp(-\beta f)]$. Suppose that φ has an analytic continuation, still denoted by φ, to the domain $\{\beta \in \mathbf{C} \mid |\beta| < \beta_0\}$. Then for $\beta \in \mathbf{C}$, $|\beta| < \beta_0$ we have $\mathbf{E}[\exp(\beta\sigma)] = \varphi(-\beta)$.

This is a standard exercise in probability theory.

[†] The original paper [DS 98, Example 5.14] contained an error in Example 14.5.23. This was pointed out by S. Biagini and M. Frittelli which is gratefully acknowledged, see [BF 04]. In their paper they adapted our example in a different way.

Claim 1 can now be improved as follows: for $-\infty < \beta \le \frac{1}{8}$ we have $\mathbf{E}[\exp(\beta\nu)] = 2^{\frac{1-\sqrt{1-8\beta}}{2}}$.

As above let us now look at $\sigma = \inf\{t \mid W_t - \frac{1}{2}t \ge \ln 2\}$ and consider the measure defined by $\frac{d\mathbf{Q}}{d\mathbf{P}} = Z_\sigma = 2\,\mathbf{1}_{\{\sigma<\infty\}}$. The extension of Girsanov's theorem for non-equivalent measures, due E. Lenglart [L 77], allows to write the Brownian motion W as $W_t = W'_t + t$, where W' is a $\mathbf{Q}$-Brownian motion and the equations hold $\mathbf{Q}$-a.s.. It follows that for $0 \le \beta \le \frac{1}{8}$

$$\mathbf{E_Q}[\exp(\beta\sigma)] = 2\mathbf{E}[\exp(\beta\sigma)\mathbf{1}_{\{\sigma<\infty\}}] = \mathbf{E}[\exp(\beta\nu)] = 2^{\frac{1-\sqrt{1-8\beta}}{2}}.$$

In other words $\mathbf{E}[\exp(\beta\sigma)\mathbf{1}_{\{\sigma<\infty\}}] = 2^{\frac{-1-\sqrt{1-8\beta}}{2}}$.

We are now ready to show:

Claim 3: $\mathbf{E_{Q_0}}[L^\gamma_{\tau\wedge\sigma}] < \infty$ for $1 \le \gamma \le \frac{1+\sqrt{2}}{2}$.

The calculations are straightforward but we prefer to give the details

$$\mathbf{E}\left[L^\gamma_{\tau\wedge\sigma} Z_{\tau\wedge\sigma}\right]$$

$$= \mathbf{E}\left[\exp\left(\gamma B_{\tau\wedge\sigma} - \frac{\gamma}{2}\tau\wedge\sigma\right)\exp\left(W_{\tau\wedge\sigma} - \frac{1}{2}\tau\wedge\sigma\right)\right]$$

$$= \mathbf{E}\left[\exp\left(\gamma B_{\tau\wedge\sigma} - \frac{\gamma^2}{2}\tau\wedge\sigma\right)\exp\left(\frac{\gamma(\gamma-1)}{2}\tau\wedge\sigma\right)\exp\left(W_{\tau\wedge\sigma} - \frac{1}{2}\tau\wedge\sigma\right)\right]$$

$$\le \mathbf{E}\left[\exp\left(\gamma B_{\tau\wedge\sigma} - \frac{\gamma^2}{2}\tau\wedge\sigma\right)\exp\left(\frac{\gamma(\gamma-1)}{2}\sigma\right)\exp\left(W_\sigma - \frac{1}{2}\sigma\right)\right]$$

$$\le \lim_{n\to\infty}\mathbf{E}\left[\exp\left(\gamma B_{\tau\wedge\sigma\wedge n} - \frac{\gamma^2}{2}\tau\wedge\sigma\wedge n\right)\exp\left(\frac{\gamma(\gamma-1)}{2}\sigma\right)\exp\left(W_\sigma - \frac{1}{2}\sigma\right)\right]$$

$$\le \lim_{n\to\infty}\mathbf{E}\left[\exp\left(\gamma B_{\tau\wedge n} - \frac{\gamma^2}{2}\tau\wedge n\right)\exp\left(\frac{\gamma(\gamma-1)}{2}\sigma\right)\exp\left(W_\sigma - \frac{1}{2}\sigma\right)\right],$$

this is seen as follows: we split into the two sets $\{\tau\wedge n \le \sigma\}$ and $\{\tau\wedge n > \sigma\} \in \mathcal{F}_\sigma$ and use the martingale property of $\exp(B_t - \frac{1}{2}t)$ for $t \le n$.

$$\le \mathbf{E}\left[\exp\left(\frac{\gamma(\gamma-1)}{2}\sigma\right)\exp\left(W_\sigma - \frac{1}{2}\sigma\right)\right] \quad \text{by independence of } B \text{ and } W!$$

For $1 \le \gamma \le \frac{1+\sqrt{2}}{2}$ we have $\frac{\gamma(\gamma-1)}{2} \le \frac{1}{8}$ and hence

$$\mathbf{E_{Q_0}}[L^\gamma_{\tau\wedge\sigma}] \le \mathbf{E}\left[\exp\left(\frac{\gamma(\gamma-1)}{2}\sigma\right)2\,\mathbf{1}_{\{\sigma<\infty\}}\right] < \infty.$$

This ends the proof of (iv) and completes the discussion of the example. $\quad\square$

We now turn to the characterisation of maximal and of attainable elements. The approach is different from the one used in Chap. 11, which was based on a change of numéraire technique. In order not to overload the statements we henceforth suppose that w is a feasible weight function.

Lemma 14.5.24. *If $g \in \mathcal{K}_w$, then there is a maximal element $h \in \mathcal{K}_w$ such that $h \geq g$.*

Proof. It is sufficient to show that every increasing sequence in $\mathcal{K}_w$ has an upper bound in $\mathcal{K}_w$. So let h_n, $h_1 = g$, be an increasing sequence in $\mathcal{K}_w$. For each n take H^n, w-admissible so that $h_n = (H^n \cdot S)_\infty$. As in the previous proof we then find, as an application of Theorem 14.5.13, that there is H^0, w-admissible such that $(H^0 \cdot S)_\infty \geq \lim_n h_n$. This concludes the proof of the lemma. $\square$

Proof of Theorem 14.5.12. If $\mathbf{E_Q}[w] < \infty$ then $H \cdot S$ is a $\mathbf{Q}$-super-martingale and hence (2) and (3) are equivalent. Also it is clear that (2) implies (1). Indeed if g is the result of a w-admissible integrand then $\mathbf{E_Q}[g] \leq 0$ for each $\mathbf{Q} \in \mathcal{M}_\sigma^e$ such that also $\mathbf{E_Q}[w] < \infty$. It follows that h is necessarily maximal.

The only remaining part is that (1) implies (2). Since always $\mathbf{E_Q}[h] \leq 0$ for $\mathbf{Q} \in \mathcal{M}_\sigma^e$ such that also $\mathbf{E_Q}[w] < \infty$, we obtain already that for measures $\mathbf{Q}$ satisfying these assumptions, h^+ is $\mathbf{Q}$-integrable. So fix such a measure $\mathbf{Q}$. Now let $w_1 = h^+ + w$. Clearly w_1 is a feasible weight function. We will work with the set $\mathcal{K}_{w_1}$. The problem is, however, that we do not (yet) know that h is still maximal in the bigger cone $\mathcal{K}_{w_1}$. From the construction of w_1 it follows that, for elements $\mathbf{Q} \in \mathcal{M}_\sigma^e$, we have $\mathbf{E_Q}[w_1] < \infty$ if and only if $\mathbf{E_Q}[w] < \infty$. Now let $g \geq h$ be the result of a w_1-admissible integrand. Hence $g = (K \cdot S)_\infty$ where K is w_1-admissible. Since $(K \cdot S)_\infty \geq g \geq h \geq -w$ and since K is w_1-admissible we have that K is already w-admissible. (Remember that $\mathbf{E_Q}[w_1] < \infty$ if and only if $\mathbf{E_Q}[w] < \infty$) From the maximality of h in $\mathcal{K}_w$ it then follows that $g = h$, i.e. h is maximal in $\mathcal{K}_{w_1}$. This can then be translated into

$$\left(\frac{h}{w_1} + L_+^\infty \right) \cap \mathcal{C}_{w_1}^\infty = \{0\}.$$

Using Yan's separation theorem (Theorem 5.2.2 above) in the same way as in the proof of Theorem 14.5.9 above, we find a measure $\mathbf{Q}_1$ such that $\mathbf{E}_{\mathbf{Q}_1}[w_1] < \infty$, $\mathbf{Q}_1 \in \mathcal{M}_\sigma^e$ and $\mathbf{E}_{\mathbf{Q}_1}[h] \geq 0$. $\square$

The following theorem generalises a result due to Ansel-Stricker and Jacka, [AS 94] and [J 92].

Theorem 14.5.25. *Let w be a feasible weight function and let $f \geq -w$. The following assertions are equivalent*

(1) *there is a measure $\mathbf{Q} \in \mathcal{M}_\sigma^e$ such that $\mathbf{E_Q}[w] < \infty$ and such that*

$$\mathbf{E_Q}[f] = \sup_{\substack{R \in \mathcal{M}_\sigma^e \\ \mathbf{E_R}[w] < \infty}} \mathbf{E_R}[f] < \infty$$

(2) *f can be hedged, i.e. there is $\alpha \in \mathbb{R}$, $\mathbf{Q} \in \mathcal{M}_\sigma^e$ such that $\mathbf{E_Q}[w] < \infty$, a w-admissible integrand H such that $H \cdot S$ is a $\mathbf{Q}$-uniformly integrable martingale, and such that $f = \alpha + (H \cdot S)_\infty$.*

Proof. Clearly (2) implies (1) by Theorem 14.5.12.

For the reverse implication take now $\mathbf{Q}$ as in (1), then the duality result (Theorem 14.5.9) gives $\alpha \in \mathbb{R}$ as well as a w-admissible integrand H such that $f \leq \alpha + (H \cdot S)_\infty$, where $\alpha = \sup_{\mathbf{R} \in \mathcal{M}^e; \mathbf{E}_\mathbf{R}[w] < \infty} [f]$. Here we use explicitly that the infimum in the duality theorem is a minimum. But then it follows from $\mathbf{E}_\mathbf{Q}[w] < \infty$ and from the equality $\mathbf{E}_\mathbf{Q}[f] = \alpha$ that $f = \alpha + (H \cdot S)_\infty$ and that $H \cdot S$ is a $\mathbf{Q}$-uniformly integrable martingale. $\square$

15

A Compactness Principle
for Bounded Sequences of Martingales
with Applications (1999)

Abstract. For $\mathcal{H}^1$-bounded sequences of martingales, we introduce a technique, related to the Kadeč-Pełczyński decomposition for L^1 sequences, that allows us to prove compactness theorems. Roughly speaking, a bounded sequence in $\mathcal{H}^1$ can be split into two sequences, one of which is weakly compact, the other forms the singular part. If the martingales are continuous then the singular part tends to zero in the semi-martingale topology. In the general case the singular parts give rise to a process of bounded variation. The technique allows to give a new proof of the optional decomposition theorem in Mathematical Finance.

15.1 Introduction

Without any doubt, one of the most fundamental results in analysis is the theorem of Heine-Borel:

Theorem 15.1.1. *From a bounded sequence $(x_n)_{n \geq 1} \in \mathbb{R}^d$ we can extract a convergent subsequence $(x_{n_k})_{k \geq 1}$.*

If we pass from $\mathbb{R}^d$ to infinite dimensional Banach spaces X this result does not hold true any longer. But there are some substitutes which often are useful. The following theorem can be easily derived from the Hahn-Banach theorem and was well-known to S. Banach and his contemporaries (see [DRS 93] for related theorems).

Theorem 15.1.2. *Given a bounded sequence $(x_n)_{n \geq 1}$ in a reflexive Banach space X (or, more generally, a relatively weakly compact sequence in a Banach space X) we may find a sequence $(y_n)_{n \geq 1}$ of convex combinations of $(x_n)_{n \geq 1}$,*

$$y_n \in \mathrm{conv}\{x_n, x_{n+1}, \dots\},$$

which converges with respect to the norm of X.

[DS 99] A Compactness Principle for Bounded Sequences of Martingales with Applications. Proceedings of the Seminar on Stochastic Analysis, Random Fields and Applications, *Progress in Probability*, vol. 45, pp. 137–173, Birkhäuser, Basel (1999).

Note — and this is a *Leitmotiv* of the present paper — that, for sequences $(x_n)_{n\geq 1}$ in a vector space, passing to convex combinations usually does not cost more than passing to a subsequence. In most applications the main problem is to find a limit $x_0 \in X$ and typically it does not matter whether $x_0 = \lim_k x_{n_k}$ for a subsequence $(x_{n_k})_{k\geq 1}$ or $x_0 = \lim_n y_n$ for a sequence of convex combinations $y_n \in \mathrm{conv}\{x_n, x_{n+1}, \ldots\}$.

If one passes to the case of non-reflexive Banach spaces there is — in general — no analogue to Theorem 15.1.2 pertaining to any bounded sequence $(x_n)_{n\geq 1}$, the main obstacle being that the unit ball fails to be weakly compact. But sometimes there are Hausdorff topologies on the unit ball of a (non-reflexive) Banach space which have some kind of compactness properties. A noteworthy example is the Banach space $L^1(\Omega, \mathcal{F}, \mathbf{P})$ and the topology of convergence in measure.

Theorem 15.1.3. *Given a bounded sequence* $(f_n)_{n\geq 1} \in L^1(\Omega, \mathcal{F}, \mathbf{P})$ *then there are convex combinations*

$$g_n \in \mathrm{conv}\{f_n, f_{n+1}, \ldots)\}$$

such that $(g_n)_{n\geq 1}$ *converges in measure to some* $g_0 \in L^1(\Omega, \mathcal{F}, \mathbf{P})$.

The preceding theorem is a somewhat vulgar version of Komlos' theorem [K 67]. Note that Komlos' result is more subtle as it replaces the convex combinations $(g_n)_{n\geq 1}$ by the Cesaro-means of a properly chosen subsequence $(f_{n_k})_{k\geq 1}$ of $(f_n)_{n\geq 1}$.

But the above *vulgar version* of Komlos' theorem has the advantage that it extends to the case of $L^1(\Omega, \mathcal{F}, \mathbf{P}; E)$ for reflexive Banach spaces E as we shall presently see (Theorem 15.1.4 below), while Komlos' theorem does not. (J. Bourgain [B 79] proved that the precise necessary and sufficient condition for the Komlos theorem to hold for E-valued functions is that $L^2(\Omega, \mathcal{F}, \mathbf{P}; E)$ has the Banach-Saks property; compare [G 79] and [S 81].)

Here is the vector-valued version of Theorem 15.1.3:

Theorem 15.1.4. *If E is a reflexive Banach space and* $(f_n)_{n\geq 1}$ *a bounded sequence in* $L^1(\Omega, \mathcal{F}, \mathbf{P}; E)$, *we may find convex combinations*

$$g_n \in \mathrm{conv}\{f_n, f_{n+1}, \ldots\}$$

and $g_0 \in L^1(\Omega, \mathcal{F}, \mathbf{P}; E)$ *such that* $(g_n)_{n\geq 1}$ *converges to* f_0 *almost surely, i.e.,*

$$\lim_{n \to \infty} \|g_n(\omega) - g_0(\omega)\|_E = 0 \qquad \textit{for a.e. } \omega \in \Omega.$$

The preceding theorem seems to be of folklore type and to be known to specialists for a long time (compare also [DRS 93]). We shall give a proof in Sect. 15.2 below.

Let us have a closer look at what is really happening in Theorems 15.1.3 and 15.1.4 above by following the lines of Kadeč and Pełczyński [KP 65].

These authors have proved a remarkable decomposition theorem which essentially shows the following (see Theorem 15.2.1 below for a more precise statement): Given a bounded sequence $(f_n)_{n\geq 1}$ in $L^1(\Omega, \mathcal{F}, \mathbf{P})$ we may find a subsequence $(f_{n_k})_{k\geq 1}$ which may be split into a *regular* and a *singular* part, $f_{n_k} = f_{n_k}^r + f_{n_k}^s$, such that $(f_{n_k}^r)_{k\geq 1}$ is uniformly integrable and $(f_{n_k}^s)_{k\geq 1}$ tends to zero almost surely.

Admitting this result, Theorem 15.1.3 becomes rather obvious: As regards the *regular part* $(f_{n_k}^r)_{k\geq 1}$ we can apply Theorem 15.1.2 to find convex combinations converging with respect to the norm of L^1 and therefore in measure. As regards the *singular part* $(f_{n_k}^s)_{k\geq 1}$ we do not have any problems as any sequence of convex combinations will also tend to zero almost surely.

A similar reasoning allows to deduce the vector-valued case (Theorem 15.1.4 above) from the Kadeč-Pełczyński decomposition result (see Sect. 15.2 below).

After this general prelude we turn to the central theme of this paper. Let $(M_t)_{t\in\mathbb{R}_+}$ be an $\mathbb{R}^d$-valued càdlàg local martingale w.r. to $(\Omega, \mathcal{F}, (\mathcal{F}_t)_{t\in\mathbb{R}_+}, \mathbf{P})$ and $(H^n)_{n\geq 1}$ a sequence of M-integrable processes, i.e., predictable $\mathbb{R}^d$-valued stochastic processes such that the integral

$$(H^n \cdot M)_t = \int_0^t H_u^n \, dM_u$$

makes sense for every $t \in \mathbb{R}_+$, and suppose that the resulting processes $((H^n \cdot M)_t)_{t\in\mathbb{R}_+}$ are martingales. The theme of the present paper is: *under what conditions can we pass to a limit H^0?* More precisely: by passing to convex combinations of $(H^n)_{n\geq 1}$ (still denoted by H^n) we would like to ensure that the sequence of martingales $H^n \cdot M$ converges to some martingale N which is of the form $N = H^0 \cdot M$.

Our motivation for this question comes from applications of stochastic calculus to Mathematical Finance where this question turned out to be of crucial relevance. For example, in chapter 9 as well as in the work of D. Kramkov ([K 96a]) the passage to the limit of a sequence of integrands is the heart of the matter. We shall come back to the applications of the results obtained in this paper to Mathematical Finance in Sect. 15.5 below.

Let us review some known results in the context of the above question. The subsequent Theorem 15.1.5, going back to the foundations of stochastic integration given by Kunita and Watanabe [KW 67], is a straightforward consequence of the Hilbert space isometry of stochastic integrands and integrals (see, e.g., [P 90, p. 153] for the real-valued and Jacod [J 79] for the vector-valued case).

Theorem 15.1.5 (Kunita-Watanabe). *Let M be an $\mathbb{R}^d$-valued càdlàg local martingale, $(H^n)_{n\geq 1}$ be a sequence of M-integrable predictable stochastic processes such that each $(H^n \cdot M)$ is an L^2-bounded martingale and such that the sequence of random variables $((H^n \cdot M)_\infty)_{n\geq 1}$ converges to a random variable $f_0 \in L^2(\Omega, \mathcal{F}, \mathbf{P})$ with respect to the norm of L^2.*

Then there is an M-integrable predictable stochastic process H^0 such that $H^0 \cdot M$ is an L^2-bounded martingale and such that $(H^0 \cdot M)_\infty = f_0$.

It is not hard to extend the above theorem to the case of L^p, for $1 < p \leq \infty$. But the extension to $p = 1$ is a much more delicate issue which has been settled by M. Yor [Y 78a], who proved the analogue of Theorem 15.1.5 for the case of $\mathcal{H}^1$ and L^1.

Theorem 15.1.6 (Yor). *Let $(H^n)_{n \geq 1}$ be a sequence of M-integrable predictable stochastic processes such that each $(H^n \cdot M)$ is an $\mathcal{H}^1$-bounded (resp. a uniformly integrable) martingale and such that the sequence of random variables $((H^n \cdot M)_\infty)_{n \geq 1}$ converges to a random variable $f_0 \in \mathcal{H}^1(\Omega, \mathcal{F}, \mathbf{P})$ (resp. $f_0 \in L^1(\Omega, \mathcal{F}, \mathbf{P})$) with respect to the $\mathcal{H}^1$-norm (resp. L^1-norm); (or even only with respect to the $\sigma(\mathcal{H}^1, BMO)$ (resp. $\sigma(L^1, L^\infty)$) topology).*

Then there is an M-integrable predictable stochastic process H^0 such that $H^0 \cdot M$ is an $\mathcal{H}^1$-bounded (resp. uniformly integrable) martingale and such that $(H^0 \cdot M)_\infty = f_0$.

We refer to Jacod [J 79, Theorème 4.63, p.143] for the $\mathcal{H}^1$-case. It essentially follows from Davis' inequality for $\mathcal{H}^1$-martingales. The L^1-case (see [Y 78a]) is more subtle. Using delicate stopping time arguments M. Yor succeeded in reducing the L^1 case to the $\mathcal{H}^1$ case. In Sect. 15.4 we take the opportunity to translate Yor's proof into the setting of the present paper.

Let us also mention in this context a remarkable result of Mémin ([M 80, Theorem V.4]) where the process M is only assumed to be a semi-martingale and not necessarily a local martingale and which also allows to pass to a limit $H^0 \cdot M$ of a Cauchy sequence $H^n \cdot M$ of M-integrals (w.r. to the semi-martingale topology).

All these theorems are *closedness results* in the sense that, if $(H^n \cdot M)$ is a *Cauchy-sequence* with respect to some topology, then we may find H^0 such that $(H^0 \cdot M)$ equals the limit of $(H^n \cdot M)$.

The aim of our paper is to prove *compactness results* in the sense that, if $(H^n \cdot M)$ is a *bounded sequence* in the martingale space $\mathcal{H}^1$, then we may find a subsequence $(n_k)_{k \geq 1}$ as well as decompositions $H^{n_k} = {}^r K^k + {}^s K^k$ so that the sequence ${}^r K^k \cdot M$ is relatively weakly compact in $\mathcal{H}^1$ and such that the singular parts ${}^s K^k \cdot M$ hopefully tend to zero in some sense to be made precise. The regular parts ${}^r K^k \cdot M$ then allow to take convex combinations that converge in the norm of $\mathcal{H}^1$.

It turns out that for *continuous* local martingales M the situation is nicer (and easier) than for the general case of local martingales with jumps. We now state the main result of this paper, in its continuous and in its general version (Theorem 15.A and 15.B below).

Theorem 15.A. *Let $(M^n)_{n \geq 1}$ be an $\mathcal{H}^1$-bounded sequence of real-valued continuous local martingales.*

Then we can select a subsequence, which we still denote by $(M^n)_{n \geq 1}$, as well as an increasing sequence of stopping times $(T_n)_{n \geq 1}$, such that $\mathbf{P}[T_n < \infty]$

tends to zero and such that the sequence of stopped processes $\left((M^n)^{T_n}\right)_{n\geq 1}$ is relatively weakly compact in $\mathcal{H}^1$.

If all the martingales are of the form $M^n = H^n \cdot M$ for a fixed continuous local martingale taking values in $\mathbb{R}^d$, then the elements in the $\mathcal{H}^1$-closed convex hull of the sequence $\left((M^n)^{T_n}\right)_{n\geq 1}$ are also of the form $H \cdot M$.

As a consequence we obtain the existence of convex combinations

$$K^n \in \mathrm{conv}\{H^n, H^{n+1}, \ldots\}$$

such that $K^n \mathbf{1}_{[0,T_n]} \cdot M$ tends to a limit $H^0 \cdot M$ in $\mathcal{H}^1$. Also remark that the remaining *singular* parts $K^n \mathbf{1}_{]T_n,\infty]} \cdot M$ tend to zero in a stationary way, i.e. for almost each $\omega \in \Omega$ the set $\{t \mid \exists n \geq n_0, K_t^n \neq 0\}$ becomes empty for large enough n_0. As a result we immediately derive that the sequence $K^n \cdot M$ tends to $H^0 \cdot M$ in the semi-martingale topology.

If the local martingale M is not continuous the situation is more delicate. In this case we cannot obtain a limit of the form $H^0 \cdot M$ and also the decomposition is not just done by stopping the processes at well-selected stopping times.

Theorem 15.B. *Let M be an $\mathbb{R}^d$-valued local martingale and $(H^n)_{n\geq 1}$ be a sequence of M-integrable predictable processes such that $(H^n \cdot M)_{n\geq 1}$ is an $\mathcal{H}^1$ bounded sequence of martingales.*

Then there is a subsequence, for simplicity still denoted by $(H^n)_{n\geq 1}$, an increasing sequence of stopping times $(T_n)_{n\geq 1}$, a sequence of convex combinations $L^n = \sum_{k\geq n} \alpha_k^n H^k$ as well as a sequence of predictable sets $(E^n)_{n\geq 1}$ such that

(1) *$E^n \subset [0, T_n]$ and T_n increases to ∞,*
(2) *the sequence $\left(H^n \mathbf{1}_{[0,T_n]\cap(E^n)^c} \cdot M\right)_{n\geq 1}$ is weakly relatively compact in $\mathcal{H}^1$,*
(3) *$\sum_{n\geq 1} \mathbf{1}_{E^n} \leq d$,*
(4) *the convex combinations $\sum_{k\geq n} \alpha_n^k H^k \mathbf{1}_{[0,T_n]\cap(E^n)^c} \cdot M$ converge in $\mathcal{H}^1$ to a stochastic integral of the form $H^0 \cdot M$, for some predictable process H^0,*
(5) *the convex combinations $V_n = \sum_{k\geq n} \alpha_n^k H^k \mathbf{1}_{]T_n,\infty[\cup E^n} \cdot M$ converge to a càdlàg optional process Z of finite variation in the following sense: a.s. we have that $Z_t = \lim_{s\searrow t;\, s\in\mathbb{Q}} \lim_{n\to\infty} (V_n)_s$ for each $t \in \mathbb{R}_+$,*
(6) *the brackets $[(H^0 - L^n) \cdot M, (H^0 - L^n) \cdot M]_\infty$ tend to zero in probability.*

If, in addition, the set

$$\{\Delta(H^n \cdot M)_T^- \mid n \in \mathbb{N};\ T \text{ stopping time}\}$$

resp.

$$\{|\Delta(H^n \cdot M)_T| \mid n \in \mathbb{N};\ T \text{ stopping time}\}$$

is uniformly integrable, e.g. there is an integrable function $w \geq 0$ such that

$$\Delta(H^n \cdot M) \geq -w \quad \text{resp.} \quad |\Delta(H^n \cdot M)| \leq w, \quad a.s.$$

then the process $(Z_t)_{t\in\mathbb{R}_+}$ is decreasing (resp. vanishes identically).

For general martingales, not necessarily of the form $H^n \cdot M$ for a fixed local martingale M, we can prove the following theorem:

Theorem 15.C. *Let $(M^n)_{n \geq 1}$ be an $\mathcal{H}^1$-bounded sequence of $\mathbb{R}^d$-valued martingales. Then there is a subsequence, for simplicity still denoted by $(M_n)_{n \geq 1}$ and an increasing sequence of stopping times $(T_n)_{n \geq 1}$ with the following properties:*

(1) *T_n increases to ∞,*
(2) *the martingales $N^n = (M^n)^{T_n} - \Delta M_{T_n} \mathbf{1}_{[\![T_n, \infty[\![} + C^n$ form a relatively weakly compact sequence in $\mathcal{H}^1$. Here C^n denotes the compensator (dual predictable projection) of the process $\Delta M_{T_n} \mathbf{1}_{[\![T_n, \infty[\![}$,*
(3) *there are convex combinations $\sum_{k \geq n} \alpha_n^k N^k$ that converge to an $\mathcal{H}^1$, martingale N^0 in the norm of $\mathcal{H}^1$,*
(4) *there is a càdlàg optional process of finite variation Z such that almost everywhere for each $t \in \mathbb{R}$: $Z_t = \lim_{s \searrow t;\, s \in \mathbf{Q}} \lim_{n \to \infty} \sum_{k \geq n} \alpha_n^k C_s^k$.*

If, in addition, the set

$$\left\{ \Delta(M^n)_T^- \mid n \in \mathbb{N};\ T \text{ stopping time} \right\}$$

resp.

$$\left\{ |\Delta(M^n)_T| \mid n \in \mathbb{N};\ T \text{ stopping time} \right\}$$

is uniformly integrable, e.g. there is an integrable function $w \geq 0$ such that

$$\Delta(M^n) \geq -w \quad resp. \quad |\Delta(M^n)| \leq w, \quad a.s.$$

then the process $(Z_t)_{t \in \mathbb{R}_+}$ is increasing (resp. vanishes identically).

Let us comment on these theorems. Theorem 15.A shows that in the continuous case we may cut off some *small* singular parts in order to obtain a relatively weakly compact sequence $((M^n)^{T_n})_{n \geq 1}$ in $\mathcal{H}^1$. By taking convex combinations we then obtain a sequence that converges in the norm of $\mathcal{H}^1$. The singular parts are small enough so that they do not influence the almost sure passage to the limit. Note that — in general — there is no hope to get rid of the singular parts. Indeed, a Banach space E such that for each bounded sequence $(x_n)_{n \geq 1} \in E$ there is a norm-convergent sequence $y_n \in \text{conv}\{x_n, x_{n+1}, \ldots\}$ is reflexive; and, of course, $\mathcal{H}^1$ is only reflexive if it is finite dimensional.

The general situation of local martingales M (possibly with jumps) described in Theorem 15.B is more awkward. As regards the convex combinations of the form $(\sum_{k \geq n} \alpha_n^k H^k \mathbf{1}_{[\![0, T_n]\!] \cap (E^n)^c} \cdot M)_{n \geq 1}$ we have convergence in $\mathcal{H}^1$ but for the *singular* parts $(V^n)_{n \geq 1}$ we cannot assert that they tend to zero. Nevertheless there is some control on these processes. We may assert that the processes $(V^n)_{n \geq 1}$ tend, in a certain pointwise sense, to a process $(Z_t)_{t \in \mathbb{R}_+}$ of integrable variation. We shall give an example (Sect. 15.3 below) which illustrates that in general one cannot do better than that. But under special assumptions, e.g., one-sided or two-sided bounds on the jumps of the processes $(H^n \cdot M)$, one may deduce certain features of the process Z (e.g., Z being monotone or vanishing identically). It is precisely this latter conclusion which has applications in Mathematical Finance and allows to give an alternative proof of Kramkov's *optional decomposition theorem* [K 96a] (see Theorem 15.5.1 below).

To finish the introduction we shall state the main application of Theorem 15.B. Note that the subsequent statement of Theorem 15.D does not use the concept of $\mathcal{H}^1(\mathbf{P})$-martingales (although the proof heavily relies on this concept) which makes it more applicable in general situations.

Theorem 15.D. *Let M be an $\mathbb{R}^d$-valued local martingale and $w \geq 1$ an integrable function.*

Given a sequence $(H^n)_{n \geq 1}$ of M-integrable $\mathbb{R}^d$-valued predictable processes such that

$$(H^n \cdot M)_t \geq -w, \qquad \text{for all } n, t,$$

then there are convex combinations

$$K^n \in \text{conv}\{H^n, H^{n+1}, \ldots\},$$

and there is a super-martingale $(V_t)_{t \in \mathbb{R}_+}, V_0 \leq 0$, such that

$$\lim_{\substack{s \searrow t \\ s \in \mathbb{Q}_+}} \lim_{n \to \infty} (K^n \cdot M)_s = V_t \quad \text{for } t \in \mathbb{R}_+, \text{ a.s.},$$

and an M-integrable predictable process H^0 such that

$$((H^0 \cdot M)_t - V_t)_{t \in \mathbb{R}_+} \quad \text{is increasing.}$$

In addition, $H^0 \cdot M$ is a local martingale and a super-martingale.

Loosely speaking, Theorem 15.D says that for a sequence $(H^n \cdot M)_{n \geq 1}$, obeying the crucial assumption of uniform lower boundedness with respect to an integrable weight function w, we may pass — by forming convex combinations — to a limiting super-martingale V in a pointwise sense and — more importantly — to a local martingale of the form $(H^0 \cdot M)$ which dominates V.

The paper is organised as follows: Sect. 15.2 introduces notation and fixes general hypotheses. We also give a proof of the Kadeč-Pełczyński decomposition and we recall basic facts about weak compactness in $\mathcal{H}^1$. We give additional (and probably new) information concerning the convergence of the maximal function and the convergence of the square function. Sect. 15.3 contains an example. In Sect. 15.4, we give the proofs of Theorems 15.A, 15.B, 15.C and 15.D. We also reprove M. Yor's Theorem 15.1.6. In Sect. 15.5 we reprove Kramkov's Optional Decomposition Theorem 15.5.1.

15.2 Notations and Preliminaries

We fix a filtered probability space $(\Omega, \mathcal{F}, (\mathcal{F}_t)_{t\in\mathbb{R}_+}, \mathbf{P})$, where the filtration $(\mathcal{F}_t)_{t\in\mathbb{R}_+}$ satisfies the *usual conditions* of completeness and right continuity. We also assume that $\mathcal{F}$ equals $\mathcal{F}_\infty$. In principle, the letter M will be reserved for a càdlàg $\mathbb{R}^d$-valued local martingale. We assume that $M_0 = 0$ to avoid irrelevant difficulties at $t = 0$.

We denote by $\mathcal{O}$ (resp. $\mathcal{P}$) the σ-algebra of optional (resp. predictable) subsets of $\mathbb{R}_+ \times \Omega$. For the notion of an M-integrable $\mathbb{R}^d$-valued predictable process $H = (H_t)_{t\in\mathbb{R}_+}$ and the notion of the stochastic integral

$$(H \cdot M)_t = \int_0^t H_u \, dM_u$$

we refer to [P 90] and to [J 79]. Most of the time we shall assume that the process $H \cdot M$ is a local martingale (for the delicacy of this issue compare [E 80] and [AS 94]) and, in fact, a uniformly integrable martingale.

For the definition of the bracket process $[M, M]$ of the real-valued local martingale M as well as for the σ-finite, non-negative measure $d[M, M]$ on the σ-algebra $\mathcal{O}$ of optional subsets of $\Omega \times \mathbb{R}_+$, we also refer to [P 90]. In the case $d > 1$ the bracket process $[M, M]$ is defined as a matrix with components $[M^i, M^j]$ where $M = (M^1, \ldots, M^d)$. The process $[M, M]$ takes values in the cone of non-negative definite $(d \times d)$-matrices. This is precisely the Kunita-Watanabe inequality for the bracket process. One can select representations so that for almost each $\omega \in \Omega$ the measure $d[M, M]$ induces a σ-finite measure, denoted by $d[M, M]_\omega$, on the Borel sets of $\mathbb{R}_+$ (and with values in the cone of non-negative definite $(d \times d)$-matrices).

For an $\mathbb{R}^d$-valued local martingale X, $X_0 = 0$, we define the $\mathcal{H}^1$-norm by

$$\|X\|_{\mathcal{H}^1} = \| \, (\operatorname{trace}([X, X]_\infty))^{\frac{1}{2}} \, \|_{L^1(\Omega, \mathcal{F}, \mathbf{P})}$$
$$= \mathbf{E}\left[\left(\int_0^\infty d\left(\operatorname{trace}([X, X]_t)\right) \right)^{\frac{1}{2}} \right] \leq \infty$$

where trace denotes the trace of a $(d \times d)$-matrix and the L^1-norm by

$$\|X\|_{L^1} = \sup_T \mathbf{E}\left[|X_T|\right] \leq \infty,$$

where $|\,.\,|$ denotes a fixed norm on $\mathbb{R}^d$, where the sup is taken over all finite stopping times T and which, in the case of a uniformly integrable martingale X, equals

$$\|X\|_{L^1} = \mathbf{E}\left[|X_\infty|\right] < \infty.$$

The Davis' inequality for $\mathcal{H}^1$-martingales ([RY 91, Theorem IV.4.1], see also [M 76]) states that there are universal constants, c_1 and c_2 (only depending on the dimension d), such that for each $\mathcal{H}^1$-martingale X we have:

$$c_1\|X^*_\infty\|_{L^1} \le \|X\|_{\mathcal{H}^1} \le c_2\|X^*_\infty\|_{L^1},$$

where $X^*_u = \sup_{t \le u}|X_t|$ denotes the maximal function.

We denote by $\mathcal{H}^1 = \mathcal{H}^1(\Omega, \mathcal{F}, (\mathcal{F}_t)_{t \in \mathbb{R}_+}, \mathbf{P})$ and $L^1 = L^1(\Omega, \mathcal{F}, (\mathcal{F}_t)_{t \in \mathbb{R}_+}, \mathbf{P})$ the Banach spaces of real-valued uniformly integrable martingales with finite $\mathcal{H}^1$- or L^1-norm respectively. Note that the space $L^1(\Omega, \mathcal{F}, (\mathcal{F}_t)_{t \in \mathbb{R}_+}, \mathbf{P})$ may be isometrically identified with the space of integrable random variables $L^1(\Omega, \mathcal{F}, \mathbf{P})$ by associating to a uniformly integrable martingale X the random variable X_∞.

Also note that for a local martingale of the form $H \cdot M$ we have the formula

$$\|H \cdot M\|_{\mathcal{H}^1} = \left\|[H \cdot M, H \cdot M]^{\frac{1}{2}}_\infty\right\|_{L^1(\Omega, \mathcal{F}, \mathbf{P})}$$

$$= \mathbf{E}\left[\left(\int_0^\infty H'_t d[M,M]_t H_t\right)^{\frac{1}{2}}\right],$$

where H' denotes the transpose of H.

We now state and prove the result of Kadeč-Pełczyński [KP 65] in a form that will be useful in the rest of our paper.

Theorem 15.2.1. (Kadeč-Pełczyński). *If $(f_n)_{n \ge 1}$ is an L^1-bounded sequence in the positive cone $L^1_+(\Omega, \mathcal{F}, \mathbf{P})$, and g is a non-negative integrable function, then there is a subsequence $(n_k)_{k \ge 1}$ as well as an increasing sequence of strictly positive numbers $(\beta_k)_{k \ge 1}$ such that β_k tends to ∞ and $(f_{n_k} \wedge (\beta_k(g+1)))_{k \ge 1}$ is uniformly integrable.*

The sequence $(f_{n_k} \wedge (\beta_k(g+1)))_{k \ge 1}$ is then relatively weakly compact by the Dunford-Pettis theorem.

Proof. We adapt the proof of [KP 65]. Without loss of generality we may suppose that the sequence $(f_n)_{n \ge 1}$ is bounded by 1 in L^1-norm but not uniformly integrable, i.e.,

$$\mathbf{E}[f_n] \le 1; \quad \delta(\beta) = \sup_n \mathbf{E}[f_n - f_n \wedge \beta(g+1)] < \infty; \quad 0 < \delta(\infty) = \inf_{\beta > 0} \delta(\beta)$$

(it is an easy exercise to show that $\delta(\infty) = 0$ implies uniform integrability). For $k = 1$ and $\beta_1 = 1$ we select n_1 so that $\mathbf{E}[f_{n_1} - f_{n_1} \wedge \beta_1(g+1)] > \frac{\delta(\infty)}{2}$. Having chosen $n_1, n_2, \ldots, n_{k-1}$ as well as $\beta_1, \beta_2, \ldots, \beta_{k-1}$ we put $\beta_k = 2\beta_{k-1}$ and we select $n_k > n_{k-1}$ so that $\mathbf{E}[f_{n_k} - f_{n_k} \wedge \beta_k(g+1)] > (1 - 2^{-k})\delta(\infty)$. The sequence $(f_{n_k} \wedge \beta_k(g+1))_{k \ge 1}$ is now uniformly integrable. To see this, let us fix K and let $k(K)$ be defined as the smallest number k such that $\beta_k > K$. Clearly $k(K) \to \infty$ as K tends to ∞. For $l < k(K)$ we then have that $f_{n_l} \wedge \beta_l(g+1) = f_{n_l} \wedge \beta_l(g+1) \wedge K(g+1)$, whereas for $l \ge k(K)$ we have

$$\mathbf{E}[f_{n_l} \wedge \beta_l(g+1) - f_{n_l} \wedge \beta_l(g+1) \wedge K(g+1)]$$
$$= \mathbf{E}[f_{n_l} - f_{n_l} \wedge K(g+1)] - \mathbf{E}[f_{n_l} - f_{n_l} \wedge \beta_l(g+1)]$$
$$\leq \delta(K) - \left(\delta(\infty) - \frac{\delta(\infty)}{2^{k(K)}}\right)$$
$$\leq \delta(\infty) - \delta(K) + \frac{\delta(\infty)}{2^{k(K)}} .$$

The latter expression clearly tends to 0 as $K \to \infty$. $\square$

Corollary 15.2.2. *If the sequence β_k is such that $f_{n_k} \wedge \beta_k(g+1)$ is uniformly integrable, then there also exists a sequence γ_k such that $\frac{\gamma_k}{\beta_k}$ tends to infinity and such that the sequence $f_{n_k} \wedge \gamma_k(g+1)$ remains uniformly integrable.*

Proof. In order to show the existence of γ_k we proceed as follows. The sequence

$$h_k = \beta_k(g+1)\mathbf{1}_{\{f_{n_k} \geq \beta_k(g+1)\}}$$

tends to zero in $L^1(\mathbf{P})$, since the sequence $f_{n_k} \wedge \beta_k(g+1)$ is uniformly integrable and $\mathbf{P}[f_{n_k} \geq \beta_k(g+1)] \leq \frac{1}{\beta_k} \to 0$. Let now α_k be a sequence that tends to infinity but so that $\alpha_k h_k$ still tends to 0 in $L^1(\mathbf{P})$. If we define $\gamma_k = \alpha_k \beta_k$ we have that

$$f_{n_k} \wedge \gamma_k(g+1) \leq f_{n_k} \wedge \beta_k(g+1) + \alpha_k h_k$$

and hence we obtain the uniform integrability of $f_{n_k} \wedge \gamma_k(g+1)$. $\square$

Remark 15.2.3. In most applications of the Kadeč-Pełczyński decomposition theorem, we can take $g = 0$. However, in Sect. 15.4, we will need the easy generalisation to the case where g is a non-zero integrable non-negative function. The general case can in fact be reduced to the case $g = 0$ by replacing the functions f_n by $\frac{f_n}{(g+1)}$ and by replacing the measure $\mathbf{P}$ by the probability measure $\mathbf{Q}$ defined as $d\mathbf{Q} = \frac{(g+1)}{\mathbf{E}[g+1]}d\mathbf{P}$.

Remark 15.2.4. We will in many cases drop indices like n_k and simply suppose that the original sequence $(f_n)_{n\geq 1}$ already satisfies the conclusions of the theorem. In most cases such passing to a subsequence is allowed and we will abuse this simplification as many times as possible.

Remark 15.2.5. The sequence of sets $\{f_n > \beta_n(g+1)\}$ is, of course, not necessarily a disjoint sequence. In case we need two by two disjoint sets we proceed as follows. By selecting a subsequence we may suppose that $\sum_{n>k}\mathbf{P}[f_n > \beta_n(g+1)] \leq \varepsilon_k$, where the sequence of strictly positive numbers $(\varepsilon_k)_{k\geq 1}$ is chosen in such a way that $\int_B f_k\, d\mathbf{P} < 2^{-k}$ whenever $\mathbf{P}[B] < \varepsilon_k$. It is now easily seen that the sequence of sets $(A_n)_{n\geq 1}$ defined by $A_n = \{f_n > \beta_n(g+1)\} \setminus \bigcup_{k>n}\{f_k > \beta_k(g+1)\}$ will do the job.

As a first application of the Kadeč-Pełczyński decomposition we prove the vector-valued Komlos-type theorem stated in the introduction:

Theorem 15.2.6. *If E is a reflexive Banach space and $(f_n)_{n\geq 1}$ a bounded sequence in $L^1(\Omega, \mathcal{F}, \mathbf{P}; E)$ we may find convex combinations*

$$g_n \in \text{conv}\{f_n, f_{n+1}, \ldots\}$$

and $g_0 \in L^1(\Omega, \mathcal{F}, \mathbf{P}; E)$ such that $(g_n)_{n\geq 1}$ converges to g_0 almost surely, i.e.,

$$\lim_{n\to\infty} \|g_n(\omega) - g_0(\omega)\|_E = 0, \qquad \text{for a.e. } \omega \in \Omega.$$

Proof. By the remark made above there is a subsequence, still denoted by $(f_n)_{n\geq 1}$ as well as a sequence $(A_n)_{n\geq 1}$ of mutually disjoint sets such that the sequence $\|f_n\|\mathbf{1}_{A_n^c}$ is uniformly integrable. By a well-known theorem on $L^1(\Omega, \mathcal{F}, \mathbf{P}; E)$ of a *reflexive space* E, [DU 77], see also [DRS 93], the sequence $(f_n \mathbf{1}_{A_n^c})_{n\geq 1}$ is therefore relatively weakly compact in $L^1(\Omega, \mathcal{F}, \mathbf{P}; E)$. Therefore (see Theorem 15.1.2 above) there is a sequence of convex combinations $h_n \in \text{conv}\{f_n \mathbf{1}_{A_n^c}, f_{n+1} \mathbf{1}_{A_{n+1}^c}, \ldots\}$, $h_n = \sum_{k\geq n} \alpha_n^k f_k \mathbf{1}_{A_k^c}$ such that h_n converges to a function g_0 with respect to the norm of $L^1(\Omega, \mathcal{F}, \mathbf{P}; E)$. Since the sequence $f_n \mathbf{1}_{A_n}$ converges to zero a.s. we have that the sequence $g_n = \sum_{k\geq n} \alpha_n^k f_k$ converges to g_0 in probability. If needed one can take a further subsequence that converges a.s., i.e., $\|g_n(\omega) - g_0(\omega)\|_E$ tends to zero for almost each ω. □

The preceding theorem allows us to give an alternative proof of [K 96a, Lemma 4.2].

Lemma 15.2.7. *Let $(N^n)_{n\geq 1}$ be a sequence of adapted càdlàg stochastic processes, $N_0^n = 0$, such that*

$$\mathbf{E}[\text{var}N^n] \leq 1, \qquad n \in \mathbb{N},$$

where $\text{var}N^n$ denotes the total variation of the process N^n.

Then there is a sequence $R^n \in \text{conv}\{N^n, N^{n+1} \ldots\}$ and an adapted càdlàg stochastic process $Z = (Z_t)_{t\in\mathbb{R}_+}$ such that

$$\mathbf{E}[\text{var}Z] \leq 1$$

and such that almost surely the measure dZ_t, defined on the Borel sets of $\mathbb{R}_+$, is the weak-star limit of the sequence dR_t^n. In particular we have that

$$Z_t = \lim_{s\searrow t} \limsup_{n\to\infty} R_s^n = \lim_{s\searrow t} \liminf_{n\to\infty} R_s^n.$$

Proof. We start the proof with some generalities of functional analysis that will allow us to reduce the statement to the setting of Theorem 15.1.4.

The space of finite measures $\mathcal{M}$ on the Borel sets of $\mathbb{R}_+$ is the dual of the space $\mathcal{C}_0$ of continuous functions on $\mathbb{R}_+ = [0, \infty[$, tending to zero at infinity. If $(f_k)_{k=1}$ is a dense sequence in the unit ball of $\mathcal{C}_0$, then for bounded sequences $(\mu_n)_{n\geq 1}$ in $\mathcal{M}$, the weak-star convergence of the sequence μ_n is equivalent to the convergence, for each k, of $\int f_k \, d\mu_n$. The mapping $\Phi(\mu) = (2^{-k} \int f_k \, d\mu)_{k\geq 1}$ maps the space of measures into the space ℓ^2. The image of a bounded weak-star-closed convex set is closed in ℓ^2. Moreover on bounded subsets of $\mathcal{M}$, the weak-star topology coincides with the norm topology of its image in ℓ^2.

For each n the càdlàg process N^n of finite variation can now be seen as a function of Ω into $\mathcal{M}$, mapping the point ω onto the measure $dN_t^n(\omega)$. Using Theorem 15.1.3, we may find convex combinations $P^n \in \operatorname{conv}\{N^n, N^{n+1}, \ldots\}$, $P^n = \sum_{k\geq n} \alpha_n^k N^k$ such that the sequence $\sum_{k\geq n} \alpha_n^k \operatorname{var}(N^k)$ converges a.s.. This implies that a.s. the sequence $P^n(\omega)$ takes its values in a bounded set of $\mathcal{M}$. Using Theorem 15.1.4 on the sequence $(\Phi(P^n))_{n\geq 1}$ we find convex combinations $R^n = \sum_{k\geq n} \beta_n^k P^k$ of $(P^k)_{k\geq n}$ such that the sequence $\Phi(dR^n) = \Phi(\sum_{k\geq n} \beta_n^k dP_t^k)$ converges a.s.. Since a.s. the sequence of measures $dR^n(\omega)$ takes its values in a bounded set of $\mathcal{M}$, the sequence $dR_t^n(\omega)$ converges a.s. weak-star to a measure $dZ_t(\omega)$. The last statement is an obvious consequence of the weak-star convergence. It is also clear that Z is optional and that $\mathbf{E}[\operatorname{var}(Z)] \leq 1$. $\qquad \square$

Remark 15.2.8. If we want to obtain the process Z as a limit of a sequence of processes then we can proceed as follows. Using once more convex combinations together with a diagonalisation argument, we may suppose that R_s^n converges a.s. for each rational s. In this case we can write that a.s. $Z_t = \lim_{s \searrow t;\, s \in \mathbf{Q}} \lim_{n \to \infty} R_s^n$. We will use such descriptions in Sections 15.4 and 15.5.

Remark 15.2.9. Even if the sequence N^n consists of predictable processes, the process Z need not be predictable. Take e.g. T a totally inaccessible stopping time and let N^n describe the point mass at $T + \frac{1}{n}$. Clearly this sequence tends, in the sense described above, to the process $\mathbf{1}_{[T,\infty[}$, i.e. the point mass concentrated at time T, a process which fails to be predictable. Also in general, there is no reason that the process Z should start at 0.

Remark 15.2.10. It might be useful to observe that if T is a stopping time such that Z is continuous at T, i.e. $\Delta Z_T = 0$, then a.s. $Z_T = \lim R_T^n$.

We next recall well-known properties on weak compactness in $\mathcal{H}^1$. The results are due to Dellacherie, Meyer and Yor (see [DMY 78]).

Theorem 15.2.11. *For a family $(M^i)_{i\in I}$ of elements of $\mathcal{H}^1$ the following assertions are equivalent:*

(1) *the family is relatively weakly compact in $\mathcal{H}^1$,*
(2) *the family of square functions $([M^i, M^i]_\infty^{\frac{1}{2}})_{i\in I}$ is uniformly integrable,*
(3) *the family of maximal functions $((M^i)_\infty^*)_{i\in I}$ is uniformly integrable.*

This theorem immediately implies the following:

Theorem 15.2.12. *If* $(N^n)_{n \geq 1}$ *is a relatively weakly compact sequence in* $\mathcal{H}^1$, *if* $(H^n)_{n \geq 1}$ *is a uniformly bounded sequence of predictable processes with* $H^n \to 0$ *pointwise on* $\mathbb{R}_+ \times \Omega$, *then* $H^n \cdot N^n$ *tends weakly to zero in* $\mathcal{H}^1$.

Proof. We may and do suppose that $|H^n| \leq 1$ and $\|N^n\|_{\mathcal{H}^1} \leq 1$ for each n. For each n and each $\varepsilon > 0$, we define E^n as the predictable set $E^n = \{|H^n| > \varepsilon\}$. We split the stochastic integrals $H^n \cdot N^n$ as $(\mathbf{1}_{E^n} H^n) \cdot N^n + (\mathbf{1}_{(E^n)^c} H^n) \cdot N^n$. We will show that the first terms form a sequence that converges to 0 weakly. Because obviously $\| (\mathbf{1}_{(E^n)^c} H^n) \cdot N^n \|_{\mathcal{H}^1} \leq \varepsilon$, the theorem follows.

From the previous theorem it follows that the sequence $(H^n \mathbf{1}_{E^n} \cdot N^n)_{n \geq 1}$ is already weakly relatively compact in $\mathcal{H}^1$. Clearly $\mathbf{1}_{E^n} \to 0$ pointwise. It follows that $F^n = \bigcup_{k \geq n} E^n$ decreases to zero as n tends to ∞. Let N be a weak limit point of the sequence $((H^k \mathbf{1}_{E^k}) \cdot N^k)_{k \geq 1}$. We have to show that $N = 0$. For each $k \geq n$ we have that $\mathbf{1}_{F^n} \cdot ((H^k \mathbf{1}_{E^k}) \cdot N^k) = (H^k \mathbf{1}_{E^k}) \cdot N^k$. From there it follows that $\mathbf{1}_{F^n} \cdot N = N$ and hence by taking limits as $n \to \infty$, we also have $N = \mathbf{1}_\emptyset \cdot N = 0$. $\qquad\square$

Related to the Davis' inequality, is the following lemma, due to Garsia and Chou, (see [G 73, pp. 34–41] and [N 75, p. 198] for the discrete time case; the continuous time case follows easily from the discrete case by an application of Fatou's lemma. The reader can also consult [M 76, p 351, (31.6)] for a proof in the continuous time case.

Lemma 15.2.13. *There is a constant c such that, for each $\mathcal{H}^1$-martingale X, we have*

$$\mathbf{E}\left[\frac{[X, X]_\infty}{X^*_\infty} \right] \leq c \, \|X\|_{\mathcal{H}^1} \, .$$

This inequality together with an interpolation technique yields:

Theorem 15.2.14. *There is a constant C such that for each $\mathcal{H}^1$-martingale X and for each $0 < p < 1$ we have:*

$$\left\| [X, X]_\infty^{\frac{1}{2}} \right\|_p \leq C \, \|X\|_{\mathcal{H}^1}^{\frac{1}{2}} \, \|X^*_\infty\|_{\frac{p}{2-p}}^{\frac{1}{2}} \, .$$

Proof. The following series of inequalities is an obvious application of the preceding lemma and Hölder's inequality for the exponents $\frac{2}{p}$ and $\frac{2}{2-p}$. The constant c is the same as in the preceding lemma.

$$\mathbf{E}\left[[X, X]_\infty^{\frac{p}{2}} \right] = \mathbf{E}\left[(X^*_\infty)^{\frac{p}{2}} \left(\frac{[X, X]_\infty}{X^*_\infty} \right)^{\frac{p}{2}} \right]$$

$$\leq \left(\mathbf{E}\left[\frac{[X, X]_\infty}{X^*_\infty} \right] \right)^{\frac{p}{2}} \left(\mathbf{E}\left[(X^*_\infty)^{\frac{p}{2-p}} \right] \right)^{\frac{2-p}{2}}$$

$$\leq c^{\frac{p}{2}} \, \|X\|_{\mathcal{H}^1}^{\frac{p}{2}} \, \|X^*_\infty\|_{\frac{p}{2-p}}^{\frac{p}{2}} \, .$$

Hence

$$\left\| [X,X]_\infty^{\frac{1}{2}} \right\|_p \leq c^{\frac{1}{2}} \|X\|_{\mathcal{H}^1}^{\frac{1}{2}} \|X_\infty^*\|_{\frac{p}{2-p}}^{\frac{1}{2}} . \qquad \square$$

Corollary 15.2.15. *If X^n is a sequence of $\mathcal{H}^1$-martingales such that $\|X^n\|_{\mathcal{H}^1}$ is bounded and such that $(X^n)^*_\infty$ tends to zero in probability, then $[X^n, X^n]_\infty$ tends to zero in probability.*

In fact, for each $p < 1$, $(X^n)^*_\infty$ as well as $[X^n, X^n]_\infty^{\frac{1}{2}}$ tend to zero in the quasi-norm of $L^p(\Omega, \mathcal{F}, \mathbf{P})$.

Proof. Fix $0 < p < 1$. Obviously we have by the uniform integrability of the sequence $\left((X^n)^*_\infty \right)^{\frac{p}{2-p}}$, that $\| (X^n)^*_\infty \|_{\frac{p}{2-p}}$ converges to zero. It then follows from the theorem that also $[X^n, X^n]_\infty \to 0$ in probability. $\qquad \square$

Remark 15.2.16. It is well-known that, for $0 \leq p < 1$, there is no connection between the convergence of the maximal function and the convergence of the bracket, [MZ 38], [BG 70], [M 94]. But as the theorem shows, for *bounded* sets in $\mathcal{H}^1$ the situation is different. The convergence of the maximal function implies the convergence of the bracket. The result also follows from the result on convergence in law as stated in [JS 87, Corollary 6.7]. This was kindly pointed out to us by A. Shiryaev. The converse of our Corollary 15.2.15 is not true as the example in the next section shows. In particular the relation between the maximal function and the bracket is not entirely symmetric in the present context.

Remark 15.2.17. In the case of *continuous* martingales there is also an inverse inequality of the type

$$\mathbf{E}\left[\frac{(X_\infty^*)^2}{[X,X]_\infty^{\frac{1}{2}}} \right] \leq c \|X\|_{\mathcal{H}^1} .$$

The reader can consult [RY 91, Example 4.17 and 4.18].

15.3 An Example

Example 15.3.1. There is a uniformly bounded martingale $M = (M_t)_{t \in [0,1]}$ and a sequence $(H^n)_{n \geq 1}$ of M-integrands satisfying

$$\|H^n \cdot M\|_{\mathcal{H}^1} \leq 1, \qquad \text{for } n \in \mathbb{N},$$

and such that

(1) for each $t \in [0,1]$ we have

$$\lim_{n \to \infty} (H^n \cdot M)_t = -\frac{t}{2} \qquad \text{a.s.}$$

(2) $[H^n \cdot M, H^n \cdot M]_\infty \to 0$ in probability.

Proof. Fix a collection $((\varepsilon_{n,k})_{k=1}^{2^{n-1}})_{n\geq 1}$ of independent random variables,

$$\varepsilon_{n,k} = \begin{cases} -2^{-n} & \text{with probability } (1-4^{-n}) \\ 2^n(1-4^{-n}) & \text{with probability } 4^{-n} \end{cases}$$

so that $\mathbf{E}[\varepsilon_{n,k}] = 0$. We construct a martingale M such that at times

$$t_{n,k} = \frac{2k-1}{2^n}, \qquad n \in \mathbb{N}, \ k = 1,\dots,2^{n-1},$$

M jumps by a suitable multiple of $\varepsilon_{n,k}$, e.g.

$$M_t = \sum_{(n,k):\, t_{n,k} \leq t} 8^{-n} \varepsilon_{n,k}, \qquad t \in [0,1],$$

so that M is a well-defined uniformly bounded martingale (with respect to its natural filtration).

Defining the integrands H^n by

$$H^n = \sum_{k=1}^{2^{n-1}} 8^n \chi_{\{t_{n,k}\}}, \qquad n \in \mathbb{N},$$

we obtain, for fixed $n \in \mathbb{N}$,

$$(H^n \cdot M)_t = \sum_{k:\, t_{n,k} \leq t} \varepsilon_{n,k},$$

so that $H^n \cdot M$ is constant on the intervals $\left[\frac{2k-1}{2^n}, \frac{2k+1}{2^n}\right[$ and, on a set of probability bigger that $1-2^{-n}$, $H \cdot M$ equals $-\frac{k}{2^n}$ on the intervals $\left[\frac{2k-1}{2^n}, \frac{2k+1}{2^n}\right[$. Also on a set of probability bigger than $1-2^{-n}$ we have that $[H^n \cdot M, H^n \cdot M]_1 = \sum_{k=1}^{2^{n-1}} 2^{-2n} = 2^{-n-1}$.

From the Borel-Cantelli lemma we infer that, for each $t \in [0,1]$, the random variables $(H^n \cdot M)_t$ converge almost surely to the constant function $-\frac{t}{2}$ and that $[H^n \cdot M, H^n \cdot M]_1$ tend to 0 a.s., which proves the final assertions of the above claim.

We still have to estimate the $\mathcal{H}^1$-norm of $H^n \cdot M$:

$$\|H^n \cdot M\|_{\mathcal{H}^1} \leq \sum_{k\geq 1}^{2^{n-1}} \|\varepsilon_{n,k}\|_{L^1}$$
$$= 2^{n-1}[2^{-n}(1-4^{-n}) + 2^n(1-4^{-n}) \cdot 4^{-n}] \leq 1. \qquad \square$$

Remark 15.3.2. What is the message of the above example? First note that passing to convex combinations $(K^n)_{n\geq 1}$ of $(H^n)_{n\geq 1}$ does not change the picture: we always end up with a sequence of martingales $(K^n \cdot M)_{n\geq 1}$ bounded in $\mathcal{H}^1$ and such that the pointwise limit equals $Z_t = -\frac{t}{2}$. Of course, the process Z is far from being a martingale.

Hence, in the setting of Theorem 15.B, we cannot expect (contrary to the setting of Theorem 15.A) that the sequence of martingales $(K^n \cdot M)_{n \geq 1}$ converges in some pointwise sense to a martingale. We have to allow that the singular parts ${}^s K^n \cdot M$ converge (pointwise a.s.) to some process Z; the crucial information about Z is that Z is of integrable variation and, in the case of jumps uniformly bounded from below as in the preceding example, decreasing.

15.4 A Substitute of Compactness for Bounded Subsets of $\mathcal{H}^1$

This section is devoted to the proof of Theorems 15.A, 15.B, 15.C, 15.D as well as Yor's Theorem 15.1.6.

Because of the technical character of this section, let us give an overview of its contents. We start with some generalities that allow the sequence of martingales to be replaced by a more suitable subsequence. This (obvious) preparation is done in the next paragraph. In Subsect. 15.4.1, we then give the proof of Theorem 15.A, i.e. the case of continuous martingales. Because of the continuity, stopping arguments can easily be used. We stop the martingales as soon as the maximal functions reach a level that is given by the Kadeč-Pełczyński decomposition theorem. Immediately after the proof of Theorem 15.A, we give some corollaries as well as a negative result that shows that boundedness in $\mathcal{H}^1$ is needed instead of the weaker boundedness in L^1. We end Subsect. 15.4.1 with a remark that shows that the proof of the continuous case can be adapted to the case where the set of jumps of all the martingales form a uniformly integrable family. Roughly speaking this case can be handled in the same way as the continuous case. Subsect. 15.4.2 then gives the proof of Theorem 15.C. We proceed in the same way as in the continuous case, i.e. we stop when the maximal function of the martingales reaches a certain level. Because this time we did not assume that the jumps are uniformly integrable we have to proceed with more care and eliminate their big parts (the *singular* parts in the Kadeč-Pełczyński decomposition). Subsect. 15.4.3 then treats the case where all the martingales are stochastic integrals, $H^n \cdot M$, with respect to a given d-dimensional local martingale M. This part is the most technical one as we want the possible decompositions to be done on the level of the integrands H^n. We cannot proceed in the same way as in Theorem 15.C, although the idea is more or less the same. Yor's theorem is then (re)proved in Subsect. 15.4.4. Subsect. 15.4.5 is devoted to the proof of Theorem 15.D. The reader who does not want to go through all the technicalities can limit her first reading to Subsects. 15.4.1, 15.4.2, 15.4.4 and only read the statements of the theorems and lemmata in the other Subsects. 15.4.3 and 15.4.5.

By $(M^n)_{n \geq 1}$ we denote a bounded sequence of martingales in $\mathcal{H}^1$. Without loss of generality we may suppose that $\|M^n\|_{\mathcal{H}^1} \leq 1$ for all n. By the Davis' inequality this implies the existence of a constant $c < \infty$ such that for all n: $\mathbf{E}[(M^n)^*] \leq c$. From the Kadeč-Pełczyński decomposition theorem we deduce the existence of a sequence $(\beta_n)_{n \geq 1}$, tending to ∞ and such that $(M^n)^* \wedge \beta_n$ is uniformly integrable. The reader should note that we replaced the original sequence by a subsequence. Passing to a subsequence once more also allows to suppose that $\sum_{n=1}^\infty \frac{1}{\beta_n} < \infty$. For each n we now define

$$\tau_n = \inf\{t \mid |M_t^n| > \beta_n\}.$$

Clearly $\mathbf{P}[\tau_n < \infty] \leq \frac{c}{\beta_n}$ for some constant c. If we let $T_n = \inf_{k \geq n} \tau_k$ we obtain an increasing sequence of stopping times $(T_n)_{n \geq 1}$ such that $\mathbf{P}[T_n < \infty] \leq \sum_{k \geq n} \frac{c}{\beta_k}$ and hence tends to zero. Let us now start with the case of continuous martingales.

15.4.1 Proof of Theorem 15.A. *The case when the martingales M^n are continuous.*

Because of the definition of the stopping times T_n, we obtain that $((M^n)^{T_n})^* \leq (M^n)^* \wedge \beta_n$ and hence the sequence $((M^n)^{T_n})_{n \geq 1}$ forms a relatively weakly compact sequence in $\mathcal{H}^1$. Also the maximal functions of the remaining parts $M^n - (M^n)^{T_n}$ tend to zero a.s.. As a consequence we obtain the existence of convex combinations $N^n = \sum_{k \geq n} \alpha_n^k (M^k)^{T_k}$ that converge in $\mathcal{H}^1$-norm to a continuous martingale M^0. We also have that $R^n = \sum_{k \geq n} \alpha_n^k M^k$ converge to M^0 in the semi-martingale topology and that $(M^0 - R^n)_\infty^*$ tends to zero in probability. From Corollary 15.2.15 in Sect. 15.2 we now easily derive that $[M^0 - R^n, M^0 - R^n]_\infty$ as well as $(M^0 - R^n)_\infty^*$ tend to zero in L^p, for each $p < 1$.

If all the martingales M^n are of the form $H^n \cdot M$ for a fixed continuous $\mathbb{R}^d$-valued local martingale M, then of course the element M^0 is of the same form. This follows from Yor's Theorem 15.1.6, stating that the space of stochastic integrals with respect to M, is a closed subspace of $\mathcal{H}^1$. This concludes the proof of Theorem 15.A. □

Corollary 15.4.1. *If $(M^n)_{n \geq 1}$ is a sequence of continuous $\mathcal{H}^1$-martingales such that*

$$\sup_n \|M^n\|_{\mathcal{H}^1} < \infty \quad and \quad M_\infty^n \to 0 \quad in \ probability,$$

then M^n tends to zero in the semi-martingale topology. As a consequence we have that $(M^n)^* \to 0$ in probability.

Proof. Of course we may take subsequences in order to prove the statement. So let us take a subsequence as well as stopping times as described in Theorem 15.A. The sequence $(M^n)^{T_n}$ is weakly relatively compact in $\mathcal{H}^1$ and since $M_{T_n}^n$ tends to zero in probability (because $\mathbf{P}[T_n < \infty]$ tends to zero and M_∞^n

tends to zero in probability), we easily see that $M_{T_n}^n$ tends to zero in L^1. Doob's maximum inequality then implies that $\left((M^n)^{T_n}\right)^*$ tends to zero in probability. It is then obvious that also $(M^n)^*$ tends to zero in probability.

Because $\left((M^n)^{T_n}\right)^*$ tends to zero in probability and because this sequence is uniformly integrable, we deduce that the sequence $(M^n)^{T_n}$ tends to zero in $\mathcal{H}^1$. The sequence M^n therefore tends to zero in the semi-martingale topology. $\qquad\square$

Remark 15.4.2. The above corollary, together with Theorem 15.2.14, show that M^n tends to zero in $\mathcal{H}^p$ (i.e., $(M^n)^*$ tends to zero in L^p) and in h^p (i.e., $[M^n, M^n]_\infty^{\frac{1}{2}}$ tends to zero in L^p) for each $p < 1$. For continuous local martingales, however, $\mathcal{H}^p$ and h^p are the same.

Remark 15.4.3. That we actually need that the sequence M^n is bounded in $\mathcal{H}^1$, and not just in L^1, is illustrated in the following *negative* result.

Lemma 15.4.4. *Suppose that* $(M^n)_{n\geq 1}$ *is a sequence of continuous, non-negative, uniformly integrable martingales such that* $M_0^n = 1$ *and such that* $M_\infty^n \to 0$ *in probability. Then* $\|M^n\|_{\mathcal{H}^1} \to \infty$.

Proof. For $\beta > 1$ we define $\sigma_n = \inf\{t \mid M_t^n > \beta\}$. Since

$$1 = \mathbf{E}\left[M_{\sigma_n}^n\right] = \beta \mathbf{P}[\sigma < \infty] + \int_{\{(M^n)^* \leq \beta\}} M_\infty^n,$$

we easily see that $\lim_{n\to\infty} \mathbf{P}[\sigma_n < \infty] = \frac{1}{\beta}$. It follows from the Davis' inequality that $\lim_{n\to\infty} \|M^n\|_{\mathcal{H}^1} \geq c \lim_{n\to\infty} \int_0^\infty \mathbf{P}[\sigma_n > \beta] \, d\beta = \infty$. $\qquad\square$

Remark 15.4.5. There are two cases where Theorem 15.A can easily be generalised to the setting of $\mathcal{H}^1$-martingales with jumps. Let us describe these two cases separately. The first case is when the set

$$\{\Delta M_\sigma^n \mid n \geq 1, \ \sigma \text{ a stopping time}\}$$

is uniformly integrable. Indeed, using the same definition of the stopping times T_n we arrive at the estimate

$$(M^n)_{T_n}^* \leq (M^n)^* \wedge \beta_n + \left|\Delta M_{T_n}^n\right|.$$

Because of the hypothesis on the uniform integrability of the jumps and by the selection of the sequence β_n we may conclude that the sequence $\left((M^n)^{T_n}\right)_{n\geq 1}$ is relatively weakly compact in $\mathcal{H}^1$. The corollary generalises in the same way.

The other generalisation is when the set

$$\{M_\infty^n \mid n \geq 1\}$$

is uniformly integrable. In this case the set

$$\{M_\sigma^n \mid n \geq 1, \ \sigma \text{ a stopping time}\}$$

is, as easily seen, also uniformly integrable. The maximal function of the stopped martingale $(M^n)^{T_n}$ is bounded by

$$\left((M^n)^{T_n}\right)^* \leq \max\left((M^n)^* \wedge \beta_n, \left|M_{T_n}^n\right|\right).$$

It is then clear that they form a uniformly integrable sequence. It is this situation that arises in the proof of M. Yor's theorem.

15.4.2 **Proof of Theorem 15.C.** *The case of an $\mathcal{H}^1$-bounded sequence M^n of càdlàg martingales.*

We again turn to the general situation. In this case we cannot conclude that the stopped martingales $(M^n)^{T_n}$ form a relatively weakly compact set in $\mathcal{H}^1$. Indeed the size of the jumps at times T_n might be too big. In order to remedy this situation we will compensate these jumps in order to obtain martingales that have *smaller* jumps at these stopping times T_n. For each n we denote by C^n the dual predictable projection of the process $(\Delta M^n)_{T_n}\mathbf{1}_{[\![T_n,\infty[\![}$. The process C^n is predictable and has integrable variation

$$\mathbf{E}[\mathrm{var}C^n] \leq \mathbf{E}[|(\Delta M^n)_{T_n}|] \leq 2c.$$

The Kadeč-Pełczyński decomposition 2.1 above yields the existence of a sequence η_n tending to ∞, $\sum_{n\geq 1}\frac{1}{\eta_n} < \infty$ and such that $(\mathrm{var}C^n) \wedge \eta_n$ forms a uniformly integrable sequence (again we replaced the original sequence by a subsequence). For each n we now define the *predictable* stopping time σ_n as

$$\sigma_n = \inf\{t \mid \mathrm{var}C_t^n \geq \eta_n\}.$$

Because the process C^n stops at time T_n we necessarily have that $\sigma_n \leq T_n$ on the set $\{\sigma_n < \infty\}$.

We remark that when X is a martingale and when ν is a predictable stopping time, then the process stopped at $\nu-$ and defined by $X_t^{\nu-} = X_t$ for $t < \nu$ and $X_t^{\nu-} = X_{\nu-}$ for $t \geq \nu$, is still a martingale.

Let us now turn our attention to the sequence of martingales

$$N^n = \left((M^n)^{T_n} - \left((\Delta M^n)_{T_n}\mathbf{1}_{[\![T_n,\infty[\![} - C^n\right)\right)^{\sigma_n-}.$$

The processes N^n can be rewritten as

$$N^n = \left((M^n)^{T_n}\right)^{\sigma_n} - (\Delta(M^n))_{\sigma_n}\mathbf{1}_{[\![\sigma_n,\infty[\![}$$
$$- (\Delta M^n)_{T_n}\mathbf{1}_{\{\sigma_n=\infty\}}\mathbf{1}_{[\![T_n,\infty[\![} + (C^n)^{\sigma_n-},$$

or which is the same:

$$N^n = (M^n)^{T_n \wedge \sigma_n} - (\Delta (M^n))_{T_n \wedge \sigma_n} \mathbf{1}_{[\![T_n \wedge \sigma_n, \infty[\![} + (C^n)^{\sigma_n -} .$$

The maximal functions satisfy

$$(N^n)^* \leq (M^n)^* \wedge \beta_n + (\operatorname{var} C^n) \wedge \eta_n$$

and hence form a uniformly integrable sequence. It follows that the sequence N^n is a relatively weakly compact sequence in $\mathcal{H}^1$. Using the appropriate convex combinations will then yield a limit M^0 in $\mathcal{H}^1$.

The problem is that the difference between M^n and N^n does not tend to zero in any reasonable sense as shown by Example 15.3.1 above. Let us therefore analyse this difference:

$$\begin{aligned} &M^n - N^n \\ &= M^n - (M^n)^{T_n \wedge \sigma_n} + (\Delta M^n)_{T_n \wedge \sigma_n} \mathbf{1}_{[\![T_n \wedge \sigma_n, \infty[\![} - (C^n)^{\sigma_n -} . \end{aligned}$$

The maximal function of the first part

$$\left(M^n - \left((M^n)^{T_n \wedge \sigma_n} \right) \right)^* ,$$

tends to zero a.s. because of $\mathbf{P}[T_n < \infty]$ and $\mathbf{P}[\sigma_n < \infty]$ both tending to zero. The same argument yields that the maximal function of the second part

$$\left((\Delta M^n)_{T_n \wedge \sigma_n} \mathbf{1}_{[\![T_n \wedge \sigma_n, \infty[\![} \right)^*$$

also tends to zero. The remaining part is $(-C^n)^{\sigma_n -}$. Applying Theorem 15.1.4 then yields convex combinations that converge in the sense of Theorem 15.1.4 to a càdlàg process of finite variation Z.

Summing up, we can find convex coefficients $(\alpha_n^k)_{k \geq n}$ such that the martingales $\sum_{k \geq n} \alpha_n^k N^n$ will converge in $\mathcal{H}^1$-norm to a martingale M^0 and such that, at the same time, $\sum_{k \geq n} \alpha_n^k C^n$ converge to a process of finite variation Z, in the sense described in Lemma 15.2.7.

In the case where the jumps ΔM^n are bounded below by an integrable function w, or more generally when the set

$$\left\{ \Delta(M^n)_\zeta^- \mid n \geq 1; \ \zeta \text{ stopping time} \right\}$$

is uniformly integrable, we do not have to compensate the negative part of these jumps. So we replace $(\Delta M^n)_{T^n}$ by the more appropriate $((\Delta M^n)_{T^n})^+$. In this case their compensators C^n are increasing and therefore the process Z is decreasing.

The case where the jumps form a uniformly integrable family is treated in the remark after the proof of Theorem 15.A. The proof of Theorem 15.C is therefore completed. $\qquad \square$

15.4.3 Proof of Theorem 15.B. *The case where all martingales are of the form* $\mathcal{M}^n = \mathcal{H}^n \cdot M$.

This situation requires, as we will see, some extra work. We start the construction as in the previous case but this time we work with the square functions, i.e., the brackets instead of the maximal functions.

Without loss of generality we may suppose that M is an $\mathcal{H}^1$-martingale. Indeed let $(\mu_n)_{n \geq 1}$ be a sequence of stopping times that localises the local martingale M in such a way that the stopped martingales M^{μ_n} are all in $\mathcal{H}^1$. Take now a sequence of strictly positive numbers a_n such that $\sum_n a_n \|M^{\mu_n}\|_{\mathcal{H}^1} < \infty$, put $\mu_0 = 0$ and replace M by the $\mathcal{H}^1$-martingale:

$$\sum_{n \geq 1} a_n \left(M^{\mu_n} - M^{\mu_{n-1}} \right).$$

The integrands have then to be replaced by the integrands

$$\sum_{k \geq 1} \frac{1}{a_k} H^n \mathbf{1}_{\llbracket \mu_{k-1}, \mu_k \rrbracket}.$$

In conclusion, we may assume w.l.g., that M is in $\mathcal{H}^1$.

Also without loss of generality we may suppose that the predictable integrands are bounded. Indeed for each n we can take κ_n big enough so that

$$\left\| \left(H^n \mathbf{1}_{\{|H^n| \geq \kappa_n\}} \right) \cdot M \right\|_{\mathcal{H}^1} < 2^{-n}.$$

It is now clear that it is sufficient to prove the theorem for the sequence of integrands $H \mathbf{1}_{\{\|H^n\| \leq \kappa_n\}}$. So we suppose that for each n we have $|H^n| \leq \kappa_n$.

We apply the Kadeč-Pełczyński construction of Theorem 15.2.1 with the function $g = (\text{trace}([M, M]_\infty))^{\frac{1}{2}}$. Without changing the notation we pass to a subsequence and we obtain a sequence of numbers β_n, tending to ∞, such that the sequence

$$[H^n \cdot M, H^n \cdot M]_\infty^{\frac{1}{2}} \wedge \beta_n \left((\text{trace}([M, M]_\infty))^{\frac{1}{2}} + 1 \right)$$

is uniformly integrable.

The sequence of stopping times T_n is now defined as:

$$T_n = \inf \left\{ t \,\middle|\, [H^n \cdot M, H^n \cdot M]_t^{\frac{1}{2}} \geq \beta_n \left((\text{trace}([M, M])_t)^{\frac{1}{2}} + 1 \right) \right\}.$$

In the general case the sequence of jumps $\Delta (H^n \cdot M^n)_{T_n}$ is not uniformly integrable and so we have to eliminate the big parts of these jumps. But this time we want to stay in the framework of stochastic integrals with respect to M. The idea is, roughly speaking, to cut out of the stochastic interval $[\![0, T_n]\!]$, the predictable support of the stopping time T_n. Of course we then have to show that these supports form a sequence of sets that tends to the empty set. This requires some extra arguments.

Since $|\Delta(H^n \cdot M)_{T_n}| \le [H^n \cdot M, H^n \cdot M]_\infty^{\frac{1}{2}}$ we obtain that the sequence

$$|\Delta(H^n \cdot M)_{T_n}| \wedge \beta_n \left((\mathrm{trace}([M,M]_\infty))^{\frac{1}{2}} + 1 \right)$$

is uniformly integrable. As in the proof of the Kadeč-Pełczyński theorem we then find a sequence $\gamma_n \ge \beta_n$ such that $\frac{\gamma_n}{\beta_n} \to \infty$ and such that the sequence

$$|\Delta(H^n \cdot M)_{T_n}| \wedge \gamma_n \left(\mathrm{trace}([M,M]_\infty)^{\frac{1}{2}} + 1 \right)$$

is still uniformly integrable. As a consequence also the sequences

$$|\Delta(H^n \cdot M)_{T_n}| \wedge \beta_n \left(\mathrm{trace}([M,M]_{T_n})^{\frac{1}{2}} + 1 \right)$$

and

$$|\Delta(H^n \cdot M)_{T_n}| \wedge \gamma_n \left(\mathrm{trace}([M,M]_{T_n})^{\frac{1}{2}} + 1 \right)$$

are uniformly integrable.

By passing to a subsequence we may suppose that

(1) the sequences β_n, γ_n are increasing,
(2) $\sum_{n\ge 1} \frac{1}{\beta_n} < \infty$ and hence $\sum \mathbf{P}[T_n < \infty] < \infty$,
(3) $\frac{\gamma_n}{\beta_n} \to \infty$,
(4) for each n we have

$$\frac{\kappa_n \beta_{n+1}(d+1)^2}{\gamma_{n+1}} \le \frac{1}{(d+1)^2} \, ,$$

which can be achieved by choosing inductively a subsequence, since $\frac{\gamma_n}{\beta_n}$ becomes arbitrarily large.

We now turn the sequence of stopping times T_n into a sequence of stopping times having mutually disjoint graphs. This is done exactly as in Subsect. 15.4.1 above. Since $\mathbf{P}[T_n < \infty]$ tends to zero, we may, taking a subsequence if necessary, suppose that

$$\lim_{n\to\infty} \mathbf{E} \left[\sup_{j\le n} [H^j \cdot M, H^j \cdot M]_\infty^{\frac{1}{2}} \mathbf{1}_{\bigcup_{k>n}\{T_k<\infty\}} \right] = 0 \, .$$

We now replace each stopping time T_n by the stopping time τ_n defined by

$$\tau_n = \begin{cases} T_n & \text{if } T_n < T_k \text{ for all } k > n \, , \\ \infty & \text{otherwise.} \end{cases}$$

For each n let $\widetilde{T}_n$ be defined as

$$\widetilde{T}_n = \begin{cases} \tau_n & \text{if } |\Delta(H^n \cdot M)_{\tau_n}| > \gamma_n \left((\text{trace}([M,M]))^{\frac{1}{2}}_{\tau_n} + 1 \right), \\ \infty & \text{otherwise.} \end{cases}$$

For each n let $\widetilde{F}^n$ be the compensator of the process $\mathbf{1}_{[\![\widetilde{T}_n,\infty[\![}$.

We now analyse the supports of the measures $d\widetilde{F}^n$. The measure $d\lambda = \sum_{n\geq1} \frac{1}{2^n} d\widetilde{F}_n$ will serve as a control measure. The measure λ satisfies $\mathbf{E}[\lambda_\infty] < \infty$ by the conditions above. Let φ^n be a predictable Radon-Nikodým derivative $\varphi^n = \frac{d\widetilde{F}^n}{d\lambda}$. It is clear that for each n we have $E^n = \{\varphi^n \neq 0\} \subset [\![0, T_n]\!]$. The idea is to show the following assertion:

Claim 15.4.6. $\sum_{n\geq1} \mathbf{1}_{E^n} \leq d$, $d\lambda$-a.s.. Hence there are predictable sets, still denoted by E^n, such that $\sum_{n\geq1} \mathbf{1}_{E^n} \leq d$ everywhere and such that $E^n = \{\varphi^n \neq 0\}$, $d\lambda$-a.s..

We will give the proof at the end of this section.

For each n we decompose the integrands $H^n = K^n + V^n + W^n$ where:

$$K^n = \mathbf{1}_{[\![0,\widetilde{T}_n]\!]}\mathbf{1}_{(E^n)^c}H^n$$
$$V^n = \mathbf{1}_{E^n}H^n$$
$$W^n = \mathbf{1}_{]\![\widetilde{T}_n,\infty[\![}H^n.$$

Since $\mathbf{P}[\widetilde{T}_n < \infty]$ tends to zero, we have that the maximal functions $(W^n \cdot M)^*_\infty$ tend to zero in probability.

We now show that the sequence $K^n \cdot M$ is relatively weakly compact in $\mathcal{H}^1$. The brackets satisfy

$$[K^n \cdot M, K^n \cdot M]^{\frac{1}{2}}_\infty \leq [H^n \cdot M, H^n \cdot M]^{\frac{1}{2}}_\infty \wedge \gamma_n \left([M,M]^{\frac{1}{2}}_\infty + 1\right)$$
$$+ [H^n \cdot M, H^n \cdot M]^{\frac{1}{2}}_\infty \mathbf{1}_{\{\widetilde{T}_n \neq T_n\}}.$$

The first term defines a uniformly integrable sequence, the second term defines a sequence tending to zero in L^1. It follows that the sequence $[K^n \cdot M, K^n \cdot M]^{\frac{1}{2}}_\infty$ is uniformly integrable and hence the sequence $K^n \cdot M$ is relatively weakly compact in $\mathcal{H}^1$.

There are convex combinations $(\alpha^k_n)_{k\geq n}$ such that $\left(\sum_k \alpha^k_n K^k\right) \cdot M$ converges in $\mathcal{H}^1$ to a martingale which is necessarily of the form $H^0 \cdot M$. We may of course suppose that these convex combinations are disjointly supported, i.e. there are indices $0 = n_0 < n_1 < n_2 < \dots$ such that α^k_j is 0 for $k \leq n_{j-1}$ and $k > n_j$. We remark that if we take convex combinations of $\left(\sum_k \alpha^k_n K^k\right) \cdot M$, then these combinations still tend to $H^0 \cdot M$ in $\mathcal{H}^1$. We will use this remark in order to improve the convergence of the remaining parts of H^n.

Let us define $L^n = \sum_k \alpha_n^k H^k$. Clearly $\|L^n \cdot M\|_{\mathcal{H}^1} \leq 1$ for each n. From Theorem 15.1.3, it follows that there are convex combinations $(\eta_n^k)_{k \geq n}$, disjointly supported, such that $\sum_k \eta_n^k [L^k \cdot M, L^k \cdot M]_\infty^{\frac{1}{2}}$ converges a.s.. Hence we have that $\sup_n \sum_k \eta_n^k [L^k \cdot M, L^k \cdot M]_\infty^{\frac{1}{2}} < \infty$ a.s.. We also may suppose that $\max_k \eta_n^k \to 0$ as $n \to \infty$. From Minkowski's inequality for the bracket it follows that also $\sup_n \left[(\sum_k \eta_n^k L^k) \cdot M, (\sum_k \eta_n^k L^k) \cdot M \right]^{\frac{1}{2}} < \infty$ a.s.. Because the convex combinations were disjointly supported we also obtain a.s. and for $R^n = \sum_k \eta_n^k \sum_j \alpha_k^j V^j$:

$$\sup_n [R^n \cdot M, R^n \cdot M]_\infty^{\frac{1}{2}} \leq \sup_n \left[\left(\sum_k \eta_n^k L^k \right) \cdot M, \left(\sum_k \eta_n^k L^k \right) \cdot M \right]_\infty^{\frac{1}{2}} < \infty.$$

From the fact that the convex combinations were disjointly supported and from $\sum_n \mathbf{1}_{E^n} \leq d$, we conclude that for each point $(t, \omega) \in \mathbb{R}_+ \times \Omega$, only d vectors $R^n(t, \omega)$ can be nonzero. Let us put $P^n = \sum_{s=2^n+1}^{s=2^{n+1}} 2^{-n} R^s$. It follows that a.s.

$$\int P^n \, d[M, M] P^n \leq d \int \left(\sum_{s=2^n+1}^{s=2^{n+1}} 2^{-2n} R^s \, d[M, M] R^s \right)$$

$$\leq d \, 2^{-n} \sum_{s=2^n+1}^{s=2^{n+1}} 2^{-n} [R^s \cdot M, R^s \cdot M]_\infty$$

$$\leq d \, 2^{-n} \sup_s [R^s \cdot M, R^s \cdot M]_\infty$$

$$\to 0.$$

If we now put $U^n = \sum_{k=2^n+1}^{k=2^{n+1}} 2^{-n} \sum_k \eta_l^k \sum_l \alpha_k^l H^l$, we arrive at convex combinations $U^n = \sum \lambda_n^l H^l$ such that

(1) the convex combinations λ_n^k are disjointly supported,
(2) $(\sum_k \lambda_n^k K^k) \cdot M \to H^0 \cdot M$ in $\mathcal{H}^1$,
(3) $\left[(\sum_k \lambda_n^k V^k) \cdot M, (\sum_k \lambda_n^k V^k) \cdot M \right]_\infty \to 0$ in probability,
(4) $\left[(\sum_k \lambda_n^k W^k) \cdot M, (\sum_k \lambda_n^k W^k) \cdot M \right]_\infty \to 0$ in probability, and even
(5) $\left((\sum_k \lambda_n^k W^k) \cdot M \right)^* \to 0$ in probability.

As a consequence we obtain that $\left[(U^n - H^0) \cdot M, (U^n - H^0) \cdot M \right]_\infty \to 0$ in probability, and hence in $L^p(\Omega, \mathcal{F}, \mathbf{P})$ for each $p < 1$.

We remark that these properties will remain valid if we take once more convex combinations of the predictable processes U^n. The stochastic integrals $(\sum_k \lambda_n^k V^k) \cdot M$ need not converge in the semi-martingale topology as the example in Sect. 15.3 shows. But exactly as in the Subsect. 15.4.2 we will show that after taking once more convex combinations, they converge in a pointwise sense, to a process of finite variation.

We consider the martingales $\left(\sum_k \lambda_n^k V^k\right) \cdot M$. For each n let D^n be the compensator of $\left(\sum_k \lambda_n^k \Delta(H^k \cdot M)_{T_k} \mathbf{1}_{[\![\widetilde{T}_k, \infty[\![}\right.$. This is a predictable process of integrable variation. Moreover $\mathbf{E}[\mathrm{var} D^n] \leq \sum_k \lambda_n^k \mathbf{E}[|\Delta(H^k \cdot M)_{T_k}|] \leq 2 \sum_k \lambda_n^k \|H^k \cdot M\|_{\mathcal{H}^1} \leq 2$. We now apply the Kadeč-Pełczyński decomposition technique to the sequence $\mathrm{var} D^n$ and we obtain, if necessary by passing to a subsequence, a sequence of numbers $\sum_n \frac{1}{\xi^n} < \infty$ such that $\mathrm{var} D^n \wedge \xi^n$ is uniformly integrable. Again we define predictable stopping times $S_n = \inf\{t \mid \mathrm{var}(D^n)_t \geq \xi^n\}$. We stop the processes at time (S_n-) since this will not destroy the martingale properties. More precisely we decompose $\sum_k \lambda_n^k V^k \cdot M$ as follows:

$$\sum_k \lambda_n^k V^k \cdot M$$

$$= \left(\sum_k \lambda_n^k V^k \cdot M - \left(\sum_k \lambda_n^k \Delta(H^k \cdot M)_{T_k} \mathbf{1}_{[\![\widetilde{T}_k, \infty[\![} - \widetilde{D}^n\right)\right)^{S_n-} \qquad \textit{first term}$$

$$+ \left(\sum_k \lambda_n^k \Delta(H^k \cdot M)_{T_k} \mathbf{1}_{[\![\widetilde{T}_k, \infty[\![} - \widetilde{D}^n\right)^{S_n-} \qquad \textit{second term}$$

$$+ \left(\left(\sum_k \lambda_n^k V^k \cdot M\right) - \left(\sum_k \lambda_n^k V^k \cdot M\right)^{S_n-}\right) \qquad \textit{third term.}$$

Since $\left([D^n, D^n]^{S_n-}\right)^{\frac{1}{2}} \leq 2 \left(\mathrm{var} D^n\right)^{S_n-} \leq \left(\mathrm{var} D^n\right) \wedge \xi^n$, we obtain that the first term defines a relatively weakly compact sequence in $\mathcal{H}^1$. Indeed, for each n we have $[\![\widetilde{T}_n]\!] \subset E^n \subset [\![0, T_n]\!]$ and hence:

$$[\textit{first term}, \textit{first term}]_\infty^{\frac{1}{2}}$$
$$\leq \sum_k \lambda_n^k [V^k \cdot M, V^k \cdot M]_{T_n-}^{\frac{1}{2}} + [D^n, D^n]_{S_n-}^{\frac{1}{2}}$$
$$\leq [H^n \cdot M, H^n \cdot M]^{\frac{1}{2}} \wedge \beta_n([M, M]_\infty + 1)$$
$$+ [H^n \cdot M, H^n \cdot M]^{\frac{1}{2}} \mathbf{1}_{\{T_n \neq \widetilde{T}_n\}} + [D^n, D^n]_\infty \wedge \xi^n .$$

It follows that the first term defines a relatively weakly compact sequence in $\mathcal{H}^1$. But the first term is supported by the set $\bigcup_{k \geq n} E^k$, which tends to the empty set if $n \to \infty$. From Theorem 15.2.12, it then follows that the sequence defined by the first term tends to zero weakly. The appropriate convex combinations will therefore tend to 0 in the norm of $\mathcal{H}^1$.

The second term splits in

$$\sum_k \lambda_n^k \Delta(H^k \cdot M)_{T_k} \mathbf{1}_{[\![\widetilde{T}_k, \infty[\![} ,$$

whose maximal functions tend to zero a.s. and the processes $(D^n)^{S_n-}$. On the latter we can apply Theorem 15.1.4, which results in convex combinations that tend to a process of finite variation. The third term has a maximal function that tends to zero since

$$\sum_n \mathbf{P}\left[\bigcup_{k\geq n}(\{T_k < \infty\} \cup \{S_n < \infty\})\right] < \infty .$$

Modulo the proof of the claim above, the proof of Theorem 15.B is complete. So let us now prove the claim.

It is sufficient to show that for an arbitrary selection of $d + 1$ indices $n_1 < \cdots < n_{d+1}$ we necessarily have that $E = \bigcap_{k\ldots d+1} E^{n_k} = \emptyset$, $d\lambda$-a.s.. For each k we look at the compensator of the processes

$$\left(\Delta(H^{n_k} \cdot M)_{T_{n_k}}\right)^+ \mathbf{1}_{[\widetilde{T}_{n_k},\infty[} \quad \text{resp.} \quad \left(\Delta(H^{n_k} \cdot M)_{T_{n_k}}\right)^- \mathbf{1}_{[\widetilde{T}_{n_k},\infty[} .$$

Let $^+E^{n_k}$ (resp. $^-E^{n_k}$) be the supports of the compensators of these processes. For each of the 2^{d+1} sign combinations $\varepsilon_k = +/-$ we look at the set $\bigcap_{k=1}^{d+1} {}^{\varepsilon_k}E^{n_k}$. If the set E is non-empty, then at least one of these 2^{d+1} sets would be non-empty and without loss of generality we may and do suppose that this is the case for $\varepsilon_k = +$ for each k.

For each k we now introduce the compensator $\widetilde{C}^k$ of the process

$$\left(\left(\text{trace}([M,M]_{T_{n_k}})\right)^{\frac{1}{2}} + 1\right) \mathbf{1}_{\{\Delta(H^{n_k}\cdot M)_{T_{n_k}}>0\}} \mathbf{1}_{[\widetilde{T}_{n_k},\infty[} .$$

The processes H^{n_k} are d-dimensional processes and hence for each (t,ω) we find that the vectors $H^{n_k}_t(\omega)$ are linearly dependent. Using the theory of linear systems and more precisely the construction of solutions with determinants we obtain $(d+1)$-predictable processes $(\alpha^k)_{k=1}^{d+1}$ such that

(1) for each (t,ω) at least one of the numbers $\alpha^k(t,\omega)$ is nonzero
(2) $\sum_k \alpha^k H^{n_k} = 0$
(3) the processes α^k are all bounded by 1.

We emphasize that these coefficients are obtained in a constructible way and that we do not need a measurable selection theorem!

We now look at the compensator of the processes

$$\Delta(H^{n_k} \cdot M)_{T_{n_l}} \mathbf{1}_{\{\Delta(H^{n_k}\cdot M)_{T_{n_k}}>0\}} \mathbf{1}_{[\widetilde{T}_{n_k},\infty[} .$$

This compensator is of the form $g^{l,k} d\widetilde{C}^l$ for a predictable process $g^{l,k}$. Because of the construction of the coefficients, we obtain that for each $l \leq d + 1$:

$$\sum g^{l,k}\alpha_n^k = 0 .$$

The next step is to show on the set $\bigcap_{k=1}^{d+1} {}^+E^{n_k}$, the matrix $\left(g^{l,k}\right)_{l,k\leq d+1}$ is non-singular. This will then give the desired contradiction, because the above linear system would only admit the solution $\alpha^k = 0$ for all $k \leq d+1$. Because of the definition of the stopping times T_{n_k} we immediately obtain that $g^{k,k} \geq \gamma_{n_k}$. For the non-diagonal elements we distinguish the cases $l < k$ and $l > k$. For $l < k$ we use the fact that on $\widetilde{T}_{n_l} < \infty$, we have that $T_{n_l} < T_{n_k}$. It follows that $|\Delta(H^{n_k} \cdot M)_{T_{n_l}}| \leq 2\beta_{n_k}((\mathrm{trace}([M,M]_{T_{n_l}}))^{\frac{1}{2}} + 1)$ and hence $|g^{l,k}| \leq \beta_{n_k}$. If $l > k$ then $|\Delta(H^{n_k} \cdot M)_{T_{n_l}}| \leq \kappa_{n_k}((\mathrm{trace}([M,M]_{T_{n_l}}))^{\frac{1}{2}}+1)$ and hence $|g^{l,k}| \leq \kappa_{n_k}$. We now multiply the last column of the matrix $g^{l,k}$ with the fraction $\frac{1}{\beta_{n_{d+1}}(d+1)^2}$ and then we multiply the last row by $\frac{\beta_{n_{d+1}}(d+1)^2}{\gamma_{n_{d+1}}}$. The result is that the diagonal element at place $(d+1, d+1)$ is equal to 1 and that the other elements of the last row and the last column are bounded in absolute value by $\frac{1}{(d+1)^2}$. We continue in the same way by multiplying the column d by $\frac{1}{\beta_{n_d}(d+1)^2}$ and the row d by $\frac{\beta_{n_d}(d+1)^2}{\gamma_{n_d}}$. The result is that the element at place (d, d) is 1 and that the other elements on row d and column d are bounded by $\frac{1}{(d+1)^2}$. We note that the elements at place $(d, d+1)$ and $(d+1, d)$ are further decreased by this procedure so that the bound $\frac{1}{(d+1)^2}$ will remain valid. We continue in this way and we finally obtain a matrix with 1 on the diagonal and with the off-diagonal elements bounded by $\frac{1}{(d+1)^2}$. By the classical theorem, due to Hadamard [G 66, Satz 1], such a matrix with dominant diagonal is non-singular. The proof of the claim is now completed and so are the proofs of the Theorems 15.A, 15.B and 15.C.   □

15.4.4 A proof of M. Yor's Theorem 15.1.6 for the L^1-convergent Case

We now show how the ideas of the proof given in [Y 78a] fit in the general framework described above. We will use the generalisation of Theorem 15.A to processes with jumps (see the remarks following the proof of Theorem 15.A). In the next theorem we suppose that M is a d-dimensional local martingale.

Theorem 15.4.7. *Let* $(H^n)_{n \geq 1}$ *be a sequence of M-integrable predictable stochastic processes such that each* $(H^n \cdot M)$ *is a uniformly integrable martingale and such that the sequence of random variables* $((H^n \cdot M)_\infty)_{n \geq 1}$ *converges to a random variable* $f_0 \in L^1(\Omega, \mathcal{F}, \mathbf{P})$ *with respect to the L^1-norm; (or even only with respect to the $\sigma(L^1, L^\infty)$-topology).*

Then there is an M-integrable predictable stochastic process H^0 such that $H^0 \cdot M$ is a uniformly integrable martingale and such that $(H^0 \cdot M)_\infty = f_0$.

Proof. If f_n converges only weakly to f_0 then we take convex combinations in order to obtain a strongly convergent sequence. We therefore restrict the proof to the case where f_n converges in L^1-norm to f_0. By selecting a subsequence we may suppose that $\|f_n\|_{L^1} \leq 1$ for each n and that $\|f_n - f_0\|_{L^1} \leq 4^{-n}$. Let N be the càdlàg martingale defined as $N_t = \mathbf{E}[f_0 \mid \mathcal{F}_t]$. From the maximal inequality for L^1-martingales it then follows that:

$$\mathbf{P}\left[\sup_t |(H^n \cdot M)_t - N_t| \geq 2^{-n}\right] \leq 2^{-n}.$$

The Borel-Cantelli lemma then implies that

$$\sup_t \sup_n |(H^n \cdot M)_t| < \infty \qquad \text{a.s.}.$$

For each natural number k we then define the stopping time T_k as:

$$T_k = \inf\{t \mid \text{ there is } n \text{ such that } |(H^n \cdot M)_t| \geq k\}.$$

Because of the uniform boundedness in t and n we obtain that the sequence T_k satisfies $\mathbf{P}[T_k < \infty] \to 0$. Also the sequence T_k is clearly increasing. For each k and each n we have that

$$\|(H^n \cdot M)^{T_k}\|_{\mathcal{H}^1} \leq k + \|(H^n \cdot M)_{T_k}\|_{L^1}.$$

Since the sequence $f_n = (H^n \cdot M)_\infty$ is uniformly integrable (it is even norm convergent), we have that also the sequence of conditional expectations, $((H^n \cdot M)_{T_k})_{n \geq 1}$ is uniformly integrable and hence the sequence $\left((H^n \cdot M)^{T_k}\right)_{n \geq 1}$ is weakly relatively compact in $\mathcal{H}^1$. Taking the appropriate linear combinations will give a limit in $\mathcal{H}^1$ of the form $K^k \cdot M$ with K^k supported by $[\![0, T_k]\!]$ and satisfying $(K^k \cdot M) = N^{T_k}$. We now take a sequence $(k_m)_{m \geq 1}$ such that $\|N_{T_{k_m}} - f_0\| \leq 2^{-m}$. If we define

$$H^0 = K^{k_1} + \sum_{m \geq 2} K^{k_m} 1_{]\!]T_{k_{m-1}}, T_{k_m}]\!]},$$

we find that $H^0 \cdot M$ is uniformly integrable and that $(H^0 \cdot M)_\infty = f_0$. $\square$

15.4.5 **Proof of Theorem 15.D**

The basic ingredient is Theorem 15.C. Exactly as in M. Yor's theorem we do not have — a priori — a sequence that is bounded in $\mathcal{H}^1$. The lower bound w only permits to obtain a bound for the L^1-norms and we need again stopping time arguments. This is possible because of a uniform bound over the time interval, exactly as in the previous part. The uniformity is obtained as in Lemma 9.4.6.

Definition 15.4.8. *We say that an M-integrable predictable process H is w-admissible for some non-negative integrable function w if $H \cdot M \geq -w$, i.e. the process stays above the level $-w$.*

Remark 15.4.9. The concept of a-admissible integrands, where $a > 0$ is a deterministic number, was used in [DS 94] (here reproduced as Chap. 9) where a short history of this concept is given. The above definition generalises the admissibility as used in Chap. 9 in the sense that it replaces a constant function by a fixed non-negative integrable function w. The concept was also used by the second named author in [S 94, Proposition 4.5].

Exactly as in Chap. 9 we introduce the cone

$$C_{1,w} = \{f \mid \text{ there is a } w\text{-admissible integrand } H \text{ such that } f \leq (H \cdot M)_\infty\}.$$

Theorem 15.4.10. *Let M be a $\mathbb{R}^d$-valued local martingale and $w \geq 1$ an integrable function.*

Given a sequence $(H^n)_{n\geq 1}$ of M-integrable $\mathbb{R}^d$-valued predictable processes such that

$$(H^n \cdot M)_t \geq -w, \qquad \text{for all } n, t,$$

there are convex combinations

$$K^n \in \text{conv}\{H^n, H^{n+1}, \ldots\},$$

and there is a super-martingale $(V_t)_{t\in\mathbb{R}_+}, V_0 = 0$, such that

$$\lim_{\substack{s \searrow t \\ s \in \mathbf{Q}_+}} \lim_{n\to\infty} (K^n \cdot M)_s = V_t \qquad \text{for } t \in \mathbb{R}_+, \text{ a.s.},$$

and an M-integrable predictable process H^0 such that

$$((H^0 \cdot M)_t - V_t)_{t\in\mathbb{R}_+} \qquad \text{is increasing.}$$

In addition, $H^0 \cdot M$ is a local martingale and a super-martingale.

Before proving Theorem 15.D we shall deduce a corollary which is similar in spirit to Theorem 9.4.2, and which we will need in Sect. 15.5 below. For a semi-martingale S we denote by $\mathcal{M}^e(S)$ the set of all probability measures $\mathbf{Q}$ on $\mathcal{F}$ equivalent to $\mathbf{P}$, such that S is a local martingale under $\mathbf{Q}$.

Corollary 15.4.11. *Let S be a semi-martingale taking values in $\mathbb{R}^d$ such that $\mathcal{M}^e(S) \neq \emptyset$ and $w \geq 1$ a weight function such that there is some $\mathbf{Q} \in \mathcal{M}^e(S)$ with $\mathbf{E}_{\mathbf{Q}}[w] < \infty$.*

Then the convex cone $C_{1,w}$ is closed in $L^0(\Omega, \mathcal{F}, \mathbf{P})$ with respect to the topology of convergence in measure.

Proof of Corollary 15.4.11. As the assertion of the corollary is invariant under equivalent changes of measure we may assume that the original measure $\mathbf{P}$ is an element of $\mathcal{M}^e(S)$ for which $\mathbf{E}_{\mathbf{P}}[w] < \infty$, i.e., we are in the situation of Theorem 15.B above. As in the proof of Theorem 15.B we also may assume that S is in $\mathcal{H}^1(\mathbf{P})$ and therefore a $\mathbf{P}$-uniformly integrable martingale.

Let

$$f_n = (H^n \cdot S)_\infty - h_n$$

be a sequence in $C_{1,w}$, where $(H^n)_{n\geq 1}$ is a sequence of w-admissible integrands and $h_n \geq 0$. Assuming that $(f_n)_{n\geq 1}$ tends to a random variable f_0 in measure we have to show that $f_0 \in C_{1,w}$.

It will be convenient to replace the time index set $[0, \infty[$ by $[0, \infty]$ by closing S and $H^n \cdot S$ at infinity, which clearly may be done as the martingale $(S_t)_{t\in\mathbb{R}_+}$

as well as the negative parts of the super-martingales $((H^n \cdot S)_t)_{t \in \mathbb{R}_+}$ are **P**-uniformly integrable. Identifying the closed interval $[0, \infty]$ with the closed interval $[0, 1]$, and identifying the processes S and $H^n \cdot S$ with process which remain constant after time $t = 1$, we deduce from Theorem 15.D that we may find $K^n \in \text{conv}\{H^n, H^{n+1}, \ldots\}$, a w-admissible integrand H^0 and a process $(V_t)_{t \in \mathbb{R}_+}$ such that

$$\lim_{\substack{s \searrow t \\ s \in \mathbf{Q}_+}} \lim_{n \to \infty} (K^n \cdot S)_s = V_t, \qquad \text{a.s. for } t \in \mathbb{R}_+$$

and

$$\lim_{n \to \infty} (K^n \cdot S)_\infty = V_\infty,$$

$$((H^0 \cdot S)_t - V_t)_{t \in \mathbb{R}_+ \cup \{\infty\}} \qquad \text{is increasing.}$$

In particular $((K^n \cdot S)_\infty)_{n \geq 1}$ converges almost surely to the random variable U_∞ which is dominated by $(H^0 \cdot S)_\infty$.

As $(f_n)_{n \geq 1}$ was assumed to converge in measure to f_0 we deduce that $f_0 \leq (H^0 \cdot S)_\infty$, i.e. $f_0 \in C_{1,w}$. $\qquad \square$

To pave the way for the proof of Theorem 15.D we start with some lemmas.

Lemma 15.4.12. *Under the assumptions of Theorem 15.D there is a sequence of convex combinations*

$$K^n \in \text{conv}\{H^n, H^{n+1}, \ldots\},$$

and a sequence $(L^n)_{n \geq 1}$ of w-admissible integrands and there are càdlàg super-martingales $V = (V_t)_{t \in \mathbb{R}_+}$ and $W = (W_t)_{t \in \mathbb{R}_+}$ with $W - V$ increasing such that

$$V_t = \lim_{\substack{s \searrow t \\ s \in \mathbf{Q}_+}} \lim_{n \to \infty} (K^n \cdot M)_s, \qquad \text{for } t \in \mathbb{R}_+, \ a.s.$$

$$W_t = \lim_{\substack{s \searrow t \\ s \in \mathbf{Q}_+}} \lim_{n \to \infty} (L^n \cdot M)_s, \qquad \text{for } t \in \mathbb{R}_+, \ a.s.$$

and such that W satisfies the following maximality condition: For any sequence $(\widetilde{L}^n)_{n \geq 1}$ of w-admissible integrands such that

$$\widetilde{W}_t = \lim_{\substack{s \searrow t \\ s \in \mathbf{Q}_+}} \lim_{n \to \infty} (\widetilde{L}^n \cdot M)_s$$

and $\widetilde{W} - W$ increasing we have that $\widetilde{W} = W$.

Proof. By Theorem 15.1.3 we may find $K^n \in \text{conv}\{H^n, H^{n+1}, \ldots\}$ such that, for every $t \in \mathbf{Q}_+$, the sequence $((K^n \cdot M)_t)_{n \geq 1}$ converges a.s. to a random variable $\widehat{V}_t$. As w is assumed to be integrable we obtain that the process $(\widehat{V}_t)_{t \in \mathbf{Q}_+}$ is a super-martingale and therefore its càdlàg regularisation,

$$V_t = \lim_{\substack{s \searrow t \\ ,s \in \mathbf{Q}_+}} \widehat{V}_s, \qquad t \in \mathbb{R}_+$$

is an a.s. well-defined càdlàg super-martingale.

Let $\mathcal{W}$ denote the family of all càdlàg super-martingales $W = (W_t)_{t \in \mathbb{R}_+}$ such that $W - V$ is increasing and such that there is a sequence $(L^n)_{n \geq 1}$ of w-admissible integrands such that

$$W_t = \lim_{\substack{s \searrow t \\ s \in \mathbf{Q}_+}} \lim_{n \to \infty} (L^n \cdot M)_s, \qquad \text{for } t \in \mathbb{R}_+$$

is a.s. well-defined.

Introducing — similarly as in [K 96a] — the order $W^1 \geq W^2$ on $\mathcal{W}$, if $W^1 - W^2$ is increasing, we may find a maximal element $W \in \mathcal{W}$, with an associated sequence $(L^n)_{n \geq 1}$ of w-admissible integrands.

Indeed, let $(W^\alpha)_{\alpha \in I}$ be a maximal chain in $\mathcal{W}$ with associated sequences of integrands $(L^{\alpha,n})_{n \geq 1}$; then $(W^\alpha_\infty)_{\alpha \in I}$ is an increasing and bounded family of elements of $L^1(\Omega, \mathcal{F}, \mathbf{P})$ and therefore there is an increasing sequence $(\alpha_j)_{j \geq 1}$ such that $(W^{\alpha_j}_\infty)_{j \geq 1}$ increases to the essential supremum of $(W^\alpha_\infty)_{\alpha \in I}$. The càdlàg super-martingale $W = \lim_{j \to \infty} W^{\alpha_j}$ is well-defined and we may find a sequence $(L^{\alpha_j, n_j})_{j \geq 1}$, which we reliable by $(L^n)_{n \geq 1}$, so that

$$W_t = \lim_{\substack{s \searrow t \\ s \in \mathbf{Q}_+}} \lim_{n \to \infty} (L^n \cdot M)_s.$$

Clearly W satisfies the required maximality condition.   □

Lemma 15.4.13. *Under the assumptions of the preceding Lemma 15.4.12 we have that for $T \in \mathbb{R}_+$, the maximal functions*

$$((L^n \cdot M) - (L^m \cdot M))^*_T = \sup_{t \leq T} |(L^n \cdot M)_t - (L^m \cdot M)_t|$$

tend to zero in measure as $n, m \to \infty$.

Proof. The proof of the lemma will use — just as in (9.4.6) and [K 96a] — the *buy low - sell high* argument motivated by the economic interpretation of L^n as trading strategies (see Remark 9.4.7).

Assuming that the assertion of the lemma is wrong there is $T \in \mathbb{R}_+$, $\alpha > 0$ and sequences $(n_k, m_k)_{k \geq 1}$ tending to ∞ such that

$$\mathbf{P}\left[\sup_{t \leq T}((L^{n_k} - L^{m_k}) \cdot M)_t > \alpha\right] \geq \alpha.$$

Defining the stopping times

$$T_k = \inf\{t \leq T \mid ((L^{n_k} - L^{m_k}) \cdot M)_t \geq \alpha\}$$

we have $\mathbf{P}[T_k \leq T] \geq \alpha$.

Define $\widehat{L}^k$ as

$$\widehat{L}^k = L^{n_k}\mathbf{1}_{[\![0,T_k]\!]} + L^{m_k}\mathbf{1}_{]\!]T_k,\infty[\![}$$

so that $\widehat{L}^k$ is a w-admissible predictable integrand.

Denote by d_k the function indicating the difference between $L^{n_k} \cdot M$ and $L^{m_k} \cdot M$ at time T_k, if $T_k < \infty$, i.e.,

$$d_k = ((L^{n_k} - L^{m_k}) \cdot M)_{T_k}\mathbf{1}_{\{T_k < \infty\}}.$$

Note that, for $t \in \mathbb{R}_+$,

$$(\widehat{L}^k \cdot M)_t = (L^{n_k} \cdot M)_t\mathbf{1}_{\{t \le T_k\}} + ((L^{m_k} \cdot M)_t + d_k)\mathbf{1}_{\{t > T_k\}}.$$

By passing to convex combinations $\sum_{j=k}^{\infty} \alpha_j \widehat{L}^j$ of $\widehat{L}^k$ we therefore get that, for each $t \in \mathbb{Q}_+$,

$$\left(\sum_{j=k}^{\infty} \alpha_j \widehat{L}^j \cdot M\right)_t = \left(\sum_{j=k}^{\infty} \alpha_j L^{n_j} \cdot M\right)_t\mathbf{1}_{\{t \le T_k\}} + \left(\sum_{j=k}^{\infty} \alpha_j L^{m_j} \cdot M\right)_t\mathbf{1}_{\{t > T_k\}} + D_t^k$$

where $(D_t^k)_{k \ge 1} = \left(\sum_{j=k}^{\infty} \alpha_j d_j\mathbf{1}_{\{t > T_k\}}\right)_{k \ge 1}$ is a sequence of random variables which converges almost surely to a random variable D_t so that $(D_t)_{t \in \mathbb{Q}_+}$ is an increasing adapted process which satisfies $\mathbf{P}[D_T > 0] > 0$ by Lemma 9.8.1.

Hence $(\widehat{L}^k)_{k \ge 1}$ is a sequence of w-admissible integrands such that, for all $t \in \mathbb{Q}_+$, $(\widehat{L}^k \cdot M)_t$ converges almost surely to $\widehat{W}_t = W_t + D_t$, and $\mathbf{P}[D_T > 0] > 0$, a contradiction to the maximality of W finishing the proof. □

Lemma 15.4.14. *Under the conditions of Theorem 15.D and Lemma 15.4.12 there is a subsequence of the sequence $(L^n)_{n \ge 1}$, still denoted by $(L^n)_{n \ge 1}$, and an increasing sequence $(T_j)_{j \ge 1}$ of stopping times, $T_j \le j$ and $\mathbf{P}[T_j = j] \ge 1 - 2^{-j}$, such that, for each j, the sequence of processes $((L^n \cdot M)^{(T_j)-})_{n \ge 1}$ is uniformly bounded and the sequence $((L^n \cdot M)^{T_j})_{n \ge 1}$ is a bounded sequence of martingales in $\mathcal{H}^1(\mathbf{P})$.*

Proof. First note that, fixing $j \in \mathbb{N}$, $C > 0$, and defining the stopping times

$$U_n = \inf\{t \mid |(L^n \cdot M)_t| \ge C\} \wedge j,$$

the sequence $((L^n \cdot M)^{U_n})_{n \ge 1}$ is bounded in $\mathcal{H}^1(\mathbf{P})$. Indeed, this is a sequence of super-martingales by [AS 94], hence

$$\mathbf{E}[|(L^n \cdot M)_{U_n}|] \le 2\mathbf{E}[((L^n \cdot M)_{U_n})_-] \le 2(C + \mathbf{E}[w]),$$

whence

$$\mathbf{E}[|\Delta(L^n \cdot M)_{U_n}|] \le 2(C + \mathbf{E}[w]) + C.$$

As the maximal function $(L^n \cdot M)_{U_n}^*$ is bounded by $C + |\Delta(L^n \cdot M)_{U_n}|$ we obtain a uniform bound on the L^1-norms of the maximal functions $((L^n \cdot$

$M)^*_{U_n})_{n\geq 1}$, showing that $((L^n \cdot M)^{U_n})_{n\geq 1}$ is a uniformly bounded sequence in $\mathcal{H}^1(\mathbf{P})$.

If we choose $C > 0$ sufficiently big we can make $\mathbf{P}[U_n < j]$ small, uniformly in n; but the sequence of stopping times $(U_n)_{n\geq 1}$ still depends on n and we have to replace it by just one stopping time T_j which works for all $(L^n{}_k)_{k\geq 1}$ for some subsequence $(n_k)_{k\geq 1}$; to do so, let us be a little more formal.

Assume that $T_0 = 0, T_1, \ldots, T_{j-1}$ have been defined as well as a subsequence, still denoted by $(L^n)_{n\geq 1}$, such that the claim is verified for $1, \ldots, j-1$; we shall construct T_j. Applying Lemma 15.4.13 to $T = j$ we may find a subsequence $(n_k)_{k\geq 1}$ such that, for each k,

$$\mathbf{P}\left[\left((L^{n_{k+1}} \cdot M) - (L^{n_k} \cdot M)\right)^*_j \geq 2^{-k}\right] < 2^{-(k+j+2)}.$$

Now find a number $C_j \in \mathbb{R}_+$ large enough such that

$$\mathbf{P}\left[(L^{n_1} \cdot M)^*_j \geq C_j\right] < 2^{-(j+1)}$$

and define the stopping time T_j by

$$T_j = \inf\left\{t \;\middle|\; \sup_k |(L^{n_k} \cdot M)_t| \geq C_j + 1\right\} \wedge j$$

so that $T_j \leq j$ and
$$\mathbf{P}[T_j = j] \geq 1 - 2^{-j}.$$

Clearly $|(L^{n_k} \cdot M)_t| \leq C_j + 1$ for $t < T_j$, whence $((L^n{}_k \cdot M)^{(T_j)-})_{k\geq 1}$ is uniformly bounded.

We have that $T_j \leq U_{n_k}$ for each k, where U_{n_k} is the stopping time defined above (with $C = C_j + 1$). Hence we deduce from the $\mathcal{H}^1(\mathbf{P})$-boundedness of $((L^{n_k} \cdot M)^{U_{n_k}})_{k\geq 1}$ the $\mathcal{H}^1(\mathbf{P})$-boundedness of $(L^{n_k} \cdot M)^{T_j}$. This completes the inductive step and finishes the proof of Lemma 15.4.14. □

Proof of Theorem 15.D. Given a sequence $(H^n)_{n\geq 1}$ of w-admissible integrands choose the sequences $K^n \in \text{conv}\{H^n, H^{n+1}, \ldots\}$ and L^n of w-admissible integrands and the super-martingales V and W as in Lemma 15.4.12. Also fix an increasing sequence $(T_j)_{j\geq 1}$ of stopping times as in Lemma 15.4.14.

We shall argue locally on the stochastic intervals $]T_{j-1}, T_j]$. Fix $j \in \mathbb{N}$ and let
$$L^{n,j} = L^n 1_{]T_{j-1}, T_j]}.$$

By Lemma 15.4.14 there is a constant $C_j > 0$ such that $(L^{n,j})_{n\geq 1}$ is a sequence of $(w + C_j)$-admissible integrands and such that $(L^{n,j} \cdot M)_{n\geq 1}$ is a sequence of martingales bounded in $\mathcal{H}^1(\mathbf{P})$ and such that the jumps of each $L^{n,j} \cdot M$ are bounded downward by $w - 2C_j$. Hence — by passing to convex combinations, if necessary — we may apply Theorem 15.B to split $L^{n,j}$ into two disjointly supported integrands $L^{n,j} = {}^r L^{n,j} + {}^s L^{n,j}$ and we may find an integrand $H^{0,j}$ supported by $]T_{j-1}, T_j]$ such that

$$\lim_{n\to\infty} \|(^rL^{n,j} - H^{0,j}) \cdot M\|_{\mathcal{H}^1(\mathbf{P})} = 0$$

$$(Z_j)_t = \lim_{\substack{q \searrow t \\ q \in \mathbf{Q}_+}} \lim_{n\to\infty} (^sL^{n,j} \cdot M)_q$$

where Z_j is a well-defined adapted càdlàg increasing process.

Finally we paste things together by defining $H^0 = \sum_{j\geq 1}^{\infty} H^{0,j}$ and $Z = \sum_{j\geq 1}^{\infty} Z_j$. By Lemma 15.4.12 we have that

$$W_t = \lim_{\substack{s \searrow t \\ s \in \mathbf{Q}_+}} (L^n \cdot M)_s$$

is a well-defined super-martingale. As

$$Z = (H^0 \cdot M) - W$$

is an increasing process and as $(H^0 \cdot M)$ is a local martingale and a super-martingale by [AS 94] we deduce from the maximality of W that $H^0 \cdot M$ is in fact equal to W. Hence $(H^0 \cdot M) - V$ is increasing and the proof of Theorem 15.D is finished.   □

15.5 Application

In this section we apply the above theorems to give a proof of the *Optional Decomposition Theorem* due to N. El Karoui, M.-C. Quenez [EQ 95], D. Kramkov [K 96a], Föllmer-Kabanov [FK 98], Kramkov [K 96b] and Föllmer-Kramkov [FK 97]. We refer the reader to these papers for the precise statements and for the different techniques used in the proofs.

We generalise the usual setting in finance in the following way. The process S will denote an $\mathbb{R}^d$-valued semi-martingale. In finance theory, usually the idea is to look for measures $\mathbf{Q}$ such that under $\mathbf{Q}$ the process S becomes a local martingale. In the case of processes with jumps this is too restrictive and the idea is to look for measures $\mathbf{Q}$ such that S becomes a *sigma-martingale*. A process S is called a $\mathbf{Q}$-sigma-martingale if there is a strictly positive, predictable process φ such that the stochastic integral $\varphi \cdot S$ exists and is a $\mathbf{Q}$-martingale. We remark that it is clear that we may require the process $\varphi \cdot S$ to be an $\mathcal{H}^1$-martingale and that we also may require the process φ to be bounded (compare Chap. 14). As easily seen, local martingales are sigma-martingales. In the local martingale case the predictable process φ can be chosen to be decreasing and this characterises the local martingales among the sigma-martingales. The concept of sigma-martingale is therefore more general than the concept of local martingale. The set $\mathcal{M}^e(S)$ denotes the set of all equivalent probability measures $\mathbf{Q}$ on $\mathcal{F}$ such that S is a $\mathbf{Q}$-sigma-martingale. It is an easy exercise to show that the set $\mathcal{M}^e(S)$ is a convex set. We suppose that this set is non-empty and we will refer to elements of $\mathcal{M}^e(S)$ as equivalent sigma-martingale measures. We refer to Chap. 14 for more details and for a discussion of the concept of sigma-martingales. We also remark that if S is a semi-martingale and if φ is strictly positive, bounded and predictable, then the sets of stochastic integrals with respect to S and with respect to $\varphi \cdot S$ are the same. This follows easily from the formula $H \cdot S = \frac{H}{\varphi} \cdot (\varphi \cdot S)$.

Theorem 15.5.1 (Optional Decomposition Theorem). *Let $S = (S_t)_{t \in \mathbb{R}_+}$ be an $\mathbb{R}^d$-valued semi-martingale, such that the set $\mathcal{M}^e(S) \neq \emptyset$, and $V = (V_t)_{t \in \mathbb{R}_+}$ a real-valued semi-martingale, $V_0 = 0$ such that, for each $\mathbf{Q} \in \mathcal{M}^e(S)$, the process V is a $\mathbf{Q}$-local super-martingale.*

Then there is an S-integrable $\mathbb{R}^d$-valued predictable process H such that $(H \cdot S) - V$ is increasing.

Remark 15.5.2. The Optional Decomposition Theorem is proved in [EQ 95] in the setting of $\mathbb{R}^d$-valued continuous processes. The important — and highly non-trivial — extension to not necessarily continuous processes was achieved by D. Kramkov in his beautiful paper [K 96a]. His proof relies on some of the arguments from Chap. 9 and therefore he was forced to make the following hypotheses: The process S is assumed to be a locally bounded $\mathbb{R}^d$-valued semi-martingale and V is assumed to be uniformly bounded from below. Later H. Föllmer and Y.M. Kabanov [FK 98] gave a proof of the Optional Decomposition Theorem based on Lagrange-multiplier techniques which allowed them to drop the local boundedness assumption on S. Föllmer and Kramkov [FK 97] gave another proof of this result.

In the present paper our techniques — combined with the arguments of D. Kramkov — allow us to abandon the one-sided boundedness assumption on the process V and to pass to the — not necessarily locally bounded — setting for the process S.

For the economic interpretation and relevance of the Optional Decomposition Theorem we refer to [EQ 95] and [K 96a].

We start the proof with some simple lemmas. The first one — which we state without proof — resumes the well-known fact that a local martingale is locally in $\mathcal{H}^1$.

Lemma 15.5.3. *For a $\mathbf{P}$-local super-martingale V we may find a sequence $(T_j)_{j \geq 1}$ of stopping times increasing to infinity and $\mathbf{P}$-integrable functions $(w_j)_{j \geq 1}$ such that the stopped super-martingales V^{T_j} satisfy*

$$|V^{T_j}| \leq w_j \quad a.s., \text{ for } j \in \mathbb{N}.$$

The next lemma is due to D. Kramkov ([K 96a, Lemma 5.1]) and similar to Lemma 15.4.12 above.

Lemma 15.5.4. *In the setting of the Optional Decomposition Theorem 15.5.1 there is a semi-martingale W with $W - V$ increasing, such that W is a $\mathbf{Q}$-local super-martingale, for each $\mathbf{Q} \in \mathcal{M}^e(S)$ and which is maximal in the following* *sense: for each semi-martingale $\widetilde{W}$ with $\widetilde{W} - W$ increasing and such that $\widetilde{W}$ is a $\mathbf{Q}$-local super-martingale, for each $\mathbf{Q} \in \mathcal{M}^e(S)$, we have $W = \widetilde{W}$.*

Proof of the Optional Decomposition Theorem 15.5.1. For the given semi-martingale V we find a maximal semi-martingale W as in the preceding Lemma 15.5.4. We shall find an S-integrable predictable process H such that we obtain a representation of the process W as the stochastic integral over H, i.e.,

$$W = H \cdot S$$

which will in particular prove the theorem.

Fix $\mathbf{Q}_0 \in \mathcal{M}^e(S)$ and apply Lemma 15.5.3 to the $\mathbf{Q}_0$-local super-martingale W to find $(T_j)_{j \geq 1}$ and $w_j \in L^1(\Omega, \mathcal{F}, \mathbf{Q}_0)$. Note that it suffices — similarly as in [K 96a] — to prove Theorem 15.5.1 locally on the stochastic intervals $]\!]T_{j-1}, T_j]\!]$. Hence we may and do assume that $|W| \leq w$ for some $\mathbf{Q}_0$-integrable weight-function $w \geq 1$. Since S is a sigma-martingale for the measure $\mathbf{Q}_0$, we can by the discussion preceding the Theorem 15.5.1, and without loss of generality, assume that S is an $\mathcal{H}^1(\mathbf{Q}_0)$-martingale. So we suppose that the weight function w also satisfies $|S| \leq w$, where $|\,.\,|$ denotes any norm on $\mathbb{R}^d$.

Fix the real numbers $0 \leq u < v$ and consider the process $^uW^v$ *starting at u and stopped at time v*, i.e.,

$$^uW^v_t = W_{t \wedge v} - W_{t \wedge u},$$

which is a $\mathbf{Q}$-local super-martingale, for each $\mathbf{Q} \in \mathcal{M}^e(S)$, and such that $|^uW^v| \leq 2w$.

Claim 15.5.5. *There is an S-integrable $2w$-admissible predictable process $^uH^v$, which we may choose to be supported by the interval $]u, v]$, such that*

$$(^uH^v \cdot S)_\infty = (^uH^v \cdot S)_v \geq f = {}^uW^v_v = {}^uW^v_\infty.$$

Assuming this claim for a moment, we proceed similarly as D. Kramkov ([K 96a, Proof of Theorem 2.1]): fix $n \in \mathbb{N}$ and denote by $\mathcal{T}(n)$ the set of time indices

$$\mathcal{T}(n) = \left\{ \frac{j}{2^n} \,\middle|\, 0 \leq j \leq n\,2^n \right\}$$

and denote by H^n the predictable process

$$H^n = \sum_{j \geq 1}^{n2^n} {}^{(j-1)2^{-n}}H^{j2^{-n}},$$

where we obtain $^{(j-1)2^{-n}}H^{j2^{-n}}$ as a $2w$-admissible integrand as above with $u = (j-1)2^{-n}$ and $v = j2^{-n}$. Clearly H^n is a $2w$-admissible integrand such that the process indexed by $\mathcal{T}(n)$

$$((H^n \cdot S)_{j2^{-n}} - W_{j2^{-n}})_{j=0,\ldots,n2^n}$$

is increasing.

By applying Theorem 15.D to the $\mathbf{Q}_0$-local martingale S — and by passing to convex combinations, if necessary — the process

$$\widetilde{W}_t = \lim_{\substack{s \searrow t \\ s \in \mathbf{Q}_+}} \lim_{n \to \infty} (H^n \cdot S)_s$$

is well-defined and we may find a predictable S-integrable process H such that $H \cdot S - \widetilde{W}$ is increasing; as $W - \widetilde{W}$ is increasing too, we obtain in particular that $H \cdot S - W$ is increasing.

As $H \cdot S \geq W$ we deduce from [AS 94] that, for each $\mathbf{Q} \in \mathcal{M}^e(S)$, $H \cdot S$ is a $\mathbf{Q}$-local martingale and a $\mathbf{Q}$-super-martingale. By the maximality condition of W we must have $H \cdot S = W$ thus finishing the proof of the Optional Decomposition Theorem 15.5.1.

We still have to prove the claim. This essentially follows from Corollary 15.4.11.

Let us define L_w^∞ to be the space of all measurable functions g such that $\frac{g}{w}$ is essentially bounded. This space is the dual of the space $L_{w^{-1}}^1(\mathbf{Q}_0)$ of functions g such that $\mathbf{E}_{\mathbf{Q}_0}[w\,|g|] < \infty$. By the Banach-Dieudonné theorem or the Krein-Smulian theorem (see Chap. 9 for a similar application), it follows from Corollary 15.4.11 that the set

$$B = \{h \mid |\varepsilon h| \leq w \text{ and } \varepsilon h \in C_{1,2w} \text{ for some } \varepsilon > 0\} \,,$$

is a weak-star-closed convex cone in L_w^∞ (the set $C_{1,2w}$ was defined in Definition 15.4.8 above). Now as easily seen, if the claim were not true, then the said function f is not in B. Since $B - L_{w\,+}^\infty \subset B$ we have by Yan's separation theorem ([Y 80]), that there is a strictly positive function $h \in L_{w^{-1}}^1$ such that $\mathbf{E}_{\mathbf{Q}_0}[hf] > 0$ and such that $\mathbf{E}_{\mathbf{Q}_0}[hg] \leq 0$ for all $g \in B$. If we normalise h so that $\mathbf{E}_{\mathbf{Q}_0}[h] = 1$ we obtain an equivalent probability measure $\mathbf{Q}$, $d\mathbf{Q} = h\,d\mathbf{Q}_0$ such that $\mathbf{E}_{\mathbf{Q}}[f] > 0$. But since S is dominated by the weight function w, we have that the measure $\mathbf{Q}$ is an equivalent martingale measure for the process S. The process W is therefore a local super-martingale under $\mathbf{Q}$. But the density h is such that $\mathbf{E}_{\mathbf{Q}}[w] < \infty$ and therefore the process ${}^uW^v$, being dominated by $2w$, is a genuine super-martingale under $\mathbf{Q}$. However, this is a contradiction to the inequality $\mathbf{E}_{\mathbf{Q}}[f] > 0$. This ends the proof of the claim and the proof of the Optional Decomposition Theorem. □

Remark 15.5.6. Let us stress out that we have proved above that in Theorem 15.5.1 for each process W with $W - V$ increasing, W a $\mathbf{Q}$-local super-martingale for each $\mathbf{Q} \in \mathcal{M}^e(S)$ and W being maximal with respect to this property in the sense of Lemma 15.5.4, we obtain the semi-martingale representation $W = H \cdot S$.

Remark 15.5.7. Referring to the notation of the proof of the optional decomposition theorem and the claim made in it, the fact that the cone B is weak-star-closed in L_w^∞ yields a duality equality as well as the characterisation of maximal elements in the set of w-admissible outcomes. These results are parallel to the results obtained in the case of locally bounded price processes. We refer to Chap. 14 for more details.

Part III

Bibliography

References

[AS 93]   J.P. Ansel, C. Stricker, (1993), *Lois de martingale, densités et décomposition de Föllmer-Schweizer*. Annales de l'Institut Henri Poincaré – Probabilités et Statistiques, vol. 28, no. 3, pp. 375–392.

[AS 94]   J.P. Ansel, C. Stricker, (1994), *Couverture des actifs contingents et prix maximum*. Annales de l'Institut Henri Poincaré – Probabilités et Statistiques, vol. 30, pp. 303–315.

[AH 95]   Ph. Artzner, D. Heath, (1995), *Approximate Completeness with Multiple Martingale Measures*. Mathematical Finance, vol. 5, pp. 1–11.

[A 97]   Ph. Artzner, (1997), *On the numeraire portfolio*. Mathematics of Derivative Securities (M. Dempster, S. Pliska, editors), Cambridge University Press, pp. 53–60.

[A 65]   R. Aumann, (1965), *Integrals of Set-Valued Functions*. Journal of Mathematical Analysis and Applications, vol. 12, pp. 1–12.

[BP 91]   K. Back, S. Pliska, (1991), *On the fundamental theorem of asset pricing with an infinite state space*. Journal of Mathematical Economics, vol. 20, pp. 1–18.

[B 00]   L. Bachelier, (1900), *Théorie de la Spéculation*. Ann. Sci. Ecole Norm. Sup., vol. 17, pp. 21–86. *English translation in:* The Random Character of stock market prices (P. Cootner, editor), MIT Press, 1964.

[B 12]   L. Bachelier, (1912), *Calcul des Probabilités*. Gauthier-Villars, Paris.

[B 14]   L. Bachelier, (1914), *Le Jeu, la Chance et le Hasard*. Ernest Flammarion, Paris.

[Be 01]   D. Becherer, (2001), *The numeraire portfolio for unbounded semimartingales*. Finance and Stochastics, vol. 5, no. 3, pp. 327–341.

[B 32]   S. Banach, (1932), *Théorie des opérations linéaires*. Monogr. Mat., Warszawa 1. *Reprint by:* Chelsea Scientific Books (1963).

[B 72]   T. Bewley, D. Heath, (1972), *Existence of equilibria in economics with infinitely many commodities*. Journal of Economic Theory, vol. 4, pp. 514–540.

[BF 02]   F. Bellini, M. Frittelli, (2002), *On the existence of minimax martingale measures*. Mathematical Finance, vol. 12, no. 1, pp. 1–21.

[B 81]   K. Bichteler, (1981), *Stochastic integration and L^p-theory of semi martingales*. Annals of Probability, vol. 9, pp. 49–89.

[BKT 98] N.H. Bingham, R. Kiesel, (1998), *Risk-Neutral Valuation*. Springer-Verlag, London.

[BF 04] S. Biagini, M. Frittelli, (2004), *On the super replication price of unbounded claims*. Annals of Applied Probability, vol. 14, no. 4, pp. 1970–1991.

[BJ 00] T. Björk (2000), *Arbitrage Theory in Continuous Time*. Oxford University Press.

[BS 73] F. Black, M. Scholes, (1973), *The pricing of options and corporate liabilities*. Journal of Political Economy, vol. 81, pp. 637–659.

[BKT 01] B. Bouchard, Y.M. Kabanov, N. Touzi, (2001), *Option pricing by large risk aversion utility under transaction costs*. Decisions in Economics and Finance, vol. 24, no. 1, pp. 127–136.

[BT 00] B. Bouchard, N. Touzi, (2000), *Explicit solution of the multivariate super-replication problem under transaction costs*. Annals of Applied Probability, vol. 10, pp. 685–708.

[B 79] J. Bourgain, (1979), *The Komlos Theorem for Vector Valued Functions*. Manuscript, Vrije Universiteit Brussel, pp. 1–12.

[BR 97] W. Brannath, (1997), *On fundamental theorems in mathematical finance*. Doctoral Thesis, University of Vienna.

[BS 99] W. Brannath, W. Schachermayer, (1999), *A Bipolar Theorem for Subsets of $L^0_+(\Omega, \mathcal{F}, P)$*. Séminaire de Probabilités XXXIII, Springer Lecture Notes in Mathematics 1709, pp. 349–354.

[B 73] D. Burkholder, (1973), *Distribution Function Inequalities for Martingales*. Annals of Probability, vol. 1, pp. 19–42.

[BG 70] D. Burkholder, R.F. Gundy, (1970), *Extrapolation and Interpolation and Quasi-Linear Operators on Martingales*. Acta Mathematica, vol. 124, pp. 249–304.

[CS 05] L. Campi, W. Schachermayer, (2005), *A Super-Replication Theorem in Kabanov's Model of Transaction Costs*. Preprint.

[C 77] C.S. Chou, (1977/78), *Caractérisation d'une classe de semimartingales*. Séminaire de Probabilités XIII, Springer Lecture Notes in Mathematics 721, pp. 250–252.

[CMS 80] C.S. Chou, P.A. Meyer, S. Stricker, (1980), *Sur les intégrales stochastiques de processus prévisibles non bornés*. In: J. Azéma, M. Yor (eds.), Séminaire de Probabilités XIV, Springer Lecture Notes in Mathematics 784, pp. 128–139.

[Cl 93] S.A. Clark, (1993), *The valuation problem in arbitrage price theory*. Journal of Mathematical Economics, vol. 22, pp. 463–478.

[Cl 00] S.A. Clark, (2000), *Arbitrage approximation theory*. Journal of Mathematical Economics, vol. 33, pp. 167–181.

[CH 89] J.C. Cox, C.F. Huang, (1989), *Optimal consumption and portfolio policies when asset prices follow a diffusion process*. Journal of Economic Theory, vol. 49, pp. 33–83.

[CH 91] J.C. Cox, C.F. Huang, (1991), *A variational problem arising in financial economics*. Journal of Mathematical Economics, vol. 20, no. 5, pp. 465–487.

[CRR 79] J. Cox, S. Ross, M. Rubinstein, (1979), *Option pricing: a simplified approch*. Journal of Financial Economics, vol. 7, pp. 229–263.

[CK 00] J.-M. Courtault, Y. Kabanov, B. Bru, P. Crépel, I. Lebon, A. Le Marchand, (2000), *Louis Bachelier: On the Centenary of "Théorie de la Spéculation"*. Mathematical Finance, vol. 10, no. 3, pp. 341–353.

[C 75]   I. Csiszar, (1975), *I-Divergence Geometry of Probability Distributions and Minimization Problems*. Annals of Probability, vol. 3, no. 1, pp. 146–158.

[C 99]   J. Cvitanic, (1999), *On minimizing expected loss of hedging in incomplete and constrained market*. SIAM Journal on Control and Optimization.

[C 00]   J. Cvitanic, (2000), *Minimizing expected loss of hedging in incomplete and constrained markets*. SIAM Journal on Control and Optimization, vol. 38, no. 4, pp. 1050–1066.

[CK 96]   J. Cvitanic, I. Karatzas, (1996), *Hedging and portfolio optimization under transaction costs: A martingale approach*. Mathematical Finance, vol. 6, no. 2, pp. 133–165.

[CPT 99]   J. Cvitanic, H. Pham, N. Touzi, (1999), *A closed-form solution to the problem of super-replication under transaction costs*. Finance and Stochastics, vol. 3, pp. 35–54.

[CSW 01]   J. Cvitanic, W. Schachermayer, H. Wang, (2001), *Utility Maximization in Incomplete Markets with Random Endowment*. Finance and Stochastics, vol. 5, no. 2, pp. 259–272.

[CW 01]   J. Cvitanic, H. Wang, (2001), *On optimal terminal wealth under transaction costs*. Journal of Mathematical Economics, vol. 35, no. 2, pp. 223–231.

[DMW 90]   R.C. Dalang, A. Morton, W. Willinger, (1990), *Equivalent Martingale measures and no-arbitrage in stochastic securities market model*. Stochastics and Stochastic Reports, vol. 29, pp. 185–201.

[D 97]   M. Davis, (1997), *Option pricing in incomplete markets*. Mathematics of Derivative Securities (M.A.H. Dempster, S.R. Pliska, editors), Cambridge University Press, pp. 216–226.

[D 00]   M. Davis, (2000), *Optimal valuation and hedging with basis risk*. System theory: modeling analysis and control (Cambridge, MA, 1999), Kluwer International Series in Engineering and Computer Science, vol. 518, pp. 245–254.

[DN 90]   M.H.A. Davis, A. Norman, (1990), *Portfolio selection with transaction costs*. Math. Operation Research, vol. 15, pp. 676–713.

[DST 01a]   M. Davis, W. Schachermayer, R. Tompkins, (2001), *Installment Options and Static Hedging*. Mathematical Finance: Trends in Mathematics (M. Kohlmann, S. Tang, editors), pp. 130–139. *Reprint:* Risk Finance, vol. 3, no. 2, pp. 46–52 (2002).

[DST 01]   M. Davis, W. Schachermayer, R. Tompkins, (2001), *Pricing, No-arbitrage Bounds and Robust Hedging of Installment Options*. Quantitative Finance, vol. 1, pp. 597–610.

[DPT 01]   G. Deelstra, H. Pham, N. Touzi, (2001), *Dual formulation of the utility maximisation problem under transaction costs*. Annals of Applied Probability, vol. 11, no. 4, pp. 1353–1383.

[D 92]   F. Delbaen, (1992), *Representing Martingale Measures when Asset Prices are Continuous and Bounded*. Mathematical Finance, vol. 2, pp. 107–130.

[De 00]   F. Delbaen, (2000), *Coherent Risk Measures*. Notes of the Scuola Normale Superiore Cattedra Galileiana, Pisa.

[DGRSSS 02]   F. Delbaen, P. Grandits, T. Rheinländer, D. Samperi, M. Schweizer, C. Stricker, (2002), *Exponential hedging and entropic penalties*. Mathematical Finance, vol. 12, no. 2, pp. 99–123.

[DKV 02]   F. Delbaen, Y.M. Kabanov, E. Valkeila, (2002), *Hedging under transaction costs in currency markets: a discrete-time model*. Mathematical Finance, vol. 12, no. 1, pp. 45–61.

[DMSSS 94] F. Delbaen, P. Monat, W. Schachermayer, M. Schweizer, C. Stricker, (1994), *Inégalité de normes avec poids et fermeture d'un espace d'intégrales stochastiques*. C.R. Acad. Sci. Paris, vol. 319, no. 1, pp. 1079–1081.

[DMSSS 97] F. Delbaen, P. Monat, W. Schachermayer, M. Schweizer, C. Stricker, (1997), *Weighted Norm Inequalities and Closedness of a Space of Stochastic Integrals*. Finance and Stochastics, vol. 1, no. 3, pp. 181–227.

[DS 93] F. Delbaen, W. Schachermayer, (1993), *Non-arbitrage and the fundamental theorem of asset pricing*. In: Abstracts of the Meeting on Stochastic Processes and Their Applications, Amsterdam, June 21-25, 1993, pp. 37–38.

[DS 94] F. Delbaen, W. Schachermayer, (1994), *A General Version of the Fundamental Theorem of Asset Pricing*. Mathematische Annalen, vol. 300, pp. 463–520. *First reprint:* The International Library of Critical Writings in Financial Economics — Option Markets (G.M. Constantinides, A.G. Malliaris, editors). *Second reprint:* Chap. 9 of this book.

[DS 94a] F. Delbaen, W. Schachermayer, (1994), *Arbitrage and free lunch with bounded Risk for unbounded continuous Processes*. Mathematical Finance, vol. 4, pp. 343–348.

[DS 95a] F. Delbaen, W. Schachermayer, (1995), *The Existence of Absolutely Continuous Local Martingale Measures*. Annals of Applied Probability, vol. 5, no. 4, pp. 926–945. *Reprint:* Chap. 12 of this book.

[DS 95b] F. Delbaen, W. Schachermayer, (1995), *The No-Arbitrage Property under a Change of Numéraire*. Stochastics and Stochastic Reports, vol. 53, pp. 213–226. *Reprint:* Chap. 11 of this book.

[DS 95c] F. Delbaen, W. Schachermayer, (1995), *Arbitrage Possibilities in Bessel processes and their relations to local martingales*. Probability Theory and Related Fields, vol. 102, pp. 357–366.

[DS 95d] F. Delbaen, W. Schachermayer, (1995), *An Inequality for the Predictable Projection of an Adapted Process*. Séminaire de Probabilités XXIX, Springer Lecture Notes in Mathematics 1613, (J. Azéma, M. Émery, P.A. Meyer, M. Yor, editors), pp. 17–24.

[DS 96] F. Delbaen, W. Schachermayer, (1996), *Attainable Claims with p'th Moments*. Annales de l'Institut Henri Poincaré – Probabilités et Statistiques, vol. 32, no. 6, pp. 743–763.

[DS 96a] F. Delbaen, W. Schachermayer, (1996), *The Variance-Optimal Martingale Measure for Continuous Processes*. Bernoulli, vol. 2, no. 1, pp. 81–105.

[DS 97] F. Delbaen, W. Schachermayer, (1997), *The Banach Space of Workable Contingent Claims in Arbitrage Theory*. Annales de l'Institut Henri Poincaré – Probabilités et Statistiques, vol. 33, no. 1, pp. 113–144. *Reprint:* Chap. 13 of this book.

[DS 98] F. Delbaen, W. Schachermayer, (1998), *The Fundamental Theorem of Asset Pricing for Unbounded Stochastic Processes*. Mathematische Annalen, vol. 312, pp. 215–250. *Reprint:* Chap. 14 of this book.

[DS 98a] F. Delbaen, W. Schachermayer, (1998), *A Simple Counter-Example to Several Problems in the Theory of Asset Pricing, which arises in many incomplete markets*. Mathematical Finance, vol. 8, pp. 1–12. *Reprint:* Chap. 10 of this book.

[DS 99] F. Delbaen, W. Schachermayer, (1999), *A Compactness Principle for Bounded Sequences of Martingales with Applications*. Proceedings of the Seminar of Stochastic Analysis, Random Fields and Applications, Progress in Probability, vol. 45, pp. 137–173. *Reprint:* Chap. 15 of this book.

[DS 99a]   F. Delbaen, W. Schachermayer, (1999), *Non-Arbitrage and the Fundamental Theorem of Asset Pricing: Summary of Main Results.* Introduction to Mathematical Finance (D.C. Heath, G. Swindle, editors), "Proceedings of Symposia in Applied Mathematics" of the AMS, vol. 57, pp. 49–58.

[DS 00]   F. Delbaen, W. Schachermayer, (2000), *Applications to Mathematical Finance.* Handbook of the Geometry of Banach Spaces (W. Johnson, J. Lindenstrauss, editors), vol. 1, pp. 367–391.

[DS 04]   F. Delbaen, W. Schachermayer, (2004), *What is a Free Lunch?* Notices of the AMS, vol. 51, no. 5, pp. 526–528.

[DSh 96]   F. Delbaen, H Shirakawa, (1996), *A Note on the No-Arbitrage Condition for International Financial Markets.* Financial Engineering and the Japanese Markets, vol. 3, pp. 239–251.

[D 72]   C. Dellacherie, (1972), *Capacités et Processus Stochastiques.* Ergebnisse der Mathematik und ihrer Grenzgebiete, vol. 67, Springer, Berlin.

[DM 80]   C. Dellacherie, P.A. Meyer, (1980), *Probabilités et Potentiel, Chapitres V à VIII.* Théorie des martingales. Hermann, Paris.

[DMY 78]   C. Dellacherie, P.A. Meyer, M. Yor, (1978), *Sur certaines propriétés des espaces de Banach* $\mathcal{H}^1$ *et BMO.* Séminaire de Probabilités XII, Springer Lecture Notes in Mathematics 649, pp. 98–113.

[D 75]   J. Diestel, (1975), *Geometry of Banach spaces — selected topics.* Springer Lecture Notes in Mathematics 485, Springer, Berlin, Heidelberg, New York.

[DRS 93]   J. Diestel, W. Ruess, W. Schachermayer, (1993), *On weak compactness in* $L^1(\mu, X)$. Proc. Am. Math. Soc., vol. 118, pp. 447–453.

[DU 77]   J. Diestel, J.J. Uhl, (1977), *Vector Measures.* Mathematical Surveys, vol. 15. Providence, R.I.: American Mathematical Society (AMS).

[D 53]   J.L. Doob, (1953), *Stochastic Processes.* Wiley, New York.

[Du 92]   D. Duffie, (1992), *Dynamic asset pricing theory.* Princeton University Press.

[DFS 03]   D. Duffie, D. Filipovic, W. Schachermayer, (2003), *Affine Processes and Applications in Finance.* Annals of Applied Probability, vol. 13, no. 3, pp. 984–1053.

[DH 86]   D. Duffie, C.F. Huang, (1986), *Multiperiod security markets with differential information; martingales and resolution times.* Journal of Mathematical Economics, vol. 15, pp. 283–303.

[DS 58]   N. Dunford, J. Schwartz, (1958), *Linear Operators. I. General theory.* Pure and Applied Mathematics, vol. 6, New York and London: Interscience Publishers.

[DR 87]   Ph. Dybvig, S. Ross, (1987), *Arbitrage.* In: J. Eatwell, M. Milgate, P. Newman (eds.), The new Palgrave dictionary of economics, vol. l, pp. 100–106, Macmillan, London.

[ET 76]   I. Ekeland, R. Temam, (1976), *Convex Analysis and Variational Problems.* North Holland, Amsterdam. *Reprint:* 1999, SIAM Classics in Applied Mathematics 38.

[E 05]   A. Einstein, (1905), *Über die von der molekularkinetischen Theorie der Wärme geforderte Bewegung von in ruhenden Flüssigkeiten suspendierten Teilchen.* Annalen der Physik, vol. IV, no. 17, pp. 549–560.

[E 81]   N. El Karoui, (1981), *Les aspects probabilistes du contrôle stochastique.*
Ecole d'Eté de Probabilités de Saint-Flour IX-1979 (P.L. Hennequin, editor), Springer Lecture Notes in Mathematics 876, pp. 74–238.

[EGR 95]   N. El Karoui, H. Geman, J.-C. Rochet, (1995), *Changes of Numéraire, Changes of Probability Measure and Option Pricing.* Journal of Applied Probability, vol. 32, no. 2, pp. 443–458.

[EQ 95]   N. El Karoui, M.-C. Quenez, (1995), *Dynamic Programming and Pricing of Contingent Claims in an Incomplete Market.* SIAM Journal on Control and Optimization, vol. 33, no. 1, pp. 29–66.

[EJ 98]   N. El Karoui, M. Jeanblanc, (1998), *Optimization of consumptions with labor income.* Finance and Stochastics, vol. 4, pp. 409–440.

[ER 00]   N. El Karoui, R. Rouge, (2000), *Pricing via utility maximization and entropy.* Mathematical Finance, vol. 10, no. 2, pp. 259–276.

[EK 99]   R. Elliott, P.E. Kopp, (1999), *Mathematics of financial markets.* Springer Finance, Springer New York.

[ELY 99]   D. Elworthy, X.-M. Li, M. Yor, (1999), *The Importance of Strictly Local Martingales; applications to radial Ornstein-Uhlenbeck processes.* Probab. Theory Relat. Fields, vol. 115, no. 3, pp. 325–355.

[E 79]   M. Émery, (1979), *Une topologie sur l'espace des semi-martingales.* In: C. Dellacherie et al. (eds.), Séminaire de Probabilités XIII, Springer Lecture Notes in Mathematics 721, pp. 260–280.

[E 80]   M. Émery, (1980), *Compensation de processus à variation finie non localement intégrables.* In: J. Azéma, M. Yor (eds.), Séminaire de Probabilités XIV, Springer Lecture Notes in Mathematics 784, pp. 152–160.

[FH 97]   B. Flesaker, L.P. Hughston, (1997), *International Models for Interest Rates and Foreign Exchange.* Net Exposure, vol. 3, pp. 55–79. *Reprinted in:* The New Interest Rate Models, L.P. Hughston (ed.), Risk Publications (2000).

[FK 98]   H. Föllmer, Y.M. Kabanov, (1998), *Optional decomposition and Lagrange multipliers.* Finance and Stochastics, vol. 2, no. 1, pp. 69–81.

[FK 97]   H. Föllmer, D. Kramkov, (1997), *Optional Decompositions under Constraints.* Probability Theory and Related Fields, vol. 109, pp. 1–25.

[FS 91]   H. Föllmer, M. Schweizer, (1991), *Hedging of Contingent Claims Under Incomplete Information.* In: M.H.A. Davis, R.J. Elliott (eds.), Applied Stochastic Analysis, Stochastic Monogr., vol. 5, pp. 389–414, Gordon and Breach, London, New York.

[FL 00]   H. Föllmer, P. Leukert, (2000), *Efficient Hedging: Cost versus Shortfall Risk.* Finance and Stochastics, vol. 4, no. 2, pp. 117–146.

[FS 86]   H. Föllmer, D. Sondermann, (1986), *Hedging of Non-redundant Contingent Claims.* Contributions to Mathematical Economics in honor of G. Debreu (Eds. W. Hildenbrand and A. Mas-Colell), Elsevier Science Publ., North-Holland, pp. 205–223.

[FWY 99]   H. Föllmer, C.-T. Wu, M. Yor, (1999), *Canonical decomposition of linear transformations of two independent Brownian motions motivated by models of insider trading.* Stochastic Processes and Their Applications, vol. 84, no. 1, pp. 137–164.

[FWY 00]   H. Föllmer, C.-T. Wu, M. Yor, (2000), *On weak Brownian motions of arbitrary order.* Annales de l'Institut Henri Poincaré – Probabilités et Statistiques, vol. 36, no. 4, pp. 447–487.

[F 90]   L.P. Foldes, (1990), *Conditions for optimality in the infinite-horizon portfolio-cum-savings problem with semimartingale investments*. Stochastics and Stochastics Reports, vol. 29, pp. 133–171.

[F 00]   M. Frittelli, (2000), *The minimal entropy martingale measure and the valuation problem in incomplete markets*. Mathematical Finance, vol. 10, no. 1, pp. 39–52.

[F 00a]   M. Frittelli, (2000), *Introduction to a Theory of Value Coherent with the No-Arbitrage Principle*. Finance and Stochastics, vol. 4, no. 3, pp. 275–297.

[G 66]   F.R. Gantmacher, (1966), *Matrizentheorie*. Springer, Berlin, Heidelberg, New York.

[G 73]   A.M. Garsia, (1973), *Martingale Inequalities, Seminar Notes on Recent Progress*. Mathematics Lecture Notes Series, W.A. Benjamin, Inc, Reading, Massachusetts.

[G 77]   R. Geske, (1977), *The valuation of corporate liabilities as compound options*. Journal of Financial and Quantitative Analysis, vol. 12, pp. 541–562.

[G 79]   R. Geske, (1979), *The valuation of compound options*. Journal of Financial Economics, vol. 7, pp. 63–81.

[GHR 96] E. Ghysets, A.C. Harvey and E. Renault, (1996), *Stochastic volatility*. G.S. Maddala and C.R. Rao (eds.), Handbook of Statistics, vol. 14, Elsevier, Amsterdam.

[G 60]   I.V. Girsanov, (1960), *On transforming a certain class of stochastic processes by absolutely continuous substitution of measures*. Theory Prob. and Appl., vol. 5, pp. 285–301.

[GK 00]   T. Goll, J. Kallsen, (2000), *Optimal portfolios for logarithmic utility*. Stochastic Processes and Their Applications, vol. 89, pp. 31–48.

[GR 01]   T. Goll, L. Rüschendorf, (2001), *Minimax and minimal distance martingale measures and their relationship to portfolio optimization*. Finance and Stochastics, vol. 5, no. 4, pp. 557–581.

[GR 02]   P. Grandits, T. Rheinländer, (2002), *On the minimal Entropy Martingale Measure*. Annals of Probability, vol. 30, no. 3, pp. 1003–1038.

[G 54]   A. Grothendieck, (1954), *Espaces vectoriels topologiques*. Sociedade de Matematica de São Paulo, São Paulo.

[G 79]   S. Guerre, (1979/80), *La propriété de Banach-Saks ne passe pas de E à $L^2(E)$ d'après J. Bourgain*. Séminaire d'Analyse Fonctionnelle, Ecole Polytechnique, Paris.

[G 05]   P. Guasoni, (2005), *No-Arbitrage with Transaction Costs, with Fractional Brownian Motion and Beyond*. Mathematical Finance, forthcoming 2006.

[HS 49]   P.R. Halmos, L.J. Savage, (1949), *Application of the Radon-Nikodým Theorem to the theory of sufficient statistics*. Ann. Math. Statist., vol. 20, pp. 225–241.

[HK 79]   J.M. Harrison, D.M. Kreps, (1979), *Martingales and Arbitrage in Multiperiod Securities Markets*. Journal of Economic Theory, vol. 20, pp. 381–408.

[HP 81]   J.M. Harrison, S.R. Pliska, (1981), *Martingales and Stochastic Integrals in the Theory of Continous Trading*. Stochastic Processes and their Applications, vol. 11, pp. 215–260.

[H 79]   O. Hart, (1979), *Monopolistic competition in a large economy with differentiated commodities*. Review of Economic Studies, vol. 46, pp. 1–30.

[HP 91]   H. He, N.D. Pearson, (1991), *Consumption and Portfolio Policies with Incomplete Markets and Short-Sale Constraints: The Finite-Dimensional Case*. Mathematical Finance, vol. 1, pp. 1–10.

[HP 91a]   H. He, N.D. Pearson, (1991), *Consumption and Portfolio Policies with Incomplete Markets and Short- Sale Constraints: The Infinite-Dimensional Case*. Journal of Economic Theory, vol. 54, pp. 239–250.

[HJM 92]   D. Heath, R. Jarrow, A. Morton, (1992), *Bond pricing and the term structure of interest rates: a new methodology for contingent claim valuation*. Econometrica, vol. 60, pp. 77–105.

[H 86]   J.R. Hicks, (1986, First Edition 1956), *A Revision of Demand Theory*. Oxford University Press, Oxford.

[HN 89]   S.D. Hodges, A. Neuberger, (1989), *Optimal replication of contingent claims under transaction costs*. Review of Futures Markets, vol. 8, pp. 222–239.

[HL 88]   C.-F. Huang, R.H. Litzenberger, (1988), *Foundations for Financial Economics*. North-Holland Publishing Co. New York.

[HS 98]   F. Hubalek, W. Schachermayer, (1998), *When does Convergence of Asset Price Processes Imply Convergence of Option Prices?* Mathematical Finance, vol. 8, no. 4, pp. 215–233.

[HS 01]   F. Hubalek, W. Schachermayer, (2001), *The Limitations of No-Arbitrage Arguments for Real Options*. International Journal of Theoretical and Applied Finance, vol. 4, no. 2, pp. 361–373.

[HK 04]   J. Hugonnier, D. Kramkov, (2004), *Optimal investment with random endowments in incomplete markets*. Annals of Applied Probability, vol. 14, no. 2, pp. 845–864.

[HKS 05]   J. Hugonnier, D. Kramkov, W. Schachermayer, (2005), *On Utility Based Pricing of Contingent Claims in Incomplete Markets*. Mathematical Finance, vol. 15, no. 2, pp. 203–212.

[HW 87]   J. Hull, A. White, (1987), *The Pricing of Options on Assets with Stochastic Volatilities*. Journal of Finance, vol. 42, pp. 281–300.

[HW 88]   J. Hull, A. White, (1988), *The Use of the Control Variate Technique in Option Pricing*. Journal of Financial and Quantitative Analysis, vol. 23, pp. 237–252.

[H 99]   J. Hull, (1999), *Options, Futures, and Other Derivatives*. 4th Edition, Prentice-Hall, Englewood Cliffs, New Jersey.

[I 44]   K. Itô, (1944), *Stochastic integral*. Proc. Imperial Acad. Tokyo, vol. 20, pp. 519–524.

[J 91]   S.D. Jacka, (1991), *Optimal stopping and the American put*. Mathematical Finance, vol. 1, pp. 1–14.

[J 92]   S.D. Jacka, (1992), *A Martingale Representation Result and an Application to Incomplete Financial Markets*. Mathematical Finance, vol. 2, pp. 239–250.

[J 93]   S.D. Jacka, (1993), *Local times, optimal stopping and semimartingales*. Annals of Probability, vol. 21, pp. 329–339.

[J 79]   J. Jacod, (1979), *Calcul Stochastique et Problèmes de Martingales*. Springer Lecture Notes in Mathematics 714, Springer, Berlin, Heidelberg, New York.

[JS 87]   J. Jacod, A. Shiryaev, (1987), *Limit Theorems for Stochastic Processes*. Springer, Berlin, Heidelberg, New York.

[J 87]   F. Jamshidian, (1987), *Pricing of Contingent Claims in the One Factor Term Structure Model*. Merrill Lynch Capital Markets, Research Paper.

[JS 00]   M. Jonsson, K.R. Sircar, (2000), *Partial hedging in a stochastic volatility environment*. Mathematical Finance, vol. 12, pp. 375–409.

[JK 95]   E. Jouini, H. Kallal, (1995), *Martingales and arbitrage in securities markets with transaction costs*. Journal of Economic Theory, vol. 66, pp. 178–197.

[JK 95a] E. Jouini, H. Kallal, (1995), *Arbitrage in securities markets with short-sales constraints*. Mathematical Finance, vol. 3, pp. 237–248.

[JK 99]   E. Jouini, H. Kallal, (1999), *Viability and Equilibrium in Securities Markets with Frictions*. Mathematical Finance, vol. 9, no. 3, pp. 275–292.

[JNS 05] E. Jouini, C. Napp, W. Schachermayer, (2005), *Arbitrage and state price deflators in a general intertemporal framework*. Journal of Mathematical Economics, vol. 41, pp. 722–734.

[K 97]   Y.M. Kabanov, (1997), *On the FTAP of Kreps-Delbaen-Schachermayer.*. (English) Y.M. Kabanov (ed.) et al., Statistics and control of stochastic processes. The Liptser Festschrift. Papers from the Steklov seminar held in Moscow, Russia, 1995-1996. Singapore: World Scientific. pp. 191–203.

[K 99]   Y.M. Kabanov, (1999), *Hedging and liquidation under transaction costs in currency markets*. Finance and Stochastics, vol. 3, no. 2, pp. 237–248.

[K 01a]   Y.M. Kabanov, (2001), *Arbitrage Theory*. Handbooks in Mathematical Finance. Option Pricing: Theory and Practice, pp. 3–42.

[KK 94]   Y.M. Kabanov, D. Kramkov, (1994), *No-arbitrage and equivalent martingale measures: An elementary proof of the Harrison–Pliska theorem*. Theory Prob. Appl., vol. 39, no. 3, pp. 523–527.

[KL 02]   Y.M. Kabanov, G. Last, (2002), *Hedging under transaction costs in currency markets: a continuous-time model*. Mathematical Finance, vol. 12, no. 1, pp. 63–70.

[KRS 02] Y.M. Kabanov, M. Rásonyi, Ch. Stricker, (2002), *No-arbitrage criteria for financial markets with efficient friction*. Finance and Stochastics, vol. 6, no. 3, pp. 371–382.

[KRS 02a] Y.M. Kabanov, M. Rásonyi, Ch. Stricker, (2003), *On the closedness of sums of convex cones in L^0 and the robust no-arbitrage property*. Finance and Stochastics, vol. 7, no. 3, pp. 403–411.

[KS 01]   Y.M. Kabanov, Ch. Stricker, (2001), *A teachers' note on no-arbitrage criteria*. Séminaire de Probabilités XXXV, Springer Lecture Notes in Mathematics 1755, pp. 149–152.

[KS 01a] Y.M. Kabanov, Ch. Stricker, (2001), *The Harrison-Pliska arbitrage pricing theorem under transaction costs*. Journal of Mathematical Economics, vol. 35, no. 2, pp. 185–196.

[KS 02]   Y.M. Kabanov, Ch. Stricker, (2002), *On the optimal portfolio for the exponential utility maximization: Remarks to the six-author paper*. Mathematical Finance, vol. 12, no. 2, pp. 125–134.

[KS 02a] Y.M. Kabanov, Ch. Stricker, (2002), *Hedging of contingent claims under transaction costs*. Sandmann, Klaus (ed.) et al., Advances in finance and stochastics. Essays in honour of Dieter Sondermann. Berlin: Springer, pp. 125–136.

[KS 03]   Y.M. Kabanov, Ch. Stricker, (2003), *On the true submartingale property d'après Schachermayer*. Séminaire de Probabilités XXXVI, Springer Lecture Notes in Mathematics 1801, pp. 413–414.

[KP 65]   M. Kadeč, A. Pełczyński, (1965), *Basic sequences, biorthogonal systems and norming sets in Banach and Fréchet spaces*. Studia Mathematica, vol. 25, pp. 297–323.

[K 00]   J. Kallsen, (2000), *Optimal portfolios for exponential Lévy processes* Mathematical Methods of Operation Research, vol. 51, no. 3, pp. 357–374.

[K 01]   J. Kallsen, (2001), *Utility-Based Derivative Pricing in Incomplete Markets*. Mathematical Finance: Bachelier Congress 2000 (H. Geman, D. Madan, S.R. Pliska, T. Vorst, editors), Springer, pp. 313–338.

[KLS 87] I. Karatzas, J.P. Lehoczky, S.E. Shreve, (1987), *Optimal portfolio and consumption decisions for a "small investor" on a finite horizon*. SIAM Journal of Control and Optimisation, vol. 25, pp. 1557–1586.

[KLS 90] I. Karatzas, J.P. Lehoczky, S.E. Shreve, (1990), *Existence and uniqueness of multi-agent equilibrium in a stochastic, dynamic consumption/investment model*. Mathematics of Operations Research, vol. 15, pp. 80–128.

[KLS 91] I. Karatzas, J.P. Lehoczky, S.E. Shreve, (1991), *Equilibrium models with singular asset prices*. Mathematical Finance, vol. 1, pp. 11–29.

[KLSX 91] I. Karatzas, J.P. Lehoczky, S.E. Shreve, G.L. Xu, (1991), *Martingale and duality methods for utility maximisation in an incomplete Market*. SIAM Journal of Control and Optimisation, vol. 29, pp. 702–730.

[KS 88]   I. Karatzas, S.E. Shreve, (1988), *Brownian motion and stochastic calculus*. Springer, Berlin, Heidelberg, New York.

[KSh 98] I. Karatzas, S.E. Shreve, (1998), *Methods of Mathematical Finance*. Springer-Verlag, New York.

[KS 96a] I. Klein, W. Schachermayer, (1996), *Asymptotic Arbitrage in Non-Complete Large Financial Markets*. Theory of Probability and its Applications, vol. 41, no. 4, pp. 927–934.

[KS 96b] I. Klein, W. Schachermayer, (1996), *A Quantitative and a Dual Version of the Halmos-Savage Theorem with Applications to Mathematical Finance*. Annals of Probability, vol. 24, no. 2, pp. 867–881.

[KPT 99] P.-F. Koehl, H. Pham, N. Touzi, (1999), *On super-replication under Transaction costs in general discrete-time models*. Theory of Probability and its Applications, vol. 45, pp. 783–788.

[K 33]   A.N. Kolmogorov, (1933), *Grundbegriffe der Wahrscheinlichkeitsrechnung*. Ergebnisse der Mathematik und ihrer Grenzgebiete 3, Springer, Berlin.

[K 67]   J. Komlos, (1967), *A generalisation of a theorem of Steinhaus*. Acta Math. Acad. Sci. Hungar., vol. 18, pp. 217–229.

[K 96a]   D. Kramkov, (1996), *Optional decomposition of supermartingales and hedging contingent claims in incomplete security markets*. Probability Theory and Related Fields, vol. 105, pp. 459–479.

[K 96b]   D. Kramkov, (1996), *On the Closure of the Family of Martingale Measures and an Optional Decomposition of Supermartingales*. Theory Probab. Appl., vol. 41, no. 4, pp. 788–791.

[KS 99]   D. Kramkov, W. Schachermayer, (1999), *The Asymptotic Elasticity of Utility Functions and Optimal Investment in Incomplete Markets*. Annals of Applied Probability, vol. 9, no. 3, pp. 904–950.

[KS 03a] D. Kramkov, W. Schachermayer, (2003), *Necessary and sufficient conditions in the problem of optimal investment in incomplete markets.* Annals of Applied Probability, vol. 13, no. 4, pp. 1504–1516.

[K 81] D.M. Kreps, (1981), *Arbitrage and Equilibrium in Economics with infinitely many Commodities.* Journal of Mathematical Economics, vol. 8, pp. 15–35.

[KW 67] H. Kunita, S. Watanabe, (1967), *On square integrable martingales.* Nagoya Mathematical Journal, vol. 30, pp. 209–245.

[K 93] S. Kusuoka, (1993), *A remark on arbitrage and martingale measures.* Publ. Res. Inst. Math. Sci., vol. 19, pp. 833–840.

[L 92] P. Lakner, (1992), *Martingale measures for a class of right-continuous processes* . Mathematical Finance, vol. 3, no. 1, pp. 43–53.

[LL 96] D. Lamberton, B. Lapeyre, (1996), *Introduction to Stochastic Calculus Applied to Finance.* Chapman & Hall, London.

[L 78] D. Lépingle, (1978), *Une inegalité de martingales.* In: C. Dellacherie et al. (eds.), Sémin. de Probab. XII. (Lect. Notes Math., vol. 649, pp. 134–137), Springer, Berlin, Heidelberg, New York.

[L 91] D. Lépingle, (1991), *Orthogonalité et équi-intégrabilité de martingales discrètes.* Séminaire de Probabilités XXVI, Springer Lecture Notes in Mathematics 1526, pp. 167–170.

[L 77] E. Lenglart, (1977), *Transformation des martingales locales par changement absolument continu de probabilités.* Zeitschrift für Wahrscheinlichkeitstheorie und verwandte Gebiete. vol. 39, pp. 65–70.

[LW 01] S.F. LeRoy, J. Werner, (2001), *Principles of Financial Economics.* Cambridge University Press.

[LS 94] S. Levental, A.V. Skorohod, (1994), *A necessary and sufficient condition for absence of arbitrage with tame portfolios.* Annals of Applied Probability, vol. 5, no. 4, pp. 906–925.

[LS 97] S. Levental, A.V. Skorohod, (1997), *On the possibility of hedging options in the presence of transaction costs.* Annals of Applied Probability, vol. 7, pp. 410–443.

[Lo 78] M. Loève, (1978), *Probability theory.* 4th edn., Springer, Berlin, Heidelberg, New York.

[L 90] J.B. Long, (1990), *The numeraire portfolio.* Journal of Financial Economics, vol. 26, pp. 29–69.

[MZ 38] J. Marcinkiewicz, A. Zygmund, (1938), *Quelques théorèmes sur les fonctions indépendantes.* Studia Mathematica, vol. 7, pp. 104–120.

[M 78a] W. Margrabe, (1978), *The value of an option to exchange one asset for another.* Journal of Finance, vol. 33, no. 1, pp. 177–186.

[M 78b] W. Margrabe, (1978), *A theory of forward and future prices.* Preprint Wharton School, University of Pennsylvania.

[M 75] A. Mas-Colell, (1975), *A model of equilibrium with differentiated commodities.* Journal of Mathematical Economics, vol. 2, pp. 263–296.

[MB 91] D.W. McBeth, (1991), *On the existence of equivalent martingale measures.* Thesis Cornell University.

[MK 69] H.P. McKean, (1969), *Stochastic Integrals.* Wiley, New York.

[M 80] J. Mémin, (1980), *Espaces de semi Martingales et changement de probabilité.* Zeitschrift für Wahrscheinlichkeitstheorie und verwandte Gebiete, vol. 52, pp. 9–39.

[M 69]   R.C. Merton, (1969), *Lifetime portfolio selection under uncertainty: the continuous-time model.* Rev. Econom. Statist., vol. 51, pp. 247–257.

[M 71]   R.C. Merton, (1971), *Optimum consumption and portfolio rules in a continuous-time model.* Journal of Economic Theory, vol. 3, pp. 373–413.

[M 73]   R.C. Merton, (1973), *The theory of rational option pricing.* Bell J. Econ. Manag. Sci., vol. 4, pp. 141–183.

[M 73b]   R.C. Merton, (1973), *An intertemporal capital asset pricing model.* Econometrica, vol. 41, pp. 867–888.

[M 76a]   R.C. Merton, (1976), *Option pricing when underlying stock returns are discontinuous.* Journal of Financial Economics, vol. 3, pp. 125–144.

[M 80a]   R.C. Merton, (1980), *On estimating the expected return on the market: an exploratory investigation.* Journal of Financial Economics, vol. 8, pp. 323–361.

[M 90]   R.C. Merton, (1990), *Continuous-Time Finance.* Basil Blackwell, Oxford.

[M 76]   P.A. Meyer, (1976), *Un cours sur les intégrales stochastiques.* In: P.A. Meyer (ed.), Séminaire de Probabilités X, Springer Lecture Notes in Mathematics 511, pp. 245–400.

[M 62]   P.A. Meyer, (1962), *A decomposition theorem for supermartingales.* Illinois J. Math., vol. 6, pp. 193–205.

[M 63]   P.A. Meyer, (1963), *Decomposition of supermartingales: the uniqueness theorem.* Illinois J. Math., vol. 7, pp. 1–17.

[M 94]   P. Monat, (1994), *Remarques sur les inégalités de Burkholder-Davis-Gundy.* Séminaire de Probabilités XXVIII, Springer Lecture Notes in Mathematics 1583, pp. 92–97.

[MR 97]   M. Musiela, M. Rutkowski, (1997), *Martingale Methods in Financial Modelling.* Springer-Verlag, Berlin.

[N 75]   J. Neveu, (1975), *Discrete Parameter Martingales.* North-Holland, Amsterdam.

[PT 99]   H. Pham, N. Touzi, (1999), *The fundamental theorem of asset pricing with cone constraints.* Journal of Mathematical Economics, vol. 31, pp. 265–279.

[PY 82]   J. Pitman, M. Yor, (1982), *A decomposition of Bessel Bridges.* Zeitschrift f. Wahrscheinlichkeit u. Verw. Gebiete, vol. 59, no. 4, pp. 425–457.

[P 86]   S.R. Pliska, (1986), *A stochastic calculus model of continuous trading: optimal portfolios.* Math. Oper. Res., vol. 11, pp. 371–382.

[P 97]   S.R. Pliska, (1997), *Introduction to Mathematical Finance.* Blackwell Publishers.

[P 90]   P. Protter, (1990), *Stochastic Integration and Differential Equations. A new approach.* Applications of Mathematics, vol. 21, Springer-Verlag, Berlin, Heidelberg, New York (second edition: 2003, corrected third printing: 2005).

[R 98]   R. Rebonato, (1998), *Interest-rate Option models.* 2nd ed., Wiley, Chichester.

[RY 91]   D. Revuz, M. Yor, (1991), *Continuous Martingales and Brownian Motion.* Grundlehren der Mathematischen Wissenschaften, vol. 293, Springer (third edition: 1999, corrected third printing: 2005).

[R 70]   R.T. Rockafellar, (1970), *Convex Analysis.* Princeton University Press, Princeton, New Jersey.

[R 93]   L.C.G. Rogers, (1993), *Notebook, private communication.* Dec. 20, 1993.

[R 94]   L.C.G. Rogers, (1994), *Equivalent martingale measures and no-arbitrage*. Stochastics and Stochastic Reports, vol. 51, no. 1–2, pp. 41–49.

[RW 00]   L.C.G. Rogers, D. Williams, (2000), *Diffusions, Markov Processes and Martingales*. Volume 1 and 2, Cambridge University Press.

[R 04]   D. Rokhlin, (2004) *The Kreps-Yan Theorem for L^∞*. Preprint.

[RS 05]   D. Rokhlin, W. Schachermayer, (2005), *A note on lower bounds of martingale measure densities*. Preprint.

[R 84]   L. Rüschendorf, (1984), *On the minimum discrimination information theorem*. Statistics and Decisions Supplement Issue, vol. 1, pp. 263–283.

[Sa 00]   D. Samperi, (2000), *Entropy and Model Calibration for Asset Pricing and Risk Management*. Preprint.

[S 65]   P.A. Samuelson, (1965), *Proof that properly anticipated prices fluctuate randomly*. Industrial Management Review, vol. 6, pp. 41–50.

[S 69]   P.A. Samuelson, (1969), *Lifetime portfolio selection by dynamic stochastic programming*. Rev. Econom. Statist., vol. 51, pp. 239–246.

[S 70]   P.A. Samuelson, (1970), *The fundamental approximation theorem of portfolio analysis in terms of means, variances, and higher moments*. Rev. Econom. Stud., vol. 37, pp. 537–542.

[S 73]   P.A. Samuelson, (1973), *Mathematics of speculative prices*. SIAM Review, vol. 15, pp. 1–42.

[SM 69]   P.A. Samuelson, R.C. Merton, (1969), *A complete model of warrant pricing that maximizes utility*. Industrial Management Review, vol. 10, pp. 17–46.

[S 81]   W. Schachermayer, (1981), *The Banach-Saks property is not L^2-hereditary*. Israel Journal of Mathematics, vol. 40, pp. 340–344.

[S 92]   W. Schachermayer, (1992), *A Hilbert space proof of the fundamental theorem of asset pricing in finite discrete time*. Insurance: Mathematics and Economics, vol. 11, no. 4, pp. 249–257.

[S 93]   W. Schachermayer, (1993), *A Counter-Example to several Problems in the Theory of Asset Pricing*. Mathematical Finance, vol. 3, pp. 217–229.

[S 94]   W. Schachermayer, (1994), *Martingale Measures for Discrete time Processes with Infinite Horizon*. Mathematical Finance, vol. 4, no. 1, pp. 25–56.

[S 00]   W. Schachermayer, (2000), *Die Rolle der Mathematik auf den Finanzmärkten*. In: "Alles Mathematik", Die Urania Vorträge auf dem Weltkongress für Mathematik, Berlin 1998 (M. Aigner, E. Behrends, editors), pp. 99–111.

[S 01]   W. Schachermayer, (2001), *Optimal Investment in Incomplete Markets when Wealth may Become Negative*. Annals of Applied Probability, vol. 11, no. 3, pp. 694–734.

[S 01a]   W. Schachermayer, (2001), *Optimal Investment in Incomplete Financial Markets*. Mathematical Finance: Bachelier Congress 2000 (H. Geman, D. Madan, S.R. Pliska, T. Vorst, editors), Springer, pp. 427–462.

[S 02]   W. Schachermayer, (2002), *No-Arbitrage: On the Work of David Kreps*. Positivity, vol. 6, pp. 359–368.

[S 03]   W. Schachermayer, (2003), *Introduction to the Mathematics of Financial Markets*. In: S. Albeverio, W. Schachermayer, M. Talagrand: Springer Lecture Notes in Mathematics 1816 — Lectures on Probability Theory and Statistics, Saint-Flour summer school 2000 (Pierre Bernard, editor), Springer Verlag, Heidelberg, pp. 111–177.

[S 03a] W. Schachermayer, (2003), *A Super-Martingale Property of the Optimal Portfolio Process*. Finance and Stochastics, vol. 7, no. 4, pp. 433–456.

[S 04]   W. Schachermayer, (2004), *The Fundamental Theorem of Asset Pricing under Proportional Transaction Costs in Finite Discrete Time*. Mathematical Finance, vol. 14, no. 1, pp. 19–48.

[S 04a] W. Schachermayer, (2004), *Utility Maximisation in Incomplete Markets*. In: Stochastic Methods in Finance, Lectures given at the CIME-EMS Summer School in Bressanone/Brixen, Italy, July 6-12, 2003 (M. Frittelli, W. Runggaldier, eds.), Springer Lecture Notes in Mathematics 1856, pp. 225–288.

[S 04b] W. Schachermayer, (2004), *Portfolio Optimization in Incomplete Financial Markets*. Notes of the Scuola Normale Superiore Cattedra Galileiana, Pisa.

[S 05]   W. Schachermayer, (2005), *A Note on Arbitrage and Closed Convex Cones*. Mathematical Finance, vol. 15, no. 1, pp. 183–189.

[ST 05] W. Schachermayer, J. Teichmann, (2005), *How close are the Option Pricing Formulas of Bachelier and Black-Merton-Scholes?* Preprint.

[Sch 99] H.H. Schaefer, (1999), *Topological Vector Spaces*. Graduate Texts in Mathematics, Springer New York.

[SH 87] M.J.P. Selby, S.D. Hodges, (1987), *On the evaluation of compound options*. Management Science, vol. 33, pp. 347–355.

[Sh 99] A.N. Shiryaev, (1999), *Essentials of Stochastic Finance. Facts, Models, Theory*. World Scientific.

[Sh 04a] S.E. Shreve, (2004), *Stochastic Calculus for Finance I: The Binomial Asset Pricing Model*. Springer Finance, Springer-Verlag, New York.

[Sh 04b] S.E. Shreve, (2004), *Stochastic Calculus for Finance II: Continuous-Time Models*. Springer Finance, Springer-Verlag, New York.

[SSC 95] H.M. Soner, S.E. Shreve, J. Cvitanic, (1995), *There is no nontrivial hedging portfolio for option pricing with transaction costs*. Annals of Applied Probability, vol. 5, pp. 327–355.

[St 70] E. Stein, (1970), *Topics in harmonic analysis*. Ann. Math., Princeton University Press.

[St 03] E. Strasser, (2003), *Necessary and sufficient conditions for the supermartingale property of a stochastic integral with respect to a local martingale*. Séminaire de Probabilités XXXVII, Springer Lecture Notes in Mathematics 1832, pp. 385–393.

[St 85] H. Strasser, (1985), *Mathematical theory of statistics: statistical experiments and asymptotic decision theory*. De Gruyter studies in mathematics, vol. 7.

[St 97] H. Strasser, (1997), *On a Lemma of Schachermayer*. Preprint, Vienna University of Economics and Business Administration.

[Str 90] C. Stricker, (1990), *Arbitrage et Lois de Martingale*. Annales de l'Institut Henri Poincaré – Probabilités et Statistiques, vol. 26, pp. 451–460.

[Str 02] Ch. Stricker, (2002), *Simple strategies in exponential utility maximization*. Preprint of the Université de Franche-Comté.

[SV 69a] D.W. Stroock, S.R.S. Varadhan, (1969), *Diffusion processes with continuous coefficients I*. Comm. Pure & Appl. Math., vol. 22, pp. 345–400.

[SV 69b] D.W. Stroock, S.R.S. Varadhan, (1969), *Diffusion processes with continuous coefficients II*. Comm. Pure & Appl. Math., vol. 22, pp. 479–530.

[T 00]   M.S. Taqqu, (2000), *Bachelier and his Times: A Conversation with Bernard Bru.* Finance and Stochastics, vol. 5, no. 1, pp. 2–32.

[T 99]   N. Touzi, (1999), *Super-replication under proportional transaction costs: from discrete to continuous-time models.* Mathematical Methods of Operation Research, vol. 50, pp. 297–320.

[W 23]   N. Wiener, (1923), *Differential space.* J. Math. Phys., vol. 2, pp. 131–174.

[W 91]   D. Williams, (1991), *Probability with Martingales.* Cambridge University Press.

[W 99a]   C.-T. Wu, (1999), *Construction of Brownian motions in enlarged filtrations and their role in mathematical models of insider trading.* Dissertation Humboldt-Universität zu Berlin.

[W 99b]   C.-T. Wu, (1999), *Brownian motions in enlarged filtrations and a case study in insider trading.* Preprint.

[WY 02]   C.-T. Wu, M. Yor, (2002), *Linear transformations of two independent Brownian motions and orthogonal decompositions of Brownian filtrations.* Publications Matemàtiques, vol. 46, pp. 237–256.

[XY 00]   J. Xia, J.A. Yan, (2000), *Martingale measure method for expected utility maximisation and valuation in incomplete markets.* Preprint.

[Y 80]   J.A. Yan, (1980), *Caractérisation d' une classe d'ensembles convexes de L^1 ou H^1.* In: J. Azéma, M. Yor (eds.), Séminaire de Probabilités XIV, Springer Lecture Notes in Mathematics 784, pp. 220–222.

[Y 05]   J.A. Yan, (2005), *A Numéraire-free and Original Probability Based Framework for Financial Markets.* In: Proceedings of the ICM 2002, vol. III, Beijing, pp. 861–874, World Scientific Publishers.

[Y 78a]   M. Yor, (1978), *Sous-espaces denses dans L^1 ou H^1 et représentation des martingales.* In: C. Dellacherie et al. (eds.), Séminaire de Probabilités XII, Springer Lecture Notes in Mathematics 649, pp. 265–309.

[Y 78b]   M. Yor, (1978), *Inégalités entre processus minces et applications.* C.R. Acad. Sci., Paris, Ser. A 286, pp. 799–801.

[Y 01]   M. Yor, (2001), *Exponential Functionals of Brownian Motion and Related Processes.* Springer Finance.

[Z 05]   G. Zitkovic, (2005), *A filtered version of the Bipolar Theorem of Brannath and Schachermayer.* Journal of Theoretical Probability, vol. 15, no. 1, pp. 41–61.